SOIL MECHANICS:
BASIC CONCEPTS AND ENGINEERING APPLICATIONS

D1375618

Soil Mechanics:

Basic Concepts and Engineering Applications

A. AYSEN

Taylor & Francis
Taylor & Francis Group

LONDON AND NEW YORK

Library of Congress Cataloging-in-Publication Data

Applied for

Hardback edition: 2002
Paperback edition: 2005

Also available: Problem Solving in Soil Mechanics, by A. Aysen, paperback edition, 2005, ISBN 0 415 38392 7

Cover design: Studio Jan de Boer, Amsterdam, The Netherlands

Published by Taylor & Francis
2 Park Square, Milton Park, Abingdon, Oxon, OX14 4RN
270 Madison Ave, New York NY 10016

Transferred to Digital Printing 2006

ISBN: 0 415 38393 5

Publisher's Note
The publisher has gone to great lengths to ensure the quality of this reprint but points out that some imperfections in the original may be apparent

Printed and bound by CPI Antony Rowe, Eastbourne

Contents

Preface

Soil Mechanics: Basic Concepts and Engineering Applications is a scholarly book designed as the main text for university students undertaking a Bachelor's Degree in Civil Engineering as well as Environmental and Agricultural Engineering. The book has a novel feature in introducing the fundamental concepts in a clear manner in each chapter and then to lead to more recent approaches with numerical analysis. As a Soil Mechanics text it is ideally suited not only to the second, third and fourth year undergraduate students, but also for post-graduates at the Master's and Doctoral levels. Practicing engineers will find the text well suited for most of their closed form calculations and in using simple numerical analysis involving the finite difference and finite element methods.

The author has extracted the most important contributions in recent publications on topics that form the cornerstone of basic Soil mechanics. For example the chapter, which includes Soil Classification, includes the Australian Standard, ASTM and BS 5930. Effective stress analysis in Chapter 2 contains the effect of negative pore pressure due to capillary rise. Similarly, potential and stream functions are introduced in the section that deals with the flow of water through soils in Chapter 3, while the chapter on shear strength of soils and failure criteria, includes, the interpretation of triaxial tests data and elementary aspects of Critical State Soil Mechanics (CSSM). The chapter on stress distribution and settlement in soils contains valuable text for the calculation of settlement in sandy and clayey soils using in-situ tests such as the Standard Penetration tests and the Cone Penetration Tests. Soil-footing interaction models also form part of this chapter. One-dimensional consolidation is treated in a comprehensive manner in Chapter 6, which will take the students and practitioners both to theoretical closed form solutions and their approximations, Skempton & Bjerrum method, and the use of finite difference techniques in solving consolidation problems in a simple manner. The chapter on limit analysis to solve stability problems deal with the upper and lower bound solutions and formulations incorporating finite element methods. This chapter then logically leads the students and the practiioners to well written chapters on earth pressure calculations, stability of slopes, and, bearing capacity calculations for shallow and deep foundations. The book thus has a well-balanced text that includes the traditional soil mechanics as covered in other similar texts and then leads the readers logically to an advanced level of handling the mechanics of soils.

SI units are used throughout and the international codes of practice are often referred to wherever necessary. The book is well suited for readers who have a good background in engineering mechanics, even without prior knowledge in soil mechanics.

A problem solving approach is adopted through all the chapters and there are 152 worked examples demonstrating the engineering applications, facilitating self-learning and creating a simulating atmosphere in acquiring soil mechanics education. Also, there are another 113 unsolved problems with answers given to the readers to gain experience and confidence in solving problems by themselves.

A.S. Balasubramaniam, B.Sc., Ph.D.
Emeritus Professor, Asian Institute of Technology, Bangkok
Professor, Griffith University, Gold Coast campus

Acknowledgements

A great debt of gratitude to my teachers Professor J.R.F. Arthur (University of London), late Professor P.W. Rowe (University of Manchester) and late Professor A.W. Bishop (University of London) for teaching me, examining me and qualifying me in the field of Soil Mechanics.

To Professor A.S. Balasubramaniam (Griffith University, Australia) for his continuous support and writing the preface. To dear friend Dr A. Kilpatrick (La Trobe University, Australia) for his help and time in proofreading of the material.

To Professor J.P. Carter (University of Sydney) for his support during my academic life. To Professor S.W. Sloan (University of Newcastle, Australia) for introducing me to the world of numerical analysis in soil mechanics during my years in the University of Newcastle. To Mr R. Fulcher for his support and friendship. To Dr H. Katebi for his constructive views and guidance. To Mr M. Conway and Mr Y. Rahimi for their valuable teaching and help in soil mechanics laboratory techniques.

I am in debt to my family, especially my wife Pari, for her unwavering support and patience during this project.

A. Aysen, M.Eng., Ph.D.
aysena@bigpond.com
March 2005

CHAPTER 1

Nature of Soils, Plasticity and Compaction

1.1   INTRODUCTION

This chapter describes the basic characteristics of soil, and physical properties that affect its engineering behaviour. For a civil engineer the geological history of a soil may not seem to be of much importance as most of the time a civil engineer is searching for a stress-strain model justified by laboratory results. However some information is necessary in order to facilitate development of the model and understand the physical and chemical behaviour of the material of interest and, consequently, a brief description of soil chemistry is included. A significant part of the chapter is devoted to the phase relationships that describe volume-mass related parameters, which control the engineering behaviour of the soil. Essential information about particle size analysis, plasticity and soil classification is provided. More details about laboratory procedures and soil classification may be obtained from the standard codes of the relevant country. The final part of the chapter discusses soil compaction and its importance to the soil.

1.2   NATURE AND CHEMISTRY OF SOILS

1.2.1   *Origin of soils, geological classification and primary minerals*

Soils are the unconsolidated layers that cover the earth's surface. For a civil engineer soil consists of particles of different size and shape with minor bonds between them forming a structure that undergoes deformation when subjected to natural or artificial forces. For an environmental engineer soil is a product of the environment, and its formation is a function of climate, organisms, parent rock or material, relief and time (Jenny, 1940). Soil mechanics is the art of applying the mechanics of solids to the soil to predict its deformation behaviour and strength. This allows the design of soil structures and the investigation of its interaction with other structures built on it.

Soils are derived from the weathering of parent rocks. Weathering or disintegration of rock is of two types: physical and chemical. The physical weathering is the action of the forces associated with wind, water, glaciers and successive freezing and thawing. The chemical agents are water, oxygen, and carbon dioxide. In the physical weathering the mineralogical composition of the parent rock does not change and the resulting soil inherits similar minerals. In the chemical weathering the alteration of the minerals of the parent rock creates new minerals. The process of soil formation is controlled by semi-independent factors of time, climate, organisms and topography. The time factor controls the equilibrium of the process with the environment, a process that may take more than a million years (FitzPatrick, 1983).

The geological classification places rocks into three major groups, viz., *igneous*, *metamorphic* and *sedimentary* rocks. Igneous rocks are those that have cooled down and crystallized from a molten material called magma. These rocks are composed of a variety of mineral crystals, the size of which depends on the rate of cooling. The igneous rocks within the earth's crust are formed by slow cooling and therefore have large crystals. They can be found on the surface of the earth if the overlying material is removed or eroded. Igneous rocks formed on the surface of the earth have been subjected to rapid cooling and therefore have small crystals. Residual soils are a result of the physical and chemical weathering of these igneous rocks and the constituent minerals are different to those of the parent rocks. Sands have quartz and muscovite, similar to quartz and muscovite of the parent rock granite. The orthoclase feldspar and plagioclase of granite break down into the clay minerals of kaolinite and montmorillonite. Metamorphic rocks originate from residual soils under extreme temperature and pressure. Both metamorphic rocks and residual soils are subjected to weathering and transportation processes that result in the formation of sedimentary deposits. Compaction of these deposits, together with cementation due to chemical reactions, produces sedimentary rocks. The sedimentary rocks are also produced directly from metamorphic rocks within a weathering, transportation, consolidation and cementation process. These sedimentary rocks could return to their origin as either sedimentary deposits or metamorphic rocks as a result of weathering or extreme temperature and pressure. This is an ongoing geological cycle through time and the residual soils and sedimentary deposits so formed are known to us as soils. The soil particles are of various sizes and are broadly categorized as clay, silt, sand, gravel and rocks. Each size fraction exhibits different mechanical properties that can be successfully predicted using the principles of *soil mechanics*.

Typical depositional environments in soil formation (Fookes & Vaughan, 1986) include: *glacial environment, lake and marine environment* and *desert environment*. For soils and soil mechanics the most significant recent geological events beside the depositional environments are rising and falling land and sea levels that result in further deposition and weathering. In a geotechnical site investigation a geotechnical or civil engineer needs the cooperation of a geologist to identify the deposition environment and the subsequent deposition and weathering cycle (Atkinson, 1993). The effect of mineralogical history on the mechanical properties is evident in clay soils where the very small particles exhibit *colloidal* properties.

Minerals are divided into two groups – primary and secondary – depending on their state in the geological cycle. Primary minerals are those who inherit the properties of parent rock and their chemical compositions remain unchanged. The common primary minerals in soils are shown in Table 1.1 (Tan, 1994). The sand and silt fraction of the most soils consists largely of primary minerals. Quartz (SiO_2) and feldspars ($MAlSi_3O_8$, $M = Na^+$, K^+ and Ca^+ cations) are the most abundant primary minerals in soils (Bohn, et al., 1985). A rock of igneous origin may have 50% to 70% quartz. Quartz is believed to have been crystallized at low temperatures (Tan, 1994) and therefore it is reasonably stable at ambient temperatures and resistant to weathering but the particles eventually become rounded in nature. Other primary minerals such as micas, and a variety of accessory minerals of primary origin, are present in soils but in smaller quantities than quartz and feldspars.

Table 1.1. Common primary minerals in soils.

Primary mineral	Chemical composition
1. Quartz	SiO_2
2. Feldspar:	
Orthoclase, microcline	$KAlSi_3O_8$
Albite(plagioclase)	$NaAlSi_3O_8$
3. Mica	
Muscovite	$H_2KAl_3Si_3O_{12}$
Biotite	$(H,K)_2(Mg,Fe)_2(Al,Fe)_2\ Si_3O_{12}$
4. Ferromagnesians	
Hornblende	$Ca,(Fe,Mg)_2\ Si_4O_{12}$
Olivine	$(Fe,Mg)_2\ SiO_4$
5. Magnesium Silicate	
Serpentine	$H_4Mg_3Si_2O_9$
6. Phosphate	
Apatite	$(Ca_3(PO_4)_2)_3.Ca(Fe,Cl)_2$
7. Carbonates	
Calcite	$CaCO_3$
Dolomite	$CaMg(CO_3)_2$

Primary minerals may also occur in the clay fraction, but these are not the major constituents of the clay soils. Secondary minerals are produced by weathering of rocks and primary minerals and have different chemical composition from their parent materials. The clay fraction of most clayey soils consists of secondary minerals.

The Jackson-Sherman weathering stages considers three stages in the weathering process (Sposito, 1989). These are classified as *early stage, intermediate stage* and *advanced stage*, a classification that indicates the intensity of leaching and oxidizing conditions. The products of the early stage consist of carbonates, sulphates and primary silicates. Intermediate stage weathering results in quartz, muscovite and secondary alumino-silicates of clay fraction of soil including mica and smectites. The advanced stage creates kaolinite that is an important clay mineral. In general clay minerals fall into intermediate and advanced stages as shown in Table 1.2. This Table shows the increasing order of persistence of the soil minerals and their occurrence in time. The size of the mineral, its shape and its hardness controls the rate of weathering and their engineering behaviour (Ollier, 1972). The large minerals are more resistant to the weathering because of their low specific surface which reduces the area of contact against the weathering agents. The clay minerals have a fine texture, with individual particles less than 0.002 mm in diameter. As a result of the large surface area associated with the small particles and masses, clay minerals develop plasticity when mixed with water. A knowledge of the crystal structure of clay minerals is necessary in order to understand their behaviour.

1.2.2   *Crystal structure of clay minerals*

In general, clay minerals are alumino-silicates made by a combination of silica and aluminium oxide units with metal ions substituted within the crystal. A schematic diagram of a silica unit is shown in Figure 1.1(a). Each silica unit consists of a tetrahedron with four oxygen atoms, O, located at the vertices of the tetrahedron equidistant from each other.

Table 1.2. Jackson-Sherman weathering stages (Sposito, 1989).

Clay minerals in soil fraction	Soil chemical and physical conditions
	Early stage
Gypsum	Very low content of water and organic matter
Carbonates	Very limited leaching
Olivine / pyroxene / amphibole	Reducing environments
Fe-bearing micas	Limited amount of time for weathering
Feldspar	
	Intermediate stage
Quartz	Retention of Na, K, Ca, Mg, Fe(II), and silica:
Dioctahedral mica / illite	Ineffective leaching and alkalinity
Vermiculite / chlorite	Igneous rock rich in Ca, Mg, Fe(II), but no
Smectites	Fe(II) oxides
	Silicates easily hydrolysed
	Flocculation of silica, transport of silica into the weathering zone
	Advanced stage
Kaoline	Removal of Na, K, Ca, Mg, Fe(II), and silica:
Gibsite	Effective leaching, fresh water
Iron oxides (geotite, hematite)	Oxidation of Fe(II), acidic compounds, low pH
Titanium oxides (anatase, rutile, ilmenite)	Dispersion of silica, Al-hydroxy polymers

A silicon atom, Si, is located inside the tetrahedron equidistant from the oxygen atoms. These units combine to form a silica sheet as shown in Figure 1.1(b). The arrangement could happen in variety of ways. Figure 1.1(b) shows an arrangement in a hexagonal pattern in which each basal oxygen atom is shared with the adjacent unit. This sharing, results a negative charge in the basic unit, which can be increased to zero if, for example, aluminium replaces silicon. Figure 1.1(c) shows the short hand symbol generally used for silica. Figure 1.2(a) shows a unit of alumina (aluminium oxide) where six hydroxyl ions surround one aluminium atom, Al, in an octahedral arrangement. The combination of alumina units in staggered rows results in the formation of an alumina sheet where each hydroxyl ion is shared by three basic units (Figure 1.2(b)).

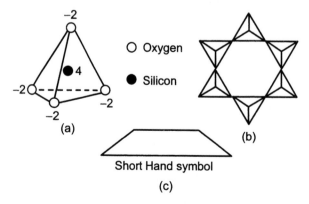

Figure 1.1. (a) A basic silica unit, (b) silica sheet, (c) short hand symbol for silica sheet.

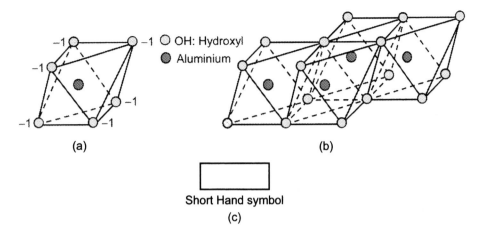

Figure 1.2. (a) A basic alumina unit, (b) alumina sheet, (c) short hand symbol for alumina sheet.

This sharing, results a positive charge in each basic unit, which can be reduced to zero if magnesium or iron replaces aluminium. Figure 1.2(c) shows the short hand symbol generally used for alumina. The basic elemental structure of a clay mineral is formed by different combinations of silica and alumina sheets. Their type and engineering characteristics depend on the strength of the bonds that connect these elemental structures.

Kaolinite. Kaolinates, with the general chemical composition of $2SiO_2Al_2O_32H_2O$, are produced from weathering of the parent rocks that have orthoclase feldspar (e.g. granite). An elementary layer of kaolinite crystal is made of one silica sheet and one alumina sheet mutually sharing the oxygen atoms between them. These layers join together to form a kaolinite particle as shown in Figure 1.3. The number of elemental layers in one stack may reach one hundred or more. The elementary layers are held together with hydrogen bonding giving a stable structure to the mineral. The hydrogen bonding is a result of the attraction forces between the oxygen atoms of the silica sheet and the hydroxyl ions of the alumina sheet.

These forces are high enough to prevent the penetration of water between the layers and consequential expansion. The mineral is moderately plastic when it is wetted and has some frictional resistance to shear.

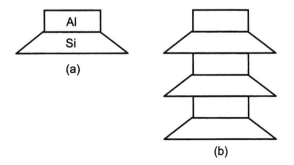

Figure 1.3. Elemental structure of kaolinite.

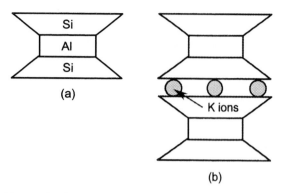

Figure 1.4. Elemental structure of illite.

Halloysite is another type of kaolinite mineral with a general chemical composition of $2SiO_2Al_2O_34H_2O$ and a tubular structure with water molecules between the elemental layers of the kaolinite crystals. As a result, in contrast to kaolinite, it is susceptible to expansion and contraction. Upon heating, halloysite is irreversibly dehydrated and is converted to metahalloysite being a precursor of kaolinite following the weathering sequence: Igneous rocks → smectite (monmorillonite) → halloysite → metahalloysite → kaolinite (Tan, 1994).

Illite. Illite is produced from the weathering of micas with the major parent mineral of muscovite. A schematic representation of illite structure is shown in Figure 1.4. Illite has a mica type structure in which an elementary layer is made from one alumina sheet sandwiched between two silica sheets. The bond between the two elementary layers is made by potassium, K^+, which joins the six oxygen atoms of the two silica sheets. This bond is not as strong as in kaolinite and, with a random staking of layers, there is more space to water to enter between the elemental layers.

Montmorillonite. Montmorillonite, with chemical composition of $4SiO_2Al_2O_3H_2O + nH_2O$, is a product of the weathering of volcanic ash in marine water under poor drainage

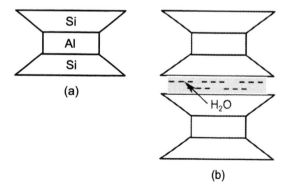

Figure 1.5. Elemental structure of montmorillonite.

conditions. Figure 1.5 shows the elemental structure of montmorillonite. Similar to illite the structure is of mica type but the bond between the elemental layers is by weak oxygen-to-oxygen links comprises of water molecules and exchangeable cations. The major characteristic of this mineral is its volume expansion or contraction due to an increase or decrease in moisture content. At high moisture contents the mineral is plastic and can be deformed easily under minor stress but hardens when dry. The type of structural units and their bonding influences the shape of the clay minerals and their surface area. As a result they are either plate shaped or tubular. The average diameter varies from 0.1 to 4 μm, while the thickness has a range of 30 Angstrom (1 Angstrom = 10^{-7} mm) to 2 μm. In general particle size decreases from kaolinite to illite and to montmorillonite. The surface area for a constant mass (specific surface) is a function of shape and size. The specific surface of montmorillonite is approximately 50 to 60 times that of kaolinite and 10 times that of illite. Due to the high adsorption of water by montmorillonite, the mixture of this group of minerals and water, into the state of a viscous fluid, is used in engineering construction as a *grouting* material to reduce void ratio and permeability of soils (ASCE, 1962).

1.2.3 *Soil organic matter*

Soil is a multiphase system containing solids, liquids, and gases in which the liquid and gas phases are essentially water and air respectively. The solid phase consists of inorganic and organic material. The organic material contains most of the carbon of the soil and usually consists of both dead and living matter. Dead organic matter is the remains of dead plant material and animal residues. It is the main source nutrients for plants and influences the engineering properties of the soil. The live microbial and plant matter comprises a few percent of the total organic matter, the average amount of which may change up to 5% to 10% (by mass) depending on the drainage conditions. In some peat soils, however, the amount of organic matter approaches 100%. According to White (1979) the amount of organic matter on the surface of soils (e.g. Australia) varies up to 16%, but decreases rapidly with depth. Organic soils and peats are recognized by their high liquid limit (the amount of water needed to turn the soil into a viscous liquid) but the range of

Table 1.3. Methods of tests for soils: chemical and electro-chemical tests.

Test	ASTM	BS	AS	AASHTO
Organic matter content	Loss on ignition: D-2974	Titration: 1377 Loss on ignition: 1377	1289.4.1.1	Titration: T194
Sulphate content		Gravitational: 1377 Ion exchange: 1377	1289.4.2.1	
pH Value	Electrometric: D-1067 Colorimetric: D-1067	Electrometric: 1377	1289.4.3.1 Colorimetric: 1289.4.3.2	
Electrical resistivity		Disc electrodes: 1377	1289.4.4.1	
Carbonate content		Titration: 1377		
Chloride content		Water-soluble and Acid-soluble: 1377		

moisture content in which the soils show plastic behaviour is small. Organic soils, especially those with a spongy structure, are also highly compressible and do not meet the settlement limitations of most structures that may be built upon them. The determination of the organic matter in a soil and its chemical composition – sulphates, carbonates and chlorides – is recommended in civil engineering construction works because of their detrimental effects on its engineering properties (Table 1.3).

1.2.4   *Electric charge in soils*

The solid phase of most soils carries a net negative charge, which is associated with the small colloidal particles of inorganic and organic origin. While the charge is negative on the surface of the clay particle; the edges carry positive or negative charges due to broken bonds. The soil particles that are responsible for cation exchange are referred to as colloid particles. A colloid is a very fine particle of organic or inorganic origin with a maximum size of 0.2 μm and a minimum size of 50 A. A permanent charge is a result of isomorphous substitution in the crystal structure of the clay minerals (e.g. substitution of the Al_3^+ ion for Si_4^+ or Mg_2^+ for Al_3^+). The resulting charge can be neutralized by the cations that are positively charged atoms present in the solution surrounding the clay particles. In illite and mica clays the potassium bond is structural rather than exchangeable (McLaren & Cameron, 1996), and the cation exchange capacity is relatively low. The unbalanced electric charge is responsible for the formation of clay structure. Surface-to-surface contact is not possible as both have net negative charges whilst the edge-to-surface contact is common as a result of the attraction between the positive and negative charges. The electric charge does not affect the engineering behaviour of coarse particles, as the ratio of surface to mass is low in comparison to that of the clay particles. A variable charge is a result of a change in pH in organic soils. A decrease or increase in pH will result a decrease or increase in negative charge. In some soils that have iron and aluminium oxides, there are some areas, along the colloid particle surface, that have a positive charge and will be neutralized by the adsorption of negatively charged anions. The quantity of exchangeable anions held is much smaller than the amount of exchangeable cations. The distribution of cations and anions from the surface of the particle towards the voids created by the arrangement of the particles has been modelled by different molecular adsorption models

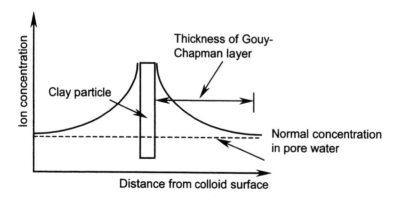

Figure 1.6. The Gouy-Chapman double layer model.

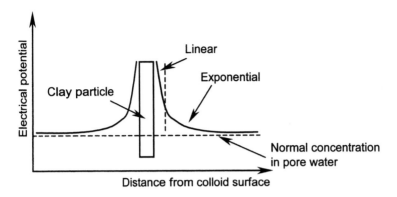

Figure 1.7. The Stern double layer model.

(Sposito, 1989). The simplest model is the Helmholtz double layer model in which it is assumed that the layer balancing the negative charges is immediately adjacent to the surface of the colloid. In the diffuse double layer, referred to as the Gouy-Chapman model, the ion concentration decreases gradually away from the surface of the colloid particle until it becomes equal to the ion concentration of the bulk solution within the voids (Figure 1.6). The forces of attraction or repulsion between electrical charges can be calculated mathematically from Coulomb's law. Combining the equation of cation attraction with the mathematical expression of diffusion, a relationship can be found between the electric potential and the distance from the colloid surface (Boltzmann equation). Thus, the characteristics of the diffuse double layer may be presented by three types of distribution that include the distribution of cations (Figure 1.6), the distribution of anions and the electric potential within the layer. The thickness of the diffuse layer depends on the type of the ion and its bulk concentration and for natural soils it may range from 1 to 20×10^{-6} mm. This model considers only the electrostatic forces of attraction and repulsion and disregards the specific forces created by the finite size of the ions. The Stern model is a modified version of the Gouy-Chapman model in which the double layer is divided into two layers of the Stern layer immediately next to the colloid surface and the Gouy-Chapman model beyond this layer (Figure 1.7). Electrical potential decreases linearly within the Stern layer and exponentially across the diffuse layer.

The investigation of the nature of the layer of water surrounding a colloid particle is important in soil chemistry, agriculture and environmental soil science from the view of exchangeable cations available for plant intake. Furthermore, the clay fraction may intercept some pollutants, carried by soil water, due to its cation exchange capacity and thus acts like a natural purifying agent. This layer is strongly attracted to the soil particle and has different properties from the water in the pore voids. In soil mechanics the study of the nature and mechanisms of held and attracted water is important, as it is responsible for the plastic behaviour of the soil. The term *adsorbed water* is used to describe this layer; whilst the water beyond the boundary of adsorbed water is called *free water*. The water particles within the adsorbed layer can move parallel to the surface of the colloid particle, but the movement in the direction perpendicular to the surface of particle is not possible or is restricted.

1.2.5   *Soil structure*

Soil structure is the three-dimensional geometrical arrangement of pores and particles of various sizes. Some geologists have referred to the structure as soil fabric to make it compatible with the term rock fabric, which describes the arrangement of mineral grains in the volume of the rock. Brewer (1964) defines fabric as the physical constitution of a soil material as expressed by the spatial arrangement of the solid particles and associated voids. The spatial arrangement of coarse material is called the soil skeleton. Brewer & Sleeman (1988) suggest that the soil structure is based on the arrangements of solids and pores in which the primary particles form a compound and the arrangement of compounds constitutes the soil structure. Although in many definitions the emphasis is on the three-dimensional nature of the soil structure, two-dimensional studies of stability problems and soil water movement have been very successful in soil mechanics. In clay soils the interparticle forces described previously have a distinct effect on the arrangement of the clay particles. If the resultant of the forces creates a net repulsion then a face-to-face arrangement called a *dispersed* structure will be created (Figure 1.8(a)). The net attraction results in an edge to face structure referred to as a *flocculated* structure (Figure 1.8(b)). Experience indicates that clay deposits that are developed in fresh or salty waters have a flocculated structure, while the transported and remoulded clays tend to have a dispersed structure. The flocculated structure in salty water is more dominant than in fresh water. In salty clays (e.g. marine clay) the concentration of cations is high, resulting in thin adsorbed layers around the surface of the particles. The attraction forces dominate the particle

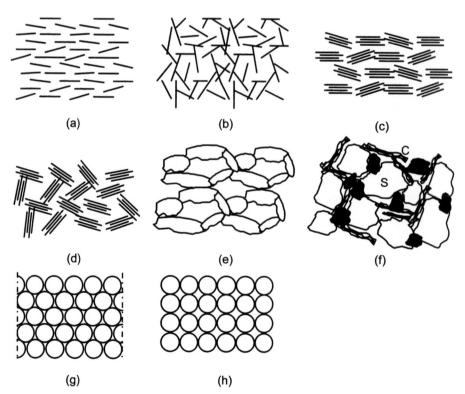

Figure 1.8. Different structures of clay, natural soil and idealization of dense and loose sand.

arrangement; thus an edge to face structure is made. A clay soil with pure water moves towards the dispersed structure. In laboratory techniques based on sedimentation, a salt, commonly sodium hexametaphosphate, is added to the suspension liquid as a dispersing agent to facilitate particle sedimentation. Sometimes the aggregations made by original mineral units make dispersed or flocculated type structures depending on the deposition environment (Figures 1.8(c) and 1.8(d)). These aggregations have face-to-face structures, as there are little or no attraction forces between mineral units. Depending on the velocity of the water in which the particles settle these aggregations, or fine silts and sands, make a honeycomb structure as shown in Figure 1.8(e). This structure contains a large amount of voids that are bridged by the assemblage of aggregations or fine silts and sand particles. A diagrammatic reconstruction of an electron microscope image of clayey sand is shown in Figure 1.8(f) (Tan, 1994). The sand particles (*S*) form the main skeleton of the soil struc-ture. Clay (*C*) and organic material (black particles) fill the voids between the sands and the remaining voids are occupied by air. This reconstruction cannot be generalized for sands and coarse materials as the process of the structural arrangement of particles is complex and must be assessed by standard tests and scanning electron micrographs. An idealization of sand or a coarse-grained soil structure is shown in Figures 1.8(g) and 1.8(h) for dense and loose structures respectively. The dense model was used by Rowe (1962) to demonstrate the stress-dilatancy in sand during shear.

1.2.6   Clay structural analysis: X-ray diffraction method

This technique was originally investigated by Whittig (1964) and later suggested by Moore (1970a, b) to the ASTM. The X-ray diffraction method has probably contributed more to the mineralogical characterization of the elemental structure (soil layer silicates) than any other single method of analysis (Bohen, et al., 1985). Optical and electron mi-croscopy methods are two other techniques used for the measurement of the fabric within a relatively small volume of soil. The optical microscopy is based on the optical birefrin-gence of the clay particles subjected to polarized light. In electron microscopy a relatively larger area is scanned by the electron beam allowing the direct observation of the material at a wide range of magnifications. In the X-ray diffraction method the material is exposed to a filtered X-ray beam as idealized by Figure 1.9. The X-ray passes into the material and causes the electrons in the atoms of the minerals to vibrate and reflect the beam through

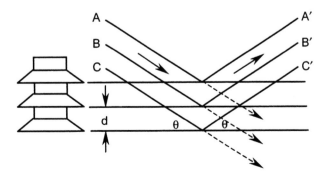

Figure 1.9. Geometric conditions for the X-ray reflection from mineral planes.

the successive planes. The method involves increasing of incidence angle and monitoring the intensity of the diffracted X-radiation until a maximum value of the diffracted intensity is achieved. The X-ray diffraction maximum is detected whenever the following equation is satisfied:

$$n\lambda = 2d \sin\theta \tag{1.1}$$

where n = a series of integers, 1, 2, 3,..., λ = known wavelength of the X-radiation, d = the repeated space between diffracting planes and θ is the angle of incidence of the X-rays. Detection of the maximum diffracted radiation intensity is carried out either by a cylindrical photographic film placed around the sample or by a rotating detector. This technique is valuable for measuring the thickness of the elemental structure as well as monitoring the interlayer spacing with any change of the external conditions. The X-ray diffraction method is also used in soil stabilization where the soil or a manufactured fine-grained material such as fly ash is stabilized by lime or cement. Development of hydration and bond formation between particles may be monitored by this technique to justify the water content used for stabilization.

1.3   MASS-VOLUME RELATIONSHIPS

1.3.1   *Basic physical properties*

A soil sample contains soil particles (solids), liquid (usually water) and air. These three phases can be visualized as three separate blocks as shown in Figure 1.10. This representation is an artificial concept and all phases are inextricably mixed. The phase diagram is dimensioned in terms of volume and mass. A completely dry soil or a fully saturated soil is a two-phase system. Volume related symbols are defined as follows:

V_a = Volume of the air within the voids between particles.
V_w = Volume of the water within the voids between particles.
V_s = Volume of the solids.
V_v = Volume of the voids within a given sample = $V_a + V_w$.
V = Total volume of the soil sample = $V_s + V_v$.
Mass related symbols are:

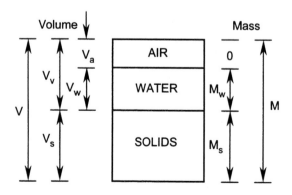

Figure 1.10. Dimensioned phase diagram.

M_a = Mass of air = 0, M_w = Mass of water, M_s = Mass of solids.
M = Total mass of soil sample = $M_w + M_s$.
The common volume related parameters are as follows:

Void ratio e. The ratio of the volume of the voids to the volume of the solids:

$$e = \frac{V_v}{V_s} \tag{1.2}$$

Specific volume v. The volume occupied by a unit volume of solids:

$$v = \frac{V}{V_s} = \frac{V_v + V_s}{V_s}, \; v = 1 + e \tag{1.3}$$

Porosity n. The ratio of the volume of voids to the total volume:

$$n = \frac{V_v}{V} \rightarrow n = \frac{V_v}{V_s + V_v} = \frac{e}{1+e} \tag{1.4}$$

Void ratio is usually expressed as a decimal whilst porosity may be expressed as a percentage.

Degree of saturation S_r. The ratio of the volume of the water to the volume of the voids:

$$S_r = \frac{V_w}{V_v} \tag{1.5}$$

Degree of saturation is usually expressed as a percentage. For a dry soil $S_r = 0$. When the volume of the voids is fully occupied by water, $V_w = V_v$ and $S_r = 1$ or 100%.

Air content A_v. The ratio of the volume of air to the total volume:

$$A_v = \frac{V_a}{V} \tag{1.6}$$

The common mass-volume related parameters are:

Moisture content or water content m or w. The moisture content or the water content is the ratio of the mass of water to the mass of the solids or dry mass of soil:

$$m \text{ or } w = \frac{M_w}{M_s} \tag{1.7}$$

This parameter is usually expressed as a percentage.

Density of solids ρ_s. The ratio of the mass of the solids to the volume of the solids:

$$\rho_s = \frac{M_s}{V_s} \tag{1.8}$$

In the SI system the preferred unit is Mg/m^3.
Numerically 1 Mg/m^3 = 1 g/cm^3 = 1 $tonne/m^3$ = 1000 kg/m^3.

Specific gravity of solids G_s. The ratio of the density of the solids to the density of water:

$$G_s = \frac{\rho_s}{\rho_w} \tag{1.9}$$

where ρ_w = density of water = 1 Mg/m^3 at 4°C. This is a dimensionless parameter within the range of 2.6 to 2.8 depending on the mineralogy of the soil.

Density of soil ρ. The ratio of the total mass to the total volume:

$$\rho = \frac{M}{V} \tag{1.10}$$

A preferred usage unit is Mg/m^3. The terms *bulk density* or *wet density* are also used. For a given void ratio, this parameter changes from a minimum at the dry state to a maximum at the saturated state:

$$\rho_d = \frac{M_s}{V} \text{ (Minimum)}, \ \rho_{sat} = \frac{M_s + V_v \rho_w}{V} \text{ (Maximum)} \tag{1.11}$$

The relationship between dry density, moisture content and bulk or wet density is:

$$\rho_d = \frac{\rho}{1 + w} \tag{1.12}$$

Density index I_D. To express the consistency states of sand and gravel, the natural void ratio is compared with the minimum and maximum void ratios obtained in the laboratory. Density index (or relative density) is defined by:

$$I_D = \frac{e_{max} - e}{e_{max} - e_{min}} \text{ (loosest) } 0 \leq I_D \leq 1 \text{ (densest)} \tag{1.13}$$

Useful equations:

Dry density expressed in terms of specific gravity and void ratio:

$$\rho_d = \frac{G_s \rho_w}{1 + e} \tag{1.14}$$

Void ratio expressed in terms of moisture content, specific gravity and degree of saturation:

$$e = \frac{w G_s}{S_r}, \ e = w G_s \text{ (for fully saturated soil)} \tag{1.15}$$

By combining Equations 1.14 and 1.15, a relationship is found to express dry density in terms of moisture content, specific gravity and degree of saturation.

$$\rho_d = \frac{G_s \rho_w}{1 + w G_s / S_r} \tag{1.16}$$

$$\rho_d = \frac{G_s \rho_w}{1 + w G_s} \text{ (for fully saturated soil)} \tag{1.17}$$

In a coordinate system with w on the horizontal axis and ρ_d on the vertical axis, this equation represents the possible states of saturated soil and is termed zero air curve.
Dry density expressed in terms of specific gravity, moisture content, and air content.

$$e = \frac{V_w + V_a}{V_s} = \frac{M_w/\rho_w + VA_v}{M_s/\rho_s} = \frac{M_w + VA_v\rho_w}{M_s/G_s} = \frac{G_sM_w + G_sVA_v\rho_w}{M_s} = wG_s + \frac{G_sA_v\rho_w}{\rho_d}.$$

Substituting for e in Equation 1.14 and rearranging,

$$\rho_d = \frac{G_s\rho_w(1-A_v)}{1+wG_s} \tag{1.18}$$

For zero air curve $A_v = 0$ this equation reduces to Equation 1.17.
Combining Equations 1.16 and 1.18 the following relationship between degree of saturation, air content and moisture content can be obtained.

$$S_r = \frac{wG_s(1-A_v)}{A_v + wG_s} \tag{1.19}$$

Relationship between e, G_s, ρ_w, and ρ_{sat}:

$$e = \frac{\rho_w G_s - \rho_{sat}}{\rho_{sat} - \rho_w}, \text{ or: } \rho_{sat} = \rho_w \frac{G_s + e}{1+e} \tag{1.20}$$

The submerged or buoyant unit weight:
$\gamma' = \gamma_{sat}(\text{saturated unit weight}) - \gamma_w(\text{unit weight of water}) = (\rho_{sat} - \rho_w)g$,
using Equation 1.20:

$$\gamma' = \gamma_w \frac{G_s - 1}{1+e} \tag{1.21}$$

Example 1.1

A soil specimen has the following properties: $G_s = 2.7$, $e = 0.6$ and $w = 14\%$. Calculate: dry density, dry unit weight, wet density, wet unit weight and degree of saturation.

Solution:

When the void ratio e is known, we calculate the volume of solids V_s and the volume of voids V_v in terms of a unit total volume ($V = 1 \text{ m}^3$).
$V_v + V_s = 1, e = V_v/V_s$ or $V_s = 1/(1+e), V_v = e/(1+e)$,
for $e = 0.6, V_s = 1/(1+0.6) = 0.625 \text{ m}^3$,
$V_v = 1 - 0.625 = 0.375 \text{ m}^3$.

Dry density is the mass of the solids that occupies a unit volume:
$\rho_d = G_s\rho_w V_s = 2.7 \times 1.0 \times 0.625 = 1.69 \text{ Mg/m}^3$.
Dry unit weight: $\gamma_d = 1.69 \times 9.81 = 16.6 \text{ kN/m}^3$.
Wet density: $\rho = \rho_d(1+w) = 1.69(1+0.14) = 1.93 \text{ Mg/m}^3$.
Wet unit weight: $\gamma = 1.93 \times 9.81 = 18.9 \text{ kN/m}^3$.
$V_w = M_w/\rho_w = 0.14M_s/\rho_w = 0.14 \times 1.69/1.0 = 0.237 \text{ m}^3$,
$S_r = V_w/V_v = 0.237/0.375 = 0.632 = 63.2\%$.

Example 1.2

A soil sample has the following properties: $G_s = 2.7$, $e = 0.7$, $S_r = 80\%$. Calculate: dry density, dry unit weight, moisture content, wet density, wet unit weight and air content.

Solution:

For $e = 0.70$, $V_s = 1/(1+0.7) = 0.588 \text{ m}^3$, $V_v = 1 - 0.588 = 0.412 \text{ m}^3$.

$\rho_d = G_s \rho_w V_s = 2.7 \times 1.0 \times 0.588 = 1.59 \text{ Mg/m}^3$, $\gamma_d = 1.59 \times 9.81 = 15.6 \text{ kN/m}^3$.

$S_r = V_w / V_v = 0.8$, $V_w / 0.412 = 0.8$, $V_w = 0.330 \text{ m}^3$, $V_a = 0.412 - 0.330 = 0.082 \text{ m}^3$.

$M_w = V_w \times \rho_w = 0.330 \times 1.0 = 0.33 \text{ Mg}$, $w = 0.33/1.59 = 0.207 = 20.7\%$.

$\rho = \rho_d (1 + w) = 1.59(1 + 0.207) = 1.92 \text{ Mg/m}^3$, $A_v = V_a / V = 0.082/1.0 = 0.082 = 8.2\%$.

Example 1.3

In a field density test the following results are obtained: $\rho = 1.95 \text{ Mg/m}^3$, $w = 0.16$. The specific gravity of solids is 2.7. Find: dry density, void ratio, air content, degree of saturation, saturated unit weight and moisture content at full saturation.

Solution:

$\rho_d = \rho/(1+w) = 1.95/(1+0.16) = 1.68 \text{ Mg/m}^3$. Assume the total volume is 1 m^3, thus:

$V_s = M_s / \rho_s = M_s / (G_s \times \rho_w) = 1.68/(2.7 \times 1.0) = 0.622 \text{ m}^3$, $V_v = 1 - 0.622 = 0.378 \text{ m}^3$.

$e = 0.378/0.622 = 0.608$. $V_w = M_w / \rho_w = (1.95 - 1.68)/1.0 = 0.27 \text{ m}^3$,

$A_v = V_a / V = (0.378 - 0.270)/1.0 = 0.108 = 10.8\%$.

$S_r = V_w / V_v = 0.270/0.378 = 0.714 = 71.4\%$.

The volume of water at the saturated state = 0.378 m^3 and its mass is 0.378 Mg. Thus,

$\rho_{sat} = 1.68 + 0.378 = 2.06 \text{ Mg/m}^3$, $\gamma_{sat} = 2.06 \times 9.81 = 20.2 \text{ kN/m}^3$.

Moisture content at full saturation = 0.378 / 1.68 = 0.225 = 22.5\%.

1.3.2 *Experimental determination of physical properties of soils*

Determination of volume-mass related parameters require the direct measurement of the total mass, mass of water, total volume, and specific gravity of solids. The volume of a specimen of an irregular shape is measured using water or mercury displacement techniques. Determination of the moisture content includes measurements of wet mass and oven-dried mass to establish the masses of the water and solid particles. In most standards the temperature of the oven for drying purposes is 105°C to 110°C, and the specimen must stay in the oven for 24 hours.

The specific gravity of solids. This parameter is measured using a simple apparatus called a *pyknometer* which is a glass jar with a conical top as shown in Figure 1.11. The cone has a hole at the top to allow the water to overflow. The test procedure described in ASTM D-854, AASHTO T100, BS 1377 and AS 1289.3.5.1, consists of the following steps:

1. Measure the mass of empty pyknometer (M_1).
2. Fill the pyknometer with water until it overflows from the hole at the top of the cone and find the total mass of the pyknometer and water (M_2).
3. Empty the pyknometer and place a sample of oven-dried soil inside the pyknometer and measure the combined mass (M_3).
4. Add water until it overflows from the top of the cone and measure the total mass of the apparatus (M_4). At this stage the jar must be agitated to remove the air bubbles. The specific gravity is the ratio of the mass of dry soil to the mass of water displaced by soil:

Figure 1.11. Pyknometer with rubber washers (Wykeham Farrance).

$$G_s = \frac{M_3 - M_1}{(M_2 - M_1) - (M_4 - M_3)} \qquad (1.22)$$

A similar method uses a narrow-necked density bottle (volumetric flask) of 25, 50 or 100 ml volume that may be agitated by a mechanical device called an *end over end shaker*.

Field density. Common methods are based on excavating a hole at the site and measuring the volume of the hole and the mass of the excavated soil. Field density is the ratio of the mass of the excavated soil to the volume of the hole. The volume of the hole is measured in a variety of ways using mostly sand or water replacement methods. In the sand replacement method two apparatus are available. In BS 1377 and BS 1924 the apparatus is an assembly of a sand pouring metal cylinder of 100 or 200 mm diameter, a metal cone and a metal tray of 300 mm square with 100 or 200 mm diameter central hole (Figure 1.12(a)). The metal tray is adjusted horizontally on the ground surface and the soil is excavated through the hole in the tray to a depth of approximately 1.5 times the diameter, with the resulting excavation being preferably cylindrical in shape. The sand-pouring cylinder is placed on the tray and fitted to the central hole. Opening the shutter system allows the sand to flow through the cone, until the hole and the cone are filled with the sand. The mass of the sand in the excavated hole is calculated by measuring the mass of the apparatus before and after the test, and knowing the mass of the sand in the cone. The volume of the hole is equal to the mass of sand in the hole divided by the pouring density of sand. To determine the mass of the sand in the cone the cylinder is filled with sufficient material and the mass of the assembly is measured. The apparatus is sanded over the hole in the tray, which is placed on a flat and level surface. When the sand stops running the tap is closed and the remaining mass of sand is measured. The difference between the two measurements before and after the operation gives the mass of the sand in the cone. The pouring density is measured by pouring the sand into a calibration cylinder (Figure 1.12(b)) of known volume. A similar method using a sand pouring can is recommended by AS 1289.5.3.2. The sand used is clean and dry, 90% passing a 1.18 mm sieve and 90% retained on a 600 μm sieve. Sometimes an upper size of 2.36 mm / 1.18 mm or a lower size of 600 μm / 300 μm is used. Care has to be taken in shielding the sand to prevent moistening and bulking that may invalidate the results.

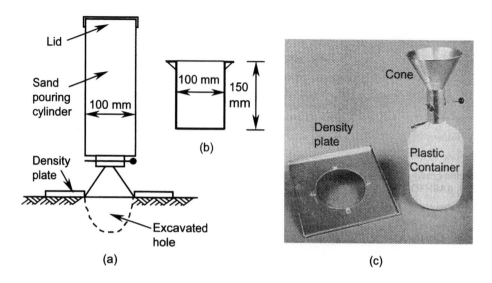

Figure 1.12. (a) Typical sand pouring apparatus, (b) calibration cylinder, (c) sand cone apparatus (Wykeham Farrance).

The second apparatus based on the sand replacement method is called *sand cone apparatus* (Figure 1.12(c)) and is used by ASTM D-1556, AASHTO T191 and AS 1289.5.3.1 standard systems. This apparatus is used in holes where the maximum particle size does not exceed 38 mm. The apparatus includes a plastic container of 5 litres capacity, a density plate (tray) with 165 mm diameter central hole and a cone with 152 mm diameter that has a shutter mechanism and threaded nozzle. According to ASTM D-1556-90 (1996) recommendations, the sand cone method is not suitable for organic, saturated, or highly plastic soils that would deform or compress during the excavation of the test hole. It may not be suitable for soils consisting of unbounded granular materials or granular soils having high void ratios that will not maintain stable sides in the test hole. The balloon density apparatus used by ASTM D-2167, AASHTO T205 and AS 1289.5.3.4, utilizes an empty rubber balloon to measure the volume of excavated soil. The balloon is slowly filled with water from a calibrated glass container until the hole is completely filled by the expanded balloon. A direct water replacement method (AS1289.5.3.5) uses a density ring on the surface and a plastic sheet to retain the water in the hole. This test is particularly appropriate for soils containing large particles as defined in AS 1289.0. In BS 1377 and BS 1924 the water displacement method is recommended only for stabilized soils. In fine-grained soils the core cutting method is used to determine the field density (ASTM D-2937, BS 1377 and AS 1289.5.3.3). The core cutter commonly has a diameter of 100 mm and is 130 mm long and is driven into the soil using a driving ram of standard weight. By measuring the mass of the soil in the core cutter and knowing its volume the field density can be calculated.

Application of nuclear gauges for the determination of field density and in-situ moisture content provides a rapid method for compaction and moisture control (ASTM D-2922, D-2950, D-3017, BS 1377 and 1924). Readouts in direct units are also provided for void ratio and degree of compaction. A nuclear moisture-density gauge operates by emitting

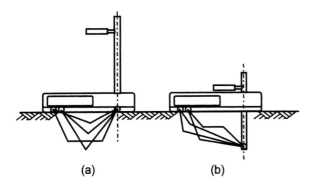

Figure 1.13. Transmission modes in nuclear moisture-density gauge.

gamma and neutron radiation from radioactive materials sealed within a capsule in the gauge. The gauge also includes a detector of the Geiger-Mueller type to collect and read gamma radiation emitted from the cesium 137 and thereby determine the field density. For neutron radiation americium 241: berylium is used in the measurement of moisture content. A material of high density absorbs the gamma radiation and acts as a radioactive shield resulting in a low reading, whilst a material of low density will give a high reading. The americium 241: berylium emits neutron radiation to the test material. Some of this radiation or high energy neutrons are moderated by the hydrogen atoms of the water in the material and the rest is detected by the Geiger-Mueller detector showing a reading on the gauge. A high reading in a specified period of time corresponds to a wet material, while a low reading corresponds to a dry material. Based on the amount of radiation detected in both cases the gauge could be calibrated either by the manufacturer or by the user. Two basic techniques, backscatter and direct transmission, are used for the determination of moisture content and density as shown in Figure 1.13. Moisture content is determined by using the backscatter mode only whilst density may be evaluated by either backscatter or direct transmission. The backscatter mode is shown in Figure 1.13(a) in which both the radiation source and detector are on the surface of the test material. Direct transmission is not possible because of the intervening shield and the detector counts only on the radiation received by reflection or radiation scattered back to the detector. Using this method depths up to 65 mm may be analysed. In the direct transmission mode (Figure 1.13(b)) a small hole is drilled in the test material to a depth of 300 mm and, whilst the radioactive source is lowered into the hole, readings are carried out at increments of 25 to 50 mm.

1.4   PARTICLE SIZE DISTRIBUTION

1.4.1   *Definition of soils and particle size distribution curve*

Figure 1.14 shows a generally used size classification for soils as proposed by MIT (Massachusetts Institute of Technology). Each group is divided into three sub-divisions of fine, medium and coarse particles. Sands and gravels are coarse-grained soils; and considered as non-cohesive (or cohesionless) materials. They are also referred to as granular materals. Clay and silt are fine-grained and are classified as cohesive materials.

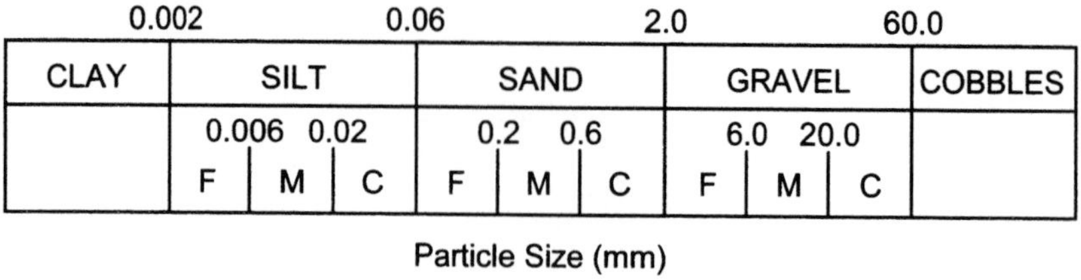

0.002
0.06
2.0
60.0
CLAY
SILT
SAND
GRAVEL
COBBLES
0.006 0.02
0.2 0.6
6.0 20.0
F M C
F M C
F M C
Particle Size (mm)

Figure 1.14. Definition of soils according to MIT.

In practice fine-grained materials are soils passing # 200 sieve, which has square openings of 0.075 mm. A particle size distribution curve describes the percentage by mass of particles of the different size ranges. The horizontal axis represents the particle size on a logarithmic scale. The vertical axis represents the percentage by weight of particles that are finer than a specific size on the horizontal axis. An example of this curve is shown in Figure 1.15. Consider point *A* on this curve with coordinates of 9.5 mm and 82%. This means that 82% by weight of particles are finer than 9.5 mm. In a typical particle size distribution test, sieves are stacked in a motorized shaker in order of decreasing aperture size and a pan at the bottom of the sieve column retains fine material that passes through all the sieves. For the material passing through sieve # 200, a sedimentation test is carried out. Figure 1.16 shows three size distribution curves: soil *W* is regarded as a well-graded soil, soil *U* is a uniformly graded soil with a size range much less than soil *W*, soil *P* is a poorly graded soil and lacks a particular range of particles. To describe the state of grading, two parameters are defined as follows:

$$\text{Coefficient of Uniformity: } C_U = \frac{D_{60}}{D_{10}} \tag{1.23}$$

$$\text{Coefficient of Curvature: } C_C = \frac{D_{30}^2}{D_{60} D_{10}} \tag{1.24}$$

where D_{10}, D_{30} and D_{60} are particle sizes corresponding to 10%, 30% and 60% finer respectively and are determined from the particle size distribution curve obtained in the

100
90
80
70
60
50
40
30
20
10
0
Percentage finer by weight
A
0.001
0.01
0.1
1
10
100
Particle size (mm)

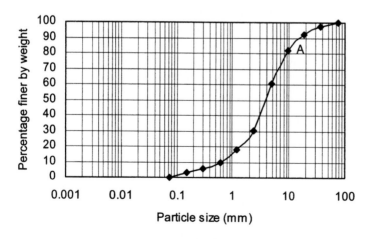

Figure 1.15. Particle size distribution curve (Example 1.4).

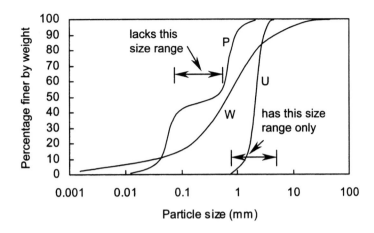

Figure 1.16. Well-graded, uniformly graded and poorly graded soils.

laboratory. The magnitude of C_U indicates the spread of grain size values; a larger range leads to larger number. According to the Unified Soil Classification, for well-graded soils, the value of C_C is between 1 and 3. For a well-graded gravel, or gravel-sand mixtures with little or no fines, the value of C_U is greater than 4. For well-graded sands, and gravelly sands with little or no fines, the value of C_U is greater than 6.

1.4.2   *Sieve and hydrometer analyses*

The standard procedure for determining the distribution of particle sizes may be found in ASTM D-422, BS 1377 and AS 1289.3.6.1. Sieve requirements are specified in ASTM E11, BS 410 and AS 1152 and are manufactured in three diameters of 200 mm, 300 mm, and 450 mm. Normally, sieves have a mesh of woven brass wire with square apertures ranging from 0.075 to 125 mm. The sieve with 0.075 mm apertures (or sieve # 200, indicating 200 openings in 25.4 mm length) is the smallest practical sieve size used in soil mechanics laboratories. Prior to sieving, soil samples must be oven-dried at 105°C to 110°C. For intermediate and fine size particles the soil sample is first submerged in a dispersing solution at least for an hour. It is then washed with water on the 0.075 mm sieve and oven dried. An analysis without washing is also permitted by standard codes, but the results are not reliable if an appreciable amount of clay and silt is present in the sample. Sieving may be done by hand or by a sieving machine. A typical motorized shaker or sieving machine is activated by electromagnetic impulses with a control panel for time adjustment and vibrating intensity. Some advanced shakers permit the adjustment of pause time between vibrations. A sieve shaker holds up to ten nesting sieves of any diameter. The particle size distribution for the fraction of the soil passing through the 0.075 mm sieve (clay and silt) is carried out using the hydrometer method specified in ASTM D-422, AASHTO T88, BS 1377 and AS 1289.3.6.2. The methods described in these standard codes are not applicable if less than 10% of the soil passes the 0.075 mm sieve. The hydrometer method is based on the measurement of the velocity of soil particles in a sedimentation solution and the dry mass of soil in the solution in different intervals of time. The velocity of falling particles and the dry mass of soil at a specific depth are measured by a hydrometer originally manufactured to measure the specific gravity of a solution. The results are combined with the Stokes' law, which gives the relationship between the velocity of a spherical particle and its diameter while settling within a solution. As a result, two relationships are constructed between the hydrometer reading and both the diameter and percentage finer than that diameter for each reading at a specific time. Stokes' law was originally presented in the following form:

$$D = \sqrt{18\eta v/(\gamma_s - \gamma_w)} \tag{1.25}$$

where D is the diameter of a spherical particle in cm, η is the dynamic viscosity of water in dyne-sec/cm^2, γ_s is the unit weight of the particle in g-force/cm^3, γ_w is the unit weight of water in g-force/cm^3, and v is the velocity of the spherical particle in cm/sec. The reading R of a hydrometer at the time t gives the amount of soil particles in g/1000 ml of solution (or equivalent suspension density) at the centre of the hydrometer volume. The distance of a reading from this centre, L (Figure 1.17(b)), represents the length covered by a particle of a specific diameter during this time. Replacing v with L/t in Equation 1.25:

$$D = \sqrt{18\eta L/t(\gamma_s - \gamma_w)} \tag{1.26}$$

Equation 1.26 may be simplified using familiar SI units:

$$D = \sqrt{L/10} \times 1.749 \sqrt{\eta/(\rho_s - \rho_w)} \times 10/\sqrt{t} \tag{1.27}$$

where D is the diameter of the soil particle in μm, ρ_s and ρ_w are in Mg/m^3, η is in $10^3 \times$ N/m$^2 \times$ seconds and t is in minutes. The effective length L is found by calibrating the hydrometer. Equation 1.27 can be written in the following form:

$$D = F_1 \times F_2 \times F_3 \tag{1.28}$$

Parameters F_1 and F_3 are calculated from the test data. Parameter F_2 depends on the properties of the soil and water and can be tabulated for a specified ρ_s (Table 1.4).

A sample calculation for $T = 15°C$: $\eta = 0.001139$ N/m$^2 \times$ s, $\rho_w = 0.9991$ Mg/m^3,

$F_2 = 1.749\sqrt{1.139/(2.7 - 0.9991)} = 1.432$.

The percentage finer than D is calculated from the following equation:

$$\text{Percentage finer} = R_C \times a/M_s \times 100\% \tag{1.29}$$

where M_s is the mass of dry soil and R_C is the corrected value of hydrometer reading for temperature, density of water, and meniscus effects. The parameter a corrects the reading due to the specific gravity of solids as the hydrometer is made (e.g. hydrometer 152H, ASTM) to measure the density of a solution in g/1000 ml for $G_s = 2.65$:

$$a = \frac{1.65 G_s}{2.65(G_s - 1)} = \frac{0.623 G_s}{G_s - 1} \tag{1.30}$$

The amount of oven-dried soil used for the test is 50 g in a clayey soil to 100 g in a silty soil and the sample may be collected from the sieve washings and then oven-dried. The test is carried out in 1000 ml sedimentation glass cylinders (Figure 1.17(a)). The soil is mixed with water and a dispersing solution (commonly sodium hexametaphosphate) and is shaken manually for about one minute. The cylinder is then placed on a firm level and vibration free surface and by immersing the hydrometer in the suspension the reading starts in logarithmic sequences of time.

Table 1.4. Values of parameter F_2 versus temperature for $\rho_s = 2.7$ Mg/m^3.

Temp.°C	F_2	Temp.°C	F_2	Temp.°C	F_2
10	1.534	17	1.395	24	1.282
11	1.512	18	1.378	25	1.267
12	1.491	19	1.361	26	1.253
13	1.471	20	1.344	27	1.239
14	1.451	21	1.328	28	1.225
15	1.432	22	1.312	29	1.212
16	1.414	23	1.297	30	1.199

An alternative method of obtaining suspension density is the use of an *Andreasen pipette* (BS 1377). The 10 ml capacity glass pipette is used to obtain a small quantity of suspension from a prescribed depth at different intervals of time to calculate the suspension density. The pipette is fixed to a carriage assembly on a vertical stand and operates with no vibration whilst the pipette is inserted and withdrawn from the liquid suspension. The cylinder used in this method is 500 ml and samples are taken from a 100 mm depth.

Example 1.4

A sieve analysis on a soil sample of 978 g has given the following results. Plot the particle size distribution curve, determine C_U and C_C and classify the soil.

Sieve size (mm)	Mass retained (g)	Sieve size (mm)	Mass retained (g)
75	0	1.18	118.5
37.5	25.5	0.6	83.1
19	52.8	0.3	39.9
9.5	97.8	0.15	24.6
4.75	207.3	0.075	27.3
2.36	298.2	Pan	3.0

Solution:

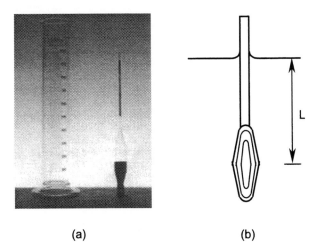

(a)   (b)

Figure 1.17. (a) Hydrometer and sedimentation cylinder (Wykeham Farrance), (b) effective depth.

The results are tabulated below and shown in Figure 1.15, from which: $D_{10} = 0.63$ mm, $D_{30} = 2.40$ mm, $D_{60} = 4.57$ mm. $C_U = D_{60} / D_{10} = 4.57/0.63 = 7.25$,

$$C_C = D_{30}^2 / (D_{60} D_{10}) = 2.40^2 / (4.57 \times 0.63) = 2.0.$$

The soil is a well-graded gravel with sand.

Sieve size (mm)	Retained (%)	Total retained (%)	Finer than (%)
75	0.00	0.000	100.000
37.5	2.607	2.607	97.393
19	5.398	8.005	91.995
9.5	10.000	18.005	81.995
4.75	21.196	39.201	60.799
2.36	30.490	69.691	30.309
1.18	12.116	81.807	18.193
0.6	8.496	90.303	9.697
0.3	4.079	94.382	5.618
0.15	2.515	96.897	3.103
0.075	2.791	99.688	0.312
Pan	0.306		

Example 1.5

For the following data with $M_s = 50$ g and $G_s = 2.7$ draw the particle size distribution curve. For the hydrometer used, L (assume corrected) $= 163 - R \times 49 / 30$ (mm).

T (C°)	Time (min.)	Reading	T (C°)	Time (min.)	Reading
18	1	45.5	18	30	27.5
18	2	41.5	20	60	26.0
18	3	38.0	22	120	25.4
18	4	35.5	24	240	25.0
18	8	32.0	19	1440	24.0
18	16	29.0			

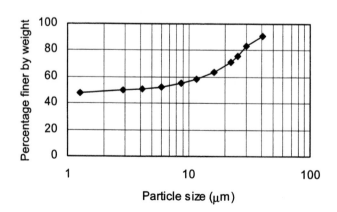

Figure 1.18. Example 1.5.

Solution:

The results are tabulated and shown in Figure 1.18. Sample calculation at $t = 30$ min.:

$L = 163.0 - 27.5 \times 49 / 30 = 118.1$ mm, $F_1 = \sqrt{118.1/10} = 3.436$,

$F_2 = 1.378$ (Table 1.4), $F_3 = 10/\sqrt{30.0} = 1.826$, $D = 3.436 \times 1.378 \times 1.826 = 8.64\,\mu m$.

From Equation 1.29: (%) finer $= 27.5 / 50 = 0.55 = 55\%$.

T (C°)	Time (min.)	Reading	L (mm)	F_1	F_2	F_3	D (μm)	(%) Finer
18	1	45.5	88.7	2.978	1.378	10.00	41.03	91.0
18	2	41.5	95.2	3.085	1.378	7.071	30.06	83.0
18	3	38.0	100.9	3.176	1.378	5.773	25.26	76.0
18	4	35.5	105.0	3.240	1.378	5.000	22.32	71.0
18	8	32.0	110.7	3.327	1.378	3.535	16.21	64.0
18	16	29.0	115.6	3.400	1.378	2.500	11.71	58.0
18	30	27.5	118.1	3.436	1.378	1.826	8.64	55.0
20	60	26.0	120.5	3.471	1.344	1.291	6.02	52.0
22	120	25.4	121.5	3.486	1.312	0.913	4.18	50.8
24	240	25.0	122.2	3.496	1.282	0.645	2.89	50.0
19	1440	24.0	123.8	3.518	1.361	0.263	1.26	48.0

1.5 INDEX PROPERTIES AND VOLUME CHANGE IN FINE GRAINED SOILS

1.5.1 *Consistency states of fine grained soils*

The volume change and flow behaviour of a fine-grained soil both depend upon its moisture content. At a high level of moisture the soil has the properties of a liquid, whilst at a low moisture content it takes on the properties of a solid. At moisture contents between these two states the soil passes from a plastic state to a semi-solid state as the moisture content decreases. The physical condition of soil-water mixture is denoted as its *consistency*. Consistency is the resistance to flow, which is related to the force of attraction between particles and is easier to feel physically than to describe quantitatively (Yong & Warkentin, 1966). Figure 1.19 shows the different consistency states of a mixture of water and fine-grained soil. The boundaries of these states, expressed in terms of moisture content, are termed the Atterberg limits (after Albert Atterberg, a Swedish agricultural scientist), and are defined as follows:

Liquid Limit: The liquid limit *LL* is the moisture content above which the soil-water mixture passes to a liquid state. At this state the mixture behaves like a viscous fluid and flows under its own weight. Below this moisture content, the mixture is in a plastic state. Any change in moisture content either side of the *LL* produces a volume change in soil.

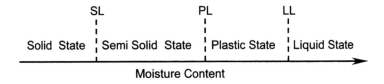

Figure 1.19. Consistency states.

Plastic Limit: The plastic limit *PL* is the moisture content above which the soil-water mixture passes to a plastic state. At this state the mixture is deformed to any shape under minor pressure. Below this moisture content, the mixture is in semi-solid state. Any change in moisture content at either side of the *PL* produces a change in volume of the soil.

Shrinkage Limit: The shrinkage limit *SL* is the moisture content above which the mixture of soil and water passes to a semi-solid state. Below this moisture content, the mixture is in a solid state. Any increase in moisture content is associated with volume change but a decrease in moisture content does not cause volume change. This is the minimum moisture content that causes full saturation of the soil-water mixture. The volume remains constant as the mixture goes through the dry state to the *SL* moving from zero saturation to 100% saturation (Figure 1.20). On the wet side of the *SL* the volume of the mixture increases linearly with increasing moisture content. On this line the mixture is fully saturated. A decrease in moisture content moves the state of the mixture along the broken line *CBA*. Using these limits the following indices are defined and used in the classification and description of fine grained-soils:

$$\text{Plasticity Index } PI = LL - PL \qquad \text{(mixture is plastic at this range)} \qquad (1.31)$$

$$\text{Activity} = PI\,/\,\text{Clay particles (\%)} \qquad (1.32)$$

$$\text{Liquidity Index } I_L = (w - PL)/\,PI \quad (w = \text{moisture content in the field}) \qquad (1.33)$$

The Atterberg limits are used extensively in the classification of fine-grained soils. Finding relationships between these limits and the engineering properties of the soil has been a matter of research for many years. Early researchers such as Terzaghi & Peck (1967) suggested the direct proportion of liquid limit and soil compressibility. Sherard (1953) reported similar behaviour while investigating the effects of index properties on the performance of earth dams. Whyte (1982) suggested a method based on extrusion for the determination of the plastic limit and found that the ratio of strength at the plastic limit to the strength at the liquid limit is approximately 70. According to Skempton & Northy (1953), however, this ratio is approximately 100. A comprehensive collection of equations relating compressibility indices and soil plasticity is reported by Bowles (1996). These relationships may be useful in guiding the engineer in the early stages of a feasibility study before conducting extensive soil exploration and strength tests.

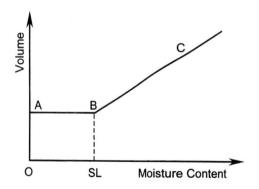

Figure 1.20. Volume-moisture content relationship.

The liquid limit for clay minerals may vary from 50% for kaolinite to 60% for illite and up to 700% for montmorillonite. Kaolinite and illite may show moderate plastic limits of 25% to 35%, whilst in montmorillonite the plastic limit can reach 100%. Note that exchangeable cations such as Na and Ca may influence clay activity, and this has to be investigated if these minerals are to be subjected to salty water.

1.5.2 *Determination of the liquid, plastic and shrinkage limits of soils*

The Casagrande method for determination of the liquid limit. The relevant apparatus is shown in Figure 1.21 and complies with ASTM D-4318, AASHTO T89, BS 1377, AS 1289.3.1.1 and AS 1289.3.1.2. It is comprised of a brass bowl (A) that is hinged to a crank, which, on rotation, causes the bowl to be lifted and dropped 10 mm onto a hard rubber base – a counter records the number of rotations (blows). A pat of soil is placed in the bowl covering approximately 2/3 of the area, and is grooved into two pieces with a standard grooving tool (B). The bowl is a part of a 54 mm diameter sphere and has a maximum depth of 27 mm. The original grooving tool was later replaced by the ASTM grooving tool (C) to overcome difficulties in grooving silty and sandy soils. The test specimen is made from 250 g of soil passing the 425 µm sieve. Water is added in increments and mixed with the soil until the soil becomes a thick homogeneous paste. It is then left at room temperature to cure for 12 hours. A part of this mixture is placed in the Casagrande cup and levelled parallel to the base to make a depth of approximately 10 mm. After closure of 10 mm of the groove as a result of the lifting and dropping of the cup, the number of blows is recorded and a sample of the mixture is taken from the centre of the closed groove for moisture content determination. The moisture content of the remaining mixture is increased or decreased (by air drying) and the test is repeated three times to give four different moisture contents and corresponding numbers of blows. These results are plotted on a semi-logarithmic scale with the number of blows on the horizontal logarithmic axis and moisture content on the vertical axis on the linear scale. A line of best fit is drawn and the moisture content corresponding to 25 blows is the liquid limit of the soil. Most of the standard codes recommend a one-point test to establish the liquid limit as a subsidiary method according to:

$$LL = w_N (N/25)^{\tan \beta} \qquad (1.34)$$

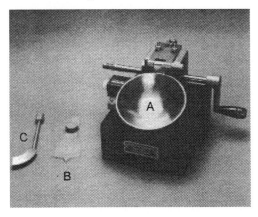

Figure 1.21. Liquid limit device with grooving tools (Wykeham Farrance).

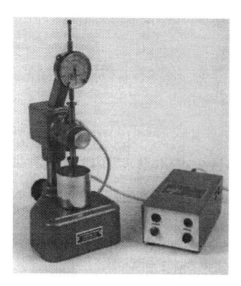

Figure 1.22. Cone penetrometer with electric timer unit (Wykeham Farrance).

where LL is the liquid limit in percentage, w_N is the moisture content in percent corresponding to N number of drops, and β is the slope of the semi-log plot.

The cone liquid limit of a soil. The cone penetration test (Figure 1.22) is described in BS 1377 and AS 1289.3.9 (allowed in 1991). In this method the liquid limit is the moisture content of a soil-water mixture placed in a standard cylinder when a standard cone penetrates 20 mm into the soil paste in five seconds after it is released freely from the surface of the sample. The 60° cone is manufactured from metal and has a height of 32 mm whilst the cylinder has a diameter of 53 mm and a height of 40 mm. The test has to be repeated at least four times using the original cured sample. Results are plotted in log-linear scale, with moisture contents on linear vertical axis and cone penetrations on logarithmic horizontal axis. A line of best fit is drawn through the data and the moisture content corresponding to a cone penetration of 20 mm is determined.

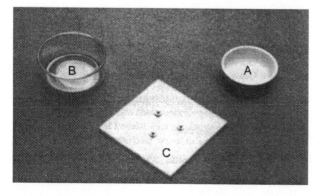

Figure 1.23. Shrinkage dish, glass cup and prong plate (Wykeham Farrance).

Determination of the plastic limit. The test procedure is described in ASTM D-4318, AASHTO T90, BS 1377 and AS 1289.3.2.1 and is performed on the material prepared for liquid limit test. The plastic limit is the moisture content at which a 3 mm diameter thread of soil-water paste shears when rolled on a glass plate with the tip of fingers, and the test is repeated several times to obtain an average value for the *PL*.

Determination of the shrinkage limit. This test is carried out according to ASTM D-427, AASHTO T92 and BS 1377. The procedure involves the measurements of mass and volume of the soil-water sample (cake) in two states of wet, close to the liquid limit (point *C* in Figure 1.20) and oven-dried (point *A* in Figure 1.20). The soil cake is prepared in multiple layers in a shrinkage dish of 45 mm diameter by 22 mm depth (dish *A* in Figure 1.23) and, after air-drying, is left in the oven until completely dry. After measuring the mass of the oven-dried soil cake it is submerged in mercury (dish *B* in Figure 1.23) and its volume is measured by mercury replacement. A prong plate *C* is necessary to level the mercury and keep the soil cake submerged in the dish. The volume of the shrinkage dish *A* is equal to the volume of the wet soil and can be measured by mercury replacement method. By knowing the mass and volume in two states of wet and dry the shrinkage limit can be calculated (Example 1.8). As the volume of the replaced mercury is measured manually, the user has to abide by appropriate safety and health practices. Whilst shrinkage limit is not used in soil classification systems, its determination is recommended to evaluate the swell potential of clayey soils. Alternatively, a shrinkage limit apparatus, which accommodates a cylindrical sample of 38 mm diameter by 76 mm long, may be used for volumetric shrinkage determination. The apparatus includes a precision micrometer and both the volume and the volume change are measured by mercury replacement. A typical apparatus is shown in Figure 1.24. For the soils with low clay contents where the plastic limit test is difficult to perform, a linear shrinkage test is recommended (BS 1377, AS 1289.3.4.1). The shrinkage moulds are semi-cylindrical troughs, 250 mm long with an internal diameter of 25 mm, made from copper, plated steel, stainless steel or brass. They are filled with

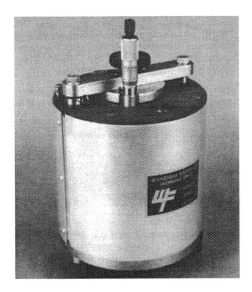

Figure 1.24. Shrinkage limit apparatus (Wykeham Farrance).

a soil-water paste, having a moisture content close or ideally equal to liquid limit. After 24 hours curing in a controlled temperature room the samples are oven-dried and the mean longitudinal shrinkage L_S is measured to the nearest millimetre. The percentage linear shrinkage LS of the specimen is defined by:

$$LS = \frac{L_S}{L} \times 100\% \tag{1.35}$$

where L is the length of the mould in mm. If the quantity of material used is limited, shorter moulds (minimum 100 mm) with a semi-circular section can be used.

In ASTM D-4943 two shrinkage factors, *volumetric shrinkage* and *shrinkage ratio*, are defined as follows.

Volumetric shrinkage is the ratio of the volume change (due to moisture content change from the liquid limit to the shrinkage limit) to the volume at the liquid limit:

$$S_V = \frac{V_{LL} - V_{SL}}{V_{LL}} \tag{1.36}$$

Shrinkage ratio is defined as the ratio of volume change (expressed as a percentage of volume of dry soil) to the moisture content change from the liquid limit to the shrinkage limit:

$$SR = \frac{\Delta V / V_{dry}}{\Delta w} = \frac{\Delta V / V_{dry}}{\Delta M_w / M_s} = \frac{\Delta V / V_{dry}}{\Delta V \times \rho_w / M_s} = \frac{M_s}{V_{dry} \times \rho_w} = \frac{\rho_d}{\rho_w} = G \tag{1.37}$$

Which is the apparent specific gravity of the soil in the dry state. Note that in the calculation of the shrinkage limit and shrinkage ratio the initial moisture content does not affect the results.

Example 1.6

The following data were obtained from a liquid limit test using the Casagrande apparatus. Determine the liquid limit of the soil.

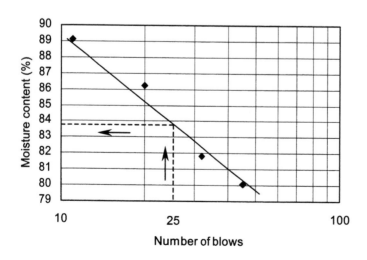

Figure 1.25. Example 1.6.

Number of blows	Moisture content (%)	Number of blows	Moisture content (%)
11	89.1	32	81.8
20	86.2	45	80.1

Solution:

The results are plotted in Figure 1.25. The line of best fit is drawn and the liquid limit is taken as the moisture content corresponding to 25 blows. Hence, $LL = 83.8\%$.

Example 1.7

Using the following data determine the cone liquid limit of the soil.

Cone penetration (mm)	Moisture content (%)	Cone penetration (mm)	Moisture content (%)
11.3	63.08	21.9	70.1
14.8	65.3	25.2	71.0

Solution:

Figure 1.26 shows the plot of moisture content versus penetration depth on a logarithmic scale. The liquid limit is taken as the moisture content corresponding to a cone penetration of 20 mm. Hence, $LL = 68.9\%$.

Example 1.8

A fully saturated sample of soil in its natural state was found to have a total mass of 204 g and a total volume of 122 ml. When the sample had been oven dried the mass was 130 g and the total volume 58 ml. Calculate the value of shrinkage limit and G_s.

Solution:

Mass of water at natural state (point C in Figure 1.20) = 204.0 − 130.0 = 74.0 g.

Volume of water at natural state = 74.0/1 (g/cm^3) = 74.0 cm^3 = 74.0 ml.

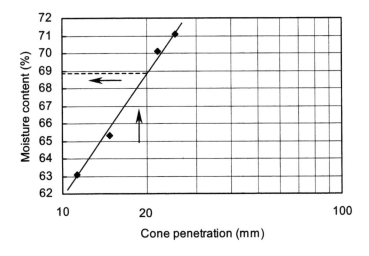

Figure 1.26. Example 1.7.

Volume change (Point C to Point A in Figure 1.20) = 122.0 − 58.0 = 64.0 ml.

Volume of water at shrinkage limit (point B in Figure 1.20) = 74.0 − 64.0 = 10 ml.

Mass of water at shrinkage limit = 10.0 g.

SL = moisture content at shrinkage limit = 10.0 / 130.0 = 0.769 = 7.7%.

$V_s = 58.0 - 10.0 = 48.0$ ml, $G_s = 130$ g/(48 ml × 1 g/ml) = 2.71.

1.6 SOIL CLASSIFICATION FOR GEOTECHNICAL PURPOSES

1.6.1 *Introduction to classification of soils*

Soil classification for engineering purposes is necessary to describe the many types of the soil that exist in nature. A classification system is a standard language (Craig, 1997) which organises the engineering knowledge of soils and is a means of communication. This section describes the Unified Soil Classification System (ASTM D-2487) which is the basis for the AS 1726-1993. The AS 1726-1993 does not preclude the usage of an alternative system and consequently references are made to BS 1377-2: 1990 where appropriate. The usage of an alternative system is useful, as the Unified Soil Classification System may not adequately describe certain types of soils such as calcareous soils, pedocretes, and laterites. The classification of Australian soils and the basic concepts of soil survey are given by Isbell (1996).

Most classification systems in soil mechanics use particle size characteristics, liquid limit and plasticity index to describe the soil and its name. An engineering soil classification system is only useful for qualitative applications. In the construction of important soil structures the classification must be supplemented by laboratory tests other than those needed for classification. The assessment is made using disturbed samples recovered from site as well as undisturbed samples from boreholes and excavations. The particle size distribution is carried out on material passing the 75 mm sieve in ASTM D-2487 and 63 mm sieve in AS 1726. Index property tests are performed on the material passing the 425 μm sieve. More often a geological investigation is needed to identify the age and type of the deposition and its environment. The soil description (using any standard system) should (a) distinguish between *composition, condition* and *structure* of the soil, (b) describe the *information* which may be obtained from a disturbed soil sample, and (c) describe the *additional condition* and *structure properties* which may only be observed in an undisturbed soil (AS 1726-1993).

1.6.2 *Unified Soil Classification System*

The Unified Soil Classification System evolved from the airfield classification system developed by Casagrande (1948). In this system two letters describe each soil. The first letter indicates the dominant size whilst the second letter describes the fines, particle size distribution, and plasticity of the soil (Table 1.5). Soils are divided into two major groups of coarse and fine-grained material. Coarse-grained soils include gravels, sands, nonplastic silts and their mixtures with more than 50% larger than the 0.075 mm sieve. Fine-grained soils include silts, clays and their mixtures with more than 50% finer than the 0.075 mm sieve.

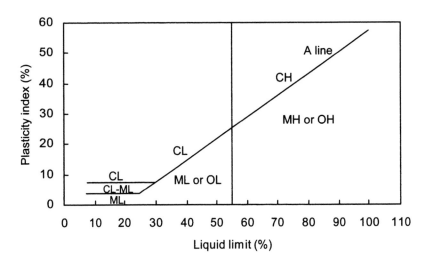

Figure 1.27. Plasticity chart (ASTM D-2487).

Table 1.5. First and second letters of group symbols.

Soil identification	First letter of group symbol	Second letter of group symbol
Coarse grained soils	*G*: gravel, *S*: sand	*W*: Well graded,
		P: Poorly graded
Fine grained soils	*M*: silt, *C*: clay	*L*: Low plasticity,
		H: High plasticity
Organic soils	*O*	*L*: Low plasticity,
		H: High plasticity
Highly organic soils	*Pt*	No second letter

Coarse-grained soils are labelled *GW, GP, GM, GC, SW, SP, SM,* and *SC* whilst fine-grained soils are labelled *ML, CL, OL, MH, CH,* and *OH*.

The name of a fine-grained soil can be determined from the plasticity chart shown in Figure 1.27 which is divided into two ranges of liquid limit of *L* for *LL* < 50% and *H* for *LL* > 50%. The equation for the *A* line which separates the clay fraction from the silt and organic matter is given by:

$$PI = 0.73(LL - 20) \tag{1.38}$$

Table 1.6 shows the Unified Soil Classification System for engineering purposes according to ASTM D-2478. The use of this standard results in a single classification group symbol and group name except when a soil has 5 to 12% fines or the state of the soil in the plasticity chart falls into the cross-hatched area. In these cases, a dual symbol is used, e.g. *GP-GM,* and *CL-ML*. Any soil on the borderline of two group symbols is identified by both group symbols separated by a slash, e.g. *SC/CL, CL/CH,* and *GM/SM*. Borderline symbols are particularly useful when the *LL* is close to 50%. These soils can have expansive characteristics and using a borderline symbol alerts the user of this potential (ASTM D-2487-1998).

Table 1.6. Unified Soil Classification System (ASTM D-2487).

Major division		Group symbol	Finer[**] than 75 μm
Coarse grained soils 50% > 75 μm	Gravelly soils 50% of coarse fraction larger than sieve No.4: 2.36 mm	*GW*	< 5%
		GP	< 5%
		GM	> 12%
		GC	> 12%
	Sandy soils 50% of coarse fraction finer than sieve No.4: 2.36 mm	*SW*	< 5%
		SP	< 5%
		SM	> 12%
		SC	> 12%
Fine grained soils 50% < 75 μm	Silts and clays *LL* < 50%	*ML*	
		CL	
		OL	
	Silts and clays *LL* > 50%	*MH*	
		CH	
		OH	
Highly organic soils		*Pt*	

[**]For soils with 5%-12% passing the No. 200 (75 μm) use a dual symbol.

1.6.3 *Description, identification and classification of soils (AS 1726)*

The Unified Soil Classification System with minor modifications is used. Coarse-grained soils are defined as materials in which more than half of material less than 63 mm is larger than 0.075 mm. The *CL* group of the original system (*LL* < 50%) has been divided into two groups of *CL* (inorganic clay of low plasticity) with *LL* < 35% and *CI* (inorganic clay of medium plasticity) with 35% < *LL* < 50%. Field identification procedures are performed on particles finer than 0.2 mm after coarse particles are removed by hand.

Dry strength (crushing characteristics). A pat of soil is made to the consistency of putty and is left in the oven or sun or air-dried until completely dry. By crushing it between the fingers, its strength can then be assessed but only after developing experience from applying the method to different soil samples with known strength characteristics.

Dilatancy (reaction to shake). A mixture of soil and water (approximately 10 cm^3) is made with sufficient moisture to make the mixture soft but not sticky. The sample is then placed in the open palm of one hand and is shaken rapidly, striking vigorously against the other hand. Water will appear on the surface of the sample, but will disappear if the

Table 1.6 (continued). Unified Soil Classification System (ASTM D-2487).

Typical names	Laboratory classification criteria	
Well graded gravels, gravel-sand mixtures Poorly graded gravels, gravel-sand mixtures	$C_U > 4$ $1 < C_C < 3$ Not meeting gradational requirements for GW	
Silty gravels, poorly graded gravel-sand-silt mixtures Clayey gravels, poorly graded gravel-sand-clay mixtures	Atterberg limits below A-line or $PI < 4$ Atterberg limits above A-line with $PI > 7$	Above A-line With $4 < PI < 7$ requires dual symbols
Well graded sands, gravelly sands Poorly graded sands, gravelly sands	$C_U > 6$ $1 < C_C < 3$ Not meeting gradational requirements for SW	
Silty sands, poorly graded sand-silt mixtures Clayey sands, poorly graded sand-clay mixtures	Atterberg limits below A-line or $PI < 4$ Atterberg limits above A-line with $PI > 7$	Above A-line with $4 < PI < 7$ requires dual symbols
Inorganic silts, very fine sands, silty or clayey fine sands with slight plasticity, rock flour Inorganic clays of low to medium plasticity, gravelly, sandy and silty clays, lean clays Organic silts and silt-clays of low plasticity	Plasticity chart Plasticity chart Plasticity chart, organic odour or colour	
Inorganic silts, micaceous or diatomaceous fine-sandy or silty soils, elastic silts, volcanic ash Inorganic clays of high plasticity Organic clays of medium to high plasticity	Plasticity chart Plasticity chart Plasticity chart, organic odour or colour	
Peat and other highly organic soils	Fibrous organic matter, will char, burn or glow	

shaking is stopped and the sample is squeezed between the fingers. The rapidity of appearance of water during the shaking process and of its disappearance during squeezing assist in identifying the character of the fines in the soil. For instance, very fine clean sands give the quickest and most distinct reaction whereas plastic clay has no reaction. Inorganic silts show a moderately quick reaction.

Toughness (consistency near plastic limit). A specimen of soil approximately 10 cm^3 in volume is moulded to the consistency of putty and then rolled out by hand on a smooth surface or between palms into a thread about 3 mm in diameter. The thread is then folded and re-rolled repeatedly. The sample gradually loses moisture and crumbles when the plastic limit is reached. The pieces of the crumbled thread are lumped together with a slight kneading action which should be continued until the lump crumbles. The tougher the thread near the plastic limit and the stiffer the lump when it finally crumbles, the more potent is the colloidal clay fraction in the soil. Weakness of the thread at the plastic limit and quick loss of coherence of the lump below the plastic limit indicate either inorganic clay of low plasticity or organic clay. Highly organic clays have a very weak and spongy feel at the plastic limit. Table 1.7 shows a guide to field identification and the application of dry strength, dilatancy and toughness concepts to the group symbols of the Unified Soil Classification System (AS 1726).

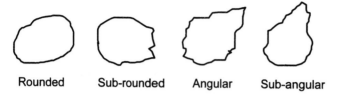

Rounded   Sub-rounded   Angular   Sub-angular

Figure 1.28. Description of particle shape.

Description of a soil. A systematic and standardized order of description similar to Table 1.8(a) (Section A2.3 of AS 1726) has been found to be suitable. Most of the descriptive terms needed in Table 1.8(a) can be found in Tables 1.6 and 1.7. For a description of colour and particle shape the following terms are used: the colour of a soil should be described in the moist condition as (modified if necessary) *pale*, *dark*, or *mottled* and black, white, grey, red, brown, orange, yellow, green and blue. Equidimensional particles may be described as *round(ed)*, *sub-rounded*, *sub-angular*, or *angular* as shown in Figure 1.28. The consistency of cohesive soils is described in terms of its undrained shear strength (Chapter 4) as shown in Table 1.8(b) (AS 1726) whilst the consistency of non-cohesive soils is described in terms of the density index according to Table 1.8(c) (AS 1726).

Table 1.7. Guide to the field identification of coarse and fine-grained soils (AS 1726).

Group symbol	Field identification of coarse grained soils		
GW	Wide range in grain size and substantial amounts of all intermediate sizes, not enough fines to bind coarse grains, no dry strength.		
GP	Predominantly one size or range of sizes with some intermediate sizes missing, not enough fines to bind coarse grains, no dry strength.		
GM	Dirty materials with excess of non-plastic fines, zero to medium dry strength.		
GC	Dirty materials with excess of plastic fines, medium to high dry strength.		
SW	Wide range in grain size and substantial amounts of all intermediate sizes, not enough fines to bind coarse grains, no dry strength.		
SP	Predominantly one size or range of sizes with some intermediate sizes missing, not enough fines to bind coarse grains, no dry strength.		
SM	Dirty materials with excess of non-plastic fines, zero to medium dry strength.		
SC	Dirty materials with excess of plastic fines, medium to high dry strength.		
	Field identification of fine grained soils		
	Dry strength	Dilatancy	Toughness
ML	None to low	Quick to slow	None
CL, CI	Medium to high	None to very slow	Medium
OL	Low to medium	Slow	Low
MH	Low to medium	Slow to none	Low to medium
CH	High to very high	None	High
OH	Medium to high	None to very slow	Low to medium
Pt	Identified by colour, odour, spongy feel and generally by fibrous texture		

Table 1.8(a). Order of description of a soil.

Order of description	Details of description
Composition	Group symbol, soil name, plasticity or particle characteristics, colour, secondary components, and minor components.
Conditions	Moisture condition (disturbed or undisturbed), consistency (undis.)
Structure	Zoning, defects, cementing (undisturbed).
Additional observations	Soil origin and other matters if significant.

Table 1.8(b). Consistency terms for cohesive soils based on undrained shear strength c_u.

Term	c_u (kPa)	Field guide to consistency
Very soft	≤ 12	Exudes between the fingers when squeezed in hand.
Soft	> 12 ≤ 25	Can be moulded by light finger pressure.
Firm	> 25 ≤ 50	Can be moulded by strong finger pressure.
Stiff	> 50 ≤ 100	Cannot be moulded by fingers; can be indented by thumb.
Very stiff	>100 ≤ 200	Can be indented by thumbnail.
Hard	>200	Can be indented with difficulty by thumbnail.

Table 1.8(c). Consistency terms for non-cohesive soils.

Term	Density index
Very loose	≤ 0.15
Loose	> 0.15 ≤ 0.35
Medium dense	> 0.35 ≤ 0.65
Dense	> 0.65 ≤ 0.85
Very dense	> 0.85

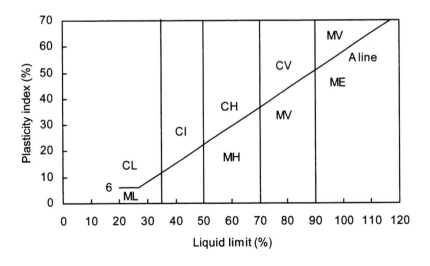

Figure 1.29. Plasticity chart according to British soil classification system.

Table 1.9. British soil classification system for engineering purposes (BS 5930).

Soil groups			Subgroups and laboratory identification				
Gravel and sand may be defined as sandy gravel and gravelly sand, etc.			Group symbol	Sub-group Symbol	Fines : % < 0.06 mm	LL %	
Coarse soils less than 35% is finer than 0.06 mm	Gravels more than 50% of coarse material is of gravel size (coarser than 2 mm)	Slightly silty or clayey gravel	G	GW GP	GW GPu, GPg	0-5	
		Silty gravel Clayey gravel	G-F	G-M G-C	GWM, GPM GWC, GPC	5-15	
		Very silty gravel Very clayey gravel	GF	GM GC	GML, etc. GCL GCI GCH GCV GCE	15-35	
	Sands more than 50% of coarse material is of sand size (finer than 2 mm)	Slightly silty or clayey sand	S	SW SP	SW SPu, SPg	0-5	
		Silty sand Clayey sand	S-F	S-M S-C	SWM, SPM SWC, SPC	5-15	
		Very silty sand Very clayey sand	SF	SM SC	SML, etc. SCL SCI SCH SCV SCE	15-35	
Fine soils more than 35% is finer than 0.06 mm	Gravelly or sandy silts and clays 35% to 65% fines	Gravelly silt Gravelly clay	FG	MG CG	MLG, etc. CLG CIG CHG CVG CEG		<35 35-50 50-70 70-90 >90
		Sandy silt Sandy clay	FS	MS CS	MLS, etc. CLS, etc.		
	Silts and clays 65% to 100% fines	Silt (M-soil) Clay	F	M C	ML, etc. CL CI CH CV CE		<35 35-50 50-70 70-90 >90
Organic soils			Descriptive letter *O* suffixed to any group or subgroup symbol				
Peat			*Pt*				

1.6.4 *Summary of British soil classification system for engineering purposes*

The British soil classification system is shown in Table 1.9. The qualifying terms are similar to the Unified Soil Classification System with more subgroup symbols for both coarse and fine soils. Poorly graded gravel and sand have been divided into two group symbols by adding *Pu* and *Pg* to represent uniform and gap-graded soils respectively. In coarse soils subgroups have been defined by adding a third letter (*M* or *C*) to well-graded or poorly graded gravel and sand. Fine soils, and the fine fraction of coarse soils, are shown by *F*, and the subgroup symbols of fines and fine fraction of coarse soils are: *L* (low plasticity), *I* (intermediate plasticity), *H* (high plasticity), *V* (very high plasticity), *E* (extremely high plasticity) and *U* (upper plasticity range) as the second or third letter (Figure 1.29 & Table 1.9). Field identification is based on particle size, grading, structure, compactness, dry strength, dilatancy, consistency and weathering (BS 5930, 1981).

1.7   COMPACTION

1.7.1   *Dry density-moisture content relationship*

Compaction is the process of reducing the air content by the application of energy to the moist soil. Compaction increases the number of particles within a specific volume thereby increasing the shear strength. Consequently, any displacement due to external loading and surface settlement thereby reduced because of the denser structure. Compaction of soil samples in the laboratory is carried out in standard cylinders. Energy is applied by a hammer of standard size and mass dropping freely from a standard height on a layer of the sample inside the cylinder. Each specimen is made in 3 or 5 layers with a specified number of blows depending on the codes and method of testing. In the field, the energy is applied by means of different types of rollers. Standard codes specify this energy per unit volume of soil. The term *compactive effort* is used to describe the energy given to a unit volume of soil. Compaction increases the number of particles per unit volume and dry density is used to indicate the degree of compaction.

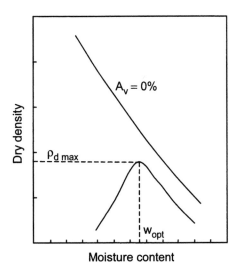

Figure 1.30. Soil behaviour under a specified compactive effort.

Compactive effort is most effective if a uniform mixture of soil and water is used. Dry density increases progressively with increasing moisture content to a maximum under a constant compactive effort. The moisture content corresponding to the maximum dry density is called the *optimum moisture content (OMC)*. Increasing the moisture content beyond this value reduces the dry density.

The common and natural behaviour of a soil-water mixture under a specified compactive effort is shown in Figure 1.30. The curve representing the experimental results is termed the *dry density-moisture content* curve. Maximum dry density and the relevant optimum moisture content are obtained from this plot.

Lambe (1958) explained the shape of the dry density-moisture content curve in terms of the development and characteristics of the diffuse double layer with increasing moisture content. At low moisture contents and on the dry side of the optimum moisture content there is insufficient water to form a double layer and the structure of the fine particles is of a flocculated type. The increase in moisture content causes the formation of the double layer and randomly distributed particles become more orientated. This produces a lubrication effect between the large particles causing them to slide over each other as the compactive effort is applied. At the maximum dry density the specimen of compacted soil has a high degree of saturation. Further increase in the moisture content has a dilution effect and dry density decreases while the degree of saturation remains approximately constant. The effects of compactive effort and moisture content are insignificant on granular soils that have little or no fines. Test points on the wet side of the *OMC* may not be established by a standard compaction test due to loss of water during compaction. A method combing vertical vibration and water gives satisfactory results. The theoretical relationship between dry density, air content, and moisture content is expressed by Equation 1.18. Figure 1.31 shows the plot of this equation for $A_v = 0\%$ (termed the *zero air curve*), 5% and 10% for a certain value of specific gravity of the solids. In practice the compactive effort cannot de - air the voids completely, thus the dry density-moisture content curve is

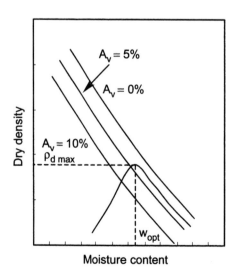

Figure 1.31. Theoretical curves and experimental compaction curve.

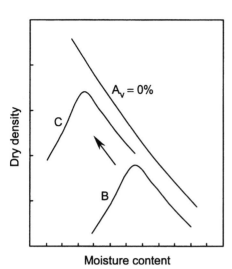

Figure 1.32. The effect of compactive effort on maximum dry density ρ_d and *OMC*.

always on the left side of the zero air curve. The descending part of the compaction curve usually has a minimum air content of 3%-5%. If the descending part intersects the zero air curve, either the test data or the specific gravity of the solids is wrong. In general the right side of the zero air curve is an impossible state for experimental data points. It is always convenient to plot the air void curves of 5%, 10%, etc. to understand the development of compaction and the progress of the degree of saturation as moisture content increases. The standard energy applied to soil specimens in the laboratory is 592 kNm per 1 m^3 of soil and was suggested by soil scientist R. R. Proctor during the 1930s. In time, the need for having a denser and stronger soil was increased and the Proctor method was modified by increasing the compactive effort to 2695 kNm per 1 m^3 of soil. Increasing the compactive effort increases the dry density and reduces the optimum moisture content, as shown in Figure 1.32. Curve *C* has a higher compactive effort than curve *B*. The increase in maximum dry density is believed to be as a result of greater orientation of the fine particles under the increased compactive effort. Figure 1.33 provides an estimation of the dry densities and optimum moisture contents for the fine and coarse-grained soils defined by the Unified Soil Classification System.

1.7.2   *Standard laboratory compaction test*

The moulds used for the compaction test, with internal diameters of 101.6 mm or 105 mm and 152.4 mm, and the hand-operated rammer are shown in Figure 1.34. Each mould has a removable collar assembly with a detachable base plate. The mould type is selected depending on the size fraction of the particles in accordance to the relevant standard code. In the standard Proctor test compaction of the prepared sample is conducted in 3 layers with a rammer of 2.5 or 2.7 kg. The number of blows per layer is 25 or 27 for the small size mould and 56 or 60 for the large size mould. This small variation in the number of blows and the mass of rammer is due to minor differences in the diameter and height of the moulds in the different codes.

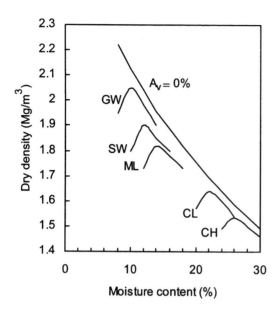

Figure 1.33. Typical compaction curves for coarse and fine grained soils.

In the modified Proctor test the layers are increased to 5 and a heavier rammer of 4.5 kg or 4.9 kg is used. Mechanical versions of hand rammers, which comply with most codes, are also available. Two common types are portable, lightweight vibrating rammer and the stationary automatic compactor. Table 1.10 shows the adopted standards for the compaction test. Details of the preparation of samples, treatment of oversize material, moisture content intervals, and plotting of the experimental results are in the designated codes in the

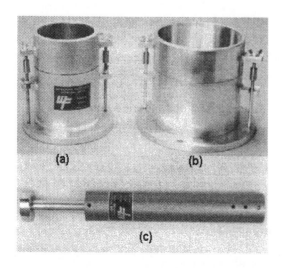

Figure 1.34. (a) Proctor mould, (b) modified Proctor mould, (c) rammer (Wykeham Farrance).

Table 1.10. Adopted standards for compaction test.

Standard designation	Mould: Diameter (mm), Height (mm), Volume (cm³)	Rammer: Drop (mm), Mass (kg)	No. of layers, No. of blows per layer
ASTM D-698	101.6, 116.5, 944	305, 2.5	3, 25
(Proctor)	152.4, 116.5, 2100	457, 4.5	5, 25
ASTM D-1557,	101.6, 116.5, 944	305, 2.5	5, 25
AASHTO T99, T180 (mod. Proc.)	152.4, 116.5, 2100	457, 4.5	5, 56
BS 1377 & BS 1924 (Proctor)	105.0, 115.5, 1000	300, 2.5	3, 27
(modified using AASHTO)	105.0, 115.5, 1000	450, 4.5	5, 27
AS 1289.5.1.1	105.0, 115.5, 1000	300, 2.7	3, 25
(standard)	152.0, 132.5, 2400		3, 60
AS 1289.5.2.1	105.0, 115.5, 1000	450, 4.9	5, 25
(modified)	152.0, 132.5, 2400		5, 100

first column of Table 1.10. The bulk density of the soil in each test is calculated by dividing the mass of the compacted soil by the volume of the mould. After removing the compacted soil from the mould, a representative sample from the whole height of the specimen is obtained for the determination of its moisture content. Dry density is calculated from Equation 1.11 or 1.12. The test is repeated 6 times over a range of moisture contents to establish the dry density-moisture content curve.

Minimum and maximum dry density of cohesionless materials. Moisture content has little or no influence on the compaction characteristics of granular soils (except when the soil is fully saturated). Their state of compaction can be obtained by relating the dry density to the minimum and maximum dry densities obtained in the laboratory. The test technique requires a vibratory rammer or table and a pouring method. Accordingly the density index (Equation 1.13) may be calculated in terms of minimum and maximum dry densities if a specified density index is required. Combining Equations 1.13 and 1.14:

$$I_D = \frac{e_{max} - e}{e_{max} - e_{min}} = \frac{\rho_{d\,max}}{\rho_d} \frac{\rho_d - \rho_{d\,min}}{\rho_{d\,max} - \rho_{d\,min}} \tag{1.39}$$

Minimum dry density is obtained using an oven-dried material and is referred to as the dry placement method. In this method a mould with known volume and a pouring device are selected according to the maximum size of the soil particles, as recommended by the relevant code. To prepare a uniform sample with minimum segregation, the mould is filled with material, using a funnel or by means of a scoop, in a steady stream, free of vibration and any other disturbances. After levelling the material the mass of the mould and its contents is measured for the calculation of dry density. The test is repeated several times until a lowest value for the mass inside the mould is obtained. Maximum dry density is determined by the wet placement method using either a vibratory rammer (BS 1377) or a vibratory table (ASTM D-4253, D-4254, and AS 1289.5.5.1). A diagrammatic section (and view) of a vibrating table permitted by AS 1289.5.5.1 is shown in Figure 1.35. Generally the apparatus comprises a vibratory table, which oscillates at a nominated rate (say 3600 vibrations per minute) and is equipped with two to five moulds with nominal volumes of 1 litre to 5 litres.

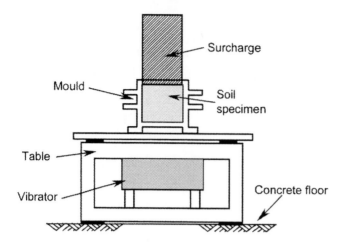

Figure 1.35. Apparatus for vibratory compaction.

A surcharge of 5 to 70 kg is necessary to provide vertical support during vibration. The saturated material is poured into the mould by means of a funnel or a scoop until it is filled and overflowing. If necessary, water is added to ensure full saturation. The mould is vibrated during the filling process with smaller amplitude of vibration. The surcharge is lowered to the surface of the sample and the vibrator is set to a required amplitude for approximately 10 minutes or until the settlement of the surface has stopped. After removing the surcharge and levelling off the material, the contents of the mould is placed in an oven to dry to a constant mass. The maximum dry density is then calculated by dividing the dry mass by the volume of the mould.

1.7.3 *Compaction in the field*

Compaction in the field may be divided into the two categories of deep and surface compaction depending on the thickness of the compacted layer. Deep compaction is based on vibro- compaction techniques, dynamic compaction and the use of explosives. A detailed study of deep compaction is beyond the scope of this book, however, due to its increasing applications in surface and near-surface compaction, a brief description of dynamic compaction is included at the end of this section. Surface compaction includes the compaction of layers in 0.15 m to 0.3 m layers, however, by using vibratory techniques and dynamic compaction, the depth of the compacted layer may be increased to 1.0 m or more in granular soils. The maximum dry density obtained in laboratory conditions is achievable in the field by using different types of compactors that employ different concepts in the application of the compactive effort. Degree of compaction is a means of comparing the field density with laboratory results and is defined as the ratio of the dry density in the field to the maximum dry density obtained in the laboratory. In most construction works the degree of compaction is specified as 95% or more, usually without any designation of the method of compaction. The bulk and dry density in the field are measured using one of the methods described in Section 1.3. If the compacted soil does not expand or consolidate after the field compaction is completed, its dry density remains constant regardless of the magnitude of its moisture content.

In the field, the moisture content may vary from zero to a maximum value, which is the intersection of a horizontal line drawn from the maximum dry density on the experimental curve with the theoretical, zero air curve. Stability studies of compacted soils show that a compacted fill may undergo compression or consolidation due to self-weight resulting a substantial surface settlement or failure (Lawton et al., 1989; Brandon et al., 1990). In order to predict the stability of a compacted fill it is necessary to perform laboratory strength and consolidation tests on the compacted specimens. Compaction equipment commonly used in the field are:

Sheep's foot roller. This roller consists of a steel drum with studs or feet 40 to 100 cm^2 in area projecting some 200 to 250 mm. It is either pushed or pulled by a tractor or is self-propelled unit. A sheep's foot roller is convenient for cohesive soils and applies compaction by a combination of tamping and kneading. Contact pressures applied by the projected studs are high and vary from 700 to 4000 kPa. The thickness of a compacted layer is of the order of 0.15 m to 0.3 m. Bond between two compacted layers is very strong due to the action of the studs. The type of the roller (light or heavy) and the number of passes must be determined by the site engineer after conducting a suitable number of field density tests.

Pneumatic tyre roller. This type of roller is either towed or is a self-propelled unit commonly equipped with an outside fitted box for weight adjustment. It usually has two axles with 2 to 5 (or more) tyres per axle and the back axle is designed to overlap the area compacted by the front axle. The wheels are sometimes equipped with a wobbling mechanism to cover the lower surfaces. The tyres have flat treads and therefore cannot apply sideways displacements to the soil particles. This compactor is suitable for many different soils and applies compaction by kneading only. Compacted layers vary from 0.15 m to 0.5 m depending on the mass of the compactor. Usually only a few passes are necessary to achieve maximum dry density.

Vibratory rollers. These are applicable to many types of soils but are primarily designed for granular materials. In the smooth drum vibrators the vibrating mechanism consists of eccentrically arranged weights (Terzaghi & Peck, 1967), which in rotation apply a vertical vibration to the drum. In pneumatic-tyred vibratory compactors the vibrating mechanism is applied to the axle of the unit. In the vibrating plate compactors a plate or plates that are in contact with the ground surface are subjected to vibration; however the depth of influence is not as great as smooth drum and pneumatic tyred vibrators. These latter two machines compact the soil to a depth of up to 1.0 m, but an effective depth of 0.3 m is generally accepted if the soil consists of silt and clay. Vibratory rollers are available in towed or self-propelled units.

Dynamic compaction. In this method a weight of 100 to 200 kN is dropped from a height of 15 to 40 m using crawler cranes. Higher weights of 500 kN may also be used if a deeper compaction is required. Establishing the number of passes and drops per pass needs a careful in-situ study. Experience has shown that granular materials require more energy than fined grained soils. In the latter the applied energy increases the pore water pressure and adequate time is needed for dissipation of any excess pore pressure. In the laboratory dynamic compaction is modelled using the centrifuge method in which the

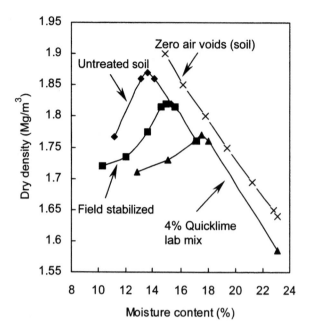

Figure 1.36. Compaction curves for a clay soil stabilized by lime (Aysen et al., 1996).

force can be increased up to $100g$ ($g = 9.81$ m/sec^2). Here dynamic compaction is provided by dropping a steel rod several mm in diameter and approximately 100 mm in length on the surface of the sample. The centrifuge machine for this purpose should be capable of accepting a 0.5 m cube of soil.

1.7.4 *Application of compaction test in soil stabilization techniques*

The standard compaction test is also used in soil stabilization to investigate the effect of the stabilizer on the maximum dry density and optimum moisture content. Depending on the type of the material and the stabilizer (e.g. lime or cement) the maximum dry density may increase or decrease slightly or moderately causing variation in the optimum moisture content in the reverse direction. In these cases the strength of the treated material is not represented by the maximum dry density and the increase in strength is mostly due to the formation of the bonds between the particles. Investigating the optimum moisture content is vital to ensure that sufficient water is available for hydration. An example of lime stabilization of clay material is shown in Figure 1.36 (Aysen et al., 1996).

Example 1.9

The following results were obtained from a standard compaction test.

Mass (g)	1768	1929	2074	2178	2106	2052	2007
w (%)	4	6	8	10	12	14	16

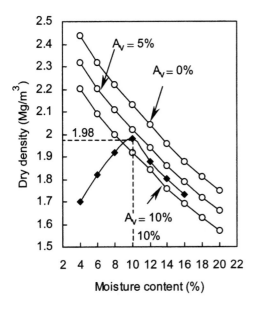

Figure 1.37. Example 1.9.

Determine the optimum moisture content and maximum dry density. Plot the curves of 0%, 5% and 10% air content and give the value of air content at the maximum dry density. The volume of the standard mould is 1000 cm³ and $G_s = 2.7$.

Solution:

Calculate dry density for each test and tabulate the results. Sample calculation for $w = 4\%$:

$\rho = M/V = 1768/1000 = 1.768 \text{ Mg/m}^3$,

$\rho_d = \rho/(1+w) = 1.768/(1+0.04) = 1.70 \text{ Mg/m}^3$.

ρ_d (Mg/m³)	1.70	1.82	1.92	1.98	1.88	1.80	1.73
w (%)	4	6	8	10	12	14	16

To plot the constant air content curves use Equation 1.18 and tabulate the results. The results are shown in Figure 1.37. It is seen that the maximum dry density is 1.98 Mg/m³ and optimum moisture content is 10%.

w (%)		4	6	8	10	12	14	16	18	20
ρ_d (Mg/m³)	$A_v = 0\%$	2.44	2.32	2.22	2.13	2.04	1.96	1.88	1.81	1.75
	$A_v = 5\%$	2.32	2.20	2.11	2.02	1.94	1.86	1.79	1.72	1.66
	$A_v = 10\%$	2.20	2.09	2.00	1.92	1.84	1.76	1.69	1.63	1.57

Calculate the air content at maximum dry density using Equation 1.18:

$\rho_d = G_s \rho_w (1 - A_v)/(1 + wG_s) = 1.98$,

$$2.7 \times 1.0(1 - A_v)/(1 + 0.1 \times 2.7) = 1.98,$$
$$A_v = 0.0686 = 6.86\%.$$

Example 1.10

The following results were obtained from a standard compaction test ($G_s = 2.65$):

Sample number	1	2	3	4	5
Moisture content (%)	16.2	16.7	19.0	20.4	21.6
Dry density (Mg/m^3)	1.580	1.620	1.647	1.605	1.566

To obtain two extra points close to the maximum dry density two standard compaction tests were performed and tabulated. (a) Determine ρ_{dmax} and w_{opt} and plot the zero air curve, (b) a 0.3 m layer of this soil is compacted in the field to its maximum dry density. After some time the natural moisture content of the layer is measured to be 16%. How much water in terms of m/m^2 is needed to make the layer fully saturated?

Sample number	6	7
Mass of mould + compacted wet soil (kg)	8.966	8.974
Mass of mould (kg)	7.0	7.0
Volume of mould (cm^3)	1000	1000
Mass of sub-sample taken from mould (g)	178.8	155.8
Mass of sub-sample after drying (g)	152.3	131.8

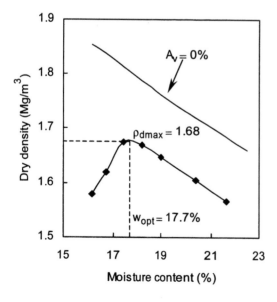

Figure 1.38. Example 1.10.

Solution:

(a) For sample number 6:

$\rho_d = \rho/(1+w) = 1.966/(1+0.174) = 1.675$ Mg/m³. Similarly for sample number 7:

$w = 0.182$, $\rho_d = 1.670$ Mg/m³. For $A_v = 0\%$ use Equation 1.17; results are tabulated below and are shown in Figure 1.38 from which $\rho_{d\max} = 1.68$ Mg/m³ and $w_{opt} = 17.7\%$.

Test points	W (%)	ρ_d (Mg/m³) $A_v = 0\%$	Test	Test points	W (%)	ρ_d (Mg/m³) $A_v = 0\%$	Test
1	16.2	1.854	1.580	3	19.0	1.762	1.647
2	16.7	1.837	1.620	4	20.4	1.720	1.605
6	17.4	1.814	1.675	5	21.6	1.685	1.566
7	18.2	1.788	1.670				

(b) $\rho_d = G_s \rho_w /(1+wG_s) = 1.68$, $2.65 \times 1.0/(1+2.65w) = 1.68 \rightarrow w = 21.8\%$.

$\Delta w = 21.8 - 16.0 = 5.8\%$, $\Delta M_w = 1.68 \times 0.058 = 0.097$ Mg/m³ or $\Delta V_w = 0.097$ m³/m³.

$\Delta V_w (0.3$ m layer$) = 0.097 \times 0.3 = 0.029$ m³/m², $h_w = 0.029/1.0 = 0.029$ m/m².

1.8 PROBLEMS

1.1 The following data are given for a specimen of clay soil: $M = 221$ g, $M_s = 128$ g, $G_s = 2.7$, $S_r = 75\%$. Determine the total volume and the porosity of the specimen.

Answers: 171.4 cm³, 72.3%

1.2 Dry soil ($G_s = 2.71$) is mixed with 16% by weight of water and compacted to produce a cylindrical sample of 38 mm diameter and 76 mm long with 6% air content. Calculate the mass of the mixed soil that will be required and its void ratio.

Answers: 177.6 g, 0.525

1.3 During a field density test 1850 g of soil was excavated from a hole having a volume of 900 cm³. The oven-dried mass of the soil was 1630 g. Determine the moisture content, dry density, void ratio and degree of saturation. $G_s = 2.71$.

Answers: 13.5%, 1.81 Mg/m³, 0.496, 73.7%

1.4 A soil specimen has a moisture content of 21.4%, void ratio of 0.72, and $G_s = 2.7$. Determine:

(a) bulk density and degree of saturation,

(b) the new bulk density and void ratio if the specimen is compressed undrained until full saturation is obtained.

Answers: 1.905 Mg/m³, 80.1%, 2.08 Mg/m³, 0.578

1.5 The moisture content of a specimen of a clay soil is 22.4%. The specific gravity of the solids is 2.71.

(a) Plot the variation of void ratio with degree of saturation and calculate the void ratio, and the dry and wet densities at 50% saturation,

(b) a sample of this soil with initial degree of saturation of 50% is isotropically compressed to achieve a void ratio of 0.55. Calculate the volume change in terms of percentage of the initial volume. How much of this volume change is due to the outward flow of water from the sample?

Answers: 1.214, 1.224 Mg/m^3, 1.50 Mg/m^3, 30.0%, 2.6%

1.6 The results of a particle size analysis are tabulated below. The total mass was 469 g. Plot the particle size distribution curve and determine the coefficient of uniformity, coefficient of curvature and soil description.

Answers: 13.3, 3.2, *GW*

Sieve size (mm)	Mass retained (g)	Sieve size (mm)	Mass retained (g)
63	0.0	4.75	50
37.5	26	2.36	137
19.0	28	1.18	46
13.2	18	0.6	31
9.5	20	0.212	34
6.7	49	0.075	30

1.7 The following data were recorded in a liquid limit test using the Casagrande apparatus. Determine the liquid limit of the soil. Classify the soil assuming $PL = 19.8\%$.

Number of blows	Mass of can (g)	Mass of wet soil + can (g)	Mass of dry soil + can (g)
8	11.80	36.05	29.18
16	13.20	34.15	28.60
27	14.10	36.95	31.16
40	12.09	33.29	28.11

Answers: 34.2%, *CL*

1.8 The recorded data in a liquid limit test using the cone penetration method are as follows. Determine the liquid limit of the soil.

Cone penetration (mm)	14.1	18.3	22.1	27.2
Moisture content (%)	28.3	42.2	52.4	63.4

Answer: 47.0%

1.9 The maximum and minimum void ratios for a sand are 0.805 and 0.501 respectively. The field density test performed on the same soil has given the following results: $\rho = 1.81$ Mg/m^3, $w = 12.7\%$. Assuming $G_s = 2.65$, compute the density index.

Answer: 0.51

1.10 The following results are obtained from a standard compaction test:

Mass of compacted soil (g)	1920.5	2051.5	2138.5	2147.0	2120.0	2081.5
Moisture content (%)	11.0	12.1	12.8	13.6	14.6	16.3

The specific gravity of the solids is 2.68, and the volume of the compaction mould is 1000 cm^3. Plot the compaction curve and obtain the maximum dry density and optimum moisture content. Plot also the 0%, 5% and 10% air void curves. At the maximum dry density, calculate the void ratio, degree of saturation and air content. If the natural moisture content in the field is 11.8%, what will be the possible maximum dry density if the soil is compacted with its natural moisture content?

Answers: 1.907 Mg/m^3, 13.0%, 0.406, 85.8%, 4.1%, 1.80 Mg/m^3

1.9 REFERENCES

ASCE. 1962. Symposium on grouting, *Journal SMFE, ASCE* 87(SM2): 1-145.

ASTM D-2487. 1998. Standard classification of soils for engineering purposes (Unified Soil Classification System), *American society for testing and materials*. West Conshohocken, PA.

Atkinson, J. 1993. *An introduction to the mechanics of soils and foundations*. London: McGraw-Hill.

Australian Standard. 1993. *Geotechnical site investigations*. 3rd edition. Australia, NSW: Standard Association of Australia.

Aysen, A., Ayers, R. & Ricks, J. 1996. *Soil stabilisation with cement and aluminium hydroxide*. Research Report No CE96/5, ISBN 0909756295. Toowoomba: The University of Southern Queensland.

Bohen, H.L., McNeal, B.L & O'Connor, G.A. 1985. *Soil chemistry*. New York: John Wiley.

Bowles, J.E. 1996. *Foundation analysis and design*. New York: McGraw-Hill.

Brandon, T.L., Duncan, J.M. & Gardener, W.S. 1990. Hydrocompression settlement of deep fills. *Journal SMFED, ASCE* 116(GT 10): 1536-1548.

Brewer, R. 1964. *Fabric and mineral analysis of soils*. New York: John Wiley & Sons.

Brewer, R. & Sleeman, J.R. 1988. *Soil structure and fabric*. Australia, Adelaide: CSIRO Div. Soils.

British Standard 5930. 1981. *Code of practice for site investigation*. London: British Standards Institution.

Casagrande, A. 1948. Classifications and identifications of soils. *Translated*: ASCE 113: 901-991.

Craig, R.F. 1997. *Soil mechanics*. 6th edition. London: E & FN SPON.

FitzPatrick, E.A. 1983. *Soils, their formation, classifications and distributions*. London: Longman.

Fookes, P.G. & Vaughan, P.R. 1986. *A handbook of engineering geomorphology*. Surry: Surry University Press.

Isbell, R.F. 1996. *The Australian soil classification*. Australia, Collingwood: CSIRO Publishing.

Jenny, H. 1941. *Factors of soil formation*. New York: McGraw-Hill.

Lambe, T.W. 1958. The engineering behavior of compacted clay. *Journal SMFE, ASCE* 84(SM2), May 1958.

Lawton, E.C., Fragaszy, R.J. & Hardcastle, J.H. 1990. Collapse of compacted clayey soils, *Journal SMFED, ASCE* 115(GT 9): 1252-1267.

McLaren, R.G. & Cameron, K.C. 1996. Iron exchange in soils and soil acidity. *Soil science: sustainable production and environmental protection*. Auckland: Oxford Press.

Moore, C.A. 1970a. Suggested method for application of X-ray diffraction of clay structural analysis to the understanding of the engineering behavior of soils. *Special procedures for testing soil and rock for engineering purposes*. Philadelphia: ASTM publication.

Moore, C.A. 1970b. Suggested techniques for measuring the fabric of engineering soils. *Special procedures for testing soil and rock for engineering purposes*. Philadelphia: ASTM publication.

Ollier, C.D. 1975. *Weathering*. London: Longman Group Ltd..

Rowe, P.W. 1962. The stress-dilatancy relation for static equilibrium of an assembly of particles in contact. *Proc. roy. soc.* A, 269: 500-527.

Sherard, J.L. 1953. The influence of soil properties and construction methods on the performance of homogeneous earth dams. *US bureau of reclamation.* Tech. Memo (645).

Skempton, A.W. & Northy, R.D. 1953. The sensitivity of clays. *Geotechnique* 3(1): 30-35.

Sposito, G. 1989. *The chemistry of soils.* Oxford: Oxford University Press.

Tan, K.H. 1994. *Environmental soil science.* New York: Marcel Dekker Inc..

Terzaghi, K. & Peck, R. B. 1967. *Soil mechanics in engineering practice.* New York: John Wiley.

White, R.E. 1979. *Introduction to the principles and practice of soil science.* Melbourne: Blackwell Scientific.

Whittig, L.D. 1964. X-ray diffraction techniques for mineral identification and mineralogical composition. In, *Methods of soil analysis, American society of agronomy monograph.* Part 1, Chapter 49.

Whyte, I.L. 1982. Soil plasticity and strength: a new approach using extrusion. *Ground engineering* 15(1): 16-24.

Yong, R.N. & Warkentin, B.P. 1966. *Introduction to soil behaviour.* New York: The Macmillan Company.

CHAPTER 2

Effective Stress and Pore Pressure in Saturated Soils

2.1 INTRODUCTION

In a saturated soil with no water movement, equilibrium exists between stresses in the solid and liquid phases. The stress related to the internal forces acting on the contact points of the particles is termed the *effective stress* whilst the stress within the liquid phase or water is termed *pore pressure*. The combination of these two stresses represents the *total stress* at a point. When a saturated soil is subjected to external forces, the state of equilibrium is altered and changes the effective stresses and pore pressures from their initial values to new values. This increase in the pore pressure, called the *excess pore pressure*, dissipates in time depending on the drainage conditions and permeability of the soil. This pressure dissipation, in turn, causes a reduction in the volume of the soil, which is predicted by a mathematical-physical model called the *consolidation analogy*.

2.2 STATE OF STRESS AT A POINT DUE TO SELF-WEIGHT

2.2.1 *The effective stress concept*

Terzaghi's concept of effective stress states that:

$$\sigma = \sigma' + u \tag{2.1}$$

where σ is the total normal stress at a point in a specified direction or plane, σ' is the effective normal stress on that plane resisted by the particles and u is the pore pressure acting on the plane. To fully understand the concept of effective stress, consider a cylinder containing coarse granular material in a dry or a wet condition subjected to a force N' (Figure 2.1(a)). Assuming there is no friction between the soil particles and the inner surface of the cylinder, the entire force N' is transmitted completely to the base. The average vertical stress at the base of the cylinder is the ratio of N' to the internal cross-sectional area of the cylinder. This stress is much less than the stresses created at the contact points. As there is no pore pressure in the system the total stress and effective stress at the base of the cylinder are identical. Consider, now, that the cylinder is filled with water until the soil is fully saturated and that a force U is applied through another piston as shown in Figure 2.1(b). The loading plate that carries force N' has a hole h for the passage of water to ensure a uniform pore pressure through the whole system. The fraction of the base area occupied by the contact points between the particles and the base may be ignored without

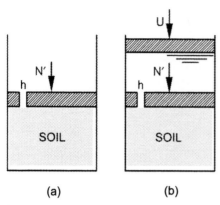

(a)   (b)

Figure 2.1. Illustration of the concept of effective stress.

significant error. If the self-weight of the components are negligible (in this example) then the total vertical stress at the bottom of the cylinder is the sum of the pore pressure caused by U and the effective vertical stress caused by N'. In this example the loading systems for the particles and the water are independent with no apparent relationship. However, in a real soil, N' and U may have the same source, changing the values of effective stresses and pore pressures as time related functions. Equation 2.1 satisfies the states of stress from the initial state through to the final state. Figure 2.2 shows the average sectional area ab that is equivalent to the area of the base of cylinder in the example above. The effective normal stress on the plane ab is equal to the sum of the components of the forces F_1 to F_7 perpendicular to the plane ab divided by the sectional area represented by ab.

2.2.2   *Effective stress and pore pressure within unconfined and confined aquifers*

A schematic illustration of the different types of ground water conditions is shown in Figure 2.3 (Bell, 1993). In an unconfined ground water the surface of the water, known as the *phreatic surface, water level* or *water table*, is in equilibrium with the atmospheric pressure; that is, the aquifer is not overlain by a material of lower permeability. In a confined aquifer both the upper and lower boundaries are confined with impermeable strata, and the water may have sufficient pressure to rise above the overlaying stratum (artesian conditions). In a leaky aquifer, a layer of low permeability material separates the confined and unconfined aquifers. This layer slowly transmits the water and creates a water table in the unconfined region and a piezometric level (the level of water rising in a piezometric standpipe) in the confined region.

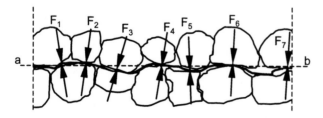

Figure 2.2. The concept of effective stress.

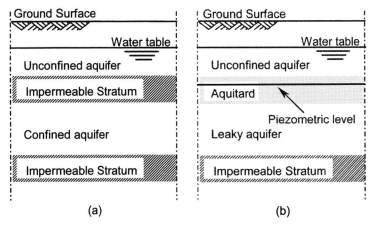

Figure 2.3. (a) Unconfined and confined aquifers, (b) leaky aquifer.

Assuming that there is no movement of water within the voids of the soil, the effective vertical stress σ'_v at depth z for three possible positions of the water table (Figure 2.4) is given by:

$$\sigma'_v = \sigma_v - u, \ \sigma'_v = \rho gz - u = \gamma_{sat} z - u.$$

If the water table is at the ground surface (Figure 2.4, case a), then

$$\sigma'_v = \gamma_{sat} z - \gamma_w z = (\gamma_{sat} - \gamma_w)z = \gamma' z \tag{2.2}$$

In case b the water table is below the ground surface and therefore

$$\sigma'_v = \gamma_{sat} z_w + \gamma_{dry}(z - z_w) - \gamma_w z_w = \gamma_{dry}(z - z_w) + \gamma' z_w \tag{2.3}$$

In case c where the water table is above the ground surface:

$$\sigma'_v = (\gamma_{sat} z + \gamma_w h_w) - \gamma_w(z + h_w) = \gamma' z,$$

which is identical to Equation 2.2 and shows that σ'_v at a point below the water table is constant regardless of the position of the water table above the ground surface.

For a soil section composed of n layers each having a thickness of h_i:

$$\sigma'_v = \Sigma \gamma z - u = \sum_{i=1}^{i=n} \gamma_{ei} h_i \tag{2.4}$$

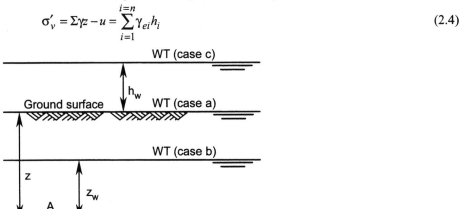

Figure 2.4. Effective stress in a saturated soil.

where γ_{ei} is the effective unit weight of each layer. In dry and partially saturated layers γ_{ei} is the dry or wet unit weight; in saturated layers with hydrostatic water pressure, γ_{ei} is the submerged unit weight. In the presence of seepage forces due to movement of the water through the voids the hydrostatic based equations become invalid and the effective stress must be evaluated using the main definition given in Equation 2.1.

Example 2.1

A soil section is comprised of two layers with the following properties: Soil 1 (0 to 3 m): $\rho_{dry} = 1.8$ Mg/m³, $\rho_{sat} = 2.0$ Mg/m³; soil 2 (3 m to 8 m): $\rho_{sat} = 2.1$ Mg/m³. The water table is 1.5 m below the ground surface. Plot the total vertical stress, pore pressure, and effective vertical stress over the soil section.

Solution:

On the ground surface, $\sigma_v = 0.0$, $u = 0.0$, $\sigma'_v = 0.0$.
At $z = 1.5$ m: $\sigma_v = 1.8 \times 9.81 \times 1.5 = 26.5$ kPa, $u = 0.0$, $\sigma'_v = 26.5$ kPa.
At $z = 3.0$ m: $\sigma_v = 1.8 \times 9.81 \times 1.5 + 2.0 \times 9.81 \times 1.5 = 55.9$ kPa,
$u = 1.0 \times 9.81 \times 1.5 = 14.7$ kPa, $\sigma'_v = 55.9 - 14.7 = 41.2$ kPa.
At $z = 8.0$ m: $\sigma_v = 2.1 \times 9.81 \times 5.0 + 55.9 = 158.9$ kPa, $u = 1.0 \times 9.81 \times 6.5 = 63.8$ kPa,
$\sigma'_v = 158.9 - 63.8 = 95.1$ kPa. The results are presented in Figure 2.5.

Example 2.2

A soil profile is shown in Figure 2.6(a). Plot the distribution of total vertical stress, pore pressure, and effective vertical stress up to a depth of 12 m.

Solution:

Calculate the dry and saturated densities in soil 1: $V_v + V_s = 1$, $e = V_v / V_s$. Thus:

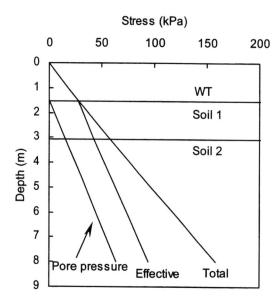

Figure 2.5. Example 2.1.

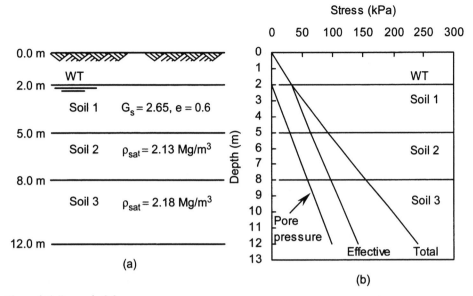

Figure 2.6. Example 2.2.

$V_s = 1/(1+e)$ and $V_v = e/(1+e)$, for $e = 0.6$: $V_s = 0.625$ m^3, $V_v = 0.375$ m^3.
$\rho_{dry} = G_s \times \rho_w \times V_s = 2.65 \times 1.0 \times 0.625 = 1.656$ Mg/m^3,
$M = 1.656$ Mg $+ 0.375$ m$^3 \times 1.0$ Mg/m$^3 = 2.031$ Mg, $\rho_{sat} = 2.031$ Mg/m^3.
The results are tabulated below and presented in Figure 2.6(b).

Depth (m)	σ_v (kPa)	u (kPa)	σ'_v (kPa)
0	0.0	0.0	0.0
2	32.5	0.0	32.5
5	92.3	29.4	62.9
8	155.0	58.9	96.1
12	240.5	98.1	142.4

Example 2.3

Resolve Example 2.2 assuming that: (a) the water table is at the ground surface, (b) the water table is 2 m above the ground surface.

Solution:

Calculations are summarized in the table below and are presented in Figure 2.7.

Depth (m)	(a)			(b)		
	σ_v (kPa)	u (kPa)	σ'_v (kPa)	σ_v (kPa)	u (kPa)	σ'_v (kPa)
0	0.0	0.0	0.0	19.6	19.6	0.0
5	99.6	49.0	50.6	119.3	68.7	50.6
8	162.3	78.5	83.8	181.9	98.1	83.8
12	247.8	117.7	130.1	267.4	137.3	130.1

Note: effective stresses at both cases are identical.

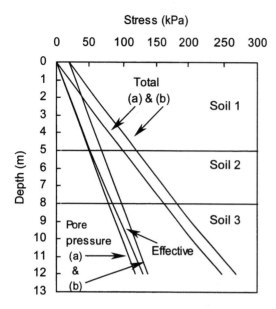

Figure 2.7. Example 2.3.

Example 2.4

The lower sand layer in the soil profile of Figure 2.8(a) is in an artesian condition. Calculate (a) σ'_v at the top and base of the clay layer, (b) the height of water in the standpipe for $\sigma'_v = 0$ at the base of the clay layer, (c) the maximum depth of the proposed excavation (Figure 2.8(b)).

Solution:

(a) At $z = 4.0$ m: $\sigma_v = (1.7 \times 2.0 + 2.1 \times 2.0) \times 9.81 = 74.6$ kPa,
$u = 1.0 \times 9.81 \times 2.0 = 19.6$ kPa, $\sigma'_v = 74.6 - 19.6 = 55.0$ kPa.

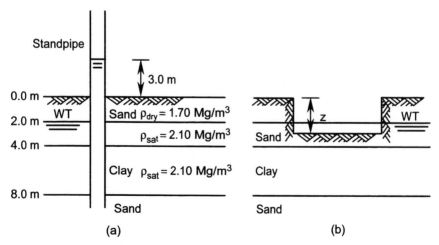

Figure 2.8. Example 2.4.

At $z = 8.0$ m: $\sigma_v = (1.7\times2.0+2.1\times2.0+2.1\times4.0)\times9.81=157.0$ kPa,
$u = 1.0\times9.81\times11.0 = 107.9$ kPa, $\sigma'_v = 157.0 - 107.9 = 49.1$ kPa.
(b) At $z = 8.0$ m, $\sigma'_v = \sigma_v - u = 0$, $\sigma_v = u = 157.0$ kPa, $u = \rho_w \times g \times h = 157.0$ kPa.
$h = 157.0 / 9.81 = 16.0$ m. Height above the ground surface $= 16.0 - 8.0 = 8.0$ m.
(c) With the arrangement of Figure 2.8(b): $\sigma'_v = \sigma_v - u = 0$, $\sigma_v = u$, assume $z > 2.0$ m,
$\sigma_v = [2.1(4.0-z)+2.1\times4]\times9.81=1.0\times11.0\times9.81=107.9$ kPa, $z = 2.76$ m > 2.0 m.

2.2.3 Negative pore pressure due to capillary rise

Capillary rise results from the combined actions of surface tension and intermolecular forces between the liquid and solids. Noting that the pressure on the water table level is zero, any water above this level must have a negative pressure. In soils a negative pore pressure increases the effective stresses and varies with the degree of saturation and by drying or wetting processes. The phenomenon can be demonstrated by immersing a glass tube of small inner diameter into water as shown in Figure 2.9. The capillary water has a hydraulic continuity with the water table and rises to a certain height where equilibrium between the weight of the water column and the vertical resultant of the surface tension forces is achieved. The height of the capillary column h_c is defined by:

$$h_c = 4T\cos\theta/(\gamma_w d) \tag{2.5}$$

where T is the surface tension (0.074 N/m at 20°C), θ is the angle of surface tension with the vertical, d is the diameter of the tube, and γ_w is the unit weight of water. The water surface in the tube (meniscus) is assumed to be a portion of a sphere. For pure water and a glass tube of very small diameter, $\theta = 0$, the meniscus is hemispherical and:

$$h_c = 4T/(\gamma_w d) \tag{2.6}$$

The maximum negative pore pressure is:

$$u_c = h_c\gamma_w = 4T/d \tag{2.7}$$

Any application of Equations 2.6 and 2.7 to a soil raises questions concerning the tube concept, the definition of d and the limitations of the equation. The capillary process starts as water evaporates from the surface of the soil. The capillary zone is comprised of a fully

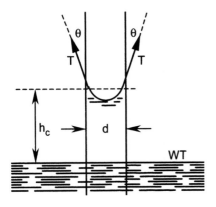

Figure 2.9. Cylindrical capillary tube.

saturated layer with a height of usually less than h_c, and a partially saturated layer overlain by wet or dry soil. In the partially saturated layer the water and air within the voids are both continuous. Terzaghi & Peck (1967) suggested that the pore pressure in this layer should be treated as a negative hydrostatic pressure with its maximum expressed by Equation 2.7. Negative pore pressure results in an increase in the effective stress and is termed *soil suction*.

The boundary between the capillary and the gravitational water is *ill defined* and may not be determined accurately (Bell, 1993). In the field, a borehole that can create a *well* is a simple way of determining the water level, but suction measuring devices are also used. The diameter of the equivalent tube d largely depends on the particle size distribution of the soil and the assumed value of $d = eD_{10}$ has been generally accepted by soil scientists. Estimation for the range of capillary rise can be made by assuming $0.2D_{10} < d < D_{10}$ (Powrie, 1997).

In granular materials (gravels and sands) the amount of capillary rise is negligible while in silty soils the water may rise up to several metres. The rise and fall of the water table and gravitational drainage from the ground surface can affect the height of capillary rise and may create different types of unsaturated zones above or close to the capillary zone. Immediately above the level h_c, the suction in the water exceeds the tensile strength of the water and generates vacuum conditions and vapour pressure (Klausner, 1991). As the elevation increases beyond the capillary rise, suction increases and the soil moisture and degree of saturation decrease. Eventually continuous air voids with atmospheric pressure evolve with typical behaviour shown in Figure 2.10. Locally created capillary menisci between the small particles causes an increase in strength due to the increase in effective stress. Simple examples include the stability of sand castles made by children at the beach or the ability of a moist sand layer to resist shear stresses induced by vehicular tyres. The apparent increase in the strength of a fine granular material disappears with full saturation or full drying. In the latter instance, the effective stress increases due to the reduction in the radius of the meniscus and the soil shrinks. The process continues until the shrinkage limit is reached and soil volume becomes constant. The strength increases as the drying process continues and reaches its maximum in full dry conditions. This behaviour is reversible in fined-grained soils (Section 1.5), that is, the soil volume increases (after the shrinkage limit), decreasing the strength of the soil.

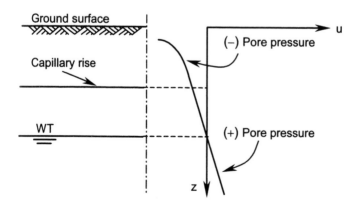

Figure 2.10. Suction profile (Klausner, 1991).

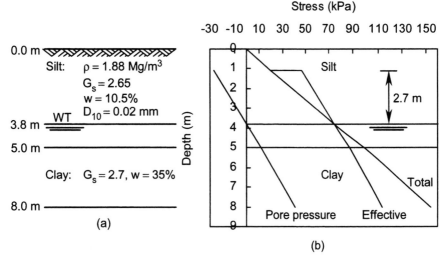

Figure 2.11. Example 2.5.

Example 2.5

For the soil profile shown in Figure 2.11(a) plot the distribution of total and effective vertical stresses and pore pressure taking account of the capillary rise in the silt layer.

Solution:

Calculate the dry and saturated densities of silt: $\rho_d = 1.88(1+0.105) = 1.701$ Mg/m^3.
Assuming $V = 1$ m^3:
$V_s = 1.701 / 2.65 = 0.642$ m^3, $V_v = 1.0 - 0.642 = 0.358$ m^3,
$\rho_{sat} = 1.701 + 0.358 \times 1.0 = 2.059$ Mg/m^3.

Calculate equivalent tube diameter: $e = 0.358/0.642 = 0.558$,
$D = e \times D_{10} = 0.558 \times 0.02 = 0.01116$ mm, $d = 1.116 \times 10^{-5}$ m.
$h_c = 4T / \gamma_w d = 4 \times 0.000074 / 9.81 \times 1.116 \times 10^{-5} = 2.7$ m.

Calculate the saturated density of clay:
e(saturated clay) $= wG_s = 0.35 \times 2.7 = 0.945$,
$V_s = 1/(1+e) = 1/(1+0.945) = 0.514$ m^3, $V_v = e/(1+e) = 0.945/(1+0.945) = 0.486$ m^3,
$\rho_{sat} = 0.514 \times 2.7 + 0.486 \times 1.0 = 1.874$ Mg/m^3.

At depth $3.8 - 2.7 = 1.1$ m:
$\sigma_v = 1.88 \times 9.81 \times 1.1 = 20.3$ kPa, $u = -1.0 \times 9.81 \times 2.7 = -26.5$ kPa,
$\sigma'_v = 20.3 + 26.5 = 46.8$ kPa.

At depth 3.8 m: $\sigma_v = 20.3 + 2.059 \times 9.81 \times 2.7 = 74.8$ kPa, $u = 0.0$ kPa, $\sigma'_v = 74.8$ kPa.
At depth 5.0 m: $\sigma_v = 74.8 + 2.059 \times 9.81 \times 1.2 = 99.1$ kPa, $u = 1.0 \times 9.81 \times 1.2 = 11.8$ kPa,
$\sigma'_v = 99.1 - 11.8 = 87.3$ kPa.

At depth 8.0 m: $\sigma_v = 99.1 + 1.874 \times 9.81 \times 3.0 = 154.2$ kPa, $u = 1.0 \times 9.81 \times 4.2 = 41.2$ kPa,
$\sigma'_v = 154.2 - 41.2 = 113.0$ kPa.
The results are plotted in Figure 2.11(b).

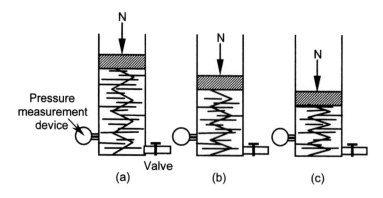

Figure 2.12. Conceptual model for consolidation.

2.3 STATE OF STRESS AT A POINT DUE TO EXTERNAL FORCES

2.3.1 *Conceptual model for consolidation*

External loading increases the total stress at every point in a saturated soil above its initial value. The magnitude of this increase depends mostly on the location of the point and is estimated from appropriate stress distribution theories based on the mechanical properties of the saturated soil (Chapter 5). In clay soils immediately after the application of the load, the pore pressure increases to almost equal the applied load and the pressurised water tends to move towards the free drained boundaries. Thus, over time, the excess pore pressure dissipates and the pore pressure approaches its initial value. The volume of the clay layer decreases to balance the volume of the transported water and remains saturated. Thus the applied load is gradually transmitted to the solid particles and the associated volume change causes the settlement of the layer. This phenomenon, known as *consolidation*, can be explained by the mathematical-physical model shown in Figure 2.12.

Consider a cylinder with a spring supporting a piston and filled with water to the top of the spring, as shown in Figure 2.12(a). The piston seals the water but has smooth contact with the walls of the cylinder. Initially a valve, which is the only escape for the water, is closed and the system is in equilibrium. There is no force in the spring and no pressure in the water. If a normal force N is applied to the piston the water will resist the entire force since, because it is incompressible, there will be no deformation in the water and consequently no deformation and therefore no force in the spring. If the valve were to be opened for a specified period of time, the pressurised water drains through the valve (Figure 2.12(b)). As a result, the spring undergoes a shortening equal to the vertical displacement of the piston and the load N is resisted by both the water and the spring. If the valve remains open until the water pressure decreases to zero (Figure 2.12(c)) the spring takes the entire force applied to the piston. The time over which the force N is transmitted from the water to the spring depends on the diameter of the valve, the volume of water inside the cylinder and the elastic characteristics of the spring. In a real soil the valve represents the voids between the solid particles and the spring represents the solid particles themselves and the load N is normally applied through a footing or a similar type of structure.

The pore pressure at every point within the volume increases initially to a maximum value equal to the increment of stress at that point caused by the applied loading (*excess pore pressure*). A gradual process of redistribution of load to the particles starts immediately after application of the external load (*consolidation*). The process is associated with volume change (equal to the drained water) and surface settlement. At the end of this process the applied load is resisted totally by the soil particles. In unloading the volume change due to consolidation is not recoverable or the rate of expansion is very small in comparison to the rate of compression. Consequently, the conceptual model shown in Figure 2.12 has to be modified to represent this behaviour.

The rate at which the total load is applied also affects the soil deformation behaviour. If the external load is applied in small increments over a long period of time and the soil has free draining boundaries, then there will be no excess pore pressure and the applied increments of load will be resisted by the solid particles. Volume change will occur in increments similar to the loading. In general this type of the loading, whether in the field or in the laboratory, is termed *drained loading*. In *undrained loading* the water in the voids cannot drain to the free boundaries and volume remain unchanged. However, in the field, the lateral boundaries are not as rigid as the walls of the cylinder model, and an element of soil will undergo vertical and horizontal deformations most probably in the reverse directions to keep the volume unchanged. The term undrained loading can also be used to describe the state of stress in the field when the load is applied very quickly and the seepage of water from the voids to the free boundaries takes place slowly. In laboratory techniques the slow application of load is assumed to be equivalent to drained loading whilst rapid load application represents undrained loading. However, the full undrained loading condition can be created regardless of the rate of load application.

The compression of a clay layer may occur if the water table is drawn downwards as a result of pumping. In this case the effective stresses increase due to the reduction in the level of the water table. However, if the pumping is stopped, the water table will recover to its initial position after time but most of the settlement that occurred in the clay layer will not be recovered. This forms the basis of a method of soil stabilization to minimize the volume change due to external loading.

Example 2.6

A layer of clay of 4 m thick is overlain by a sand layer of 5 m, the top of which is the ground surface. The clay overlays an impermeable stratum. Initially the water table is at the ground surface but it is lowered 4 metres by pumping. Calculate σ'_v at the top and base of the clay layer before and after pumping.

For sand $e = 0.45$, $G_s = 2.6$, S_r (sand, after pumping) = 50%. For clay $e = 1.0$, $G_s = 2.7$.

Solution:

Calculate the density of sand at $S_r = 50\%$ and 100%:

For $V = 1$ m^3,

$V_s = 1/(1+e) = 1/(1+0.45) = 0.690$ m^3,

$V_v = e/(1+e) = 0.45/(1+0.45) = 0.31$ m^3.

$\rho_{50\%sat} = 0.690 \times 2.6 + 0.310 \times 0.5 \times 1.0 = 1.949$ Mg/m^3.

$\rho_{sat} = 0.690 \times 2.6 + 0.310 \times 1.0 = 2.104$ Mg/m^3.

Using similar calculation for $e = 1$, ρ_{sat} (clay) = 1.850 Mg/m^3.

At the top of the clay layer before pumping:

$\sigma_v = 2.104 \times 9.81 \times 5.0 = 103.2$ kPa, $u = 1.0 \times 9.81 \times 5.0 = 49.0$ kPa,

$\sigma_v' = 103.2 - 49.0 = 54.2$ kPa.

At the base of the clay layer before pumping:

$\sigma_v = 103.2 + 1.850 \times 9.81 \times 4.0 = 175.8$ kPa, $u = 1.0 \times 9.81 \times 9.0 = 88.3$ kPa,

$\sigma_v' = 175.8 - 88.3 = 87.5$ kPa.

At the top of the clay layer after pumping:

$\sigma_v = 1.949 \times 9.81 \times 4.0 + 2.104 \times 9.81 \times 1.0 = 97.1$ kPa, $u = 1.0 \times 9.81 \times 1.0 = 9.8$ kPa,

$\sigma_v' = 97.1 - 9.8 = 87.3$ kPa.

At the base of the clay layer after pumping:

$\sigma_v = 97.1 + 1.850 \times 9.81 \times 4.0 = 169.7$ kPa, $u = 1.0 \times 9.81 \times 5.0 = 49.0$ kPa,

$\sigma_v' = 169.7 - 49.0 = 120.7$ kPa.

$\Delta\sigma'_v$ (at the top) $= 87.3 - 54.2 = 33.1$ kPa.

$\Delta\sigma'_v$ (at the base) $= 120.7 - 87.5 = 33.2$ kPa.

The increase in the effective vertical stress throughout the clay layer is uniform.

2.3.2 *The case of partially saturated soil*

Bishop (1959) extended Equation 2.1 to represent the state of stress in an unsaturated soil:

$$\sigma = \sigma' + u_a - \chi(u_a - u_w) \tag{2.8}$$

where u_a is the pore air pressure, u_w is the pore water pressure and χ is a constitutive parameter depending on the degree of saturation. Note that $(u_a - u_w)$ represents the soil suction. To explain the physical meaning of χ we rearrange the terms in Equation 2.8:

$$\sigma = \sigma' + \chi u_w + (1 - \chi)u_a \rightarrow u = \chi u_w + (1 - \chi)u_a \tag{2.9}$$

It can be seen that χ is the fraction of the unit area occupied by water, and $1 - \chi$ represents the area of the air. The magnitude of χ can be determined experimentally, however a linear relationship between degree of saturation and χ may be adopted between the following limits: $S_r = 0$, $\chi = 0$, $S_r = 1$, $\chi = 1$. As the pore water and pore air are assumed to have an interface due to surface tension, then the pore water pressure must be always less than the pore air pressure. Therefore the reliability of Equations 2.9 reduces with a decreasing degree of saturation; however, for higher degrees of saturation which are on the wet side of the optimum moisture content, Equations 2.9 work well.

Fredlund (1973 and 1979) and Fredlund & Morgenstern (1977) investigated the stress state in an unsaturated soil assuming four independent phases viz., solids, pore air, pore water and the air-water interface. Three stress variables of $\sigma - u_a$, $\sigma - u_w$, and $u_a - u_w$, were introduced to express the state of stress in a three-dimensional system of which only two are independent, as adding the first variable to the third variable will yield the second variable. The strength parameters of the unsaturated soil were then defined differently to those of saturated soils in classical soil mechanics. The conditions of transition to the saturated case were successfully applied in the proposed models.

2.4   PROBLEMS

2.1   For the soil profile shown in Figure 2.13(a) plot the variation of total vertical stress, pore pressure and effective vertical stress and indicate their values on the boundaries of each layer.

Answers: $\sigma'_v = 31.5$ kPa, 58.4 kPa, 92.3 kPa

2.2   For the given soil profile of Figure 2.13(b) calculate the effective vertical stress at a depth of 7.5 m.

Answer: 57.2 kPa

2.3   A clay layer of 4 m thick with $\rho_{sat} = 2$ Mg/m³ is overlain by a 4 m sand with $\rho_{sat} = 1.9$ Mg/m³ and $\rho_{dry} = 1.65$ Mg/m³, the top of this layer being the ground surface. The water table is located 2 m below the ground surface. The clay layer is underlain by a sand stratum that is in artesian conditions with the water level in a standpipe being 4 m above the ground surface.

Calculate the effective vertical stresses at the top and the base of the clay layer. If the dry sand is excavated, in what depth the effective stress at the bottom of the clay layer will become zero?

Answers: 50.0 kPa, 30.4 kPa, 1.9 m

2.4   A clay layer 10 m thick has a density of 1.75 Mg/m³ and is underlain by sand. The top of the clay is the ground surface. An excavation in the clay layer failed when the depth of the excavation reached to 6.5 m from the ground surface.

Calculate the depth of water in a standpipe sunk to the sand layer.

Answer: 3.875 m

2.5   A stratum of soil is 15 m thick and its top surface is the ground surface. Formulate the effective vertical stress within the layer if:

(a) the water table is at the ground surface,

(b) the water table is lowered 3 m by pumping.

$\rho_{sat} = 2$ Mg/m³ and $\rho_{dry} = 1.65$ Mg/m³.

Answers: 9.81 z kPa, 9.81 z + 19.13 kPa (z is in m)

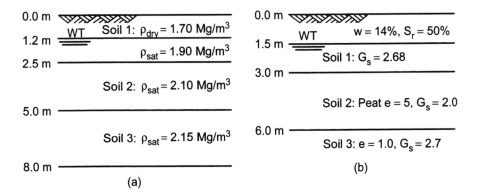

Figure 2.13. Problems 2.1 & 2.2.

2.5 REFERENCES

Bell, F.G. 1993. *Engineering geology*. Oxford: Blackwell Scientific Publications.

Bishop, A.W. 1959. The principal of effective stress. From a lecture in Oslo, Norway 1955. Reprinted in *Teknisk Ukeblad* (39): 859-863.

Fredlund, D.G. 1973. Discussion of the second technical session. *Proc. 3rd intern. conf. on expansive soils* 1: 71-76.

Fredlund, D.G. 1979. Second Canadian geotechnical colloquium: Appropriate concepts and technology for unsaturated Soils, *Canadian geotechnical journal* (16): 121-139.

Fredlund, D.G. & Morgenstern, N.R. 1977. Stress state variables for unsaturated Soils. *Journal SMFED, ASCE* 103 (GT5): 447-466.

Klausner, Y. 1991. *Fundamentals of continuum mechanics of soils*. London: Springer-Verlag.

Powrie, W. 1997. *Soil mechanics-concepts and applications*. London: E & FN Spon.

Terzaghi, K. & Peck, R. B. 1967. *Soil mechanics in engineering practice*. New York: John Wiley.

CHAPTER 3

The Movement of Water through Soil

3.1 INTRODUCTION

This chapter investigates the flow of water through interconnected pores between soil particles in both one and two dimensions. Often, the flow of ground water through a confined or unconfined porous material can be represented as a uniform flow in a plane. The term *seepage flow* is used to describe the movement of water through or under soil structures to wells, drains and reservoirs. Generally, the Reynolds number defined in fluid mechanics is less than 1, and therefore the flow is laminar and Darcy's law is valid. Any departure from laminar flow is not serious as long as the Reynolds number is below 10, the upper limit for the validity of Darcy's law. A simplified theoretical and numerical approach is developed and applied to practical examples to obtain *flow nets* that describe the seepage flow. This is followed by descriptions of test methods (laboratory and in-situ) used to obtain the *coefficient of permeability* (or *hydraulic conductivity*) that controls the velocity and the volume of water flowing per unit of time.

3.2 PRINCIPLES OF FLOW IN POROUS MEDIA

3.2.1 *Darcy's law*

Figure 3.1 shows the flow of water through a tube of soil of length L and cross-sectional area A. Although water moves through the interconnected voids of the soil, it is convenient to define the velocity as if the water flows through the whole section. Whilst this imaginary velocity is many times less than the actual velocity, the flow rate calculated using this velocity is real. The water level at its entry point B into the soil tube is higher than the water level at the exit point C. This shows the dissipation of potential energy as the water flows across the length L. The water level at each point is represented by its distance h from an arbitrary datum and is called *head* or *total head*. According to Bernoulli's definition of head:

$$h = z + \frac{u}{\gamma_w} + \frac{v^2}{2g} \tag{3.1}$$

where z is the positional head (vertical coordinate of the point), u is the water pressure or pore pressure, γ_w is the unit weight of water, v is the velocity and $v^2/2g$ is the velocity head. In classical soil mechanics, the velocity head term is ignored due to low velocities and therefore:

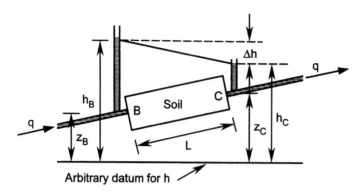

Figure 3.1. Idealized diagram for flow of water through a tube of soil.

$$h = z + \frac{u}{\gamma_w} \tag{3.2}$$

The *hydraulic gradient* within the length L is a dimensionless parameter and is defined as the rate of change in total head (or head loss) Δh over the length L:

$$i = \frac{\Delta h}{L} \tag{3.3}$$

In steady flow conditions the hydraulic gradient along a finite length is assumed constant, otherwise it is a point-related parameter that defines the reduction of total head per unit length of flow across the specified direction. If the head loss for an increment of ΔL is Δh then:

$$i = \frac{\Delta h}{\Delta L} \tag{3.4}$$

If the total head is expressed as a function of L, then Equation 3.4 is replaced by:

$$i_L = -\frac{\partial h}{\partial L} \tag{3.5}$$

The negative sign implies that the velocity is in the direction of decreasing total head. As long as the calculations are based on absolute values and there is no coordinate system involved in the seepage problem, Equations 3.3 and 3.4 may be used. In two or three-dimensional coordinate systems the hydraulic gradient is defined by Equation 3.5. Darcy, a French hydraulic scientist of 19th century, stated that the flow rate through a porous medium is directly proportional to the head loss and inversely proportional to the length of flow path:

$$v = ki \tag{3.6}$$

where k is the coefficient of permeability of the material and is determined experimentally. The quantity of water that flows in a unit of time through an area of A or *flow rate* is then given by:

$$q = Av = Aki \tag{3.7}$$

The coefficient of permeability has the dimension of length per unit time and decreases as the particle size decreases. An estimate of the actual velocity v_s, referred to as the *seepage velocity*, can be made by considering the following equation:

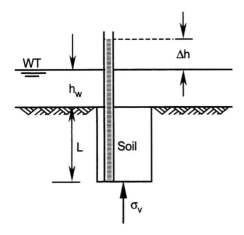

Figure 3.2. The concept of seepage pressure.

$$q = Av = A_v v_s$$

where A_v is the cross-sectional area of the pore voids. Multiplying the both sides of this equation by length L we have:

$$ALv = A_v L v_s \rightarrow V \times v = V_v \times v_s \rightarrow v_s = \frac{V}{V_v} v = \frac{v}{n} \text{ or:}$$

$$v_s = \frac{v}{n} = \frac{v(1+e)}{e} \qquad (3.8)$$

where n is the porosity and e is the void ratio of the soil.

3.2.2   *Seepage pressure and critical hydraulic gradient*

In an upward one-dimensional flow the effective vertical stress at the base of a column of soil reduces by an amount equal to $\gamma_w \, \Delta h$. Figure 3.2 shows an upward flow through a soil column due to a higher head in the underlying soil. Using Equation 2.1, the effective vertical stress at the base of the soil column is:

$$\sigma'_v = \sigma_v - u = (\gamma_{sat} L + \gamma_w h_w) - \gamma_w (L + h_w + \Delta h),$$

$$\sigma'_v = \sigma_v - u = (\gamma_{sat} - \gamma_w) L - \gamma_w \Delta h \qquad (3.9)$$

If the water level at the column base were equal to the water level at the top of the column, there would be no upward movement of water and σ'_v would be equal to the first term of Equation 3.9. Considering a soil column of unit area the reduction in σ'_v per unit volume or *seepage pressure* is:

$$j = \frac{\gamma_w \Delta h}{L} = i \gamma_w \qquad (3.10)$$

where i is the hydraulic gradient. The first term of Equation 3.9 represents the submerged weight of the soil column of length L with unit cross-sectional area. When the seepage pressure multiplied by the volume of the soil column becomes equal to the submerged weight of the soil column, σ'_v at the base of the column reduces to zero. The soil particles loose their contact and the column is subjected to heave and consequently fails. This state is called the *quick condition* and defines a critical hydraulic gradient given by:

$$\sigma'_v = L(\gamma_{sat} - \gamma_w) - \gamma_w \Delta h = 0,$$

$$i_c = \frac{\gamma_{sat} - \gamma_w}{\gamma_w} = \frac{\gamma'}{\gamma_w} \tag{3.11}$$

Substituting γ' from Equation 1.21 we have:

$$i_c = \frac{\gamma'}{\gamma_w} = \frac{(G_s - 1)}{1 + e} \tag{3.12}$$

It should be noted that, at zero effective vertical stress, the strength of the soil depends entirely on its cohesion and therefore the quick conditions for clays may not obey Equation 3.12. In sands the strength and stability of an element is provided by the forces at the contact points of the particles. Thus, in the $\sigma'_v = 0$ condition, the sand element is no longer stable. The state of sand in this condition is similar to that of a slurry liquid with a density greater than that of water and is sometimes referred to as *quicksand*. In the case of seepage in the direction of gravity, it may be shown that the effective vertical stress increases from its static value by an amount equal to $\gamma_w \Delta h$. In general the seepage force acts in the direction of flow and may increase or decrease σ'_v depending on the angle between the direction of flow and gravity.

3.3 PERMEABILITY

3.3.1 *Coefficient of permeability*

The coefficient of permeability k is determined by experimental methods and varies from 1 m/s in gravels to less than 10^{-9} m/s in clays (Table 3.1). The magnitude of k depends on the size of the particles, their shape and orientations, and the structure of the soil. The viscosity of the water also has some effect on k as it varies with change in temperature. Whilst many mathematical expressions have been proposed for the coefficient of permeability in terms of the physical properties of soils, experimental determination of k is always recommended.

Table 3.1. Range of values for the coefficient of permeability.

k(m/s)											
1	10^{-1}	10^{-2}	10^{-3}	10^{-4}	10^{-5}	10^{-6}	10^{-7}	10^{-8}	10^{-9}	10^{-10}	10^{-11}

Clean gravels	Clean gravels, clean sand and gravel	Very fine sands, organic and inorganic silts, mixture of sand, silt and clay	Clays

Well drained soils	Poorly drained soils	Practically impervious

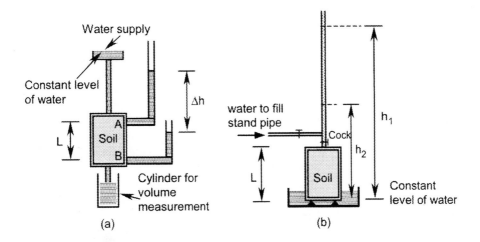

Figure 3.3. Laboratory permeability tests: (a) constant head test, (b) falling head test.

An empirical equation proposed by Hazen a century ago relates the coefficient of permeability of fine-graded granular materials to their particle size distribution characteristics by:

$$k = CD_{10}^2 \qquad (3.13)$$

where k is in m/s and D_{10} = diameter corresponding to 10% finer in mm and C is a coefficient that varies from 0.005 for silt and well-graded sands to 0.012 for uniform sands.

3.3.2 *Laboratory permeability tests*

Two tests commonly conducted in the laboratory are the constant head test and the falling head test.

Constant head test. A diagrammatic view of the test arrangement is shown in Figure 3.3(a). This test is most suitable for granular soils (gravels and sands) where k is smaller than 10^{-7} m/s. The soil specimen is formed in a cylindrical mould (75 or 100 mm diameter) and water is passed through the sample from a constant head tank. Two outlets, or pressure take-off points, A and B separated vertically by a distance L (≈ 100 mm) show the water levels at these outlets. Longer permeability cells with 3 or 6 outlets are also used. Two filters using coarse-grained soils are placed at the top and the bottom of the cell to ensure uniform flow across the soil sample. Before any readings are taken, a specified time must elapse to ensure steady flow conditions are achieved. The discharged water Q is collected and measured by one of two methods, either the volume over a given time t or the time for a nominated volume. The difference Δh between the water levels at A and B is recorded, and the rate of flow is the ratio of discharged water Q to the elapsed time t. Using Equation 3.7:

$$q = \frac{Q}{t} = Av = Aki = Ak\frac{\Delta h}{L}, \text{ and hence}$$

$$k = \frac{QL}{\Delta h \times At} = \frac{qL}{\Delta h \times A} \tag{3.14}$$

To ensure laminar flow conditions a hydraulic gradient of 0.2 to 0.3 for loose material and 0.3 to 1.0 for dense material is used. The constant head is approximately 120 mm above the top surface of the sample and is applied through the base of the cell to ensure full saturation.

Falling head test. This method is used for fine-grained soils such as silts and clays with a coefficient of permeability between 10^{-7} and 10^{-9} m/s. A schematic view of the apparatus is shown in Figure 3.3(b). The permeability cell is a cylinder of approximately 100 mm in diameter and a height of 130 mm. A filter layer of coarse-grained material is positioned at the top of the sample to ensure uniform laminar flow. A perforated disc and a porous plate with a wire gauze rest at the base of the cell and are held between the top and bottom plates of the cell by retaining ties. A capillary tube is fitted to the top of the cell and the assembly is placed in a soaking tank. If the constant water level at the tank is lower than the height of the cylinder, precautions have to be made to ensure full saturation of the soil specimen in the cell. At the start of the test and at $t = 0$ the head is equal to h_1. The valve on the vertical capillary tube is opened and the time t for the head to fall to h_2 is recorded. The head h corresponding to time t is reduced by dh due to an increase in time of dt. Applying Darcy's law and noting the volume of water discharged from the capillary tube is equal to the volume of flow through the sample we have:

$$q = Av = Aki = Ak \frac{h}{L} = -a \frac{dh}{dt}$$

where a is the internal sectional area of the capillary tube and the negative sign shows that the increase in time causes a decrease in the head. Integrating this continuity condition, we have:

$$\int_0^t \frac{kA}{aL} dt = \int_{h_1}^{h_2} (-\frac{dh}{h}) \rightarrow k = 2.3 \frac{aL}{At} \log \frac{h_1}{h_2} \tag{3.15}$$

Recorded data now are substituted into Equation 3.15 to determine the coefficient of permeability. To ensure optimum accuracy, various standpipe sizes can be used. Soils of very low permeability are sealed inside the permeability cell (using a thin layer of wax to the internal sides of the mould) to prevent seepage and piling of water between the mould and specimen. Details of both techniques may be found in ASTM D-2434-68 (2000), BS 1377 and AS 1289.6.7.1 & 2 (1999). Some important notes with regard to sample preparation, surcharge and calculations are as follows:

1. Remoulded specimens are compacted with a specified compactive effort and must represent the field condition. The size of the permeability cell depends on the maximum particle size of the soil and is selected according to code requirements.

2. If necessary a confining axial load or surcharge is applied to the sample to represent the field stress conditions.

3. To include the effect of temperature the following equation may be used:

$$k_T = k_\theta \frac{\eta_\theta}{\eta_{20}} \tag{3.16}$$

where θ is the temperature of the outflow water in degrees Celsius, k_T and k_θ are the coefficients of permeability at $20°C$ and at $\theta°C$, η_{20} and η_θ are the dynamic viscosities of water at $20°C$ and $\theta°C$ respectively.

Constant head method using a flexible wall permeameter. For soils with $k < 10^{-9}$ m/s, a constant head method, incorporating a flexible wall permeameter, is used (ASTM D-5084 (2000), BS 1377, AS 1289.6.7.3 (1999)). In this test remoulded or undisturbed specimens are tested under a specified effective stress to simulate the field condition. A cylindrical soil specimen, protected by a thick rubber membrane, is tested in a triaxial cell (Chapter 4). The cell pressure is increased gradually causing both the effective stress and the pore pressure to increase within the sample. A back pore pressure is applied through the top of the specimen and the pore pressure is monitored at the base of the specimen to ensure that the effective stress does not exceed the required final value. Both the back pore pressure and cell pressure are increased incrementally until the required cell pressure and effective stress are achieved. At this stage the base of the specimen is under constant head and the fall in the head at the top of the specimen can be measured to establish the head loss. The rate of flow is measured over a sufficient period of time, until the coefficient of permeability calculated using Equation 3.14 is constant.

Permeability test using Rowe type consolidation cell. A diagrammatic section of a Rowe type consolidation cell is shown in Figure 3.4 and is the development of an original design by Professor P W Rowe at Manchester University UK. Whilst the apparatus is designed for a consolidation test, it can be used to determine the coefficient of permeability for both vertical and radial flow. The test is carried out under a specified effective vertical stress but the effective horizontal stress, which is the reaction of the rigid wall to the applied effective vertical stress, is unknown. This, in many ways, simulates the consolidation process under structures build upon saturated clay soils. The cell is commercially available in three diameters of 75 mm, 150 mm and 250 mm and complies with BS 1377. It is manufactured from aluminium alloy and consists of a cylinder, and top and base plates. The soil specimen within the cell is hydraulically loaded through a diaphragm that ensures a uniform distribution of total vertical stress at the top surface of the sample. Under a specified effective vertical stress different drainage conditions vertically upwards or downwards or radially outwards or inwards can be applied.

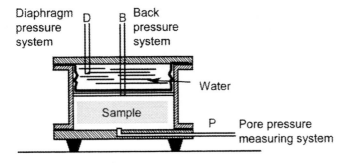

Figure 3.4. Rowe type hydraulic consolidation cell.

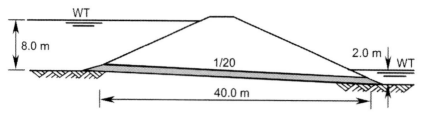

Figure 3.5. Example 3.1.

Permeability tests are carried out in the 250 mm diameter cells. Both methods of constant head and falling head can be performed depending on the type of soil. In the vertical permeability test two porous drainage discs are placed at the top and base of the sample. For radial flow porous plastic liners are placed around the sample with a central porous hole filled with sand. The diameter of the hole is approximately 5% of the sample diameter. Categories of the sample include undisturbed, remoulded, compacted soil and undisturbed prepared from remoulded specimens. Reference may be made to Head (1986) for a detailed test procedure.

Example 3.1

An impermeable dam is constructed on 0.5 m thick layer of silty sand as shown in Figure 3.5. Estimate the flow rate per metre width of the dam in m^3/day. $k = 2 \times 10^{-4}$ m/s.

Solution:

Assume a constant hydraulic gradient along the silty-sand layer: $i = \dfrac{\Delta h}{\Delta L}$.

Assume the toe of the dam as the datum for total heads:
h_1 = total head at the upstream side $= 8.0 + 40.0 \times (1/20) = 8.0 + 2.0 = 10.0$ m.
h_2 = total head at the downstream side $= 2.0$ m.
$\Delta h = h_1 - h_2 = 10.0 - 2.0 = 8.0$, $L = \sqrt{40.0^2 + 2.0^2} = 40.05$ m,

$i = \dfrac{\Delta h}{L} = \dfrac{8.0}{40.05} = 0.1997$.

$q = Aki = (0.5 \times 1.0) \times 2.0 \times 10^{-4} \times 0.1997 \times 3600 \times 24 = 1.73$ m^3/day.

Example 3.2

A constant head permeability cell has an internal diameter of 75 mm and three tapped outlets A, B, and C at a 100 mm vertical pitch. During three sets of tests on a specimen of sand the following data were recorded. The flow was upward and each test was conducted with a different constant head. Determine k in m/s (average of six values).

Test no.	Q (ml/10 minutes)	Head (mm) above datum		
		A	B	C
1	281.5	139.0	100.0	60.0
2	422.5	208.5	150.5	91.5
3	591.0	292.0	210.0	127.0

Solution:

A = area of cell = $75.0^2 \times \pi / 4 \times 10^{-6} = 4.4179 \times 10^{-3}$ m^2.

Using Equation 3.14 for each test:

Test 1: $q = 281.5 / 10.0 = 28.15$ ml/min $= 28.15 \times 10^{-6} / 60.0 = 0.469 \times 10^{-6}$ m^3/s,

(a) $k = 0.469 \times 10^{-6} \times 0.1 / [(0.139 - 0.100) \times 4.4179 \times 10^{-3}] = 2.72 \times 10^{-4}$ m/s,

(b) $k = 0.469 \times 10^{-6} \times 0.1 / [(0.100 - 0.060) \times 4.4179 \times 10^{-3}] = 2.65 \times 10^{-4}$ m/s.

Test 2: $q = 422.5 / 10.0 = 42.25$ ml/min $= 42.25 \times 10^{-6} / 60.0 = 0.704 \times 10^{-6}$ m^3/s,

(a) $k = 0.704 \times 10^{-6} \times 0.1 / [(0.2085 - 0.1505) \times 4.4179 \times 10^{-3}] = 2.75 \times 10^{-4}$ m/s,

(b) $k = 0.704 \times 10^{-6} \times 0.1 / [(0.1505 - 0.0915) \times 4.4179 \times 10^{-3}] = 2.70 \times 10^{-4}$ m/s.

Test 3: $q = 591.0 / 10.0 = 59.10$ ml/min $= 59.10 \times 10^{-6} / 60.0 = 0.985 \times 10^{-6}$ m^3/s,

(a) $k = 0.985 \times 10^{-6} \times 0.1 / [(0.292 - 0.210) \times 4.4179 \times 10^{-3}] = 2.72 \times 10^{-4}$ m/s,

(b) $k = 0.985 \times 10^{-6} \times 0.1 / [(0.210 - 0.127) \times 4.4179 \times 10^{-3}] = 2.68 \times 10^{-4}$ m/s.

Average $k = 2.70 \times 10^{-4}$ m/s.

Example 3.3

A falling head test was conducted in a cell of 100 mm diameter and 127 mm high. The recorded data were: $h_1 = 560$ mm, $h_2 = 465$ mm, $t = 1081$ s, and the diameter of the vertical standpipe was 7 mm. Determine (a) k in m/s, (b) the time required for a similar head and drop if a standpipe 4 mm in diameter was used. Include the effect of capillary rise.

Solution:

(a) $A = 0.1^2 \times \pi / 4 = 7.854 \times 10^{-3}$ m^2, $a = 0.007^2 \times \pi / 4 = 3.848 \times 10^{-5}$ m^2.

From Equation 3.15:

$k = 2.3 \times 3.848 \times 10^{-5} \times 0.127 / (7.854 \times 10^{-3} \times 1081.0) \log(560.0 / 465.0) = 1.069 \times 10^{-7}$ m/s.

Correction for capillary rise: $h_c = 4T / \gamma_w d = 4 \times 0.000074 / 9.81 \times 0.007 \times 1000 = 4.3$ mm.

$h_1 = 560.0 - 4.3 = 555.7$ mm, $h_2 = 465.0 - 4.3 = 460.7$ mm.

$k = 2.3 \times 3.848 \times 10^{-5} \times 0.127 / (7.854 \times 10^{-3} \times 1081.0) \log(555.7 / 460.7) = 1.078 \times 10^{-7}$ m/s.

(b) $h_c = 4T / \gamma_w d = 4 \times 0.000074 / 9.81 \times 0.004 \times 1000 = 7.5$ mm,

$h_1 = 560.0 - 7.5 = 552.5$ mm, $h_2 = 465.0 - 7.5 = 457.5$ mm.

For $d = 4$ mm, $a = 1.257 \times 10^{-5}$ m^2. Using Equation 3.15:

$t = 2.3 \times 1.257 \times 10^{-5} \times 0.127 \times \log(552.5 / 457.5) / (7.854 \times 10^{-3} \times 1.069 \times 10^{-7}) = 358$ s.

From Equation 3.15 it can be shown that in the absence of capillary rise:

$t_1 / t_2 = D_1^2 / D_2^2$, $1081.0 / t_2 = 7.0^2 / 4.0^2 \rightarrow t_2 = 353$ s.

Example 3.4

A permeability test is arranged according to Figure 3.6. The cell has an inner diameter of 100 mm and a total length of 300 mm. Two soil specimens of equal length are fitted into the cell as shown. Soil 1 has a coefficient of permeability of 6.5×10^{-5} m/s. The total heads at points A and B (h_A and h_B) have been recorded 650 mm and 320 mm respectively. The amount of discharged water is 210 ml in 5 minutes. Calculate the coefficient of permeability of soil 2.

Solution:

$A = 7.854 \times 10^{-3}$ m^2, $L = 0.3 / 2 = 0.15$ m, using Equation 3.14 for soil 1:

$\Delta h = QL / kAt = 210.0 \times 10^{-6} \times 0.15 / (6.5 \times 10^{-5} \times 7.854 \times 10^{-3} \times 5.0 \times 60) = 0.2057$ m.

For soil 2: $\Delta h = 0.65 - 0.2057 - 0.32 = 0.1243$ m,

$k = QL / \Delta h A t = 210.0 \times 10^{-6} \times 0.15 / (0.1243 \times 7.854 \times 10^{-3} \times 5.0 \times 60) = 1.075 \times 10^{-4}$ m/s.

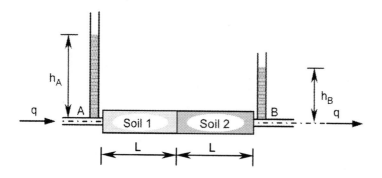

Figure 3.6. Example 3.4.

3.3.3 *In-situ permeability tests*

The main disadvantage of a laboratory permeability test is the small size of the sample. It is inappropriate to use only laboratory results to study flow conditions in the field. These results must be supported by in-situ tests before selecting an average k for the conditions involved in the project. In-situ permeability tests require the construction of test and observation wells both of which are expensive. Moreover, the complete hydraulic behaviour of any aquifer may need to be known. A comprehensive pumping test provides the necessary information about permeability characteristics and is a reliable basis for the prediction of ground water behaviour. In-situ tests are normally based either on the constant or variable head methods. The flow rate and the position of the piezometric level in the aquifer must also be determined which could be unconfined, confined or a combination of both. Care has to be taken to avoid disturbance, which may happen if aquifers are connected through the test wells and boreholes. A typical section of a modern test well is shown in Figure 3.7 (AS 2368). The well is sometimes protected against instability of its sides by casing but in dense and consolidated soils there may not be a need for casing. Water is extracted from the screened interval(s) that may also be perforated or unlined.

Pumping test in an unconfined aquifer. An idealized section of an unconfined aquifer with a pumping well (radius r_w) and two observation wells is shown in Figure 3.8. A pump is selected in accordance with the relevant code with regard to the maximum discharge and is mounted at the top of the pumping well to discharge a specified flow rate of q. As a result of pumping, the water level in the well is lowered (*drawdown*) creating a localized hydraulic gradient as shown in Figure 3.8. This causes the water to flow into the well in an axisymmetric condition making a cone type potentiometric surface around the well. The shape of the inverted cone (*cone of depression*) depends on the quantity of the discharge and hydraulic properties of the aquifer. After equilibrium is reached, the water levels in the two observation wells are measured using manual, automatic, electronic recorders or pressure gauges and the total heads of z_1 and z_2 are established. The flow passing through any cylindrical soil volume of radius r and height of z is equal to the discharged flow:

$$q = Av = Aki = (2\pi rz)k(\frac{dz}{dr}), \quad q\frac{dr}{r} = 2\pi kzdz,$$

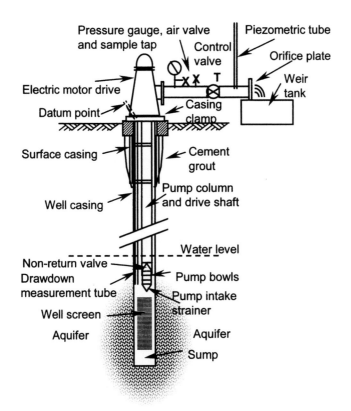

Figure 3.7. Typical test well in an unconfined aquifer.

$$q\int_{r_1}^{r_2} \frac{dr}{r} = 2\pi k \int_{z_1}^{z_2} z\,dz \rightarrow k = \frac{q}{\pi(z_2^2 - z_1^2)} \ln\frac{r_2}{r_1}$$ (3.17)

From Equation 3.17 it can be seen that for multiple observation wells with a constant flow rate the ratio of $a = \ln(r_2/r_1)$ to $b = (z_2^2 - z_1^2)$ remains constant for each pair of observation wells.

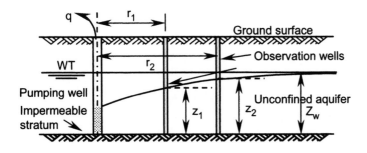

Figure 3.8. Pumping well in an unconfined aquifer.

This implies that, in a coordinate system representing a and b, the test data must be averaged by a line of best fit. The slope of this line with b axis gives the best average value for a/b to be used in Equation 3.17. Rearranging this equation, the piezometric level z, in terms of r and one of observation well data, may be expressed in the following form:

$$z = \sqrt{z_1^2 + \frac{q}{\pi k}\ln\frac{r}{r_1}}$$

(3.18)

The *radius of influence* r_o defines the point(s) where the piezometric level fully recovers to its original value of z_w. Substituting z_w in Equation 3.18, r_o can be expressed as:

$$r_o = r_1 \exp\frac{\pi k(z_w^2 - z_1^2)}{q}$$

(3.19)

Accordingly, the theoretical drawdown in the well is obtained by replacing r_w for r in Equation 3.18 and subtracting the result from the original piezometric level z_w.

$$D_w = z_w - \sqrt{z_1^2 + \frac{q}{\pi k}\ln\frac{r_w}{r_1}}$$

(3.20)

Note that in the application of the Darcy's law we assumed that the water moves horizontally towards the centreline of the well and perpendicular to the soil cylinder between the two observation wells. Thus vertical components of the velocity, which may be of significance close to the well, are ignored. Furthermore, due to the head loss caused by screens and gravel packs in the bottom of the well, the observed water level in the well may be lower than the water level immediately behind the well casing. Correction factors for the drawdown in the test well are available from standard codes (e.g. BS 1377). These reduce the actual drawdown to a more realistic value which is used to assess the overall behaviour of the aquifer.

Pumping test in a confined aquifer. Similarly, for a confined aquifer of thickness D (Figure 3.9) subjected to a discharge flow of q, Darcy's law can be applied to obtain k:

$$q = Av = Aki = (2\pi rD)k(\frac{dz}{dr}) \rightarrow q\frac{dr}{r} = 2\pi kDdz \rightarrow q\int_{r_1}^{r_2}\frac{dr}{r} = 2\pi kD\int_{z_1}^{z_2}dz,$$

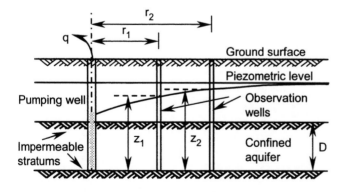

Figure 3.9. Pumping well in a confined aquifer.

$$k = \frac{q \ln(r_2 / r_1)}{2\pi D(z_2 - z_1)} \qquad (3.21)$$

By rearranging Equation 3.21, the piezometric level, radius of influence and drawdown in the well are expressed by Equations 3.22 to 3.24 respectively.

$$z = z_1 + \frac{q}{2\pi Dk} \ln \frac{r}{r_1} \qquad (3.22)$$

$$r_o = r_1 \exp[\frac{2\pi Dk}{q}(z_w - z_1)] \qquad (3.23)$$

$$D_w = z_w - [z_1 + \frac{q}{2\pi Dk} \ln \frac{r_w}{r_1}] \qquad (3.24)$$

In both unconfined and confined aquifers it is useful to estimate the radius of influence r_o where the localized water level recovers to its original level ($z = z_w$). The drawdown in the well may be approximated from Equations 3.20 and 3.24 without the need for an observation well if an arbitrary value of $r_o = 300$ m to 350 m is substituted in these equations (Roberson et al., 1997). However, the result must be interpreted with care and considered only as a preliminary step towards the hydraulic study of the aquifer. Another approximation reported by Linsley et al. (1992) is to assume that the radius of influence is proportional to the flow rate. In this case, by substituting two sets of data comprising two flow rates and two drawdowns into Equations 3.20 or 3.24, an estimate of the coefficient of permeability and drawdowns for specific flow rates can be obtained. In confined aquifer if, as a result of pumping, the water level is lowered to the aquifer zone, then a modification in the flow rate equation is necessary as the flow in zone adjacent to the well transfers to the unconfined state.

Equation 3.21 may also be applied to the Rowe type cell for the determination of the radial coefficient of permeability. In this case D is the height of the sample and r_2 and r_1 represent the radii of the sample and the central hole respectively. The term $z_2 - z_1$ is replaced by $\Delta p / \rho_w g$ where Δp is the recorded pressure difference between the porous boundary and the central hole, ρ_w is the density of water and $g = 9.81$ m^2/s.

Conventional slug test. This in-situ test is described in ASTM D-4044 (1996) and AS 2368 (1990) and is categorized as a *rapid displacement method.* The test has achieved

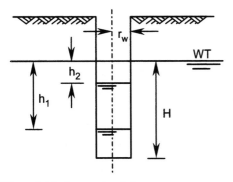

Figure 3.10. Conventional slug test.

widespread acceptance due to its simplicity, as there is neither a need for a pump installation or for observation wells. The test studies the rapid displacement of the water level in the well by instantaneous removal or injection of water. In a conventional slug test the water within the well is rapidly removed by bailing and is then allowed to recover. Records of the response of the well are obtained by measuring the recovered water level at different intervals of time. Figure 3.10 shows two positions of the recovered water level in a well over a time interval of *t*.

Auger hole test. This test is similar to the slug test but the removal of water is carried out in an unlined cylindrical hole drilled with an auger. The results may be affected by the instability of the hole's internal walls. However, problems due to well screen and gravel pack that influence the slug test results are non-existent.

Piezometer test. This test is similar to auger hole test with the difference that a piezometer (a rigid pipe approximately 110 mm in diameter) is inserted into the aquifer. Advantages of this method over the auger hole test include the possibility of testing a particular zone of the aquifer and improved stability of the hole.

Pneumatic displacement test. In this test drawdown in the water level in the well is achieved by increasing the air pressure above the water. After sealing the top of the well, the air pressure is increased by means of an air compressor. The volume of water associated with the downward displacement of the water level is absorbed by the aquifer. Constant air pressure is maintained until equilibrium is reached. The pressure is then suddenly released and the displacement of the water level is measured at various time intervals as in the slug test.

Tracer test. In this method the time interval for a trace (a dye or a salt) to travel between two observation wells is measured. The average velocity is calculated by dividing the distance between the two observation wells by the measured time. By measuring the water levels in the test wells, the average hydraulic gradient can then be calculated. The coefficient of permeability is the ratio of the velocity to the hydraulic gradient. This result has to be considered as an approximation due to limitations in the field as well in the theory. The direction of the flow has to be established by constructing multiple observation wells, which in turn increases the cost of the operation. Moreover, the existence of low or high permeability layers or of lenses may affect the results. For more information about the pumping tests and investigation of the aquifer response, reference may be made to BS 5930 (1981) and AS 2368 (1990).

Example 3.5

A pump test was carried out in an unconfined layer of sand 15 m thick and overlying an impermeable stratum. The water table was 2 m below the ground surface. The discharge after achieving equilibrium was measured 1.1 m^3/min whilst the drawdown in the two observation wells at 20 m and 40 m from the pumping well were measured at 0.80 m and 0.55 m respectively. Estimate the average k and r_o.

Solution:

$r_1 = 20.0$ m, $z_1 = 15.0 - 2.0 - 0.80 = 12.2$ m.

$r_2 = 40.0$ m, $z_2 = 15.0 - 2.0 - 0.55 = 12.45$ m.

$q = 1.1$ m^3/min $= 0.018333$ m^3/s. From Equation 3.17:

$k = 0.018333 \times \ln(40.0/20.0)/[\pi(12.45^2 - 12.20^2)] = 6.56 \times 10^{-4}$ m/s.

Using Equation 3.19:

$r_o = 20.0 \exp[\pi \times 6.56 \times 10^{-4}(13.0^2 - 12.2^2)/0.018333] = 193$ m.

Example 3.6

A pump test was carried out in an unconfined aquifer of $k = 3 \times 10^{-6}$ m/s with a flow rate of 20 m^3/hour. The radius of the well is 0.4 m and the aquifer has a depth of 80 m above an impermeable stratum. The drawdown in an observation well at a distance of 150 m from the well is 2.5 m. Calculate the radius of influence and the depth of water in the well.

Solution:

$z_1 = 80.0 - 2.5 = 77.5$ m, $r_1 = 150$ m, $r_w = 0.4$ m, $q = 20$ m^3/hour $= 5.555 \times 10^{-3}$ m^3/s.

From Equations 3.19 and 3.20:

$r_o = 150.0 \times \exp[\pi \times 3 \times 10^{-6}(80.0^2 - 77.5^2)/5.555 \times 10^{-3}] = 292.5$ m.

$D_w = 80.0 - \sqrt{77.5^2 + 5.555 \times 10^{-3} \ln(0.4/150.0)/(\pi \times 3 \times 10^{-6})} = 29.9$ m.

Example 3.7

A well is constructed to fully penetrate a confined aquifer of thickness of 25 m. The water level at two observation wells 40 m and 150 m from the well are 1.1 m and 0.4 m below the original piezometric level respectively. Determine the value of k. $q = 2.4$ m^3/min.

Solution:

$q = 2.4/60 = 0.04$ m^3/s, $z_1 = z_w - 1.1$, $z_2 = z_w - 0.4 \rightarrow z_2 - z_1 = 0.7$ m.

From Equation 3.21:

$k = 0.04 \ln(150.0/40.0)/(2\pi \times 25.0 \times 0.7) = 4.8 \times 10^{-4}$ m/s.

Example 3.8

An unconfined aquifer has a thickness of 40 m. A well of diameter 0.3 m is constructed for a pump test. For flow rates of 43 m^3/hour and 135 m^3/hour the actual drawdowns in the well are measured 3.7 m and 16.1 m respectively. Estimate the drawdown in the well for a flow rate of 60 m^3/hour assuming the radius of influence is proportional to the flow rate.

Solution:

Using Equation 3.20 for tests 1 and 2:

(a) $3.7 = 40.0 - \sqrt{40.0^2 + (43.0/3600.0)\ln(0.15/r_{o1})/(\pi \times k)}$,

$16.1 = 40.0 - \sqrt{40.0^2 + (135.0/3600.0)\ln(0.15/r_{o2})/(\pi \times k)}$.

But $r_{o2} = (135.0/43.0) r_{o1} = 3.139 r_{o1}$, thus:

(b) $16.1 = 40.0 - \sqrt{40.0^2 + (135.0/3600.0)\ln(0.15/3.139 r_{o1})/(\pi \times k)}$.

Solve two equations of (a) and (b) for r_{o1} and k:

$r_{o1} = 184.5$ m, $r_{o2} = 184.5 \times 3.139 = 579.2$ m, $k = 9.58 \times 10^{-5}$ m/s. For $q = 60.0$ m^3/hour:

$r_{o3} = (60.0 / 135.0) \times 579.2$ or $(60.0 / 43.0) \times 184.5 = 257.4$ m.

$$D_w = 40.0 - \sqrt{40.0^2 + (60.0/3600.0)\ln(0.15/257.4)/(\pi \times 9.58 \times 10^{-5})} = 5.54 \text{ m.}$$

3.3.4 *Permeability of stratified soils*

Consider the stratified soil section shown in Figure 3.11. The average values of k in the x and z directions are estimated assuming that the water moves either parallel or perpendicular to the layers. When the seepage is parallel to z:

$$q = Av = Aki = Ak_{z1}\frac{\Delta h_1}{z_1} = Ak_{z2}\frac{\Delta h_2}{z_2} = \cdots = Ak_{zn}\frac{\Delta h_n}{z_n}$$

where Δh_1, Δh_2, and Δh_n are the losses in the total head as the water passes through the layers of thickness of z_1, z_2, ..., and z_n. The total loss is equal to the sum of head losses, hence:

$\Delta h = \Delta h_1 + \Delta h_2 + \cdots + \Delta h_n$, or:

$$\Delta h = \frac{qz}{Ak_z} = \frac{qz_1}{Ak_{z1}} + \frac{qz_2}{Ak_{z2}} + \cdots + \frac{qz_n}{Ak_{zn}} = \frac{q}{A}\left(\frac{z_1}{k_{z1}} + \frac{z_2}{k_{z2}} + \cdots + \frac{z_n}{k_{zn}}\right),$$

where $z = z_1 + z_2 + \cdots + z_n$ and k_z is the average for the stratified soil in z direction, thus:

$$k_z = \frac{z}{\dfrac{z_1}{k_{z1}} + \dfrac{z_2}{k_{z2}} + \cdots + \dfrac{z_n}{k_{zn}}} \tag{3.25}$$

Note that the xz coordinate is selected to match the geometry of the layered structure and the z-axis may not be in the direction of gravity. For the seepage in the x direction the total flow rate is the sum of flow rates passing through each layer:

$q = q_1 + q_2 + \cdots + q_n$, or: $Ak_x i = A_1 k_{x1} i + A_2 k_{x2} i + \cdots + A_n k_{xn} i$,

where $A = z \times 1 = (z_1 + z_2 + \cdots + z_n) \times 1$, $A_1 = z_1 \times 1, \ldots,$ and $A_n = z_n \times 1$.

Substituting A values into the equation above, we obtain:

$$k_x = \frac{z_1 k_{x1} + z_2 k_{x2} + \cdots + z_n k_{xn}}{z} \tag{3.26}$$

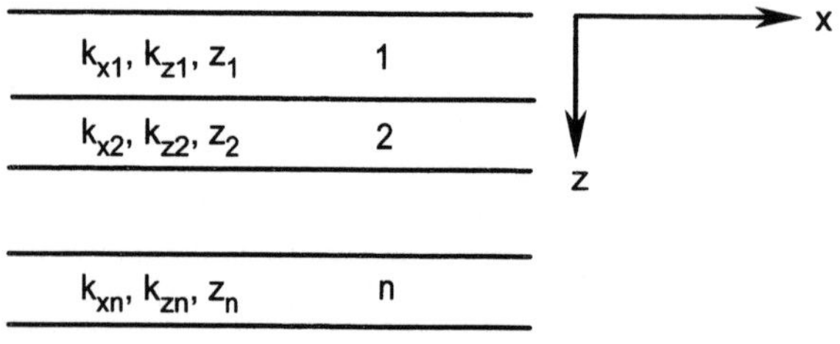

Figure 3.11. Analysis of average horizontal and vertical coefficient of permeability in stratified soil.

Example 3.9

The profile of a stratified soil contains three horizontal layers each having a thickness of 0.3 m. The top of the first layer is the ground surface. The k values are 10^{-2} m/s for the top, 2.5×10^{-4} m/s for the middle and 4.0×10^{-5} m/s for the bottom layer. There is 0.2 m of water above the surface of the soil. The section is drained at the base level. Calculate the flow rate per m^2 of the stratified soil in litre/hour. If the overflow mechanism of the constant head is deleted, what will be the time required for the water level to drop to the ground surface.

Solution:

Using Equation 3.25:

$k_z = 0.9/(0.3/10^{-2} + 0.3/2.5 \times 10^{-4} + 0.3/4.0 \times 10^{-5}) = 1.03 \times 10^{-4}$ m/s.

$q = Aki = 1.0 \times 1.03 \times 10^{-4} \times (0.2 + 0.9)/0.9 \times 1000 \times 3600 = 453.2$ l/hour.

Using Equation 3.15 and assuming that, $h_1 = 1.1$ m, $h_2 = 0.9$ m, $L = 0.9$ m and

$A = a = 1$ m^2: $k = 1.03 \times 10^{-4} = 2.3(aL/At)\log(h_1/h_2) = 2.3(0.9/t)\log(1.1/0.9)$,

$t = 1751$ s $= 29$ min., 11 s.

3.4   FLOW NETS

3.4.1   *Flow nets in an isotropic material*

A flow net is a graphical representation of two-dimensional seepage and consists of two groups of curves of *flow lines* and *equipotential lines* (Figure 3.12). A flow line in a steady flow condition represents a curve on which a particle of water commences its movement from a known head (point S, Figure 3.12) and terminates at a specified lower head (point F, Figure 3.12). The tangent at a point on a flow line represents the direction of the velocity defined by Darcy's law. Thus the velocity at each point is proportional to the hydraulic gradient which in turn depends on the coordinate of the point. As a result the average velocity on a flow line becomes dependent on the length of the flow line. A shorter flow line means a higher hydraulic gradient for a specified loss of head, resulting in a higher velocity. Referring to Figure 3.12 it can be seen that the velocity in the vicinity of the base of the dam is much higher than the velocity immediately above the imperme-

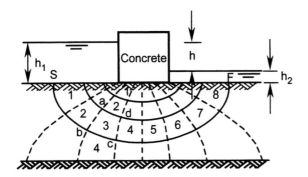

Figure 3.12. Flow and equipotential lines under an impermeable dam.

able stratum. An equipotential line is a curve on which the total head is constant (water rises to a constant level) and whose magnitude ranges between the highest and lowest total heads related to the seepage problem. The flow lines are everywhere normal to the equipotential lines, and thus the velocity at every point is perpendicular to the equipotential line passing through that point. From the definition of flow and equipotential lines we can see that their numbers are infinite, however a finite number of flow and equipotential lines (*flow net*) can give valuable information on seepage characteristics such as velocity and flow rate. In steady flow there is no flow occurring across the flow lines and the rate of flow between two flow lines is constant. Figure 3.13 is an example of a soil element bounded with two flow lines and two equipotential lines. The flow net is constructed in a way that the average distance between two flow lines within an element is equal to the average distance between two equipotential lines (l_i in Figure 3.13). Furthermore, the head loss between successive equipotential lines is equal. The flow lines selected to represent the flow net are those in which the amount of flow between two pair of flow lines is equal. Two flow lines represent a *lane* (Wu, 1966) with a variable section and the total flow rate is the flow rate corresponding to one lane multiplied by the number of lanes. The flow rate passing through an element *abcd* (Figure 3.12) is the constant flow between two flow lines of q_i and q_{i+1} and is given by:

$$\Delta q = Av = Aki = (l_i \times 1) \times k \times \frac{\Delta h}{l_i} = k\Delta h = k\frac{h_1 - h_2}{N_d} = k\frac{h}{N_d},$$

where N_d is the number of equal drops in total head and h is the total loss due to the seepage. If the total number of flow lanes is N_f the total flow rate is:

$$q = N_f \Delta q = kh\frac{N_f}{N_d} \tag{3.27}$$

Flow nets are constructed either analytically or by numerical methods. Both techniques are based on the solution of the continuity equation (Section 3.5). In seepage problems with simple boundary conditions the flow net is constructed by sketching, as demonstrated in Figure 3.14. This may be done by hand or by computer graphics in accordance with the following:

1. The boundary between soil and water is an equipotential line.

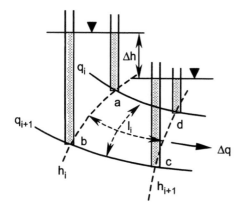

Figure 3.13. Flow and equipotential lines.

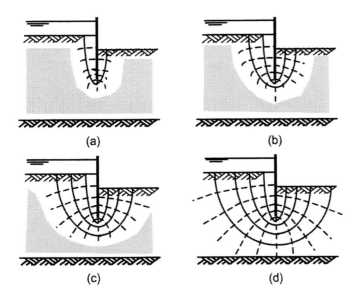

(a) (b)

(c) (d)

Figure 3.14. Step by step construction of a flow net.

2. Any boundary between the soil and an impermeable material is a flow line. The impermeable material is either an impermeable stratum(s) of the aquifer or the base of the foundations and sheet piles that touch the aquifer.

3. Sketching may start by constructing the first flow lane as shown in Figure 3.14(a). The flow lane is divided into a number of squares and equipotential lines are projected outwards into the second flow lane.

4. The second flow line is drawn to form the squares of the second flow lane.

5. The procedure is continued until the last flow lane is formed. This lane must consist of squares and satisfy the boundary conditions; otherwise the first flow line has to be repositioned and the whole procedure repeated.

Example 3.10

A sheet piling system with its corresponding flow net is shown in Figure 3.15. (a) Estimate the flow rate in m^3/day per 1 m run of piling, (b) for the element A with $l = 1.5$ m calculate the average velocity and the effective vertical stress, (c) determine the magnitude of the effective vertical stress at the base and at the right-hand side of the sheet pile, (d) calculate the factor of safety against the quick condition, defined as the ratio of the existing hydraulic gradient along the downstream face of sheet pile to the critical hydraulic gradient defined by Equation 3.11. $k = 0.02$ mm/s, $\gamma_{sat} = 20$ kN/m^3.

Solution:

(a) $N_f = 5$ and $N_d = 11$, $h = 3.0$ m,

$$q = \frac{khN_f}{N_d} = \frac{(0.02/1000) \times 24 \times 3600 \times 3.0 \times 5}{11} = 2.356 \text{ m}^3/\text{day}.$$

(b) $v = ki = 0.02 \times [(3.0/11)/1.5] \times 3600 = 13.1$ mm/hour.

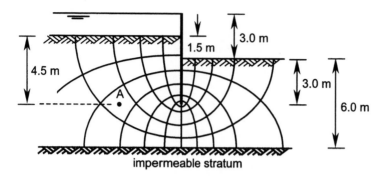

Figure 3.15. Example 3.10.

Head loss at the centre of the element $= 2.5$ drops $\times 3.0 / 11 = 0.682$ m,
$h_w =$ depth of A from water surface $-$ loss of head $= (4.5 + 1.5) - 0.682 = 5.318$ m,
$u = 5.318 \times 9.81 = 52.2$ kPa, $\sigma_v = 4.5 \times 20.0 + 1.5 \times 9.81 = 104.7$ kPa,
$\sigma'_v = 104.7 - 52.2 = 52.5$ kPa.
(c) Head loss at the base of the sheet pile $=$
6 (number of equal drops) $\times (3.0 / 11) = 1.636$ m,
Height of water at this point $=$ depth from the water surface $-$ loss of head
$= (4.5 + 1.5) - 1.636 = 4.364$ m, $u = 4.364 \times 9.81 = 42.8$ kPa.
$\sigma_v = 3 \times 20.0 = 60.0$ kPa, $\sigma'_v = 60.0 - 42.8 = 17.2$ kPa.
(d) $i = (4.364 - 3.0) / 3.0 = 0.455$, $i_c = \gamma' / \gamma_w = (20.0 - 9.981) / 9.81 = 1.039$.
$F = 1.039 / 0.455 = 2.28$.
The factor of safety against the quick condition is also estimated by evaluation of the stability of a prism of soil D (embedment depth) by $D / 2$ (width) by unit thickness attached to the back of the sheet pile. The factor of safety is defined as the effective weight of the prism divided by the upward force resulting from the seepage pressure at the base of the prism.

3.4.2 Flow nets in an anisotropic material

Consider an anisotropic soil with horizontal and vertical coefficients of permeability of k_x and k_z respectively. The material can be treated as an isotropic soil by assuming an isotropic coefficient of permeabilty of:

$$k_i = \sqrt{k_x k_z} \qquad (3.28)$$

and using a transformed scale of:

$$x' = x\sqrt{k_z / k_x} \qquad (3.29)$$

Considering an element of soil in x direction, having the length of l and cross-sectional area of A, the flow rate in the x direction in the untransformed state is:
$q_x = A k_x i = A k_x \Delta h / l$. In the transformed state:

$$q_{x'} = A k_i i = A k_i \frac{\Delta h}{l \sqrt{k_z / k_x}} = A \sqrt{k_x k_z} \frac{\Delta h}{l \sqrt{k_z / k_x}} = A k_x \frac{\Delta h}{l} = q_x.$$

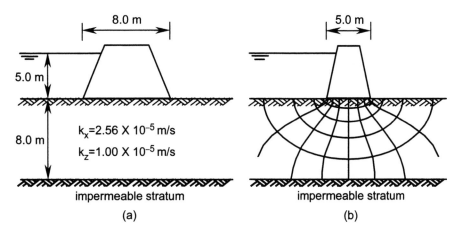

Figure 3.16. Example 3.11.

Example 3.11

A concrete dam has a base length of 8 m and retains 5 m of water as shown in Figure 3.16(a). The water level on the downstream side is at the ground surface. Under the dam there exists an 8 m thick layer of anisotropic permeable soil with the k values shown in the Figure 3.16(a). Calculate the flow rate under the concrete dam in l/day.

Solution:

The scale factor in the x direction is:

$$\sqrt{k_z / k_x} = \sqrt{1.0\times10^{-5}/2.56\times10^{-5}} = 0.625.$$

The length of the transformed base is: $8.0 \times 0.625 = 5.0$ m.
The equivalent isotropic coefficient of permeability is calculated using Equation 3.28:

$$k_i = \sqrt{1.0\times10^{-5}\times2.56\times10^{-5}} = 1.6\times10^{-5} \text{ m/s}.$$

Figure 3.16(b) shows the transformed section of the soil-dam, with the flow net drawn by sketching in which: $N_d = 8$, $N_f = 5$.
Using Equation 3.27: $q = khN_f / N_d = 1.6\times10^{-5}\times5.0\times1000\times3600\times24/8 = 4320$ l/day.

3.4.3   Control of the flow rate under impermeable dams

The flow rate in the soil under an impermeable dam (Figure 3.17(a)) and the factor of safety against the quick condition may be controlled by introducing horizontal or vertical impermeable blankets and a toe filter. A horizontal impermeable blanket may contain a thin layer of compacted clay constructed on the upstream side (Figure 3.17(b)). This increases the length of the flow line immediately under the dam, projecting all the flow lines to the left of the blanket. As a result, the number of equipotential lines increases, thereby reducing the flow rate (Equation 3.27) to a required value. It may be shown that if the number of equipotential lines is kept the same as in the unmodified section, then the number of flow lines and the flow rate will both decrease. In addition, the pore pressure at the toe decreases, improving the overall stability of the dam and the factor of safety against

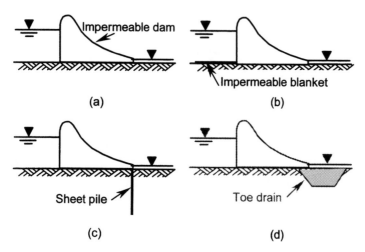

(a)   (b)

(c)   (d)

Figure 3.17. Methods to control the flow rate and quick condition.

the quick condition. An example of a vertical impermeable blanket is a sheet pile system driven into the soil, either at the toe or at another position within the base of the dam (Figure 3.17(c)). Sheet piles are more effective in increasing the factor of safety but less effective in decreasing the flow rate. A toe drain shown in Figure 3.17(d) may also be used to significantly increase the factor of safety but it will result in a higher flow rate that may or may not be desirable for the project. A combination of the above methods may be used to optimise the length of the base which is designed mainly to provide stability against hydrostatic and gravity forces.

Example 3.12

Resolve Example 3.11 with a horizontal impermeable blanket of 8 m length added to the system at the upstream side.

Solution:

The transformed length of the horizontal blanket is: $8.0 \times 0.625 = 5.0$ m.

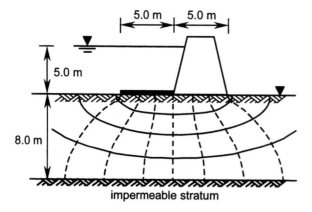

Figure 3.18. Example 3.12.

Figure 3.18 shows the flow net drawn schematically for the transformed section. The number of equipotential lines were kept the same: $N_d = 8$, and $N_f = 4$.

Using Equation 3.27:

$q = 1.6 \times 10^{-5} \times 5.0 \times 1000 \times 3600 \times 24 \times 4/8 = 3456$ l/day.

The flow rate is reduced by: $(4320 - 3456) / 4320 = 0.2 = 20\%$.

Example 3.13

For the concrete dam shown in Figure 3.19 construct the flow net under the dam, calculate the flow rate (in m^3/day) and the distribution of uplift pressure on the base of the dam. The base is 1 m below the original ground surface and a sheet pile 7 m long is driven into the soil at the upstream side.

$h = 9$ m, $k = 4 \times 10^{-5}$ m/s, $\gamma_{sat} = 21$ kN/m^3.

Solution:

The flow net is shown in Figure 3.19, from which: $N_d = 10$ and $N_f = 4$.
Using Equation 3.27:

$q = 4.0 \times 10^{-5} \times 9.0 \times 3600 \times 24 \times 4/10 = 12.44$ m^3/day (per metre run).

The total head across the length of the base is formulated as follows:
$h = 10.0 + 1.0$ (depth of foundation) $- 0.9 n_d$, where n_d is the number of equal drops of $9.0 / 10 = 0.9$ m up to the point of interest on the base. The distance of the intersection points of equipotential lines from the left corner of the dam and the relevant heads and water pressures are presented in the table below. For point (1) we may use interpolation or, conservatively, simply accept the higher magnitude corresponding to $n_d = 3$.

Point	1	2	3	4	5	6	7
x (m)	0.00	4.75	9.75	15.00	19.25	22.50	25.00
h (m)	8.3	7.4	6.5	5.6	4.7	3.8	2.9
u (kPa)	81.4	72.6	63.8	54.9	46.1	37.3	28.5

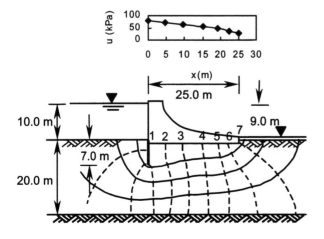

Figure 3.19. Example 3.13.

The diagram of uplift pressure is also shown in Figure 3.19. The average hydraulic gradient between the base and ground surface on the downstream side is:

$$i = \frac{\Delta h}{l} = \frac{0.9}{1.0} = 0.9, \; i_c = \frac{\gamma'}{\gamma_w} = \frac{21.0 - 9.81}{9.81} = 1.14.$$

Factor of safety against the quick condition $= 1.14 / 0.9 = 1.27$.

3.5 MATHEMATICS OF THE FLOW IN SOIL

3.5.1 *Continuity equation for a steady flow*

Figure 3.20 shows a three-dimensional element of soil in which the total head of water is h. Assuming a steady state flow, the quantity of water entering the element parallel to one of the axes (e.g. x-axis) is: $q_x = v_x A_x$, where v_x is the component of velocity in the x direction and $A_x = dy \times dz$ is the sectional area of the element perpendicular to the direction of flow. Hence:

$$v_x = k_x i_x = -k_x \frac{\partial h}{\partial x}, \; v_y = k_y i_y = -k_y \frac{\partial h}{\partial y}, \; v_z = k_z i_z = -k_z \frac{\partial h}{\partial z} \tag{3.30}$$

The negative sign is arbitrary and causes the value of h to decrease in the direction of the velocity. Substituting v_x into the flow rate equation we obtain:

$$q_x = -k_x \frac{\partial h}{\partial x} dy \, dz \tag{3.31}$$

The amount of flow that exits the element in the x direction is $q_x + dq_x$ in which dq_x is the variation of the flow rate in the x direction and is expressed by:

$$dq_x = -k_x \frac{\partial^2 h}{\partial x^2} dx \, dy \, dz \tag{3.32}$$

Assuming a zero volume change, the flow rate entering the element is equal to the flow rate exiting the element. Therefore, the continuity condition can be written as:

$$q_x + q_y + q_z = q_x + dq_x + q_y + dq_y + q_z + dq_z,$$

$$dq_x + dq_y + dq_z = \Delta_{volume} = 0 \tag{3.33}$$

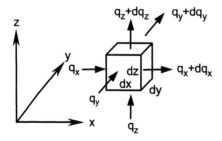

Figure 3.20. Steady flow through a three-dimensional element.

Substituting Equation 3.32 into the continuity condition we have:

$$k_x \frac{\partial^2 h}{\partial x^2} + k_y \frac{\partial^2 h}{\partial y^2} + k_z \frac{\partial^2 h}{\partial z^2} = 0 \tag{3.34}$$

In two-dimensional flow, where flow in, say, the y direction is zero, the corresponding term in the continuity equation vanishes. Equation 3.34 therefore becomes:

$$k_x \frac{\partial^2 h}{\partial x^2} + k_z \frac{\partial^2 h}{\partial z^2} = 0, \text{ or } \frac{\partial^2 h}{\partial x^2} + \frac{\partial^2 h}{\partial z^2} = 0 \qquad (\text{for } k_x = k_z) \tag{3.35}$$

The solution to this differential equation yields the total head within the seepage zone. It is seen that the distribution of the total head is independent of the coefficient of permeability (when $k_x = k_z$) and depends entirely on the geometry of the problem. The continuity condition expressed by Equation 3.35 may be arranged in terms of the components of the velocity by substituting Equations 3.30, resulting in:

$$\frac{\partial v_x}{\partial x} + \frac{\partial v_z}{\partial z} = 0 \tag{3.36}$$

Equation 3.35 (or 3.36) is the classical differential equation of flow in two dimensions and is referred to as the *Laplace equation*. Traditional methods of solving this equation include direct integration and numerical analysis. Direct integration is carried out only for seepage problems with simple boundary conditions by using complex functions. Whilst there are an infinite number of solutions to the Laplace equation, the boundary conditions control the selection of the function. For one-dimensional flow, in the x direction:

$$\frac{\partial^2 h}{\partial x^2} = 0 \tag{3.37}$$

Equation 3.37 may be integrated twice to give the following linear relationship for the total head: $h = C_1 x + C_2$, where C_1 and C_2 are integration constants found by substituting two known boundary conditions into the linear equation. These known boundary conditions may be selected as the total heads h_1 and h_2 corresponding to two sections with $x_1 = 0$ and $x_2 = L$ where L is the distance between the two sections. Substituting these values into the linear equation of total head we get:

$$C_1 = -\frac{h_1 - h_2}{L}, C_2 = h_1.$$

The total head between two sections of x_1 and x_2 is represented by:

$$h = -\frac{x(h_1 - h_2)}{L} + h_1 \tag{3.38}$$

By differentiating Equation 3.38 and applying Darcy's law, the velocity is found to be:

$$v_x = k_x i_x = -k_x \frac{\partial h}{\partial x} = k_x \frac{h_1 - h_2}{L} = k_x \frac{\Delta h}{L} \tag{3.39}$$

This shows that the velocity is constant between sections 1 and 2 and is consistent with the basic assumption stated previously (Figure 3.1).

3.5.2   *Potential and stream functions*

Equations 3.30 represent the projections of the velocity in a three-dimensional system and imply that the velocity is perpendicular to the constant *h* passing through the point. The analytical solution of two-dimensional flow through an isotropic material expressed by Equation 3.35 is obtained by the introduction of two functions Φ and Ψ to satisfy Equations 3.30.

$$v_x = -k\frac{\partial h}{\partial x} = \frac{\partial \Phi}{\partial x} = \frac{\partial \Psi}{\partial z}, \quad v_z = -k\frac{\partial h}{\partial z} = \frac{\partial \Phi}{\partial z} = -\frac{\partial \Psi}{\partial x} \tag{3.40}$$

where Φ and Ψ are called *potential function* and *stream function* respectively. Substituting Equations 3.40 into 3.35 we obtain:

$$\frac{\partial^2 \Phi}{\partial x^2} + \frac{\partial^2 \Phi}{\partial z^2} = 0 \tag{3.41}$$

The solution of Equation 3.41 will yield a result for $\Phi(x, z)$, which can be expressed as follows based on the definition of Φ:

$$\Phi(x,z) = -kh(x,z) + C \tag{3.42}$$

where C is a constant depending on the boundary conditions. For different but constant values of Φ, such as Φ_1, Φ_2, and Φ_3, Equation 3.42 represents a series of curves for which the potential function and total head are constant i.e. equipotential lines. From the definition of the function Ψ it is seen that it is tangential to the velocity and represents the direction of the flow and flow lines. From Equations 3.40:

$$\frac{\partial \Phi}{\partial x} = \frac{\partial \Psi}{\partial z}, \quad \frac{\partial \Phi}{\partial z} = -\frac{\partial \Psi}{\partial x} \tag{3.43}$$

These are called the Cauchy-Riemann equations whose solution yields a series of stream functions or flow lines for constant values of ψ_1, ψ_2, ...assigned for $\psi(x, z)$. Taking the derivative of the first term of Equation 3.43 with respect to *z* and the second term with respect to *x* and adding the results, then:

$$\frac{\partial^2 \Psi}{\partial x^2} + \frac{\partial^2 \Psi}{\partial z^2} = 0 \tag{3.44}$$

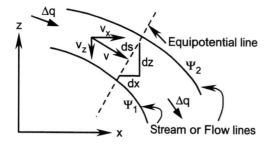

Figure 3.21. Calculation of the quantity of flow between two flow lines.

which means that the stream functions are in the form of the Laplace equation. Figure 3.21 shows two flow lines represented with Ψ_1 and Ψ_2. At an arbitrary point within these flow lines the velocity v is shown with its projections in the xz plane. With the arrangement shown for the positive variation of x and z along the equipotential line, one of the projections of the velocity is in the negative direction of the corresponding plane axis. The flow dq corresponding to the length of ds on the equipotential line is:

$$dq = v \times ds = v_x \times dz - v_z \times dx.$$

Substituting velocities from Equations 3.40 we obtain:

$$dq = \frac{\partial \Psi}{\partial x} dx + \frac{\partial \Psi}{\partial z} dz = d\Psi, \text{ or}$$

$$\Delta q = \int_{\Psi_1}^{\Psi_2} d\Psi = \Psi_2 - \Psi_1 \tag{3.45}$$

This shows that the flow rate between two successive flow lines is equal to the difference between the stream functions. If the stream function is known in terms of x and z then the velocity components can be calculated using Equation 3.40. Conversely, with known functions for the velocity components, the stream function can be obtained by:

$$\Psi = \int \frac{\partial \Psi}{\partial x} dx + \int \frac{\partial \Psi}{\partial z} dz + C \tag{3.46}$$

A useful result is obtained by substituting Equations 3.43 into Equation 3.41:

$$\frac{\partial^2 \Psi}{\partial z \partial x} - \frac{\partial^2 \Psi}{\partial x \partial z} = 0 \tag{3.47}$$

This indicates that a stream function obtained from Equation 3.46 automatically satisfies the continuity equation. Equations 3.35 (or 3.41) and 3.44 are a set of differential equations for which closed form solutions are available only in very simple cases. Analytical solutions are obtained by using the theory of complex variables. For example, for an impermeable dam with a base length of b, and a total head loss of h, resting on a permeable layer, similar to Figure 3.12 but with infinite depth, Wu (1966) reported the following closed form solutions for potential and stream functions:

$$\frac{x^2}{(b\cos\frac{\pi\Phi}{h})^2} - \frac{z^2}{(b\sin\frac{\pi\Phi}{h})^2} = 1,$$

$$\frac{x^2}{(b\cosh\frac{\pi\Psi}{h})^2} + \frac{z^2}{(b\sinh\frac{\pi\Psi}{h})^2} = 1 \tag{3.48}$$

In general, the solution of two-dimensional flow is achieved by numerical techniques such as *finite element* and *finite difference* methods. In the finite element method the seepage zone is divided into discrete elements with common nodes. The parameter of interest (h, Φ or ψ) is assumed to vary within each element according to a linear or a non-linear function.

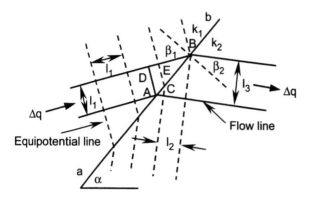

Figure 3.22. Deflection of flow lines at the boundary between two soils.

A solution is obtained for specified boundary conditions that covers the entire seepage zone. In the finite difference method the seepage zone is divided into a network of nodes usually known as grid points. The partial derivatives at each point are replaced with a finite difference approximation that results in a linear algebraic equation.

With known boundary conditions, and application of the continuity condition at the boundaries with unknown numerical values for the specified function (h, Φ or ψ), a unique solution can be obtained for grid point values. The variation of h, Φ or ψ is assumed to be linear between grid points, which is a major disadvantages compared to the finite element method. However, the finite difference method is easy to formulate and a useful alternative to the analytical solution.

3.5.3 *The transfer conditions*

The line ab in Figure 3.22 represents the boundary between two soils with different k values. The flow lines at the k_1 side are deflected as they enter into the k_2 side. This is represented by different values of β, the angle between a flow line and the direction normal to the boundary. The flow net at the k_1 side is comprised of squares of length l_1. At the k_2 side the squares change into rectangles l_2 by l_3 which represent the distance between the equipotential lines and flow lines respectively. Consider the case where one flow line and one equipotential line intersect at a point B located on the boundary. According to the continuity condition the flow rate inside a flow lane at the k_1 side is equal to the flow rate in the deflected lane at the k_2 side.

$$\Delta q_1 = Aki = (l_1 \times 1)k_1 \frac{\Delta h}{l_1} = k_1 \Delta h, \ \Delta q_2 = Aki = (l_3 \times 1)k_2 \frac{\Delta h}{l_2}.$$

From the geometry of Figure 3.22, l_2 and l_3 can be expressed in terms of β_1, β_2 and l_1:

$$l_2 = BC \sin\beta_2 = l_1 \frac{\sin\beta_2}{\sin\beta_1}, \ l_3 = AB\cos\beta_2 = l_1 \frac{\cos\beta_2}{\cos\beta_1}.$$

Substituting l_2 and l_3 in the flow equations and equating Δq_1 by Δq_2 we obtain:

$$\frac{k_1}{\tan\beta_1} = \frac{k_2}{\tan\beta_2} \rightarrow k_1 \tan\beta_2 = k_2 \tan\beta_1 \tag{3.49}$$

3.5.4 *Numerical analysis of two-dimensional flow using the finite difference method*

In this method two-dimensional flow, represented by the continuity Equations 3.35 or 3.41 or the differential equation of stream function in Equation 3.44, is approximated by finite difference equations. In this section we develop numerical values for the solution of the Laplace equation for total head given in Equation 3.35. Let Δx and Δz be the change in the coordinate of a point represented by i and j as shown in Figure 3.23(a). Restricting the Taylor's series to a finite number of terms, the total head at points $(i + 1, j)$ and $(i − 1, j)$ representing $+\Delta x$ and $-\Delta x$ respectively, are:

$$h_{i+1,j} = h_{i,j} + \Delta x (\frac{\partial h}{\partial x})_{i,j} + \frac{\Delta x^2}{2!} (\frac{\partial^2 h}{\partial x^2})_{i,j} + \frac{\Delta x^3}{3!} (\frac{\partial^3 h}{\partial x^3})_{i,j} + \frac{\Delta x^4}{4!} (\frac{\partial^4 h}{\partial x^4})_{i,j} \quad (3.50)$$

$$h_{i-1,j} = h_{i,j} - \Delta x (\frac{\partial h}{\partial x})_{i,j} + \frac{\Delta x^2}{2!} (\frac{\partial^2 h}{\partial x^2})_{i,j} - \frac{\Delta x^3}{3!} (\frac{\partial^3 h}{\partial x^3})_{i,j} + \frac{\Delta x^4}{4!} (\frac{\partial^4 h}{\partial x^4})_{i,j} \quad (3.51)$$

Adding Equations 3.50 and 3.51, we obtain:

$$(\frac{\partial^2 h}{\partial x^2})_{i,j} = \frac{h_{i+1,j} + h_{i-1,j} - 2h_{i,j}}{(\Delta x)^2} \quad (3.52)$$

Similarly, for flow in the z direction, we have:

$$(\frac{\partial^2 h}{\partial z^2})_{i,j} = \frac{h_{i,j+1} + h_{i,j-1} - 2h_{i,j}}{(\Delta z)^2} \quad (3.53)$$

Substituting Equations 3.52 and 3.53 into the continuity Equation 3.35:

$$k_x \frac{h_{i+1,j} + h_{i-1,j} - 2h_{i,j}}{(\Delta x)^2} + k_z \frac{h_{i,j+1} + h_{i,j-1} - 2h_{i,j}}{(\Delta z)^2} = 0 \quad (3.54)$$

For convenience we assume $\Delta x = \Delta z$, and $k_x = k_z$ and therefore:

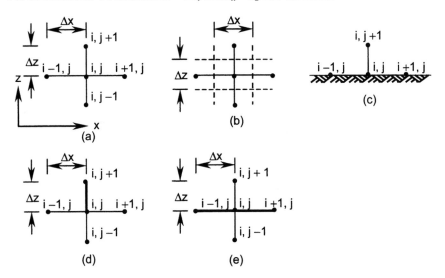

Figure 3.23. The basic finite difference grid.

$$h_{i,j} = \frac{h_{i+1,j} + h_{i-1,j} + h_{i,j+1} + h_{i,j-1}}{4} \tag{3.55}$$

This equation shows that, in a square grid, the total head at every grid point is the average of the total heads at the four adjacent grid points. It also implies that the total head at every point in the grid represents the average total head within a flow lane of thickness Δx or Δz, and that the point is located on the centre line of the flow lane (Figure 3.23(b)). This means that if we apply Darcy's law in the form $q = Aki$, and the continuity condition in the form $\Delta q = 0$ (at every point) we can obtain Equation 3.55. Note that in this case, the hydraulic gradient is the first derivative of the total head expressed in finite difference form. Equation 3.55 can be modified (using the concept of flow lane and continuity condition) for the case when the point (i, j) is located on an impermeable boundary (Figure 3.23(c)):

$$h_{i,j} = \frac{h_{i+1,j} + h_{i-1,j} + 2h_{i,j+1}}{4} \tag{3.56}$$

If the point (i, j) is located at the base of a vertical impermeable boundary with negligible thickness (e.g. sheet pile), as illustrated in Figure 3.23(d), the total head at the grid point $(i, j + 1)$ can be taken as the average of the total heads at the left and right of the boundary:

$$h_{i,j} = \frac{h_{i+1,j} + h_{i-1,j} + 0.5[(h_{i,j+1})_L + (h_{i,j+1})_R] + h_{i,j-1}}{4} \tag{3.57}$$

If the point (i, j) is located on the boundary of two soils with different permeabilities, and the boundary is parallel to the horizontal axis (Figure 3.23(e)), then:

$$h_{i,j} = \frac{1}{4}(h_{i+1,j} + h_{i-1,j} + \frac{2k_B}{k_T + k_B}h_{i,j+1} + \frac{2k_T}{k_T + k_B}h_{i,j-1}) \tag{3.58}$$

where k_T and k_B are coefficients of permeability at the top zone and the bottom zone of the boundary respectively. For each unknown head at an interior or a boundary grid point, there is a condition expressed by Equations 3.55 to 3.58. Thus, the total number of linear equations and unknowns are equal, and a unique solution can be obtained. Traditionally, the set of equations was solved using a relaxation method, which included the readjustment of assumed values of the unknowns to satisfy the appropriate condition at each grid point. However, with the improvement of computational tools, a direct solution is now preferred.

Example 3.14

Figure 3.24(a) shows a section of a concrete dam, with an 8 m base, resting on a permeable layer of thickness 8 m. Taking the impermeable stratum as the datum for total heads, calculate the total heads at the grid points using the finite difference method and construct the equipotential line for a head loss of 1.25 m. Repeat the solution using the relaxation method by a refined grid (occupying the same area) with $\Delta x = \Delta z = 2$ m.

Solution:

A mesh of $\Delta x = \Delta z = 4$ m is constructed to evaluate the total heads at the grid points. Since the flow net in the permeable layer is symmetric, only one-half of the section is considered.

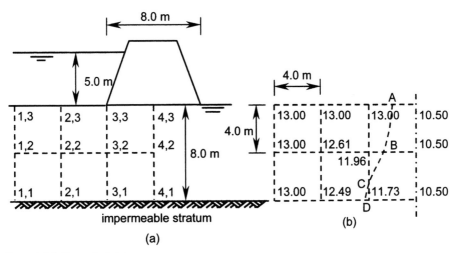

Figure 3.24. Example 3.14.

Boundary conditions: $h_{1,3} = h_{2,3} = h_{3,3} = 5.0 + 8.0 = 13.0$ m.

For a symmetrical flow net:

$h_{4,1} = h_{4,2} = h_{4,3} = 2.5 + 8.0 = 10.5$ m.

Assume at the left boundary (grid points 1,1-1,2 and 1,3) the flow rate is zero and the pore pressures at these points are hydrostatic: $h_{1,1} = h_{1,2} = h_{1,3} = 13.0$ m.

Applying Equation 3.55 at the interior grid points 2,2 and 3,2 and Equation 3.56 at the grid points 2,1 and 3,1 on the impermeable boundary, and substituting the known boundary heads:

$$-h_{2,1} + 4h_{2,2} - h_{3,2} = 26.0, \quad -h_{2,2} - h_{3,1} + 4h_{3,2} = 23.5,$$

$$4h_{2,1} - 2h_{2,2} - h_{3,1} = 13.0, \quad -h_{2,1} + 4h_{3,1} - 2h_{3,2} = 10.5.$$

Solving for the unknowns: $h_{2,1} = 12.49$ m, $h_{2,2} = 12.61$ m, $h_{3,1} = 11.73$ m, $h_{3,2} = 11.96$ m.

The total head for the equipotential line with a 1.25 m loss is $13.0 - 1.25 = 11.75$ m. The dashed line in Figure 3.24(b) shows the equipotential line for $h = 11.75$ m, where the locations of points A, B, C and D have been established by linear interpolation between the two grid points at both sides of the point of interest. For the finer grid the first estimations of the total head at the grid points are shown in Figure 3.25 (first row). The first iteration is applied using Equation 3.55 for the interior points and Equation 3.56 for the points on the impermeable boundary. These results are presented in the second row whilst the third row shows the results of the fourth and final iteration. The equipotential line corresponding to $h = 11.75$ m is represented by the solid line, while the dashed line represents the previous analysis with the coarse grid. Sample calculations for the first iteration for grid points 3,2 (interior) and 3,1 (on the impermeable boundary) are:

$$h_{3,2} = \frac{h_{4,2} + h_{2,2} + h_{3,3} + h_{3,1}}{4} \rightarrow h_{4,2} + h_{2,2} + h_{3,3} + h_{3,1} - 4h_{3,2} = 0.0.$$

But as the selected values are not the exact values, a residual R is defined according to:

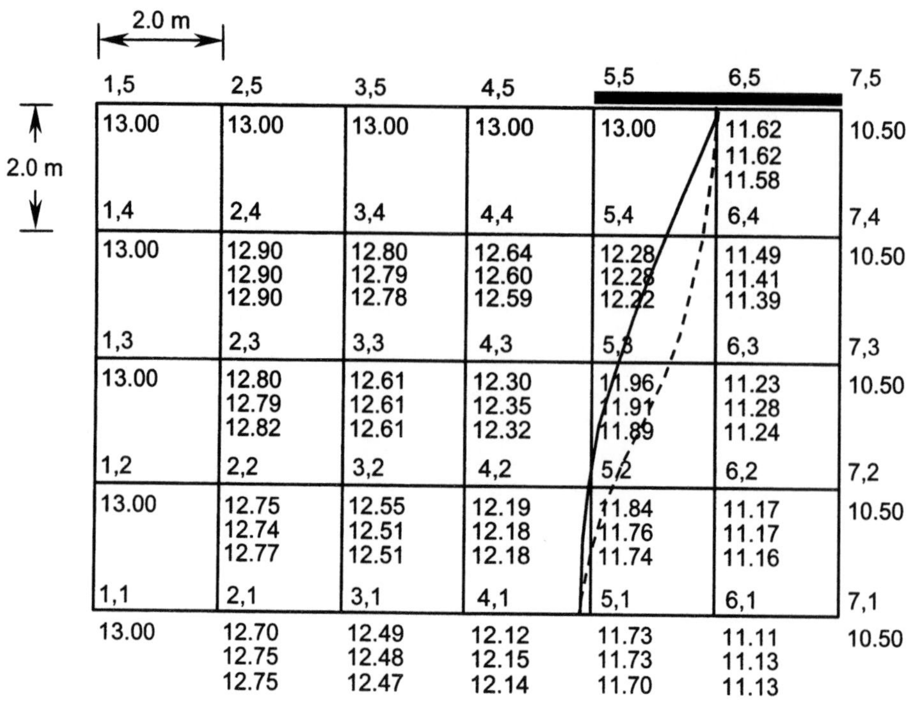

	1,5	2,5	3,5	4,5	5,5	6,5	7,5
	13.00	13.00	13.00	13.00	13.00	11.62 / 11.62 / 11.58	10.50
	1,4	2,4	3,4	4,4	5,4	6,4	7,4
	13.00	12.90 / 12.90 / 12.90	12.80 / 12.79 / 12.78	12.64 / 12.60 / 12.59	12.28 / 12.28 / 12.22	11.49 / 11.41 / 11.39	10.50
	1,3	2,3	3,3	4,3	5,3	6,3	7,3
	13.00	12.80 / 12.79 / 12.82	12.61 / 12.61 / 12.61	12.30 / 12.35 / 12.32	11.96 / 11.91 / 11.89	11.23 / 11.28 / 11.24	10.50
	1,2	2,2	3,2	4,2	5,2	6,2	7,2
	13.00	12.75 / 12.74 / 12.77	12.55 / 12.51 / 12.51	12.19 / 12.18 / 12.18	11.84 / 11.76 / 11.74	11.17 / 11.17 / 11.16	10.50
	1,1	2,1	3,1	4,1	5,1	6,1	7,1
	13.00	12.70 / 12.75 / 12.75	12.49 / 12.48 / 12.47	12.12 / 12.15 / 12.14	11.73 / 11.73 / 11.70	11.11 / 11.13 / 11.13	10.50

2.0 m (horizontal), 2.0 m (vertical)

Figure 3.25. Example 3.14.

$R_{3,2} = h_{4,2} + h_{2,2} + h_{3,3} + h_{3,1} - 4h_{3,2} = 12.19 + 12.75 + 12.61 + 12.49 - 4 \times 12.55 = -0.16$ m.

The corrected value of $h_{3,2}$ to satisfy the continuity condition is:

$h_{3,2} = h_{3,2} + R_{3,2} / 4 = 12.55 + (-0.16)/4 = 12.51$ m.

For grid point 3,1:

$h_{3,1} = (h_{4,1} + h_{2,1} + 2h_{3,2})/4,$

$R_{3,1} = h_{4,1} + h_{2,1} + 2h_{3,2} - 4h_{3,1} = 12.12 + 12.7 - 2 \times 12.55 - 4 \times 12.49 = -0.04$ m,

$h_{3,1} = h_{3,1} + R_{3,1}/4 = 12.49 + (-0.04)/4 = 12.48$ m.

More precise values may be obtained if the left boundary is moved further to the left in order to justify the assumptions made on the total heads for the points on this boundary.

3.6 SEEPAGE THROUGH EARTH DAMS

3.6.1 *Entry and exit conditions*

An idealized section of a typical earth dam is shown in Figure 3.26. It is constructed on an impermeable base and the seepage zone is limited to the embankment soil only. The height of the water at the upstream side is h whilst at the downstream side, the water table is assumed to be on the ground surface.

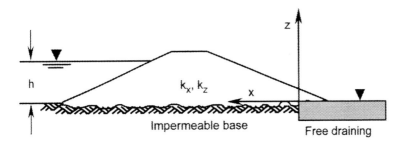

Figure 3.26. An idealized section of a permeable earth dam.

A toe drain may be constructed at the downstream side to control the uppermost flow line or *phreatic surface*. Its length is designed to stop the phreatic surface from exiting at the downstream slope. The main objective is to establish the phreatic surface and the flow net within the dam in order to calculate the flow rate as well as the parameters necessary for a stability calculation, such as pore pressure and seepage pressure. Since the coordinate system shown in Figure 3.26 does not agree with the conventions of the previous section, the relevant equations are modified for this arrangement. The problem involves three cases of transfer conditions:

1. Entry of water from the $k = \infty$ zone (water) to the soil with coefficient of permeability of k.
2. Possible exit of water from the soil to the air ($k = \infty$) at the downstream slope.
3. Entry of water from the soil to the free draining at toe with $k = \infty$.

The entry and exit conditions for a boundary between a soil of permeability of k and a zone of infinite permeability may be established by constructing a flow net around the boundary similar to Figure 3.22. However, in this section, a complete proof of the transfer conditions is not presented.

Figure 3.27(a) shows a case in which water enters from the upstream face to the soil. As the upstream face is an equipotential line, the flow line must start perpendicular to this face. The flow line rises up a little above the normal line to the face as shown in the figure. If there is a gravel wall between the water and the earth dam, the phreatic surface will be tangent to the water level in the gravel and will drop downwards, as shown in Figure 3.27(b). Note that the downstream side of the gravel wall does not represent an equipotential line. Figure 3.28(a) shows the entry condition from the soil to the air at the downstream face of the earth dam where the phreatic surface is tangent to the slope.

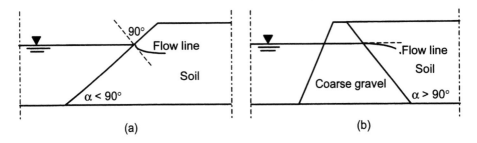

Figure 3.27. Entry conditions from the water to the soil.

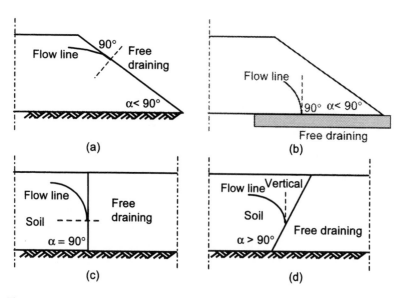

Figure 3.28. Entry conditions from the soil to the air and drain.

Figures 3.28(b)-3.28(d) show the exit conditions from the soil to a drain with a horizontal, a vertical and an inclined surface respectively. In all cases the flow line is tangent to a vertical line irrespective of the slope of the surface of the drain.

3.6.2 *The equation of the phreatic surface: the basic parabola*

A traditional (*Casagrande*) method defines the phreatic surface as a parabola with its focus located at the origin O of the xz coordinate system (Figure 3.29). This *basic parabola* can be defined mathematically if the coordinate of one point on the boundary (or within the seepage zone) is known, and is shown as curve FMC in Figure 3.29. Experimental investigations have shown that the intersection point of the basic parabola and the water surface, point F, is located such that $FA = 0.3\ EA$, which means that the coordinates of point F are known. The parabola has to be corrected at point A to meet the requirements of the entry conditions.

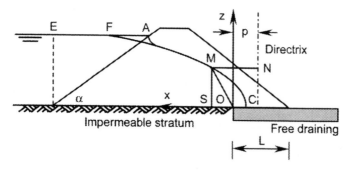

Figure 3.29. Casagrande method to establish the phreatic surface in an earth dam.

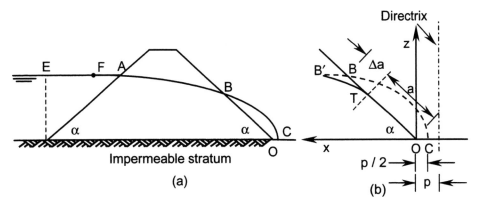

Figure 3.30. Correction of the phreatic surface to satisfy the exit conditions.

The distance of any point M on the basic parabola from the focus is equal to the distance of this point from the directrix, which is located at an unknown distance of p from the z-axis.

$$MO = MN \rightarrow \sqrt{x^2 + z^2} = x + p \rightarrow x = (z^2 - p^2)/2p \tag{3.59}$$

By substituting the coordinates of point F in Equation 3.59, the value of p so obtained is:

$$p = \sqrt{x_F^2 + z_F^2} - x_F \tag{3.60}$$

The flow rate is estimated by constructing the flow net schematically as explained earlier. An alternative solution is to assume a constant hydraulic gradient in the vertical sections:

$$q = Aki = (z \times 1)k \frac{dz}{dx}.$$

From Equation 3.59 $\dfrac{dz}{dx} = \dfrac{p}{z}$ and:

$$q = kp \tag{3.61}$$

The horizontal length of the toe drain L must be sufficient to allow the basic parabola to be located inside the earth dam. It is calculated by intersecting the equation of the basic parabola with the equation of the downstream face $z = \tan\alpha\,(x + L)$ and seeking the condition in which both functions become tangent to each other:

$$L = \frac{p}{2}(1 + \cot^2 \alpha) \tag{3.62}$$

In the absence of a toe drain the basic parabola intersects the downstream face at point B as shown in Figure 3.30(a). In reality, the phreatic surface must be tangent to the downstream face at point T with a distance a from the origin O. In the Casagrande method, the correction length of Δa (Figure 3.30(b)) is found from Figure 3.31 which is based on experimental results. Note that the distance $OB = \Delta a + a$ can be easily established, as the equations of the basic parabola and the downstream face are both known.

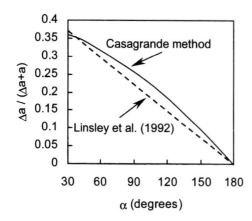

Figure 3.31. Correction of the phreatic surface.

Table 3.2. $\Delta a / (\Delta a + a)$ in terms of angle α.

α (degrees)	30	60	90	120	150	180
$\Delta a / (\Delta a + a)$	0.36	0.32	0.26	0.18	0.10	0

Table 3.2 shows the approximate values of $\Delta a / (\Delta a + a)$ for given angles of α based on Casagrade's experimental work. An estimation of $\Delta a / (\Delta a + a)$ can be made by assuming the following linear relationship of the form reported by Linsley et al. (1992):

$$\frac{\Delta a}{\Delta a + a} = \frac{180 - \alpha}{400} \tag{3.63}$$

(Figure 3.31). The top flow line can now be corrected by sketching a smooth curve tangent to the basic parabola at B' and downstream face at T (Figure 3.30(b)).

An alternative solution, called the Dupuit method, defines the correction curve as follows:
(a) At any vertical section the hydraulic gradient defined by dz / dx has a constant value.
(b) The curve passes through point F, which is already defined.
(c) The curve is tangent to the downstream face.
It can be shown that the first condition yields the correction curve as a parabola in the form: $x = C_1 z^2 + C_2$.
The integration constants C_1 and C_2 are found by substituting the second and third conditions in the above equation. Hence, the equation of the flow line adjacent to the downstream face is:

$$x = \frac{\cos\alpha(z^2 - z_F^2)}{2a\sin^2\alpha} + x_F \tag{3.64}$$

The coordinates of point T are $x = a \cos\alpha$ and $z = a \sin\alpha$. Substituting these values into Equation 3.64 two answers are obtained, the positive value of which is the distance a.

$$a = \frac{x_F}{\cos\alpha} - \sqrt{\frac{x_F^2}{\cos^2\alpha} - \frac{z_F^2}{\sin^2\alpha}} \tag{3.65}$$

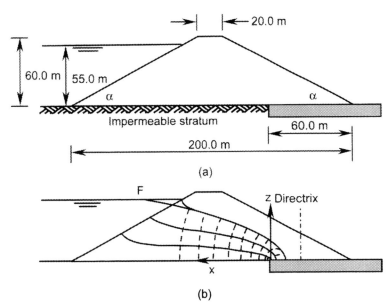

Figure 3.32. Example 3.15.

The flow rate can be obtained by calculating dz / dx from Equation 3.64 and substituting this result and $z_T = a \sin\alpha$ into the flow equation:

$$q = Aki = (z_T \times 1)k \frac{dz}{dx} \rightarrow q = ka \sin\alpha \tan\alpha \qquad (3.66)$$

Example 3.15

A vertical section of a large earth dam is shown in Figure 3.32(a). Estimate the flow rate by sketching the flow net and use the Dupuit assumption. $k_x = k_z = 3 \times 10^{-6}$ m/s.

Solution:

$\tan\alpha = 60.0/[(200.0 - 20.0)/2] = 0.666 \rightarrow \alpha = 33.69°$. Referring to Figure 3.29:

$EA = 55.0/\tan 33.69° = 82.50$ m, $x_F = 200.0 - 82.5 + 0.3 \times 82.5 - 60.0 = 82.25$ m,

$z_F = 55.0$ m.

From Equation 3.60: $p = \sqrt{82.25^2 + 55.0^2} - 82.25 = 16.69$ m.

The equation of the basic parabola is: $x = (z^2 - p^2)/2p = (z^2 - 278.71)/33.38$.

Figure 3.32(b) shows the flow net within the earth dam in which $N_f = 3$ and $N_d = 10$.

$q = khN_f / N_d = 3.0 \times 10^{-6} \times 24 \times 3600 \times 55.0 \times 3/10 = 4.28$ m^3/day.

When the Dupuit condition applies:

$q = kp = 3 \times 10^{-6} \times 16.69 \times 24 \times 3600 = 4.33$ m^3/day.

Example 3.16

For the earth dam section shown in Figure 3.33 calculate the values of a and Δa and the flow rate across the dam in m^3/day. $k_x = k_z = 3 \times 10^{-6}$ m/s.

Solution:

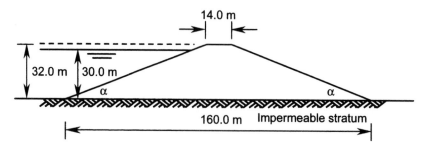

Figure 3.33. Example 3.16.

$\tan \alpha = 32.0/[(160.0-14.0)/2] = 0.43836 \rightarrow \alpha = 23.67°$. Referring to Figure 3.29:

$EA = 30.0/\tan 23.67° = 68.44$ m, $x_F = 160.0 - 68.44 + 0.3 \times 68.44 = 112.09$ m,

$z_F = 30.0$ m. From Equation 3.60:

$$p = \sqrt{112.09^2 + 30.0^2} - 112.09 = 3.94 \text{ m}.$$

The equation of the basic parabola is: $x = (z^2 - p^2)/2p = (z^2 - 3.94^2)/7.88$.

The length $OB = \Delta a + a$ is calculated by substituting:

$x_B = OB \cos 23.67°$ and $z_B = OB \sin 23.67°$ into the equation of the basic parabola; this results in $OB = 46.90$ m. Using Equation 3.65:

$$a = 112.09/\cos 23.67° - \sqrt{112.09^2/\cos^2 23.67° - 30.0^2/\sin^2 23.67°} = 25.46 \text{ m}.$$

$\Delta a = 46.90 - 25.46 = 21.44$ m.

$q = ka \sin \alpha \tan \alpha = 3.0 \times 10^{-6} \times 25.46 \times \sin 23.67° \times \tan 23.67° \times 3600 \times 24 = 1.16 \text{ m}^3/\text{day}.$

3.6.3 Design considerations to control the flow rate

Different types of earth dams are shown in Figure 3.34 (Wilson & Marsal, 1979; Roberson, et al., 1997). The major factors involved in the control of the flow rate are the coefficient of permeabilities of the dam material and the base soil, and the height of the water at the upstream side. Horizontal or inclined drains, or a combination of both, can be used to control the downstream seepage, as shown in Figure 3.34(a). A traditional section includes an internal core constructed from compacted clay material. This core reduces the seepage and has back drains to remove water to the downstream side (Figure 3.34(b)). In the case of a permeable base, the length of the flow line is increased by constructing impermeable blankets or by developing the central core to the permeable base to form a cut-off trench (Figure 3.34(c)). Sometimes the horizontal and vertical impermeable blankets are combined to make a cut-off trench, as shown in Figure 3.34(d). Alternatively, the horizontal blanket may be projected upwards along the slope at the upstream side (Figure 3.34(e)); asphaltic cement is sometimes used for this purpose (Linsley, et al., 1992). In recent years synthetic fibres have been used to provide drainage as well as impermeable blankets. Synthetics have also been used as a reinforcement to provide stability to the upstream and downstream slopes. If modifications such as these are employed, particular care has to be taken in the evaluation of the overall factor of safety of the structure. Any increase in the pore pressure will tend to reduce the factor of safety against shear failure.

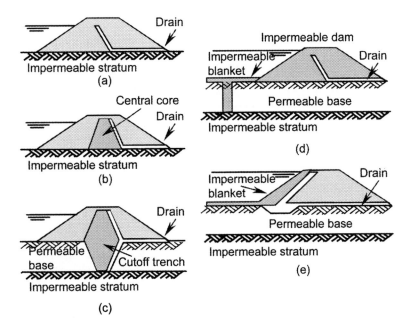

Figure 3.34. Different types of earth dams.

3.7 PROBLEMS

3.1 In a constant head permeability test, a cylindrical sample 100 mm in diameter and 150 mm high is subjected to an upward flow of 540 ml/min. The head loss over the length of the sample is 360 mm. Calculate the coefficient of permeability in m/s.

Answer: $k = 4.8 \times 10^{-4}$ m/s

3.2 In a laboratory falling head test, the recorded data are: diameter of the tube = 20 mm, diameter of the cell = 100 mm, length of the sample = 1000 mm. The head measured from the top level of the sample dropped from 800 mm to 600 mm within

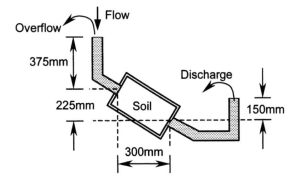

Figure 3.35. Problem 3.3.

1 hour and the temperature of the water was 30°C. Calculate the coeeficient of permeability at 20°C.

$\eta = 1.005 \times 10^{-3}$ N.s/m^2 (at 20° C), $\eta = 0.801 \times 10^{-3}$ N.s/ m^2 (at 30°C).

Answer: $k = 2.54 \times 10^{-6}$ m/s

3.3 For the test arrangement shown in Figure 3.35, calculate the volume of water discharged in 20 minutes. The cross sectional area of the soil is 4000 mm^2 and $k = 4.0$ mm/s.

Answer: $Q = 23.04\,\text{l}$

3.4 A long trench is excavated parallel to a river, as shown in Figure 3.36. The soil profile consists of a permeable soil of thickness D confined between two impermeable layers. Initially the water level in the trench is the same as that in the river. Water is pumped out of the trench at a flow rate of q. (a) Formulate q in terms of the geometrical parameters shown in Figure 3.36, (b) for $z_w = 7$ m, $D = 5$ m, L (the average horizontal distance between the trench and the river's slope) = 100 m, and $k = 4 \times 10^{-5}$ m/s, calculate the flow rate corresponding to a drawdown of $D_w = 2$ m, (c) calculate q when the water table in the trench is 2 m below the permeable layer.

Answers: $q = kD(z_w - D)/l$ (for full flow),

$q = k[2D(z_w - D) + D^2 - (z_w - D_w)^2]/2L$ (water table in the trench in D zone),

0.346 and 0.622 m^3/day/metre run of trench.

3.5 A well of diameter 0.3 m is constructed to the full depth of an unconfined aquifer of thickness of 150 m. The water table is 10 m below the ground surface. A pumping test of 12 m^3/hour has resulted a drawdown of 10 m. Assuming $r_o = 400$ m, calculate the coefficient of permeability of the aquifer. If the flow rate increases to 18 m^3/day, and in the absence of any other data, what will be the best estimate for the drawdown in the well?

Answers: 3.1×10^{-6} m/s, 15.3 m

3.6 A pumping test carried out in a 50 m thick confined aquifer (well dia. = 0.6 m) results in a flow rate of 600 l/min. The thickness of the impermeable layer above the aquifer is 20 m and the original water level in the well is 2 m below the ground surface (which is also the top of the impermeable layer). Drawdowns in two observation wells located 50 m and 100 m from the well are 3 and 1 m respectively.

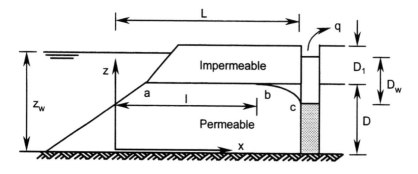

Figure 3.36. Problem 3.4.

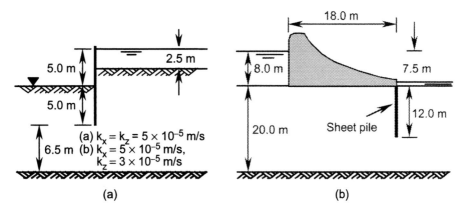

Figure 3.37. (a) Problem 3.8, (b) Problem 3.9.

Calculate: (a) the coefficient of permeability of the aquifer, (b) the drawdown in the well, (c) the radius of influence.

Answers: 1.1×10^{-5} m/s, 17.8 m, 141.3 m

3.7 A soil profile consists of three layers with the properties shown in the table below. Calculate the coefficients of permeability parallel and normal to the stratum.

Layer	Thickness (m)	k_x (parallel, m/s)	k_z (normal, m/s)
1	3.0	2.0×10^{-6}	1.0×10^{-6}
2	4.0	5.0×10^{-8}	2.5×10^{-8}
3	3.0	3.0×10^{-5}	1.5×10^{-5}

Answers: $k_x = 9.6 \times 10^{-6}$ m/s, $k_z = 6.1 \times 10^{-8}$ m/s

3.8 For the sheet pile system shown in Figure 3.37(a), calculate the flow rate in m^3/day by constructing the flow net in the two conditions (a) and (b) shown.

Answers: 9.82 m^3/day, 7.60 m^3/day

3.9 A concrete dam retains 8 m of water, as shown in Figure 3.37(b). Calculate the flow rate in m^3/day by constructing the flow net under the dam. $k = 5 \times 10^{-5}$ m/s.

Answer: 12.15 m^3/day

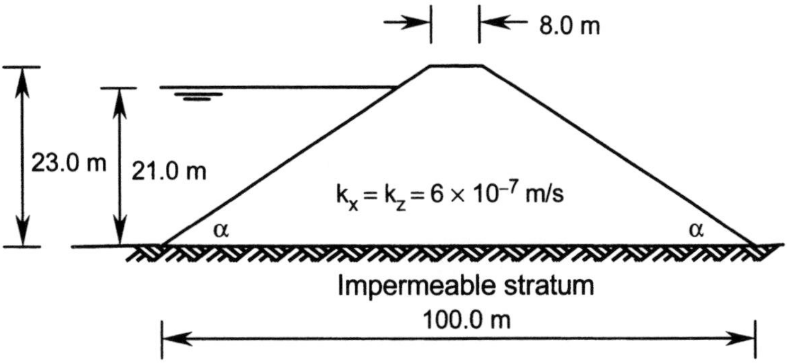

Figure 3.38. Problems 3.10 & 3.11.

3.10 For the earth dam section shown in Figure 3.38, calculate the flow rate in m^3/day.
Answer: 0.18 m^3/day

3.11 For the earth dam section shown in Figure 3.38, calculate the minimum length of the toe drain required to ensure that the phreatic surface becomes tangent to the downstream face.
Answer: 8.65 m

3.8 REFERENCES

AS 2368. 1990. *Test pumping of water wells*. NSW, Australia: Standard Association of Australia.

AS 1289.6.7.1. 1999. *Methods of testing soils for engineering purposes: Soil strength and consolidation tests-Determination of permeability of a soil-Constant head method for a remoulded specimen*. NSW, Australia: Standard Association of Australia.

AS 1289.6.7.2. 1999. *Methods of testing soils for engineering purposes: Soil strength and consolidation tests-Determination of permeability of a soil-Falling head method for a remoulded specimen*. NSW, Australia: Standard Association of Australia.

AS 1289.6.7.3. 1999. *Methods of testing soils for engineering purposes: Soil strength and consolidation tests-Determination of permeability of a soil-Constant head method using a flexible wall permeameter*. NSW, Australia: Standard Association of Australia.

ASTM D-5084. 2000. *Standard test methods for measurement of hydraulic conductivity of saturated porous materials using a flexible wall permeameter*. West Conshohocken, PA: ASTM.

ASTM D-2434-68. 2000. *Standard test method for permeability of granular soils (constant head)*. West Conshohocken, PA: ASTM.

ASTM D-4044. 1996. *Standard test method (field procedure) for instantaneous change in head (Slug): Tests for determining hydraulic properties of aquifers*. West Conshohocken, PA: ASTM.

BS 5930. 1981. *Code of practice for site investigation*. London: British Standard Institution.

Head, K.H. 1986. *Manual of soil laboratory testing*. London: Pentech Press.

Linsley, R.K., Franzini, J.B., Freyberg, D.L. & Tchobanoglous, G. 1992. *Water-resources engineering*. New York: McGraw-Hill.

Roberson, J.A., Cassidy, J.J. & Chaudhry, M.H. 1997. *Hydraulic engineering*. New york: John Wiley & Sons.

Wilson, S.D. & Marsal, R.J. 1979. *Current trends in design and construction of embankment dams*. New York: ASCE.

Wu, T.H. 1966. *Soil mechanics*. Boston: Allyn & Bacon.

CHAPTER 4

Shear Strength of Soils and Failure Criteria

4.1 INTRODUCTION

This chapter describes the shear strength characteristics of soils and related failure criteria. Soils fail either in tension or in shear; however, in the majority of soil mechanics problems, only failure in shear requires consideration. The shear strength along any plane is mobilized by cohesion and by the angle of internal friction, collectively referred to as *shear strength parameters*. If, at a point along a specified plane, the shear stress becomes equal to the peak shear strength along that plane, then the soil will fail at that point and large shear strains will result. A part of the soil on one side of this plane, called the *failure plane*, will slide relative to the other side, bringing about collapse of the soil structure. A failure criterion is a mathematical relationship between the state of peak stress and the shear strength parameters. When combined with the principles of solid mechanics, a failure criterion can be used in different types of stability problems to predict (collapse) loads that cause the failure. The post peak strength behaviour of the soil may be predicted by using the critical state models that are now common for specific types of soils.

4.2 MOHR-COULOMB FAILURE CRITERION

4.2.1 *The Mohr's circle of stress*

Figure 4.1(a) shows the sign convention for two-dimensional stresses. From equilibrium of the forces acting on blocks 1 or 2 (Figure 4.1(b)), the normal and shear stresses on any plane P, with angle α to the x-axis, are determined by the following equations:

$$\sigma = \frac{\sigma_z + \sigma_x}{2} + \frac{\sigma_z - \sigma_x}{2} \cos 2\alpha + \tau_{xz} \sin 2\alpha \tag{4.1}$$

$$\tau = \frac{\sigma_z - \sigma_x}{2} \sin 2\alpha - \tau_{xz} \cos 2\alpha \tag{4.2}$$

Combining these equations to eliminate the parameter α, we obtain:

$$(\sigma - \frac{\sigma_z + \sigma_x}{2})^2 + \tau^2 = (\frac{\sigma_z - \sigma_x}{2})^2 + \tau_{xz}^2,$$

which is the equation of a circle in the σ-τ coordinate system (Figure 4.1(c)). This circle is called *Mohr's circle of stress* whose centre and radius are defined by s and t respectively:

$$s = \frac{\sigma_1 + \sigma_3}{2} = \frac{\sigma_z + \sigma_x}{2} \tag{4.3}$$

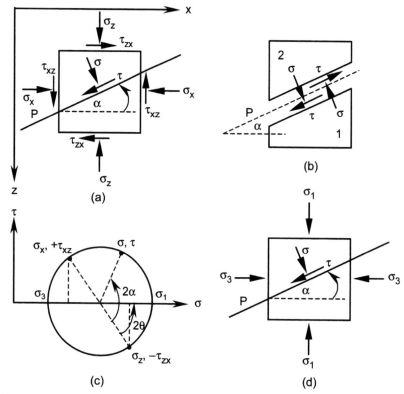

Figure 4.1. Mohr's circle of stress.

$$t = \sqrt{(\frac{\sigma_z - \sigma_x}{2})^2 + \tau_{xz}^2}$$ (4.4)

The major and minor principal stresses are given by:

$$\sigma_1 = s + t, \quad \sigma_3 = s - t$$ (4.5)

whilst the angle of the major principal plane to the x direction is defined by:

$$\theta = \frac{1}{2} \tan^{-1}(\frac{2\tau_{xz}}{\sigma_z - \sigma_x})$$ (4.6)

The minor principal plane is defined by $\theta + 90°$. Equations 4.1 and 4.2 show that the state of stress in any specified direction α is found by knowing the normal and shear stresses in two reference directions. If the σ_1 and σ_3 stress directions are taken as reference axes (Figure 4.1(d)), then:

$$\sigma = \frac{\sigma_1 + \sigma_3}{2} + \frac{\sigma_1 - \sigma_3}{2} \cos 2\alpha$$ (4.7)

$$\tau = \frac{\sigma_1 - \sigma_3}{2} \sin 2\alpha$$ (4.8)

Example 4.1

An element of soil is subjected to the two-dimensional stresses shown in Figure 4.2(a).

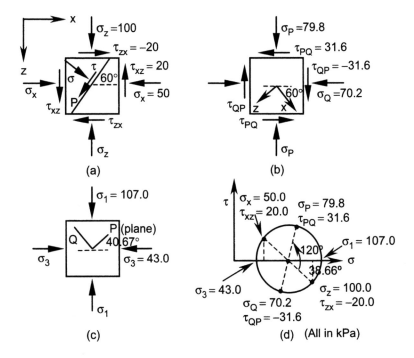

Figure 4.2. Example 4.1.

Determine (a) the normal and shear stresses on the P and Q planes which are orthogonal, (b) the magnitudes and directions of the major and minor principal stresses, (c) the corresponding α values of P and Q if the principal axes are taken as the reference axes.

Solution:

(a) Using Equations 4.1 and 4.2:

$\sigma_P = (100.0 + 50.0)/2 + [(100.0 - 50.0)/2]\cos 120.0° + 20.0\sin 120.0° = 79.8$ kPa,

$\tau_P = [(100.0 - 50.0)/2]\sin 120.0° - 20.0\cos 120.0° = 31.6$ kPa.

$\sigma_Q = (100.0 + 50.0)/2 + [(100.0 - 50.0)/2]\cos 300.0° + 20.0\sin 300.0° = 70.2$ kPa,

$\tau_Q = [(100.0 - 50.0)/2]\sin 300.0° - 20.0\cos 300.0° = -31.6$ kPa.

Figure 4.2(b) shows the normal and shear stresses on the PQ planes.

(b) From Equations 4.3 and 4.4:

$$s = (100.0 + 50.0)/2 = 75.0 \text{ kPa}, \quad t = \sqrt{(\frac{100.0 - 50.0}{2})^2 + 20.0^2} = 32.0 \text{ kPa},$$

$\sigma_3 = s - t = 75.0 - 32.0 = 43.0$ kPa, $\sigma_1 = s + t = 75.0 + 32.0 = 107.0$ kPa.

The direction of major principal plane is:

$\theta = 1/2[\tan^{-1}(2\tau_{xz}/\sigma_z - \sigma_x)] = 1/2[\tan^{-1}(2 \times 20.0/100.0 - 50.0)] = 19.33°$.

(c) In σ_1 and σ_3 coordinate system the angle α for the plane P is:

$\alpha_P = -19.33° + 60.0° = 40.67°$.

A summary of the results is shown in Figures 4.2(c) and 4.2(d).

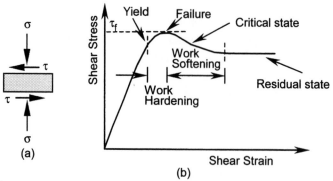

Figure 4.3. Stress-strain behaviour of a soil element subjected to simple shear.

4.2.2 *Stress-strain models for soils*

Figure 4.3 shows an element of soil subjected to a variable shear stress τ and a constant normal stress σ. The deformation behaviour of this element is represented by a plot of shear stress versus shear strain. At small shear stresses the response of the element is linear until the *yield point* is reached; beyond this point plastic straining begins to occur (*work hardening*). After the yield point the shear stress continues to increase and eventually reaches a peak value (or *failure point*), that is, the maximum shear strength of the element corresponding to the applied normal stress. Further shear deformation causes the shear stress to decease (*work softening*) to an ultimate value representing a critical state in which the material fails while maintaining constant volume. Failure planes may develop before or after the critical state leading to a constant shear strength called the residual strength. However, this state is achievable only with very large strains. In overconsolidated soils, failure planes may develop immediately after the peak point and the critical state may not be obtained and the state of the element moves towards the residual. Normally consolidated soil may not have a peak strength and the critical state may be obtained at relatively low strains. Consider, now, a soil mass subjected to a system of external loading which increases from an initial value until the mass fails. Every element of this mass goes through the same deformation pattern shown in Figure 4.3; however, the shear strength at a given point depends on the magnitude of the load transmitted to this point. Due to the complexity of the behaviour shown in Figure 4.3, some idealizations must be made so that predictions of shear strength and the maximum value of the external loading can be made. A few of the more common idealizations are shown in Figure 4.4.

4.2.3 *Mohr-Coulomb failure criterion*

Coulomb (1776) suggested that the shear strength of a soil along the *failure plane* could be described by:

$$\tau_f = c + \sigma \tan \phi \tag{4.9}$$

where τ_f is the absolute value of the shear strength on the failure plane (corresponding to the peak point in Figure 4.3(b)), σ is the stress normal to the plane, c is the cohesion and ϕ is the angle of internal friction of the soil (or internal friction angle). The two parameters c and ϕ are called *shear strength parameters*.

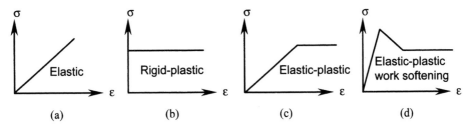

Figure 4.4. Common models for soil behaviour.

To understand the concept behind Equation 4.9, consider the two blocks *A* and *B* (Figure 4.5) of unit area that are in contact with each other and subjected to the normal and shear stresses shown. The contact surface is not smooth and contains frictional asperities. Under a constant normal stress, the shear stress is increased from zero to the maximum τ_f, forcing the two blocks to slide along their contact area. When $\sigma = 0$, the shear stress has to be mobilized to a maximum value of *c* to make the sliding possible. If the friction angle between blocks *A* and *B* is ϕ then, for the values of $\sigma > 0$, τ has to be increased to overcome the resistance to sliding $\sigma \times \tan\phi$ caused by friction. Consequently, the summation of *c* and $\sigma \times \tan\phi$ represents the maximum shear stress needed to slide the two blocks on the plane of contact. In a real soil, if a predetermined sliding surface is forced to occur, the soil below and the soil above the failure plane will not act as rigid materials but will deform, causing a volume change around the sliding surface. Moreover, the internal friction angle is not the same as the friction angle between solid particles as it depends mostly on the interlocking mechanism between the solid particles. In a coordinate system with σ plotted horizontally and τ vertically, Equation 4.9 is represented by the line shown in Figure 4.6(a). This equation was originally written in terms of total stress and was only partially successful in predicting the shear strength of real soils. Coulomb's failure criterion was subsequently redefined as:

$$\tau_f = c' + \sigma' \tan \phi' \tag{4.10}$$

where τ_f is the absolute value of the shear strength, σ' is the effective normal stress, c' is the effective cohesion and ϕ' is the effective angle of internal friction of the soil (Figure 4.6(b)). In both the total and effective stress conditions, the shear stress is taken solely by the particles since the liquid in the voids, which is normally water, has no resistance to shear. The tensile strength of soil is commonly ignored and therefore cohesion is the minimum shear strength at zero normal stress. Coulomb's criterion does not apply in the case of a normal tensile stress.

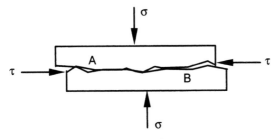

Figure 4.5. Mechanical concept of sliding.

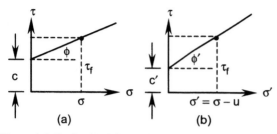

Figure 4.6. Coulomb's failure criterion.

The shear strength parameters corresponding to the total and effective states have different values. However, in the case of clean sand, it is reasonable to assume $c' = c = 0$ and $\phi' = \phi$. The magnitude of ϕ' or ϕ in a sandy soil varies from 25° to 45° depending on the soil type and density whereas, in a clay soil, it may vary from 0 to 25° depending on the soil type, moisture content, and drainage conditions. In undrained conditions the cohesion (c_u) varies from 10 kPa to 100 kPa or more, depending on the density of the soil.

Coulomb's failure criterion is also used to represent the residual strength and the subscript r is used with each term in Equations 4.9 or 4.10 to identify them as the shear strength parameters in this state. Figure 4.7 shows the total and effective states of stress at the peak strength (failure point) represented by Mohr's circles. It is apparent that the shear stress at every plane in the total stress Mohr's circle is the same as in the effective stress Mohr's circle. The difference between the normal stresses in two perpendicular directions in the total and effective states are equal:

$$\sigma_z - \sigma_x = (\sigma'_z + u) - (\sigma'_x + u) = \sigma'_z - \sigma'_x.$$

Thus, the radius of both the total and the effective Mohr's circles are identical. The horizontal distance of the two circles is equal to pore pressure u. Any point F on the failure envelope represents the normal and shear stresses on a failure plane at a specified point in soil. These stresses must also satisfy the equilibrium conditions at the point, which is represented by Mohr's circle of stress. This implies that, at failure, Mohr's circle of stress must be tangent to the line expressed by Equation 4.9 (or 4.10). This condition, referred to as the Mohr-Coulomb failure criterion, is shown in Figure 4.7. The angle between the failure plane F and the major principal plane is $\alpha = (90° + \phi')/2 = 45° + \phi'/2$. Due to

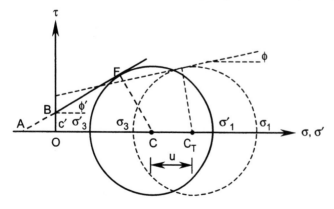

Figure 4.7. Mohr-Coulomb failure criterion.

symmetry, there will be another failure plane at angle $-\alpha$ from the σ'_1 plane. From Figure 4.7, a relationship between the state of stress (σ'_1 and σ'_3) or (σ'_z, σ'_x and τ_{xz}) and the shear strength parameters c' and ϕ' may be formulated by equating the radius of Mohr's circle to the distance of the centre of the circle from the failure envelope:

$$t' = \sqrt{[(\sigma_z - \sigma_x)/2]^2 + \tau_{xz}^2} = OB\cos\phi' + OC\sin\phi'.$$

Considering that $\sigma_z - \sigma_x = \sigma'_z - \sigma'_x$, $OC = (\sigma'_z + \sigma'_x)/2$ and $OB = c'$ then:

$$(\sigma'_z - \sigma'_x)^2 + (2\tau_{xz})^2 = [2c'\cos\phi' + (\sigma'_z + \sigma'_x)\sin\phi']^2 \tag{4.11}$$

This equation can be expressed in terms of the principal stresses σ'_1 and σ'_3:
$\sigma'_1 - \sigma'_3 = 2c'\cos\phi' + (\sigma'_1 + \sigma'_3)\sin\phi'$, or:

$$\sigma'_1 = \sigma'_3 \frac{1+\sin\phi'}{1-\sin\phi'} + 2c'\frac{\cos\phi'}{1-\sin\phi'} \rightarrow \sigma'_1 = \sigma'_3 \frac{1+\sin\phi'}{1-\sin\phi'} + 2c'\sqrt{\frac{1+\sin\phi'}{1-\sin\phi'}}, \text{ or: } \tag{4.12}$$

$$\sigma'_1 = \sigma'_3 \tan^2(45° + \phi'/2) + 2c'\tan(45° + \phi'/2) \tag{4.13}$$

$$\sigma'_3 = \sigma'_1 \tan^2(45° - \phi'/2) - 2c'\tan(45° - \phi'/2) \tag{4.14}$$

Example 4.2

At a point in a soil mass, the total vertical and horizontal stresses are 240 kPa and 145 kPa respectively whilst the pore pressure is 40 kPa. Shear stresses on the vertical and horizontal planes passing through this point are zero. Calculate the maximum excess pore pressure to cause the failure of the point. What is the magnitude of the shear strength on the plane of failure? The effective shear strength parameters are: $c' = 10$ kPa, $\phi' = 30°$.

Solution:

Check if the initial state of stress is below the failure envelope:
$\sigma'_1 = 240.0 - 40.0 = 200.0$ kPa, $\sigma'_3 = 145.0 - 40.0 = 105.0$ kPa. Referring to Figure 4.7:
$CF = OB\cos\phi' + OC\sin\phi' = 10.0\cos 30.0° + [(200.0 + 105.0)/2]\sin 30.0° = 84.9$ kPa,
$t' =$ radius of Mohr's circle $= (200.0 - 105.0)/2 = 47.5$ kPa < 84.9 kPa.
Hence, the initial state of stress is below the failure envelope.
$\sigma_1 = 240.0$ kPa, $\sigma'_1 = 240.0 - 40.0 - u_e$, $\sigma'_1 = 200.0 - u_e$
$\sigma_3 = 145.0$ kPa, $\sigma'_3 = 145.0 - 40.0 - u_e$, $\sigma'_3 = 105.0 - u_e$
where u_e is the excess pore pressure. Substituting the major and minor effective principal stresses into Equation 4.13 and noting that $\alpha = 45° + \phi'/2 = 60°$:
$200.0 - u_e = (105.0 - u_e)\tan^2 60.0° + 2\times10.0\times\tan 60.0° \rightarrow u_e = 74.8$ kPa.
$\sigma'_1 = 200.0 - u_e = 200.0 - 74.8 = 125.2$ kPa, $\sigma'_3 = 105.0 - u_e = 105.0 - 74.8 = 30.2$ kPa.
Using Equations 4.7 and 4.8:
$\sigma' = (125.2 + 30.2)/2 + [(125.2 - 30.2)/2]\cos 120° = 53.9$ kPa,
$\tau_f = 53.9\tan 30° + 10.0 = 41.1$ kPa, or: $\tau_f = [(125.2 - 30.2)/2]\sin 120.0° = 41.1$ kPa.

Example 4.3

A cubic sample of clay soil is subjected to the plane strain loading of: $\sigma'_z = \sigma'_x = 80$ kPa, τ_{xz}(initial) $= 0$. The shear stress τ_{xz} is gradually increased until the sample fails. If the shear strength parameters are $c' = 40$ kPa and $\phi' = 15°$, determine (a) the maximum shear stress applied on the four faces of the sample, (b) the shear stress on the failure plane, (c) the magnitudes of the major and minor principal stresses, (d) the minimum value of $\sigma'_z = \sigma'_x$ to avoid any tensile stresses in the sample and the corresponding magnitude of τ_{xz}.

Solution:

(a) At the initial state when $\tau_{xz} = 0$, Mohr's circle of stress is represented by a point with $\sigma'_z = \sigma'_x = 80$ kPa on the horizontal axis. Using Equation 4.11:

$(2\tau_{zx})^2 = [2 \times 40.0 \cos 15.0° + (80.0 + 80.0) \sin 15.0°]^2 \rightarrow \tau_{zx} = 59.3$ kPa.

(b) Using Equation 4.2 with the sign convention of Figure 4.1(c):

$2\alpha = 90.0° + 2(45.0° + 15.0°/2) = 195.0°$, $\tau_f = -59.3 \cos 195.0° = 57.3$ kPa.

(c) $\sigma'_1 = 80.0 + 59.3 = 139.3$ kPa, $\sigma'_3 = 80.0 - 59.3 = 20.7$ kPa.

(d) From Equation 4.13 with $\sigma'_3 = 0$:

$\sigma'_1 = 2 \times 40.0 \tan(45° + 15°/2) = 104.2$ kPa, $\sigma'_z = \sigma'_x = (104.2 + 0.0)/2 = 52.1$ kPa.
The radius of Mohr's circle $= \tau_{xz} = (104.2 - 0.0)/2 = 52.1$ kPa.

Example 4.4

An element of soil 8 m below the ground surface is adjacent to a long vertical retaining wall where the water table is 2 m below the ground surface. The retaining wall is subjected to a horizontal displacement towards the outside of the wall until the soil behind the wall fails. Assuming that shear stresses cannot develop on the surface of the retaining wall, determine: (a) the total pressure applied to the retaining wall through the element, (b) the direction of the failure plane, (c) the shear strength on the failure plane passing through the point. Properties of the soil are:

$\rho_{dry} = 1.8$ Mg/m^3, $\rho_{sat} = 2.0$ Mg/m^3, $c' = 10$ kPa, and $\phi' = 30°$.

Solution:

(a) There is no shear stress on the vertical plane, thus there will be no shear stress on the horizontal plane passing through the point. The mechanism of the failure indicates that the vertical plane is the minor principal plane:

$\sigma_1 = (2.0 \times 1.8 + 6.0 \times 2.0) \times 9.81 = 153.0$ kPa, $u = 6.0 \times 1.0 \times 9.81 = 58.9$ kPa,

$\sigma'_1 = 153.0 - 58.9 = 94.1$ kPa. Substituting σ'_1 in Equation 4.14:

$\sigma'_3 = 94.1 \tan^2(45.0° - 30° / 2) - 2 \times 10.0 \times \tan(45.0° - 30.0° / 2) = 19.8$ kPa,

$\sigma_3 = $ total pressure to the retaining wall through the element $= 19.8 + 58.9 = 78.7$ kPa.

(b) $2\alpha = 2(45.0° + 30.0°/2) = 120.0°$, $\alpha = 60.0°$. This means that the soil behind the retaining wall could slide downwards at an angle of $90.0° - 60.0° = 30.0°$ to the wall.

(c) Substituting $2\alpha = 120°$ in Equations 4.7 and 4.8:

$\sigma' = (94.1 + 19.8)/2 + [(94.1 - 19.8)/2] \cos 120.0° = 38.4$ kPa,

$\tau_f = 38.4 \tan 30° + 10.0 = 32.2$ kPa, or: $\tau_f = [(94.1 - 19.8)/2] \sin 120.0° = 32.2$ kPa.

4.3   LABORATORY SHEAR STRENGTH TESTS

4.3.1   *Direct shear, simple shear and residual shear tests*

In the direct shear test points on the failure envelope are determined directly by applying a constant normal stress to the sample and measuring the shear strength on the predetermined failure plane. A diagrammatic section of a direct shear test apparatus, called the shear box (or split shear box), is shown in Figure 4.8(a). This box is made from metal and has two halves that can move horizontally relative to each other. In this way the soil specimen inside the box is sheared in a horizontal direction. The normal load N is applied vertically to the loading platen by means of a hanger with dead weights. Horizontal displacement is applied by a motorized system causing shear stress on the horizontal shear plane. Thin metal plates, either grooved or solid with or without porous stones, are placed at the top and the bottom of the specimen to ensure a uniform distribution of shear stress and to control the drainage conditions. In plan, the box is square with a side of 60 mm, 64 mm or 100 mm whilst its total thickness of 20 mm is shared between the upper and the lower boxes. Circular sections are also manufactured with diameters the same as the square boxes. Soils with larger size aggregates are tested in a shear box apparatus that can accommodate specimens up to 300 mm square. These larger shear boxes have recently been recognized by some standards (e.g. BS 1377). In a traditional shear box apparatus the vertical and horizontal loads and displacements are measured mechanically; recent improvements include electromechanical measurement using an electronic recording system that accepts analogue inputs. Figure 4.8(b) shows the mechanical concept of a simple shear apparatus, referred to as a research shear apparatus (Powrie, 1997), whose advanced design include the electromechanical measurements of forces and displacements and includes a pore pressure port. Square or circular shear boxes are also used to obtain the residual strength that is achievable only with very large horizontal displacements. Such large displacements are provided by reversing the direction of the shear displacement and

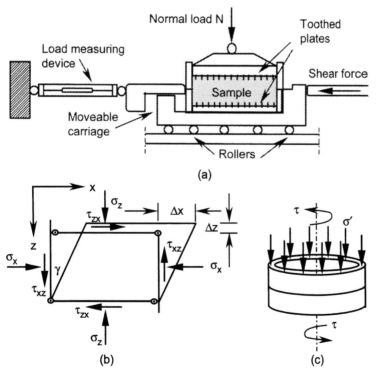

Figure 4.8. (a) Diagrammatic section of a shear box, (b) simple shear mechanism, (c) ring shear mechanism.

repeating the forward and reverse displacements until the residual strength is constant. Based on the mechanism shown in Figure 4.8(c), a ring residual shear apparatus has been developed that accommodates 5 mm thick ring samples with inner and outer diameters of 70 mm and 100 mm respectively. The sample is compressed vertically between porous bronze loading platens by means of a counterbalance system providing the normal stress. A rotation is applied to the base plate to shear the sample, forming a shear surface close to the upper platen that is artificially roughened to prevent slip at the platen-soil interface.

Soil specimens are either undisturbed or remoulded, depending on the type of the material. Undisturbed specimens are prepared from fine-grained soils. However, in a coarse-grained soil, the test sample is remoulded and particular care is taken to achieve similar physical properties to that of the undisturbed soil. In order to construct the failure envelope, a test has to be repeated several times, each with different normal stress and, ideally, identical samples. Measurements recorded in a direct shear test include shear force, horizontal relative displacement between the two split halves (or shear displacement) and vertical displacement of the sample. The stress data is presented in (σ or σ'), τ coordinate system whilst the volume change is represented by volumetric strain, vertical displacement of force N, void ratio, or specific volume. Specific volume v is a dimensionless parameter and represents a volume in which the solids occupy a unit volume. From a phase diagram it can be shown that $v = 1 + e$. Any of the volume-related parameters may be plotted against shear displacement to evaluate volume change properties. The volumetric strain, void ratio or specific volume cannot be calculated accurately as the distribution of volume change within the thickness of the sample is not uniform but concentrated mostly above and below the failure plane.

The application of the direct shear test in the assessment of the strength properties of soils in plane strain conditions has been recommended by many standard codes, symposia and textbooks, e.g. the *Geotechnique* symposium of 1987 drew special attention to this application (Jewell, 1989). Disadvantages of the conventional shear box have been highlighted by early researchers including Terzaghi & Peck (1948) and Roscoe (1953). These include the effects of progressive shear failure initiated from the ends, the unknown stress distribution within the box and the creation of tensile zones under light normal loads. Arthur & Aysen (1977) showed that the volume change is concentrated above and below the failure plane within a zone approximately 6 mm thick for a medium size sand. Results obtained from X-ray radiography showed that failure is progressive and starts from the ends and develops towards the centre of the sample. A non-uniform distribution of shear stress has been identified as one source of inaccuracy in the direct shear test and results in an underestimate of the angle of internal friction. The author believes that the major reason for the non-uniform stress distribution is due to the lack of friction at the ends, which causes the failure plane to divert from the predetermined horizontal plane and creates a retaining wall failure pattern in the sample. A detailed analysis of the direct shear test can be found in Arthur et al. (1977 and 1988). Jewell (1989) recommended that the internal friction angle measured in a split shear box should be specified as the direct shear angle rather than the plane strain shear angle. However, this method of analysis does not include the effects of deposition direction, which changes the resulting internal friction angle within a range of 6° in medium size sand (Arthur & Aysen, 1977; Wong & Arthur, 1985).

Example 4.5

From the following direct shear test data determine: (a) the internal friction angles at peak and residual states, and (b) the principal stresses and their directions inside the box at the peak strength when the normal stress is 100 kPa. The shear box is 64 mm square in plan.

Vertical load (kg)	Shear force (N) at peak	Shear force (N) at residual
50	399.8	228.7
100	801.9	457.4
150	1214.0	686.1

Solution: (a) Calculate normal and shear stresses and tabulate the results:

Normal stress (kPa)	Shear strength (kPa) at peak	Shear strength (kPa) at residual
119.7	97.6	55.8
239.5	195.8	111.7
359.2	296.4	167.5

The results are plotted in Figure 4.9, from which $\phi' = 39.25°$ and $\phi'_r = 25.00°$.
(b) $2\alpha = (90.0° + 39.25°) = 129.25°$. Using Equations 4.7 and 4.8:
$$100.0 = (\sigma'_1 + \sigma'_3)/2 + [(\sigma'_1 - \sigma'_3)/2]\cos 129.25°,$$
$$100.0 \times \tan 39.25° = [(\sigma'_1 - \sigma'_3)/2]\sin 129.25°.$$
Solving for σ'_1 and σ'_3: $\sigma'_1 = 272.2$ kPa, $\sigma'_3 = 61.2$ kPa, $\alpha = 129.25° / 2 = 64.62°$.

4.3.2 Triaxial compression test

Figure 4.10 shows a schematic view of a triaxial cell and a cylindrical soil sample. The soil sample is protected by a thin rubber membrane and is under pressure from water that occupies the volume of the cell. This confining pressure enforces a condition of equality on two of the total principal stresses i.e. $\sigma_2 = \sigma_3$. Vertical stress is applied via a loading ram, and therefore, the total major principal stress σ_1 is the sum of the confining pressure and the *deviator stress* applied through the ram. In a traditional triaxial compression test, the confining pressure σ_3 is kept constant whilst the major principal stress σ_1 is increased incrementally by loading ram until the sample fails. The loading ram has a deformation-control mechanism in which the horizontal surface of the sample is subjected to a con-trolled rate of deformation; the associated force in the ram is measured by means of prov-ing rings or load cells. The rate of applied deformation is specified by the relevant code to

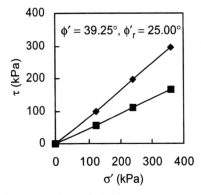

Figure 4.9. Example 4.5.

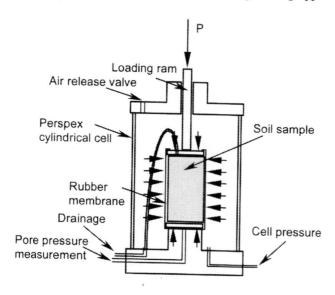

Figure 4.10. Schematic view of a triaxial cell.

avoid generation of any pore pressure in drained tests. Facilities to measure the pore pressure and volume change at any stage of the test are available. To eliminate any end effects on the results, the height of a specimen is twice its diameter. Specimen diameters are normally either 38 or 100 mm, however some cells have been manufactured to accommodate larger diameter specimens. Specimens are either undisturbed or remoulded depending on the type of the material. To construct the failure envelope for a soil, a test has to be performed several times with different confining pressures using, ideally, identical samples. Recorded measurements include deviator stress at different stages of the test, particularly at failure (peak deviator stress) and the critical state (ultimate deviator stress), vertical displacement of the ram, volume change and pore pressure. Stress data is presented in (σ or σ')-τ coordinate system, as shown in Figure 4.11, and for each test, a Mohr's circle of effective stresses at the peak is drawn. The common tangent to the circles, which is obtained graphically or by simple mathematics, represents the failure envelope.

The stress data may be presented in a $t' = (\sigma'_1 - \sigma'_3) / 2$ and $s' = (\sigma'_1 + \sigma'_3) / 2$ coordinate system (or in their equivalent total stresses), as shown in Figure 4.12. This is most convenient if a stress path method is used. In this presentation the failure envelope passes

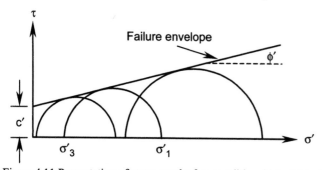

Figure 4.11 Presentation of stress results from traditional triaxial test.

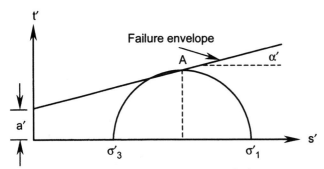

Figure 4.12. Presentation of effective stresses in t'- s' coordinate system.

through the maximum shear stress point (A) of each circle. The parameters ϕ' and c' can be related to the equivalent shear strength parameters of α' and a' according to:

$$\phi' = \sin^{-1}(\tan\alpha'), \; c' = a'/\cos\phi' \qquad (4.15)$$

An alternative presentation of the stress state for both the peak and critical states is in the $q' = \sigma'_1 - \sigma'_3$ and effective mean stress $p' = (\sigma'_1 + 2\sigma'_3)/3$ coordinate system (or in their equivalent total stresses). From the definitions of p' and q' we obtain:

$$\sigma'_3 = p' - q'/3, \; \sigma'_1 = p' + 2q'/3 \qquad (4.16)$$

Substituting σ'_3 and σ'_1 into the Mohr-Coulomb failure criterion (Equation 4.12) we obtain the following linear relationship between q' and p':

$$q' = p'[6\sin\phi'/(3-\sin\phi')] + 6c\cos\phi'/(3-\sin\phi') \qquad (4.17)$$

Equation 4.17 represents the Mohr-Coulomb failure criterion in the q', p' coordinate system. The stress-strain model and volume change properties are investigated by plotting the deviator stress and volume change against the axial strain. It is sometimes convenient to normalize the stress data by dividing the deviator stress by σ'_3, or q' by p'. If the cohesion (either effective or total) is zero then the stress-strain curves for different confining pressures all fall close to a single curve. From the Equation 4.13 we can see that, in the absence of any cohesion, the ratio of σ'_1 to σ'_3 is constant and is equal to $\tan^2(45° + \phi'/2)$. The cross sectional area of the sample, which varies during a test, is given by:

$$A = A_0(1-\varepsilon_V)/(1-\varepsilon_1) \qquad (4.18)$$

where A_0 is the initial cross sectional area, $\varepsilon_V = \Delta V / V_0$ (ΔV being volume change and V_0 being the initial volume) is the volumetric strain and $\varepsilon_1 = \Delta L / L_0$ (ΔL being the axial deformation and L_0 being initial height) is the axial strain. In undrained conditions, the volume change is zero, and therefore Equation 4.18 becomes:

$$A = A_0/(1-\varepsilon_1) \qquad (4.19)$$

4.3.3 Common types of triaxial test

Different types of triaxial tests have been developed in order to provide a comprehensive description of the behaviour of the soil under different drainage conditions and loading rates; these are:

Drained test: There is no condition on the degree of saturation and drainage allowed during an experiment. Incremental loads are applied slowly to avoid building up excessive pore pressures.

Consolidated-Undrained test: In this test, the soil sample is fully saturated and is consolidated under a specified constant total stress during which drainage is permitted (to facilitate the consolidation). The applied pressure normally represents the in-situ conditions. After completion of consolidation the sample is sheared in undrained conditions.

Unconsolidated-Undrained test: Here, the sample is fully saturated and no drainage permitted during the test. Thus, the average moisture content remains constant and no volume change occurs during the experiment. In consolidated-undrained and unconsolidated-undrained tests a back pressure system may be used to increase the initial pore pressure to insure full saturation. The magnitude of the back pressure does not affect the test results.

Example 4.6

A consolidated-undrained triaxial test gave the following data (columns 1 to 4): Diameter of the sample = 38 mm, height of the sample = 76 mm. Pore pressures at peak points are 10.0, 61.6, 113.2 kPa for σ_3 = 100, 200, 300 kPa respectively. Determine: (a) the deviator stress-axial strain curve and Modulus of Elasticity of the soil, and (b) the shear strength parameters (effective and total).

Axial deformation (mm)	Applied axial force (N) σ_3=100 (kPa)	σ_3=200 (kPa)	σ_3=300 (kPa)	Axial strain	Deviator stress (kPa) σ_3=100 (kPa)	σ_3=200 (kPa)	σ_3=300 (kPa)
0.00	0.0	0.0	0.0	0.0000	0.0	0.0	0.0
0.95	126.3	155.7	183.7	0.0125	110.0	135.6	160.0
1.90	218.9	304.1	353.2	0.0250	188.2	261.4	303.6
2.85	277.7	415.4	532.1	0.0375	235.7	352.5	451.6
3.80	308.8	475.8	622.4	0.0500	258.7	398.6	521.3
4.75	328.9	499.0	661.2	0.0625	271.9	412.5	546.6
5.70	332.3	512.9	689.3	0.0750	271.0	418.3	562.2
6.65	335.6	518.6	701.9	0.0875	270.0	417.3	564.7
7.60	338.1	523.0	708.7	0.1000	268.3	415.0	562.4
8.55	341.3	528.9	715.7	0.1125	267.1	413.9	560.1
9.50	344.2	534.6	724.0	0.1250	265.6	412.5	558.6

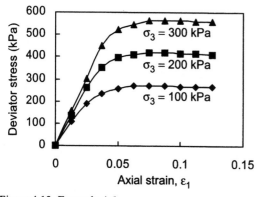

Figure 4.13. Example 4.6.

Solution:

(a) Deviator stresses are calculated from the corrected sectional areas and tabulated (columns 6 to 8 in the above table). Sample calculation for axial deformation of 4.75 mm:
$\varepsilon_1 = \Delta L / L_0 = 4.75 / 76.0 = 0.0625$, $A_0 = (\pi \times 38.0^2 / 4) \times 10^{-6} = 1.134115 \times 10^{-3}$ m^2.
$A = 1.134115 / (1 - 0.0625) = 1.209723 \times 10^{-3}$ m^2.

At $\sigma_3 = 100$ kPa deviator stress $\sigma'_1 - \sigma'_3 = (328.9/1000)/1.209723 \times 10^{-3} = 271.9$ kPa,
similarly at $\sigma_3 = 200$ kPa $\sigma'_1 - \sigma'_3 = 412.5$ kPa and at $\sigma_3 = 300$ kPa $\sigma'_1 - \sigma'_3 = 546.6$ kPa.

Figure 4.13 shows the variation of deviator stress against axial strain.

At $\sigma_3 = 100$ kPa the initial Modulus of Elasticity E_s is the best tangent to the curve from the origin or is approximately equal to: $E_s = 110.0 / 0.0125$ (second row) $= 8800$ kPa.
At $\sigma_3 = 200$ kPa, $E_s = 10848$ kPa, and at $\sigma_3 = 300$ kPa, $E_s = 12800$ kPa.

(b) Calculate σ'_1 and σ'_3 at failure for each test:

Test 1: $\sigma'_3 = 100.0 - 10.0 = 90.0$ kPa, $\sigma'_1 = 100 + 271.9 - 10.0 = 361.9$ kPa.
Test 2: $\sigma'_3 = 200.0 - 61.6 = 138.4$ kPa, $\sigma'_1 = 200 + 418.3 - 61.6 = 556.7$ kPa.
Test 3: $\sigma'_3 = 300.0 - 113.2 = 186.8$ kPa, $\sigma'_1 = 300 + 564.7 - 113.2 = 751.5$ kPa.

Mohr's circles of stress are shown in Figure 4.14 from which: $c' = 0$ and $\phi' = 37°$.

σ_1 values at failure for tests 1, 2, and 3 respectively: $\sigma_1 = 100 + 271.9 = 371.9$ kPa, $200 + 418.3 = 618.3$ kPa, $300 + 564.7 = 864.7$ kPa. Mohr's circles of stress are shown in Figure 4.14 from which: $c = 40$ kPa, and $\phi = 25°$. If the effective stresses are presented in the t and s' coordinate system, then, from Figure 4.14, we obtain $a' = 0$ and $\alpha' = 31°$.

Example 4.7

The following results were obtained from consolidated-undrained tests on specimens of a saturated clay. Determine the shear strength parameters (effective and total).

σ_3 (kPa)	$\sigma'_1 - \sigma'_3$ (kPa) at peak	u (kPa) at peak
100	137	28
200	210	86
300	283	147

Solution: Calculate σ'_1 and σ'_3 for each test:

Test 1: $\sigma'_1 = 100.0 + 137.0 - 28.0 = 209.0$ kPa, $\sigma'_3 = 100.0 - 28.0 = 72.0$ kPa.
Test 2: $\sigma'_1 = 200.0 + 210.0 - 86.0 = 324.0$ kPa, $\sigma'_3 = 200.0 - 86.0 = 114.0$ kPa.

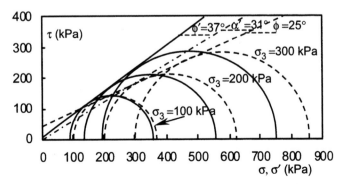

Figure 4.14. Example 4.6.

Test 3: $\sigma'_1 = 300.0 + 283.0 - 147.0 = 436.0$ kPa, $\sigma'_3 = 300.0 - 147.0 = 153.0$ kPa.

These results represent three Mohr's circles in the σ'-τ coordinate system. The graphical method may be replaced by the following simplified mathematical analysis:

Using the Mohr-Coulomb failure criterion of Equation 4.13:

(1) $209.0 = 72.0 \tan^2(45° + \phi'/2) + 2c' \tan(45°+ \phi'/2)$
(2) $324.0 = 114.0 \tan^2(45° + \phi'/2) + 2c' \tan(45° + \phi'/2)$
(3) $436.0 = 153.0 \tan^2(45° + \phi'/2) + 2c' \tan(45° + \phi'/2)$

This produces three equations with two unknown values of c' and ϕ'. Solving (1) and (2), (1) and (3), and (2) and (3) for c' and ϕ', and taking the average, we obtain $c' = 1.6$ kPa $\approx$ 0.0 and $\phi' = 28.3°$.

For the total shear strength parameters:

Test 1: $\sigma_1 = 100.0 + 137.0 = 237.0$ kPa, $\sigma_3 = 100.0$ kPa.
Test 2: $\sigma_1 = 200.0 + 210.0 = 410.0$ kPa, $\sigma_3 = 200.0$ kPa.
Test 3: $\sigma_1 = 300.0 + 283.0 = 583.0$ kPa, $\sigma_3 = 300.0$ kPa.

Substituting the test results into the Mohr-Coulomb failure criterion of Equation 4.13 and solving for c and ϕ, we obtain $c = 24.3$ kPa and $\phi = 15.5°$.

Example 4.8

The effective shear strength parameters of a fully saturated soil are $c' = 20$ kPa and $\phi' = 25°$. In an unconsolidated and undrained triaxial test, the cell pressure was 200 kPa and deviator stress at failure was 107 kPa. Compute the pore pressure at failure.

Solution:

$\sigma'_1 = 200.0 - u + 107.0 = 307.0 - u$, $\sigma'_3 = 200.0 - u$, substituting into Equation 4.13:
$307.0 - u = (200.0 - u) \tan^2(45.0° + 25.0°/2) + 2 \times 20.0 \times \tan(45.0° + 25.0°/2)$,
$u = 169.8$ kPa.

4.3.4 Unconfined compression test

Unconfined compression is a special case of triaxial compression where $\sigma_3 = 0$. However, rather than using a triaxial cell, the test is carried out in a mechanical apparatus specially manufactured for this purpose. Soil samples are cylindrical with a diameter of 38 mm and a height of 76 mm. This test is suitable only for fully saturated clays where ϕ_u is zero and the undrained cohesion can be found by constructing only one Mohr's circle. Unconfined compressive strength (*UCS*, σ_1 in Figure 4.15) is frequently used in stabilization studies to

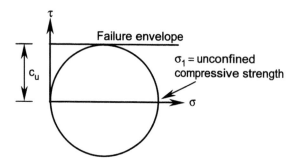

Figure 4.15. Unconfined compression test.

ascertain the proportion of stabilizer required to satisfy a specific strength criterion. Samples used in this case are 100 mm in diameter and 200 mm high.

4.3.5 Stress path method

The presentation of stress data in either the t'-(s' or s) or the q'-(p' or p) coordinate systems facilitates the monitoring of the stress increments and pore pressure variation during a triaxial test, which is a considerable advantage. Consider a triaxial test specimen of normally consolidated clay (point C in Figure 4.16(a)) that is consolidated under a confining pressure σ_3. The confining pressure is greater than the preconsolidation pressure (maximum effective stress under which the soil has been consolidated in the field) to ensure that the soil will behave as a normally consolidated soil. At point C, $s' = s = \sigma_3$ and the pore pressure is zero. The progress of the undrained triaxial test is shown by path CE for the effective stresses and path CT for the total stresses. Both paths terminate on the corresponding failure envelopes at points E and T. These points are the topmost points of the effective and total Mohr' circles respectively. As a result, the angle of line CT with the horizontal is always 45°. In a drained test the stress path coincides with CT but terminates on the effective failure envelope. In the q'-(p' or p) coordinate system the slope of the effective stress path (in drained conditions) or total stress path (in undrained conditions) is 3 vertical to 1 horizontal, as given by:

$$\text{Slope} = q'/(p'-\sigma_3') = (\sigma_1' - \sigma_3')/[(\sigma_1' + 2\sigma_3')/3 - \sigma_3'] = 3.$$

The horizontal distance between CE and CT represents the pore pressure according to:

$$s - s' = (\sigma_1 + \sigma_3)/2 - (\sigma_1' + \sigma_3')/2 = (\sigma_1' + u + \sigma_3' + u)/2 - (\sigma_1' + \sigma_3')/2 = u.$$

Figure 4.16(b) shows a typical stress path for an overconsolidated clay subjected to a consolidated-undrained triaxial test.

Example 4.9

The results of consolidated-undrained triaxial tests on three identical samples of clay are tabulated below. Determine: (a) the effective and total shear strength parameters using the t'-(s', s) coordinate system, (b) the stress path for both effective and total stresses.

Solution:

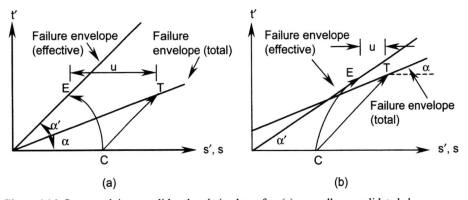

Figure 4.16. Stress path in consolidated-undrained test for: (a) normally consolidated clay, (b) overconsolidated clay.

Axial strain	$\sigma_3 = 150$ kPa			$\sigma_3 = 300$ kPa			$\sigma_3 = 450$ kPa		
	t'	u	s'	t'	u	s'	t'	u	s'
0.00	0.0	0.0	150.0	0.0	0.0	300.0	0.0	0.0	450.0
0.02	50.0	45.0	155.0	85.0	95.0	290.0	142.0	140.0	452.0
0.04	84.0	47.6	186.4	135.0	110.0	325.0	276.0	160.0	566.0
0.06	124.0	35.8	238.2	164.0	110.0	354.0	382.0	144.0	688.0
0.08	153.0	22.0	281.0	234.0	96.0	438.0	430.0	115.0	765.0
0.0972	-	-	-	-	-	-	440.6	101.4	789.2
0.10	176.0	12.0	314.0	271.0	81.0	490.0	440.0	101.0	789.0
0.12	192.0	2.0	340.0	296.0	66.0	530.0	440.0	90.0	800.0
0.14	200.0	−6.0	356.0	314.8	52.0	562.8	434.0	80.0	804.0
0.1512	-	-	-	319.2	46.2	573.0	-	-	-
0.1545	204.6	−12.6	367.2	-	-	-	-	-	-
0.16	204.0	−13.0	367.0	318.0	42.0	576.0	426.0	72.0	804.0
0.18	202.0	−17.0	369.0	308.0	31.3	576.7	416.0	66.0	800.0
0.20	195.0	−20.0	365.0	292.0	23.6	568.4	406.0	56.0	800.0
0.22	188.0	−23.0	361.0	279.0	17.5	561.5	400.0	52.0	798.0
0.24	184.0	−26.0	360.0	266.0	14.0	552.0	396.6	50.0	796.6
0.26	180.0	−28.0	358.0	-	-	-	394.0	48.0	796.0
0.28	179.0	−30.0	359.0	-	-	-	391.0	47.0	794.0
0.30	178.0	−32.0	360.0	-	-	-	388.0	47.0	791.0
0.32	177.0	−32.5	359.5	-	-	-	384.0	46.5	787.5
0.34	174.0	−32.6	356.6	-	-	-	380.0	46.5	783.5
0.36	171	−32.7	353.7	-	-	-	378.0	46.2	781.8

The recorded data includes deviator stresses (twice of t') and pore pressures at specified strains, except at peak points where the strains have been calculated from the recorded axial deformation. The values of $s' = (\sigma'_1 + \sigma'_3) / 2$ are then computed and tabulated. Check s' at peak point for $\sigma_3 = 150$ kPa: $t' = (\sigma'_1 - \sigma'_3) / 2 = 204.6$ kPa, $u = -12.6$ kPa, $s = (\sigma_1 + \sigma_3) / 2 = (\sigma_1 - \sigma_3 + 2\sigma_3) / 2 = t' + \sigma_3 = 204.6 + 150 = 354.6$ kPa, $s' = s - u = 354.6 - (-12.6) = 367.2$ kPa.
Plotting the points in the t'-(s', s) coordinate system (Figure 4.17(a)) we have:
Total Stresses: $\alpha = 23.8°$, $a = 48.5$ kPa. Substituting these values into Equation 4.15: $\phi = 26.2°$, $c = 54.0$ kPa. Effective Stresses: $\alpha' = 29.1°$, $a' = 0.0$, from Equation 4.15: $\phi' = 33.8°$, $c' = 0.0$. The stress paths are shown in Figure 4.17(b).

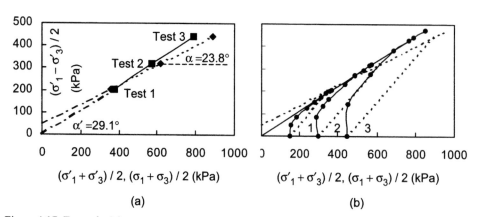

(a)   (b)

Figure 4.17. Example 4.9.

4.3.6   *Pore pressure coefficients A and B*

Pore pressure coefficients relate the applied load to the excess pore pressure in undrained loading. In an axisymmetric stress system where the loading is controlled by variation of σ_1 and σ_3, two coefficients are sufficient to construct the relationship. Pore pressure coefficients are computed from the triaxial test results and may be applied to a field situation to predict excess pore pressures created by undrained loading. Figure 4.18(a) shows an element of elastic soil under total principal stresses of σ_1, σ_2, and σ_3, with an initial pore pressure of u_i. All three principal stresses are increased equally by $\Delta\sigma$, causing an increase of Δu in pore pressure (Figure 4.18(b)). The change in effective stress in all three directions is: $\Delta\sigma' = \Delta\sigma - \Delta u$. In an elastic element subjected to these equal stress increments in the three directions, the principal strains are defined by:

$$\varepsilon_1 = \varepsilon_2 = \varepsilon_3 = \frac{\Delta\sigma'}{E_s}(1-2\mu) \tag{4.20}$$

where E_s is the Modulus of Elasticity of the soil and μ is Poisson's ratio. The volumetric strain and volume change are given by:

$$\varepsilon_V = \frac{\Delta V}{V} = \varepsilon_1 + \varepsilon_2 + \varepsilon_3 = \frac{3\Delta\sigma'}{E_s}(1-2\mu) \rightarrow \Delta V = V\frac{3\Delta\sigma'}{E_s}(1-2\mu) \tag{4.21}$$

The volume change in the pore water due to excess pore pressure Δu is obtained from the fundamental definition of volumetric strain for pore water:

$$\frac{\Delta V_w}{V_w} = \frac{\Delta u}{K} \rightarrow \Delta V_w = V_w\frac{\Delta u}{K} \tag{4.22}$$

where V_w and K are the volume and the bulk modulus of the pore water respectively. In full saturation, the volume of pore water is equal to the total volume of the element V multiplied by the porosity n. Therefore:

$$\Delta V_w = nV\frac{\Delta u}{K} \tag{4.23}$$

In undrained conditions no drainage is allowed, and consequently, the volume change of the soil skeleton and pore water must be equal. By equating Equations 4.21 and 4.23 we can express the excess pore pressure Δu in terms of the total incremental stress $\Delta\sigma$:

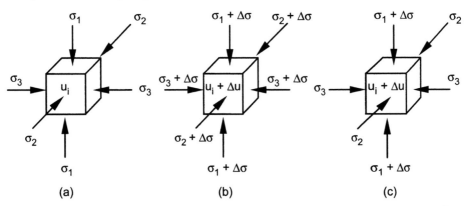

Figure 4.18. (a) Soil element, (b) excess pore pressure due to equal increase in σ_1, σ_2, and σ_3, (c) excess pore pressure due to increase in σ_1.

$$\Delta u = \frac{1}{1 + \dfrac{nE_s}{3K(1-2\mu)}} \Delta\sigma = B\Delta\sigma \qquad (4.24)$$

where B is termed the pore pressure coefficient. Figure 4.18(c) shows an elastic soil element in which only the major principal stress σ_1 is subjected to an increase of $\Delta\sigma$. Following the development described above, we obtain:

$$\Delta u = \frac{1}{3[1 + \dfrac{nE_s}{3K(1-2\mu)}]} \Delta\sigma = \frac{1}{3}B\Delta\sigma \qquad (4.25)$$

Considering the test path in a traditional triaxial compression test we can combine the two cases to yield:

$$\Delta u = B[\Delta\sigma_3 + (\Delta\sigma_1 - \Delta\sigma_3)/3] \qquad (4.26)$$

where $\Delta\sigma_3$ represents the change in the confining pressure and $\Delta\sigma_1 - \Delta\sigma_3$ is the change in the deviator stress. However, it is customary in soil mechanics practice to employ Equation 4.26 for the determination of the excess pore pressure, where both pore pressure coefficients A and B are determined from data recorded during a triaxial test:

$$\Delta u = B[\Delta\sigma_3 + A(\Delta\sigma_1 - \Delta\sigma_3)] \qquad (4.27)$$

The parameters A and B are generally different from their theoretical values as the soil is not a perfectly elastic material and the response depends on the level of the applied stresses. The magnitude of parameter B depends on the degree of saturation, and since the bulk modulus of pore water is very high, B must theoretically be unity for saturated soils. However, in partially saturated soils and in saturated soils with air bubbles in the pore water, B decreases to values less than unity. The magnitude of the parameter A depends on the test path and stress level, and diverges from its theoretical value having positive values for normally consolidated soils to negative values for overconsolidated soils.

Example 4.10

Using the data of Example 4.9, calculate the pore pressure coefficient A at every stage of the three tests and show its variation with axial strain. Assume $B = 1$.

Solution:

Strain	A $\sigma_3 = 150$ (kPa)	$\sigma_3 = 300$ (kPa)	$\sigma_3 = 450$ (kPa)	Strain	A $\sigma_3 = 150$ (kPa)	$\sigma_3 = 300$ (kPa)	$\sigma_3 = 450$ (kPa)
0.00	0.000	0.000	0.000	0.16	−0.032	0.066	0.084
0.02	0.450	0.559	0.493	0.18	−0.042	0.051	0.079
0.04	0.283	0.407	0.290	0.2	−0.051	0.040	0.069
0.06	0.144	0.335	0.188	0.22	−0.061	0.031	0.065
0.08	0.072	0.205	0.134	0.24	−0.071	0.026	0.063
0.0972	-	-	0.115	0.26	−0.078	-	0.061
0.10	0.034	0.149	0.115	0.28	−0.084	-	0.060
0.12	0.005	0.111	0.103	0.3	−0.090	-	0.060
0.14	−0.015	0.082	0.092	0.32	−0.092	-	0.060
0.1512	-	0.072	-	0.34	−0.094	-	0.061
0.1545	−0.030	-	-	0.36	−0.096	-	0.061

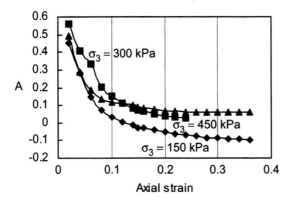

Figure 4.19. Example 4.10.

The results are tabulated and shown in Figure 4.19.

Example 4.11

Data obtained from a consolidated-undrained triaxial test are: $\sigma_3 = 300$ kPa, u (before consolidation) = 276 kPa, u (at failure) = 108 kPa, $\sigma_1 - \sigma_3$ (at failure) = 475 kPa. Calculate the pore pressure coefficients A and B.

Solution:

Using Equation 4.27 at the consolidation stage: $276.0 = B\,[300.0 + A\,(0)]$, $B = 0.92$,
At failure: $108.0 = B\,[0 + A \times 475.0] = 0.92 \times A \times 475.0$, $A = 0.247$.

4.3.7   In-situ shear strength: shear vane test

In soft and saturated clays where an undisturbed specimen is difficult to obtain, the undrained shear strength is measured using a shear vane test. A diagrammatic view of the apparatus is shown in Figure 4.20. It consists of four thin mutually perpendicular blades

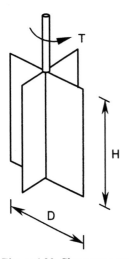

Figure 4.20. Shear vane apparatus.

connected to a central rod where the height H is twice the overall diameter D. Commonly used diameters are 38, 50 and 75 mm. The vane is pushed into the soil either at the ground surface or at the bottom of a borehole until totally embedded in the soil (at least 0.5 m). A torque T is applied and the vane is rotated at the slow rate of 9° per minute. As a result, shear stresses are mobilized on all surfaces of a cylindrical volume of the soil generated by the rotation. The maximum torque is measured by a suitable instrument and equals the moment of the mobilized shear stress about the central axis of the apparatus. From this, and after suitable rearrangement, we obtain:

$$c_u = \frac{T}{\pi D^2 (H/2 + D/6)} \tag{4.28}$$

If the test is carried out with a rapid rotation, then the result may indicate the remoulded undrained shear strength and could be used as a measure of sensitivity. The sensitivity in terms of comparison between undisturbed and remoulded samples is defined as the ratio of c_u obtained from the standard rate to the c_u obtained from the rapid test.

4.4 STRESS-STRAIN BEHAVIOUR OF SANDS AND CLAYS

4.4.1 *Stress-strain behaviour of dense and loose sands*

Figure 4.21 shows typical results of direct shear tests in dense and loose sands and, based on these results, the following observations can be made. In dense sand the shear strength increases with the shear displacement (relative displacement between the lower and the upper box) to a maximum value of $\tau_{max} = \tau_f$, and then decreases to a critical value of τ_{cr} that is approximately a constant for each type of sand. The term *critical state* is used to describe the state of the soil at this point. At the critical state the sample is sheared while having a constant volume. If further shear displacement is applied, the shear strength decreases to a residual value τ_r and shear planes will be fully developed within the sample. Initially, the volume of the specimen decreases slightly, and then increases with further shear displacement. This dilation can be explained by plotting the vertical movement (z) of the loading plate, or the normal force N of Figure 4.8(a), against the shear displacement (x). The rate of volume change is represented by the dilation angle ψ, which is the slope ($\Delta z/\Delta x$) of the resulting curve. The upper half of the shear box diverges from its horizontal

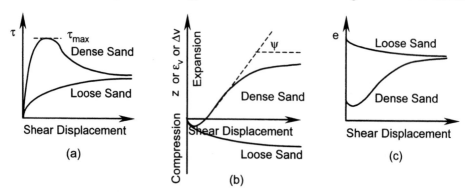

Figure 4.21. Direct shear test results for dense and loose sands.

direction to follow the dilation angle. As the increasing shear displacement moves the sample towards its residual state, the dilation angle decreases to zero. Dilation is attribute to interlocking of the sand particles that have to be subsequently loosened to allow the shear plane to develop. This (volume change) behaviour may be illustrated by plotting void ratio (or specific volume) against shear displacement (Figure 4.21(c)). The void ratio of the sample increases from its initial low value to a higher critical value, which is approximately a constant for each type of sand. Note that in the calculation of the void ratio (while shearing) the thickness of the specimen is assumed to be the thickness of the failure zone. As the failure zone is more concentrated immediately above and below the horizontal failure plane, standard computations may underestimate the critical void ratio for dense sands. The magnitude of the critical void ratio, or the equivalent internal friction angle, depends on the level of normal stress applied to the sample. When the normal stress is increased to very high levels, the internal friction angle corresponding to the peak strength decreases and approaches the critical value, thereby creating a two-stage Mohr-Coulomb failure criterion. In loose sand, the shear strength increases with the shear displacement and, without passing through a peak point, approaches its critical value at which it remains approximately constant. The volume of the specimen gradually decreases to a certain value and remains approximately constant thereafter.

Similar patterns of behaviour for dense and loose sands are observed in triaxial compression tests. However, the angle of internal friction related to the peak point is 4°- 6° less than that obtained in a modified direct shear test (Arthur & Aysen, 1977) in which the modification facilitates the generation of shear stresses at the end planes. This difference is mostly as a result of axisymmetric conditions in the triaxial test and the angle between direction of the deposition of the particles with the principal stress directions.

Example 4.12

The results of a direct shear test on a sample of dense medium size dry sand are tabulated below. The sample reached equilibrium under a normal stress of 100 kPa and its initial void ratio under this load was 0.51. If the height of the sample was 20 mm, (a) plot the variation in shear stress and vertical displacement against the shear displacement, (b) plot the variation in void ratio against the shear displacement for:
h = thickness of the sample = 20 mm, h = thickness of the failure zone = 10 mm.

Reading points	Shear disp.(mm)	Vertical disp.(mm)	Shear stress (kPa)	Reading points	Shear disp.(mm)	Vertical disp.(mm)	Shear stress (kPa)
1	0.25	0.00	25	11	2.75	1.01	60
2	0.50	0.08	50	12	3.00	1.11	56
3	0.75	0.19	68	13	3.25	1.20	54
4	1.00	0.29	78	14	3.50	1.28	51
5	1.25	0.39	83	15	3.75	1.34	50
6	1.50	0.50	84	16	4.00	1.39	49
7	1.75	0.60	80	17	4.25	1.42	48
8	2.00	0.71	74	18	4.50	1.45	48
9	2.25	0.81	69	19	4.75	1.47	48
10	2.50	0.91	64	20	5.00	1.48	48

Solution:

(a) Figures 4.22(a) and 4.22(b) show the variation in shear stress and vertical displacement against shear displacement from which:

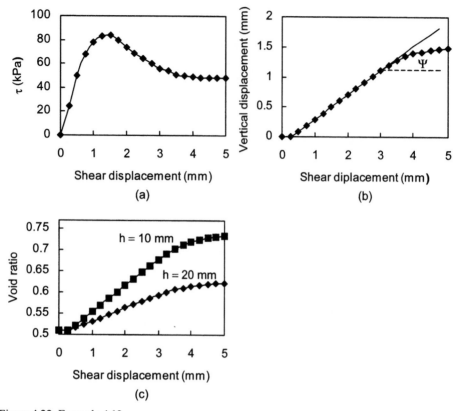

Figure 4.22. Example 4.12.

$\tan\phi$ (or ϕ') = (maximum shear strength = 84 kPa) / (normal stress = 100 kPa) = 0.84, $\phi = 40°$. From Figure 4.22(b) and for reading points 4 and 12:
$\tan\psi = (1.11\ mm - 0.29\ mm) / (3.0\ mm - 1.0\ mm) = 0.41$, $\psi = 22.3°$.
(b) Sample calculation for shear displacement of 1 mm:
Void ratio at any stage: $e = e_0 + \Delta e = e_0 + zA / h_sA = e_0 + z / h_s$
where e_0 is the initial void ratio, z is the vertical displacement, A is the cross-sectional area of the shear box, and h_s is the equivalent height (or thickness) of the solids.

Reading points	Void ratio $h = 20$ mm	Void ratio $h = 10$ mm	Reading points	Void ratio $h = 20$ mm	Void ratio $h = 10$ mm
-	0.510	0.510	11	0.586	0.662
1	0.510	0.510	12	0.594	0.678
2	0.516	0.522	13	0.600	0.691
3	0.524	0.539	14	0.607	0.703
4	0.532	0.554	15	0.611	0.712
5	0.539	0.569	16	0.615	0.720
6	0.548	0.585	17	0.617	0.724
7	0.555	0.600	18	0.619	0.729
8	0.564	0.617	19	0.621	0.732
9	0.571	0.632	20	0.622	0.733
10	0.579	0.647			

Calculation of h_s: $e_0 = 0.51 = V_v / V_s$ or $V_v = 0.51 V_s$. Substitute V_v in: $V = V_s + V_v$:

$V_s = V / 1.51 = A$ (area) $\times h_s \rightarrow h_s = V / (1.51 A) = h \times A / 1.51 A = h / 1.51$,

where h is the height or thickness of the sample. For $h = 20$ mm we have:

$e = e_0 + z / h_s = 0.51 + (1.51 / 20) z = 0.51 + 0.0755 z$.

If we assume the volume change is concentrated within the 10-mm zone centred on the failure plane, then: $e_{(h = 10\,mm)} = 0.51 + 0.151 z$.

For a shear displacement of 1.0 mm, $z = 0.29$ mm:

$e = 0.51 + 0.0755 \times 0.29 = 0.532$, $e_{(h = 10\,mm)} = 0.51 + 0.151 \times 0.29 = 0.554$.

Void ratios for $h = 20$ mm and $h = 10$ mm are tabulated and are shown in Figure 4.22(c).

4.4.2   *Stress-strain behaviour of clay*

Figures 4.23 and 4.24 present the results of drained and consolidated-undrained triaxial tests on fully saturated clay soils. In drained tests, a normally consolidated clay specimen contracts under increasing deviator stress until a constant volume is reached. The constant volume under which shear planes start to generate is called the critical volume. At this stage, the angle of internal friction is slightly greater than the residual internal friction angle but equal to or less than that of the peak point (if it exists). In general the failure envelope passes through the origin. An overconsolidated clay, however, expands as it is sheared and, after passing a peak point, approaches a constant volume. For this soil, the failure envelope indicates the presence of some cohesion, the magnitude of which depends on the type of soil. The magnitude of the expansion depends on the overconsolidation ratio (the maximum effective stress once applied to the soil in the past, or preconsolidation pressure divided by the present effective stress) and increases as the overconsolidation ratio increases. In determining the magnitude of the confining pressure applied in a triaxial test on samples of both normally consolidated and overconsolidated clays, consideration must be given to the field conditions. Care must be taken in constructing the failure envelope to distinguish test points that are situated to the left or right of the preconsolidation pressure. When the test points are scattered on both sides of the preconsolidation pressure, the best fit for the failure envelope may not be the best answer.

In the consolidated-undrained test, water is not permitted to flow in to or out of the sample, and positive or negative pore pressures are generated as the sample is sheared.

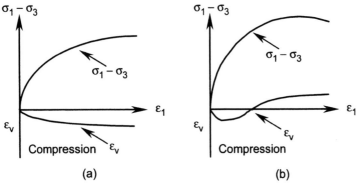

Figure 4.23. Drained triaxial tests on cohesive soils: (a) normally consolidated clay, (b) overconsolidated clay.

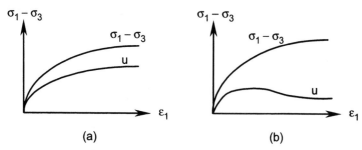

Figure 4.24. Consolidated-undrained triaxial tests on: (a) normally consolidated clay, (b) overconsolidated clay.

In a normally consolidated clay an increase in deviator stress creates a tendency for the soil to contract and push water out of the sample, but as flow is not permitted, positive pore pressure is generated in the pore water. In an overconsolidated clay, however, after an initial increase in pore pressure, the soil tends to expand and therefore draws water in, but, as water is prevented from flowing, a decrease in pore pressure will be produced that may develop into a negative pore pressure. The magnitude of this negative pore pressure depends on the overconsolidation ratio and increases (in negative values) as the overconsolidation ratio increases. Similarities between the stress-strain behaviour of loose sand and normally consolidated clay on the one hand, and between dense sand and oveconsolidated clay on the other hand, suggest that there may exist a unique relationship relating the state of the stress to the volume change properties, and a general failure criterion for soils. In an unconsolidated-undrained triaxial test carried out on a fully saturated clay, water flow is prevented in to or out of the sample at every stage of the test. Any applied confining pressure σ_3 is then resisted solely by the pore water and the incremental effective stress remains zero. The increase in deviator stress is then resisted by the undrained cohesion c_u that is an inherent strength characteristic of the soil. Frictional resistance cannot be mobilized, as the effective stress remains zero. Consequently, the failure envelopes for both total and effective stresses are the same, as is shown in Figure 4.25. This means that the strength is the same regardless of the magnitude of the applied confining pressure.

Example 4.13

Using the data of Example 4.9 plot the variation of deviator stress and pore pressure against axial strain. Show the stress - strain behaviour in the normalized coordinate system

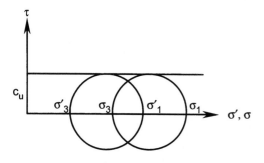

Figure 4.25. Unconsolidated-undrained triaxial test on a saturated clay.

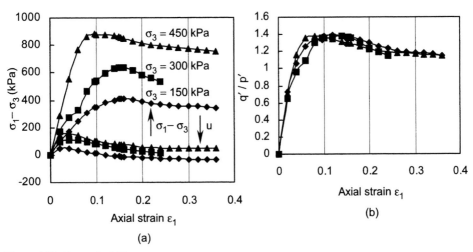

Figure 4.26. Example 4.13.

of ε_1, q' / p' and discuss the results.

Solution:

The variations of deviator stress and pore pressure against axial strain is shown in Figure 4.26(a). Figure 4.26(b) shows the normalized stress-strain behaviour in ε_1, q' / p' coordinate system. The parameter q' is calculated by multiplying t' by 2; the parameter p' is calculated from Equation 4.16. The generation of negative pore pressure in the first sample – with $\sigma_3 = 150$ kPa, a peak strength ϕ' of 33.8°, a residual ϕ' of 28.9° (calculations not shown), zero effective cohesion, and the pattern of stress paths shown in Figure 4.17(b) – shows that the soil is a lightly to medium overconsolidated fine silt.

4.5   CRITICAL STATE THEORY

4.5.1   *Isotropic compression*

From the discussion of Section 4.4 it is evident that, as a soil is sheared, the elastic-plastic deformations transfer the state of the soil into a condition of perfect plasticity in which plastic shearing continues with constant effective stress and volume. In soil mechanics this condition is known as *critical state*. Critical state theory proposes a relationship between the state of stress and the specific volume v while shearing from an initial state to the critical state.

The consolidation stage in consolidated-undrained and drained triaxial tests is represented by point A in the p', v coordinate system, as shown in Figure 4.27(a). At this point, $p'_A = (\sigma'_1 + 2\sigma'_3) / 3 = \sigma'_3 = \sigma_3$. Any further increase in the confining pressure will induce a reduction in volume, which is represented by the point B. The state of the sample subjected to an isotropic compression or confining pressure σ_3 is defined by a relationship between v and p' called the normal compression line (*NCL*), as shown in Figure 4.27(a). This normal compression line is established by means of a triaxial compression test. If at any point on the *NCL*, say point B, the sample is unloaded (arc BC), the volume recovers

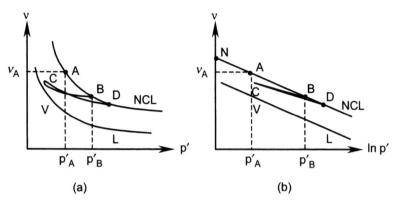

Figure 4.27. Isotropic compression in v, p' and $v, \ln p'$ coordinate systems.

slightly and the sample expands. Reloading causes compression and the state of the soil moves towards the normal compression line on arc CD. The state of the soil on BC or CD is termed as overconsolidated. The deformation behaviour of the soil along these lines can be assumed to be elastic whereas, on the normal compression line, the soil deformation behaviour is plastic. It is seen that the NCL is a boundary to this state, and that any condition to the left of the NCL when moving from A to D cannot exist.

Figure 4.27(b) represents the normal compression line in $v, \ln p'$ coordinate system. Experimental results show that the normal compression line can be approximated without serious error by a line defined by:

$$v = N - \lambda \ln p' \qquad (4.29)$$

where λ is the slope of the normal compression line and N is the magnitude of v at $p' = 1$ kPa. The expansion line BC and recompression line CD can be idealized by just one line with a slope of κ, which implies that during unloading and reloading, expansion and recompression are equal and points B and D coincide. The equation of this line may be presented in a similar way as Equation 4.29:

$$v = v_\kappa - \kappa \ln p' \qquad (4.30)$$

where $\kappa = -(v_C - v_D)/(\ln p'_C - \ln p'_D)$ and v_κ is the magnitude of v at $p'_C = 1$ kPa.

Note that v_κ is not a constant for the soil and its magnitude depends on the magnitude of p'_D. Alternatively:

$$(v - v_C)/(\ln p' - \ln p'_C) = -\kappa, \text{ or: } (v - v_D)/(\ln p' - \ln p'_D) = -\kappa \qquad (4.31)$$

For clay soils the normal compression line is established at relatively low to medium levels of stress. For sand, the stress level should be more than 700 kPa in order to approach the relevant NCL (Atkinson & Bransby, 1978). Recompression at a moderate stress is represented by a horizontal line that indicates no volume change during this process. At the same time, there is no relationship between the state of stress and the initial volume. For example, a very loose sample of sand that is prepared by pouring the sand into a container, could turn into a dense sand if it is submerged in water and the water is allowed to drain. Whilst the process induces no serious stresses in the sand particles, there is a marked difference in the void ratios. In general, the behaviour of sand at higher stresses is similar to the behaviour of clay subjected to lower isotropic compression.

One-dimensional consolidation is represented by the line *VL* in Figure 4.27(b). The slope of this line is similar to the slope of the normal compression line and it is convenient to assume that they are equal. As a result, the equation of the one-dimensional consolidation line is:

$$v = \bar{n} - \lambda \ln p' \tag{4.32}$$

where λ is the slope of the normal compression line and $\bar{n}$ is the magnitude of v in an oedometer test with $p' = 1$ kPa. Consolidation test results are often presented in a e, log σ' coordinate system where e is the void ratio and σ' is the effective vertical stress applied to the sample. The compression index C_c is the slope of the linear portion of e-log σ' plot and it can be shown that $C_c = \ln (10) \times \lambda = 2.3025 \lambda$. The slopes of the expansion and recompression lines in this case have the same ratio (2.3025) to the expansion and recompression lines in normal compression. These lines represent elastic deformation which is recoverable while the deformation on e-log σ' line is plastic and unrecoverable.

An element of soil in the field may undergo elastic and plastic deformation depending on the level of the loading, the coordinate of the element, its initial state of stress and specific volume. Critical state theory combines the parameters, λ, κ, N, and n with the state of stress and gives an elastic-plastic model for the soil behaviour while moving from a given initial state to the critical state.

Example 4.14

Results obtained from an isotropic compression test on a specimen of saturated clay are tabulated below. The volume of the sample and the moisture content at the end of the test are 76.80 ml and 30.6% respectively whilst the specific gravity of solids is 2.7. Calculate λ, N, and κ and plot the normal compression line. What are the overconsolidation ratios at test points 1 and 7?

Test points	Confining pressure (kPa)	Volume change (ml)	Test points	Confining pressure (kPa)	Volume change (ml)
1	15	0.00	5	400	14.25
2	50	1.75	6	50	11.20
3	100	5.50	7	15	9.40
4	200	9.90			

Solution:

Calculate the void ratio and specific volume corresponding to each loading:

For a saturated soil the void ratio is (Equation 1.15): $e = w G_s = 0.306 \times 2.7 = 0.826$, thus $v_7 = 1 + e_7 = 1.826$. Calculate the volume of solids:

$V_v + V_s = 76.8$ ml, $e_7 = V_v / V_s = 0.826$ or $V_s = 76.8 / (1 + 0.826) = 42.06$ ml.

The void ratio and specific volume corresponding to each load is calculated as follows:

$e_6 = e_7 + \Delta V / V_s = 0.826 + (9.4 - 11.20) / 42.06 = 0.783$, $v_6 = 1.783$,

$e_5 = e_7 + \Delta V / V_s = 0.826 + (9.4 - 14.25) / 42.06 = 0.711$, $v_5 = 1.711$, and similarly:

$v_4 = 1.814$, $v_3 = 1.919$, $v_2 = 2.008$, $v_1 = 2.050$.

The results are shown in Figure 4.28. Points 1 and 2 make the recompression line, while points 3, 4, and 5 represent the normal compression line. The line of best fit gives: $\lambda = 0.15$, $\kappa = 0.035$. Substituting the λ and v values of points 3, 4, and 5 into Equation 4.29 we obtain an average value of 2.61 for N.

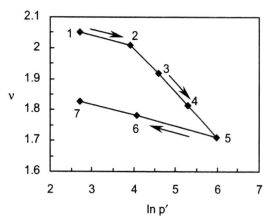

Figure 4.28. Example 4.14.

In order to calculate the overconsolidation ratio at point 1, we need to know the maximum isotropic effective stress (p'_c) that has been applied to the soil in the past. The preconsolidation pressure is the effective stress at the intersection point of line 1-2 with the *NCL*.

Equation of line 1-2: $v - 2.05 = -0.035 (\ln p' - \ln 15.0)$,

Equation of *NCL*: $v = 2.61 - 0.15 \ln p'$. Solving for p':

$p' = 57.1$ kPa, $OCR = 57.1 / 15.0 = 3.8$. At point 7: $OCR = 400.0 / 15.0 = 26.7$.

4.5.2 *The critical state line*

The critical state line (*CSL*) is a spatial curve in the v, p', q' coordinate system (Figure 4.29) and is located on a surface called the state boundary surface (*SBS*). This surface intersects the v-p' plane at the normal compression line. It bounds all possible states during loading and relates the state of stress to the specific volume at any stage of loading while moving the state of the soil from a specified initial conditions towards the critical state line. Therefore, if the equation of the *SBS* is known then evaluation of the elastic and plastic strains is possible at any stage of the loading. In this book we will concentrate on the basics of the critical state theory in terms of the definition of the *CSL* and its application. Calculation of strains, and distinguishing elastic strains from plastic strains, that require the derivation of *SBS* and other mathematical operations are beyond the scope of this book. Based on the critical state theory a comprehensive model for lightly consolidated clay was developed at Cambridge University during the 1960s, and subsequently became known as original Cam clay model (Roscoe & Schofield, 1963). This model proposed a mathematical equation for state boundary surface that included a logarithmic spiral-shaped yield surface in the p', q' coordinate system. This yield surface was later modified (Roscoe & Burland, 1968) to an elliptical surface, and this model became known as modified Cam clay model. Carter et al. (1982) extended the modified Cam clay model to predict the behaviour of a soil under the cyclic loading. The Cam clay based models have been applied successfully to geotechnical engineering problems using numerical methods. Applications of the finite element method to the critical state theory can be found in many texts and research papers (e.g. Britto & Gunn, 1987).

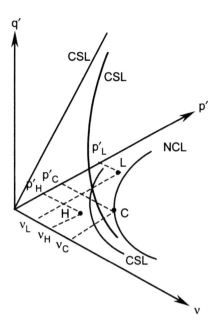

Figure 4.29. Critical state line in v, p', q' coordinate system.

Projection of the *CSL* onto the p'-q' plane (Figure 4.29) defines the state of stress at the critical state and is a line that passes through the origin. The gradient of this line is defined as *M*, which is one of the critical state parameters of the soil, and its equation is:

$$q' = Mp' \tag{4.33}$$

This means the strength of the soil depends solely on the normal effective stress and that it has no cohesion. Thus the critical friction angle can be found by equating *M* to the corresponding term in Equation 4.17. The volumetric behaviour at the critical state is defined by the projection of the *CSL* onto the v-p' plane (Figures 4.29 and 4.30(a)). The initial position is located on the v-p' plane shown by the points *C*, *L*, and *H*. Point *C* is on the *NCL* and represents a normally consolidated soil. Point *B* is located between the *NCL* and the projection of the *CSL*, and represents lightly consolidated soil. Point *H* is located between the v axis and the projection of the *CSL*, and represents a heavily consolidated soil.

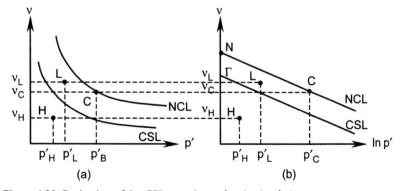

Figure 4.30. Projection of the *CSL* onto the v-p' and v-ln p' planes.

Depending on the stress path during loading, the initial point will move towards the critical state line while being inside or on the state boundary surface. The projection of the *CSL* onto the v, ln p' coordinate system (Figure 4.30(b)) can be approximated by a line parallel to the *NCL* and located on the permitted side of it:

$$v = \Gamma - \lambda \ln p' \tag{4.34}$$

where Γ is the critical specific volume at $p' = 1$ kPa and λ is the slope of the line equal to the slope of the *NCL*.

4.5.3 Triaxial tests on normally consolidated saturated clay

The stress paths for drained and undrained tests on normally consolidated clay are shown in Figure 4.31. Point C represents the initial state under a confining pressure σ_3. The stress path CD of the drained test is on a plane that passes through point C. It is perpendicular to the $p'-q'$ plane and has a gradient of 3 vertical to 1 horizontal with the v-p' plane. Point D represents the critical state and is the intersection point of the plane described above with the *CSL*. The stress path is on the state boundary surface and is defined by its projections on $p'-q'$ and v-p' (or v-ln p') planes. Projections of the stress path onto the $p'-q'$ and v-p' (or v-ln p') planes are p'_C-$D_{p'q'}$ and C-$D_{p'v}$ respectively. The stress path CU of an undrained test is on a plane that passes through point C and is perpendicular to the v-p' plane, making an angle of 90° with the v axis to ensure a constant volume during the test. Point U represents the critical state and is the intersection point of this plane with the *CSL*.

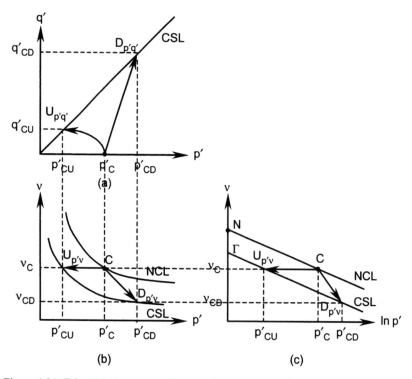

Figure 4.31. Triaxial tests on normally consolidated saturated clay.

The path CU is located on the surface of the state boundary, meaning that similar paths initiating from the different points on the normal compression line construct the surface of the state boundary. The projections of the stress path on p'-q' and v-p' (or v-ln p') planes are p'_C-$U_{p'q'}$ and C-$U_{p'v}$ respectively. As the effective stress path is on a constant v plane therefore its projection on the p'-q' plane is identical to its position in the space. In the case of the original Cam clay model the following state boundary surface is used:

$$q' = Mp'(1+\frac{\Gamma-v-\lambda\ln p'}{\lambda-\kappa})$$ (4.35)

This surface meets the v-p' plane on the normal compression line $v = N - \lambda \ln p'$ and is tangent to the q'-v plane along the v axis where $q' = p' = 0$. Equation 4.35 implies that:

$$N-\Gamma=\lambda-\kappa$$ (4.36)

The equation of the undrained stress path can be obtained by intersecting the *SBS* with a constant specific volume plane:

$$q' = Mp'\frac{\lambda}{\lambda-\kappa}\ln(\frac{p'_C}{p'})$$ (4.37)

where p'_C is the initial mean effective stress on the normal compression line (isotropic compression or isotropic consolidation pressure) corresponding to the constant v plane. For most normally consolidated clays Equation 4.37 represents the projection of the state boundary surface on the p'-q' plane. This surface (Equation 4.37) is called the Roscoe surface (after Professor K H Roscoe, Cambridge University, UK). The relationship between the mean effective stress at the critical state p'_{CU} and the initial mean effective stress p'_C can be established by combining Equations 4.37 (with replacing p'_{CU} for p') and 4.36:

$$\ln(\frac{p'_C}{p'_{CU}}) = \frac{N-\Gamma}{\lambda}$$ (4.38)

Undrained shear strength c_u. The undrained shear strength at the critical state can be determined from the following equation:

$$c_u = (\sigma_1 - \sigma_3)/2 = (\sigma'_1 - \sigma'_3)/2 = q'/2 = (Mp'_{CU})/2$$ (4.39)

where q' is the deviator stress at the critical state. Substituting p'_{CU} from Equation 4.38 the undrained shear strength can be related to the isotropic compression or isotropic consolidation pressure p'_C according to:

$$c_u = (Mp'_C/2)\exp[(\Gamma-N)/\lambda]$$ (4.40)

As the specific volume is a function of the isotropic compression p'_C, then Equation 4.40 can be written in terms of specific volume by replacing p'_C from Equation 4.29. Hence:

$$c_u = (M/2)\exp[(\Gamma-v)/\lambda]$$ (4.41)

In a fully saturated clay the void ratio $e = wG_s$ (w is the moisture content and G_s is the specific gravity of solids). Therefore $v = 1 + e = 1 + wG_s$, and substituting for v in Equation 4.41 we obtain:

$$c_u = (M/2)\exp[(\Gamma-1-wG_s)/\lambda]$$ (4.42)

The undrained shear strength may be expressed in terms of the one-dimensional consolidation parameters p'_{1C} (one-dimensional consolidation pressure) and $\bar{n}$ by combining Equations 4.32 and 4.41:

$$c_u = (Mp'_{1C}/2)\exp[(\Gamma - \bar{n})/\lambda] \qquad (4.43)$$

Example 4.15

Two triaxial compression tests were carried out on two identical specimens of a particular clay whose diameter was 38 mm and height 76 mm. The first sample was isotropically consolidated under 300 kPa, and then subjected to drained test. The critical state was achieved under a deviator stress of 360 kPa and the volume change was 4.4 ml. After completion of the test, the sample was oven dried and the mass was measured as 145.8 g. The second specimen was isotropically consolidated under 300 kPa, and then subjected to an undrained test. The critical state was achieved under a deviator stress of 152 kPa. Determine the values of M, λ, Γ, N and the ultimate pore pressure generated in the second specimen. $G_s = 2.72$.

Solution:

$p' = (\sigma'_1 + 2\sigma'_3)/3 = (\sigma'_1 - \sigma'_3 + 3\sigma'_3)/3 = \sigma'_3 + q'/3$.
p'(drained test) $= 300.0 + 360.0/3 = 420.0$ kPa, $M = q'/p' = 360.0/420.0 = 0.857$.
Initial volume of the specimen: $V_0 = 7.6 \times (3.8)^2 \times \pi/4 = 86.193$ ml,
Volume at the end of the test $= 86.193 - 4.4 = 81.793$ ml.
Volume of solids $= 145.8/(2.72 \times 1.0) = 53.603$ ml,
Volume of voids $= 81.793 - 53.603 = 28.19$ ml.
Void ratio at critical state $= 28.19/53.603 = 0.526$.
Specific volume at critical state $= v_f = 1 + 0.526 = 1.526$. The point with $v = 1.526$ and $p' = 420$ kPa is on the projection of the CSL onto the p'-v plane.
Initial void ratio $= 0.526 + 4.4/53.603 = 0.608$.
$v_0 = 1 + 0.608 = 1.608$ (at the start of both drained and undrained tests but is constant during the undrained test).
At the end of the undrained test: $p' = q'/M = 152.0/0.857 = 177.4$ kPa.
The point with $v = 1.608$ and $p' = 177.4$ kPa is on the projection of the CSL onto the v-p' plane. Having two points on this line the equation of the CSL in the v, $\ln p'$ coordinate system can be found:
$\lambda = (1.608 - 1.526)/(\ln 420.0 - \ln 177.4) = 0.095$.
Substituting the coordinates of one of these points into Equation 4.34:
$1.608 = \Gamma - 0.095 \ln 177.4$, $\Gamma = 2.1$. Therefore the equation of the CSL in the v-p' plane is:
$v = 2.1 - 0.095 \ln p'$. The initial states of both samples ($p' = 300$ kPa, $v = 1.608$) are on the normal compression line. Using Equation 4.29: $1.608 = N - 0.095 \ln 300.0$, $N = 2.15$. Hence the equation of the NCL is: $v = 2.15 - 0.095 \ln p'$.
Using Equation 4.16: $\sigma'_3 = 177.4 - 152.0/3 = 126.7$ kPa, $u = 300.0 - 126.7 = 173.3$ kPa.

4.5.4 *Triaxial tests on lightly overconsolidated saturated clay*

Projections of the stress paths on the p'-q', and v-p' (or v-ln p') planes for two specimens of lightly overconsolidated clay (point L) are shown in Figure 4.32. The stress paths in the space are located inside the state boundary surface and in the undrained test the stress path

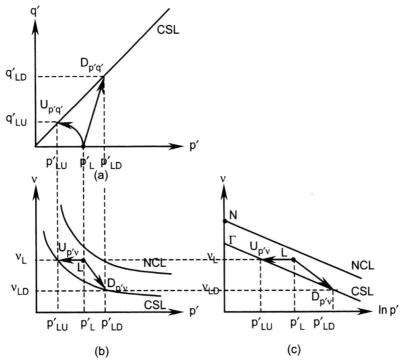

Figure 4.32. Triaxial tests on lightly overconsolidated saturated clay.

is identical to its projection onto the p'-q' plane. If the initial position of point L is on the projection of the *CSL* onto the v-p' plane, then the undrained stress path on the p'-q' plane is a vertical line having a constant effective mean stress equal to p'_{LU}. The preconsolidation pressure for this initial state is on the normal compression line and it can be shown that in the original Cam clay model the ratio of the effective mean stress at this point to the initial effective mean stress at L is equal to e (the base of natural logarithms) = 2.72. This means that the overconsolidation ratio *OCR* at this point is 2.72. Note that at any point on the normal compression line *OCR* = 1.

Example 4.16

A specimen of the soil of Example 4.15 is consolidated under 500 kPa and then unloaded to 200 kPa. The volumetric strain due to expansion is 2.64%. Determine the range of the consolidation pressures to prepare samples of void ratio of 0.62 and the formulation of the unloading.

Solution:

(a) Calculation of κ: from Equation 4.29: $v_{500} = 2.15 - 0.95\ln 500.0 = 1.560$.

$\varepsilon_V = -(v_{200} - v_{500})/v_{500} = -(v_{200} - 1.560)/1.560 = -0.0264 \rightarrow v_{200} = 1.601$.

$\kappa = -(1.601 - 1.560)/(\ln 200.0 - \ln 500.0) = 0.045$. Check if $N - \Gamma = \lambda - \kappa$:

$2.15 - 2.1 = 0.095 - 0.045 = 0.05$, thus the Cam clay model can be applied.

The initial states of the specimens are within p'_A to p'_B, as shown in Figure 4.33.

Applying Equation 4.34: $1.62 = 2.1 - 0.095\ln p'_A \rightarrow p'_A = 156.4$ kPa.

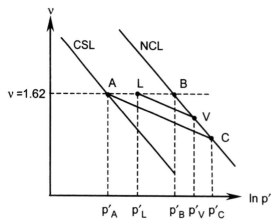

Figure 4.33. Example 4.16.

From Equation 4.29: $1.62 = 2.15 - 0.095 \ln p'_B \rightarrow p'_B = 264.8$ kPa.

To find the mean effective stress at point C (Figure 4.33) we construct two equations based on the definition of κ and λ:

$-\kappa = -0.045 = (1.62 - v_C)/(\ln 156.4 - \ln p'_C)$,

$-\lambda = -0.095 = (1.62 - v_C)/(\ln 264.8 - \ln p'_C)$, solving for v_C and p'_C.

$p'_C = 425.3$ kPa (note that $425.3 / 156.4 = 2.719 =$ base of the natural log), $v_C = 1.575$.

To determine the amount of load after unloading from *NCL*, we draw line LV parallel to AC, where L is the state of a lightly overconsolidated specimen.

$-\kappa = -0.045 = (v_L - v_V)/(\ln p'_L - \ln p'_V)$, replacing v_V from *NCL*:

$-0.045 = [1.62 - (2.15 - 0.095 \ln p'_V)]/(\ln p'_L - \ln p'_V)$, after rearranging and simplifying:

$\ln p'_L = 11.77 - 1.11 \ln p'_V \rightarrow p'_L = 130323/p'^{1.11}_V$,

where: 156.4 kPa $\leq p'_L \leq 264.8$ kPa and 264.8 kPa $\leq p'_V \leq 425.3$ kPa.

4.5.5   Triaxial tests on heavily overconsolidated saturated clay

Projections of the stress paths onto the p'-q', and v-p' (or v-ln p') planes for two specimens of heavily overconsolidated clay (point H) are shown in Figure 4.34. The stress path of the drained test on the p'-q' plane has a slope of 3 vertical to 1 horizontal, and passes beyond the critical state line (point D) approaching point L which represents the peak deviator stress. Ideally, if straining is continued, the state of the soil must descend to point D. However, any additional strain beyond L may produce failure planes, thereby preventing attainment of the critical state. The line SU is a part of the state boundary surface corresponding to the initial specific volume and is called the Hvorslev surface (after Professor M J Hvorslev). This surface is the envelope for the peak strength of drained triaxial results having the same specific volume at peak irrespective of the stress conditions. With this definition a series of tests with different void ratios at failure (peak) will yield parallel envelopes. The ideal path for the undrained test is to approach the intersection of the Roscoe surface and the *CSL* on the Hvorslev surface. However, the development of failure planes may terminate the path at the left of point U and hence the critical state may not be achieved. The equation of the Hvorslev surface in the v, p', q' coordinate system is:

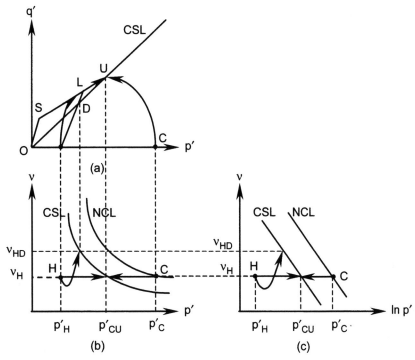

Figure 4.34. Triaxial tests on heavily overconsolidated saturated clay.

$$q' = Hp' + (M - H)\exp[\lambda(\Gamma - v)] \tag{4.44}$$

where H is the slope of the Hvorslev surface on the p'-q' plane. The line OS (with a slope of 3 vertical to 1 horizontal) represents a drained test with zero confining pressure. This line eliminates the tensile stress on the Hvorslev surface (no tension cut off) and thus the complete section of the state boundary surface on the p'-q' plane is $OSUC$.

In order to evaluate the overconsolidation ratio for a constant specific volume, consider Figure 4.35 in which L represents an overconsolidated clay and OL is the expansion line. The two points C and O are on the normal compression line, thus:

$$v_C - v_O = \lambda(\ln p'_O - \ln p'_C) \tag{4.45}$$

On the expansion line the change in specific volume is defined by:

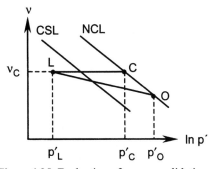

Figure 4.35. Evaluation of overconsolidation ratio.

$$v_L - v_O = v_C - v_O = \kappa(\ln p'_O - \ln p'_L) \tag{4.46}$$

Equating equations 4.45 and 4.46 and considering the definition of the overconsolidation ratio *OCR*, we find:

$$OCR = \frac{p'_O}{p'_L} = (\frac{p'_O}{p'_C})^{\lambda/\kappa} = \exp(\frac{v_C - v_O}{\kappa}) \tag{4.47}$$

4.5.6 *Three-dimensional and normalized state boundary surface*

Figure 4.36(a) shows the three-dimensional state boundary surface in the v, p', q' coordinate system. The section *OSUC* of Figure 4.34(a) can be normalized in terms of the preconsolidation pressure to represent the unique section of the state boundary surface in the $p'/p'_C - q'/p'_C$ plane as shown in Figure 4.36(b). During the progress of the drained test, the stress path intersects different sections with constant v, each having a different p'_C. The state of the stress at each intersection must be normalized against the corresponding preconsolidation pressure, and it can be shown that the resulting path will be the same as the normalized Rosoce surface for the undrained condition. To describe samples that lie between the *NCL* and the *CSL* on the v-p' plane, the term *wet of critical* is used. The maximum difference in the void ratio within this area is $N - \Gamma$ and, clearly, samples in this area have a greater moisture content than samples below the *CSL* that are *dry of critical*. The states within each group have similar stress strain patterns.

Example 4.17

The data obtained from triaxial tests on two specimens of normally consolidated clay are shown in table below (q', u for undrained, q', ε_V for drained) where the last row represents the critical state.

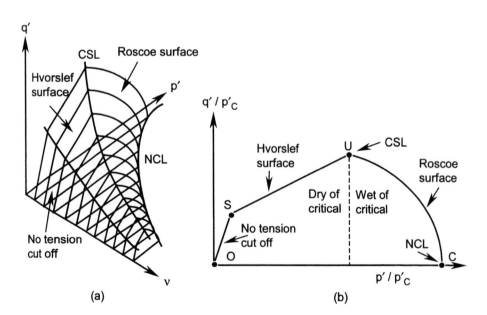

Figure 4.36. (a) Three-dimensional state boundary surface, (b) normalized state boundary surface.

Test points	Undrained: $\sigma_3 = 200$ kPa			Drained: $\sigma_3 = 150$ kPa		
	q' (kPa)	u (kPa)	p' (kPa)	q' (kPa)	$\varepsilon_{V\%}$	p' (kPa)
1	0.0	0.0	200.0	0.0	0.00	150.0
2	27.8	29.3	179.9	15.6	1.10	155.2
3	52.4	57.4	160.0	33.9	2.31	161.3
4	73.2	84.4	140.0	55.1	3.57	168.4
5	82.1	97.3	130.1	80.3	4.99	176.8
6	89.9	110.0	119.9	112.2	6.63	187.4
7	96.4	122.1	110.0	149.5	8.42	199.8
8	101.7	133.9	100.0	200.0	10.53	216.7
9	103.9	140.2	94.4	260.4	12.74	236.8

The moisture contents at the end of the tests are 31% and 24.21% for the undrained and drained tests respectively. (a) Plot the stress paths for both undrained and drained tests in the p'-q' plane, (b) evaluate the normal compression and critical state parameters, (c) plot the normalized stress paths for both samples on p' / p'_C - q' / p'_C plane. $G_s = 2.71$.

Solution:

(a) From Equation 4.16: p' (undrained test) $= 200.0 - u + q' / 3$.
For the drained test: $p' = 150.0 + q' / 3$. The values of p' for both tests are tabulated above and both stress paths are shown in Figure 4.37(a).
(b) In the undrained test and at the critical state $M = q' / p' = 103.9 / 94.4 = 1.1$.
Check M from the drained test results: $M = q' / p' = 260.4 / 236.8 = 1.1$.
Calculate the final void ratios for both tests:
e_9 (undrained) $= w\,G_s = 0.31 \times 2.71 = 0.840$, e_9 (drained) $= 0.2421 \times 2.71 = 0.656$; thus:
v_9 (undrained) $= 1.840$, v_9 (drained) $= 1.656$.
The initial void ratio (drained) can be calculated using the definition of volumetric strain:
$\varepsilon_V = \Delta V / V_0 = -(v_9 - v_1)/v_1 = (v_1 - 1.656)/v_1 = 0.1274 \rightarrow v_1 = 1.898$.
Similarly, the specific volumes at the other test points can be calculated. Using Equation 4.29 for the initial states we construct two equations with the two unknowns λ and N:
$1.840 = N - \lambda \ln 200.0, 1.898 = N - \lambda \ln 150.0$. Solving for λ and N: $\lambda = 0.2, N = 2.9$.
The equation of the *NCL* is: $v = 2.9 - 0.2 \ln p'$.
Both final states are located on the *CSL*, hence:

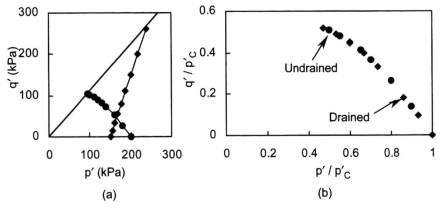

(a)   (b)

Figure 4.37. Example 4.17.

$1.840 = \Gamma - 0.2 \times \ln 94.4 \rightarrow \Gamma = 2.75$.

The equation of the *CSL* on the v-p' plane is: $v = 2.75 - 0.2 \ln p'$.

(c) In the undrained test the p' and q' values are divided by 200 (kPa) to normalise the stress state. In the drained test, the equation of the *NCL* ($v = 2.9 - 0.2 \ln p'$) is used to calculate the preconsolidation pressures on the constant v curves (test points) which intersect the drained stress path while the sample progresses from the initial state towards the critical state. To normalise the stress state, all p' and q' values are divided by the preconsolidation pressure. The results are shown in table below and the normalized stress paths are shown in Figure 4.37(b).

Sample calculation for test point 5 in the undrained and drained conditions:

$p' / p'_C = 130.1 / 200.0 = 0.650$, $q' / p'_C = 82.1 / 200.0 = 0.411$.

$\varepsilon_V = \Delta V / V_0 = -(v_5 - v_1) / v_1 = (1.898 - v_5) / 1.898 = 0.0499 \rightarrow v_5 = 1.803$,

$v = 2.9 - 0.2 \ln p'_C = 1.803 \rightarrow p'_C = 241.1 \, \text{kPa}$.

$p' / p'_C = 176.8 / 241.1 = 0.733$, $q' / p'_C = 80.3 / 241.1 = 0.333$.

Test points	Undrained: $\sigma_3 = 200$ kPa			Drained: $\sigma_3 = 150$ kPa		
	$q' / 200$	$p' / 200$	v	p'_C (kPa)	q' / p'_C	p' / p'_C
1	0.000	1.000	1.898	150.0	0.000	1.000
2	0.139	0.900	1.877	166.5	0.094	0.932
3	0.262	0.800	1.854	186.8	0.181	0.863
4	0.366	0.700	1.830	210.6	0.262	0.800
5	0.411	0.650	1.803	241.1	0.333	0.733
6	0.449	0.599	1.772	281.4	0.399	0.666
7	0.482	0.550	1.738	333.6	0.448	0.599
8	0.508	0.500	1.698	407.4	0.491	0.532
9	0.520	0.472	1.656	502.7	0.518	0.471

4.6 PROBLEMS

4.1 At a point 15 m below the ground surface, the relationship between the effective vertical stress σ'_z and the effective lateral stress σ'_x is: $\sigma'_x = \sigma'_z (1 - \sin \phi')$.
If the water table is 2 m below the ground surface, calculate the normal and shear stresses on the two perpendicular planes P and Q where the angle α for the P plane is $45° + \phi' / 2$. $c' = 0$, $\phi' = 40°$, $\rho_{dry} = 1.7$ Mg/m^3 and $\rho_{sat} = 1.95$ Mg/m^3.

Answers: $\sigma_P = 72.9$ kPa, $\tau_P = 38.0$ kPa, $\sigma_Q = 136.8$ kPa, $\tau_Q = -38.0$ kPa

4.2 At a point within a soil mass, the effective lateral and shear stresses are 100 kPa and 50 kPa respectively. Calculate the effective vertical stress to cause the failure of the point. $c' = 0$, $\phi' = 30°$.

Answer: $\sigma'_z = 233.3$ kPa

4.3 An element of soil in xz plane is subjected to σ'_z, σ'_x, and τ_{xz}. Assuming that: $\tau_{xz} = 0.306 \, \sigma'_x$, calculate the ratio σ'_x / σ'_z at failure. $c' = 0$, $\phi' = 36°$.

Answer: 0.271

4.4 The results of a direct shear test on a specimen of dry sand are as follows: normal stress = 96.6 kPa, shear stress at failure = 67.7 kPa. By means of a Mohr's circle of stresses, find the magnitude and directions of the principal stresses acting on a soil element within the zone of failure.

Answers: $\sigma_1 = 226.7$ kPa, $\sigma_3 = 61.4$ kPa, $\alpha = 62.5°$ (from failure plane)

4.5 Data obtained from three drained triaxial tests (on identical samples) are as follows. Determine c' and ϕ'.

Test no.	σ_3 (kPa)	$\sigma_1 - \sigma_3$ (kPa) at peak
1	50	191
2	100	226
3	150	261

Answers: $c' = 60$ kPa, $\phi' = 15°$

4.6 The results of three consolidated-undrained triaxial tests on identical specimens of a particular soil are:

Test no.	σ_3 (kPa)	$\sigma_1 - \sigma_3$ (kPa) at peak	u (kPa) at peak
1	200	244	55
2	300	314	107
3	400	384	159

Determine c' and ϕ'. What would be the expected pore pressure at failure for $\sigma_3 = 100$ kPa?

Answers: $c' = 10$ kPa, $\phi' = 25°$, $u = 2.6$ kPa

4.7 The results of drained and consolidated-undrained triaxial tests on two samples of a normally consolidated clay are shown below:

Type of the test	σ_3 (kPa)	$\sigma_1 - \sigma_3$ (kPa) at peak
Drained	300	650
Consolidated-undrained	200	250

Determine: (a) ϕ' from the drained test, (b) ϕ from the consolidated-undrained test, (c) the pore pressure in the consolidated-undrained test at failure.

Answers: $31.3°$, $22.6°$, 84.6 kPa

4.8 A soil has the following properties: $n = 0.38$, $E_s = 10$ MPa, $\mu = 0.3$. The bulk modulus of the pore water is 2200 MPa. Estimate the pore pressure coefficient B.

Answer: 0.9986

4.9 An unconfined compression test has given a *UCS* value of 126.6 kPa. The effective shear strength parameters are: $c' = 25$ kPa, $\phi' = 30°$. Assuming the pore pressure parameter $A = -0.09$, calculate the initial pore pressure in the sample.

Answer: -8.6 kPa

4.10 The values of the critical state parameters for a particular type of clay are: $N = 2.1$, $\lambda = 0.087$, $\Gamma = 2.05$, $M = 0.95$. Two samples of this soil are consolidated under a confining pressure of 300 kPa. One sample has been subjected to a drained triaxial test whilst the second sample has been sheared in an undrained condition. Determine: (a) the deviator stress at the critical state for both the drained and undrained tests, (b) the pore pressure in the undrained test at the critical state, (c) the volumetric strain in the drained test when the sample approaches the critical state.

Answers: 417.0 kPa, 160.4 kPa, 184.6 kPa, 5.19%

4.11 In a drained triaxial test carried out on a sample of the clay of Problem 4.10, the sample was first consolidated under a confining pressure of 400 kPa. It was then unloaded to 300 kPa and, after equilibrium was reached, it was sheared in a drained condition. If the κ value is 0.037, calculate the volumetric strain at failure.

Answer: 4.32%

4.12 The critical state parameters of a soil are: $M = 0.857$, $\lambda = 0.095$, $N = 2.1$, $\Gamma = 2.05$, $\kappa = 0.045$. Specimens of this soil have been consolidated and unloaded to obtain an initial void ratio of 0.62. (a) If the specimens are subjected to an undrained triaxial test, find the minimum overconsolidation ratio $(OCR = m)$ above which the pore pressure at the critical state becomes negative, (b) calculate the volumetric strains for three specimens of $OCR = 1$, $OCR = m$ (as defined above) and $OCR = 8$ that are subjected to drained tests.

Answers: 5.15, 5.1% (compression), 0%, 1.36% (expansion)

4.7 REFERENCES

Arthur, J. R. F. & Aysen, A. 1977. Ruptured sand sheared in plane strain. *Proc. intern. conf. SMFE* 1: 19-22. Tokyo.

Arthur, J. R. F., Dalili, A. & Dunstan, T. 1988. Discussion on the engineering application of direct and simple shear testing. *Geotechnique* 27(1): 53-74.

Arthur, J. R. F., Dunstan, T., Al-Ani, Q. & Aysen, A. 1977. Plastic deformation in granular media. *Geotechnique* 38(1): 140-144.

Atkinson, J. H. & Bransby, P. L. 1978. *The mechanics of soils*. UK: McGraw-Hill.

Britto, A. & Gunn, M.J. 1987. *Critical state soil mechanics via finite elements*. Chichester, UK: Ellis Horwood.

Carter, J.P., Booker, J.I. & Wroth, C.P. 1982. A critical state soil model for cyclic loadings. In G.N. Pande & O.C. Zienkiewicz (eds), *Soil mechanics-transient and cyclic loads*: 219-252. New York: John Wiley & Sons.

Jewell, R. A. 1989. Direct shear tests on sands. *Geotechnique* 39(2): 309-322.

Powrie, W. 1997. *Soil mechanics-concepts and applications*. London: E & FN Spon.

Roscoe, K. H. 1953. An apparatus for the application of simple shear to soil samples. *Proc. 3rd intern. conf. SMFE* 1: 186-191. Switzerland.

Roscoe, K.H. & Burland, J.B. 1968. On the generalized stress-strain behaviour of wet clay. In J. Heyman & F.A. Leckie (eds), *Engineering plasticity*: 535-639. Cambridge, UK: Cambridge University Press.

Roscoe, K.H. & Schofield, A.N. 1963. Mechanical behaviour of an idealized wet clay. *Proc. European conf. on soil mechanics and foundation engineering* 1: 47-54. Wiezbaden.

Terzaghi, K. & Peck, R. B. 1948. *Soil mechanics in engineering practice*. New York: John Wiley & Sons.

Wong, R. K. S. & Arthur, J. R. F. 1985. Induced and inherent anisotropy in sands. *Geotechnique* 35(4): 471-481.

CHAPTER 5

Stress Distribution and Settlement in Soils

5.1 INTRODUCTION

This chapter describes the stress distribution and the calculation of settlement within an idealized elastic soil mass due to applied external and internal loading. The elastic properties include the *Modulus of Elasticity E_s* and *Poisson's ratio* μ. For soils where the compressibility characteristics are non-uniform and depend upon the state of the stress (e.g. sands), the concept of a compressibility index m_v is introduced. The coefficient of (volume) compressibility is defined as the ratio of the strain increment to the stress increment in a specified direction caused by external or internal loading. The soil is assumed to be an ideal semi-infinite homogenous elastic material obeying Hooke's law, and the elastic stress-strain model shown in Figure 4.4(a) is employed to determine the stress distribution within the soil. Furthermore, the soil is assumed to be weightless which means the stress field resulting from the solution of the equilibrium and compatibility conditions satisfies only the boundary conditions. Nevertheless, with a known stress field the strains, and consequently the deformation of the soil layer and surface settlement, can be evaluated.

When the compressibility index concept is used, a mean stress increment within a zone of influence is estimated by in-situ testing. Surface settlement can be accurately predicted even though the stress distribution in the soil remains undefined.

In an ideal semi-infinite elastic soil the stress field under a specified loading is independent of the Modulus of Elasticity, but does depends on Poisson's ratio. However, in most of the loading types described in this chapter, the vertical stress (parallel to the load) is also independent of Poisson's ratio. The magnitude of the Modulus of Elasticity affects the prediction of elastic deformations and surface settlements. Moreover, the reliability of the predicted stresses depends upon the stress-strain model, and the accuracy of the elastic parameters. Unfortunately, the elastic parameters determined in the laboratory are not representative of field conditions due to differences in the density, moisture content and stress history.

Stresses and settlements within a soil mass are caused by both external and internal loading. External loading includes vertical loads applied on the ground surface or near the ground surface. Internal loading is applied inside the soil mass away from the ground surface (e.g. piles) and may include a vertical concentrated force or a distributed shear stress, or a combination of both. In the case of internal loading, only axisymmetric conditions are considered in this chapter. Surface loads are transmitted through a footing, and the stress distribution will depend on the interaction model selected for the contact area between the soil and the footing. In general, the contact pressure between a footing and the soil is

151

somewhat indeterminate and depends on the rigidity of the footing and the type of soil. If the thickness of the footing satisfies a rigidity criterion, then the contact pressure can be assumed uniform or linear, depending on the location of the resultant of the applied forces. If, for a given plan size of footing, the thickness does not satisfy the rigidity criterion, then the solution for the contact pressure is obtained by assuming the footing acts as an elastic beam resting on an elastic foundation. In this chapter, the resulting stress field within a soil is determined assuming a certain stress distribution at the boundary without taking account of the interaction model. Elastic settlements are predicted for two extreme conditions of footing stiffness viz., flexible and perfectly rigid.

5.2 FUNDAMENTAL EQUATIONS OF ELASTICITY

5.2.1 *Stresses and equilibrium*

In two-dimensional space, the state of stress is represented by the three stress components, σ_x, σ_z, and τ_{xz}, as shown in Figure 5.1(a). Partial derivatives define the change in stress in the specified direction. For convenience, the positive sense of the z-axis is selected in the direction of gravity. Equating the sum of the forces in the x and z directions to zero, two corresponding differential equations of equilibrium for the element are obtained:

$$\frac{\partial \sigma_x}{\partial x} + \frac{\partial \tau_{zx}}{\partial z} = 0, \quad \frac{\partial \sigma_z}{\partial z} + \frac{\partial \tau_{xz}}{\partial x} = \gamma \qquad (5.1)$$

where γ is the unit weight of the soil. A two-dimensional stress field describing the three unknown components of stress at every point in the soil cannot be created by considering only the two equations of equilibrium. That is, whilst these equations are necessary, they are not sufficient, and a third condition is needed to relate the state of stress to the elastic

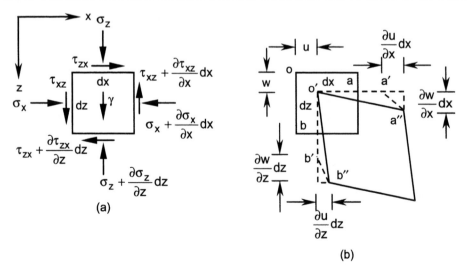

Figure 5.1. (a) Two-dimensional stresses on an element, (b) two-dimensional displacements and strains.

behaviour of the soil. Compatibility of deformations or strains to keep the body intact creates the third condition and, in an elastic analysis, this condition is based on Hooke's law that defines the proportional relationship between stresses and strains. Furthermore, the stress field must satisfy the stress boundary conditions within the equilibrium equations. In three-dimensional space there are six components of stress that satisfy the three differential equations of equilibrium:

$$\frac{\partial \sigma_x}{\partial x} + \frac{\partial \tau_{yx}}{\partial y} + \frac{\partial \tau_{zx}}{\partial z} = 0, \quad \frac{\partial \sigma_y}{\partial y} + \frac{\partial \tau_{zy}}{\partial z} + \frac{\partial \tau_{xy}}{\partial x} = 0,$$

$$\frac{\partial \sigma_z}{\partial z} + \frac{\partial \tau_{xz}}{\partial x} + \frac{\partial \tau_{yz}}{\partial y} = \gamma \tag{5.2}$$

It can be seen that at least three independent conditions of compatibility of deformations or strains are needed to define a three-dimensional stress field.

5.2.2 Strains and compatibility

Figure 5.1(b) shows the deformation of a two-dimensional element in x, z coordinate system. Two parameters u (in the x direction) and w (in the z direction) represent the displacement of point o to a new position o'. The two sides of the element (oa and ob) undergo axial deformation as well as rotation. These deformations, and the total change in the original 90° angle, are described by axial strains in the x and z directions (ε_x and ε_z) and the shear strain (γ_{zx}):

$$\varepsilon_x = \frac{\partial u}{\partial x}, \quad \varepsilon_z = \frac{\partial w}{\partial z}, \quad \gamma_{zx} = \frac{\partial w}{\partial x} + \frac{\partial u}{\partial z} \tag{5.3}$$

The compatibility condition in two-dimensional space is expressed by:

$$\frac{\partial^2 \varepsilon_x}{\partial z^2} + \frac{\partial^2 \varepsilon_z}{\partial x^2} - \frac{\partial^2 \gamma_{zx}}{\partial x \partial z} = 0 \tag{5.4}$$

In three-dimensional space the following relationships between strains and displacements are additional to Equation 5.3:

$$\varepsilon_y = \frac{\partial v}{\partial y}, \quad \gamma_{xy} = \frac{\partial v}{\partial x} + \frac{\partial u}{\partial y}, \quad \gamma_{yz} = \frac{\partial w}{\partial y} + \frac{\partial v}{\partial z} \tag{5.5}$$

where v is the displacement in the y direction. In an ideal elastic material Hooke's law relates the axial strains to the normal stresses by the following linear equations:

$$\varepsilon_x = \frac{1}{E_s}(\sigma_x - \mu\sigma_y - \mu\sigma_z), \quad \varepsilon_y = \frac{1}{E_s}(\sigma_y - \mu\sigma_z - \mu\sigma_x),$$

$$\varepsilon_z = \frac{1}{E_s}(\sigma_z - \mu\sigma_x - \mu\sigma_y) \tag{5.6}$$

where E_s is the Modulus of Elasticity and μ is Poisson's ratio. The values of E_s and μ depend on the drainage conditions (undrained or drained) and are estimated from experimental stress-strain relationships obtained in appropriate laboratory triaxial tests, or from in-situ loading tests. It may be shown that the shear strains are proportional to shear stresses and are given by:

$$\gamma_{xy} = \frac{\tau_{xy}}{G}, \gamma_{yz} = \frac{\tau_{yz}}{G}, \gamma_{zx} = \frac{\tau_{zx}}{G} \rightarrow G = \frac{E_s}{2(1+\mu)} \qquad (5.7)$$

where G is the shear modulus. The volumetric strain is approximated by the sum of the axial strains:

$$\varepsilon_V = \frac{\Delta V}{V} = \varepsilon_x + \varepsilon_y + \varepsilon_z = \frac{1-2\mu}{E_s}(\sigma_x + \sigma_y + \sigma_z) \qquad (5.8)$$

In this equation the stress terms may be replaced by the incremental stresses as indicated in Equation 4.21. In undrained loading no volume change occurs and therefore:

$$\varepsilon_V = \frac{\Delta V}{V} = 0, \frac{1-2\mu}{E_s}(\sigma_x + \sigma_y + \sigma_z) = 0, \mu = 0.5 \qquad (5.9)$$

Substituting Equation 5.9 into 5.7 we obtain:

$$E_s = 3G \text{ or } E_{su} = 3G \qquad (5.10)$$

where E_{su} is the Modulus of Elasticity in undrained conditions. Since the shear stresses are resisted solely by the soil particles and their magnitudes are independent of the drainage conditions (or pore pressures), it is reasonable to accept that, in an isotropic elastic soil, the shear modulus for drained and undrained conditions are alike.

Example 5.1

A 200 mm cube of sand is subjected to plane strain ($\varepsilon_y = 0$) loading with $\sigma_x = 150$ kPa (compression), $\sigma_z = 800$ kPa (compression) and $\tau_{xz} = 0$. The applied stresses have caused 6.4 mm compression and 0.6 mm expansion at the z and x directions respectively. Determine: (a) E_s and μ for the sand, (b) the normal stress in the x direction to recover the 0.6 mm expansion and overall compression in the z direction.

Solution:

(a) Substituting $\varepsilon_y = 0$ in Equation 5.6:

$\sigma_y = \mu(\sigma_z + \sigma_x) = \mu(800.0 + 150.0) = 950.0\mu$.

Replacing σ_y in ε_x and ε_z (Equation 5.6), we obtain two equations with the two unknowns E_s and μ:

$800.0 - \mu(150.0 + 950\mu) = (6.4/200.0)E_s = 0.032E_s$,

$150.0 - \mu(800.0 + 950\mu) = -(0.6/200.0)E_s = -0.003E_s$. Solving for E_s and μ:

$E_s = 22592$ kPa $= 22.6$ MPa, $\mu = 0.2165$. $\sigma_y = 950.0\mu = 205.7$ kPa.

(b) Plane strain conditions apply in both x and y directions ($\varepsilon_x = \varepsilon_y = 0, \sigma_x = \sigma_y$):

$\varepsilon_x = (\sigma_x - \mu\sigma_y - 800.0\mu)/E_s = 0.0$,

$\sigma_x - 0.2165\sigma_x - 800.0 \times 0.2165 = 0.0 \rightarrow \sigma_x = \sigma_y = 221.0$ kPa.

$\varepsilon_z = [800.0 - 0.2165 \times (221.0 + 221.0)]/22592 = 0.03117$,

$\Delta z = 0.03117 \times 200$ mm $= 6.23$ mm.

5.3 STRESS DISTRIBUTION DUE TO EXTERNAL AND INTERNAL LOADING

5.3.1 *Types of loading*

The state of stress within an elastic soil can be studied in two- and three-dimensional systems. Stresses generated under the following loading conditions are examined in three-dimensional space:

1. A concentrated vertical load applied at the ground surface.
2. Loaded areas using concentrated loads and the concept of superposition.
3. A circular loaded area and the construction of influence charts.
4. A rectangular loaded area.
5. A concentrated line load of finite length.

Two-dimensional states of stress are examined for the following loading conditions:

6. A concentrated line load of infinite length.
7. A uniformly or linearly loaded infinitely long strip.

In all cases the soil is assumed to be a semi-infinite elastic, homogeneous, isotropic and weightless material (the weight does not affect the pattern of stress distribution).

5.3.2 *Concentrated vertical load*

More than a century ago Boussinesq gave solutions for the stress distribution within an idealized elastic material under a concentrated vertical load applied to a horizontal ground surface. Body forces were not included in the analysis, and the stress field described only the effect of the external concentrated load. The stress field satisfied the three conditions of equilibrium, compatibility and the stress boundary conditions. Although independent of the Modulus of Elasticity it does depend on the magnitude of Poisson's ratio. Boussinesq's stress components may be presented in variety of coordinate systems (Jumikis, 1969; Poulos & Davis, 1974).

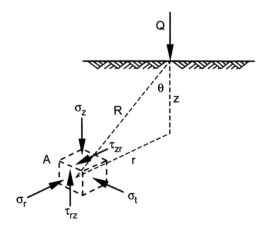

Figure 5.2. Stress components at a point due to a concentrated load.

However only the cylindrical coordinate system (z, r) and the combination of cylindrical and polar coordinate systems (θ, z) have been considered here. Equations 5.11-5.14 define the normal and shear stresses in the cylindrical coordinate system shown in Figure 5.2.

$$\sigma_z = \frac{Q}{2\pi} \frac{3z^3}{(r^2+z^2)^{5/2}} \tag{5.11}$$

$$\sigma_r = \frac{Q}{2\pi} [\frac{3r^2 z}{(r^2+z^2)^{5/2}} - \frac{1-2\mu}{r^2+z^2+z(r^2+z^2)^{1/2}}] \tag{5.12}$$

$$\sigma_t = -\frac{Q}{2\pi}(1-2\mu)[\frac{z}{(r^2+z^2)^{3/2}} - \frac{1}{r^2+z^2+z(r^2+z^2)^{1/2}}] \tag{5.13}$$

$$\tau_{rz} = \frac{Q}{2\pi} \frac{3r\,z^2}{(r^2+z^2)^{5/2}} \tag{5.14}$$

Equations 5.15-5.18 represent the same stresses in terms of the angle θ and depth z.

$$\sigma_z = \frac{Q}{2\pi z^2}(3\cos^5\theta) \tag{5.15}$$

$$\sigma_r = \frac{Q}{2\pi z^2}[3\sin^2\theta\cos^3\theta - \frac{(1-2\mu)\cos^2\theta}{1+\cos\theta}] \tag{5.16}$$

$$\sigma_t = -\frac{Q}{2\pi z^2}(1-2\mu)(\cos^3\theta - \frac{\cos^2\theta}{1+\cos\theta}) \tag{5.17}$$

$$\tau_{rz} = \frac{Q}{2\pi z^2}(3\sin\theta\cos^4\theta) \tag{5.18}$$

A common representation of the vertical stress component in terms of the dimensionless parameter r/z and an influence factor I_q is:

$$\sigma_z = \frac{Q}{z^2}\frac{3}{2\pi[(r/z)^2+1]^{5/2}} = \frac{Q}{z^2}I_q, \quad I_q = \frac{3}{2\pi[(r/z)^2+1]^{5/2}} \tag{5.19}$$

General patterns of the distribution of vertical stress component at sections for $r = 0$, constant z, and constant r are shown in Figures 5.3(a), 5.3(b) and 5.3(c) respectively. When r is constant, it is a requirement of the boundary conditions that the vertical stress component is zero on the ground surface, increases to a maximum at a specified depth and decreases as z increases. It may be shown that the r/z ratio is a constant for the maximum stress point regardless of the distance from the applied concentrated load.

Equations 5.16 and 5.17 represent the horizontal stresses in the radial and tangential directions. Depending on the limiting value of the angle θ these stresses are either compressive or tensile. For the horizontal radial stress σ_r, the limiting value depends on the magnitude of Poisson's ratio. For example, for $\mu = 0.2$ it may be shown that the limiting value of θ is 79.9°. For angles less than 79.9°, σ_r is compressive. When $\theta = 79.9°$, σ_r is zero whilst for $\theta > 79.9°$, σ_r becomes tensile. Consequently, for $\mu = 0.2$ the horizontal radial stress outside a vertical cone with a central angle of $2 \times 79.9° = 159.8°$ is tensile.

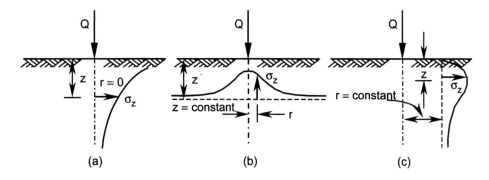

Figure 5.3. Stress distribution under a concentrated load.

As the value of μ increases the corresponding central angle of the vertical cone increases until, at $\mu = 0.5$, the central angle becomes $180°$. Similarly, it may be shown that the horizontal tangential stress σ_t for all values of $\mu < 0.5$ is tensile inside a vertical cone with a central angle of $2 \times 51.83° \approx 103.7°$ and compressive outside this cone.

The majority of the deformation occurs in the z direction. Early attempts to compare the calculated and measured vertical stresses were described in the technical reports of the Corps of Engineers in the USA (Wu, 1966). These studies show that the Boussinesq's stress distribution may still be applied with reasonable accuracy when estimating the stress conditions in a soil near the loading area.

Example 5.2

Under a concentrated vertical load of 100 kN, determine:
(a) the distribution of σ_z on horizontal planes at depths of 1 m, 2 m, and 3 m below the ground surface – vary r/z from 0.0 to 3.0 in 0.25 increments,
(b) the distribution of σ_z on the vertical planes at 1 m, 2 m, and 3 m from the applied concentrated load and the position of the maximum vertical stress on these planes – vary z in 0.5 m increments.

Solution:

Using Equation 5.19 the vertical stress component for both cases are calculated and tabulated in the following tables and are shown in Figure 5.4. In order to locate the depth corresponding to the maximum vertical stress component, we set the derivative of the vertical stress (Equation 5.11) to zero:

r/z	σ_z (kPa)			r/z	σ_z (kPa)		
	$z = 1.0$ m	$z = 2.0$ m	$z = 3.0$ m		$z = 1.0$ m	$z = 2.0$ m	$z = 3.0$ m
0.00	47.75	11.94	5.30	1.75	1.43	0.36	0.16
0.25	41.03	10.26	4.56	2.00	0.85	0.21	0.09
0.50	27.33	6.83	3.04	2.25	0.53	0.13	0.06
0.75	15.64	3.91	1.74	2.50	0.34	0.08	0.04
1.00	8.44	2.11	0.94	2.75	0.22	0.05	0.02
1.25	4.54	1.13	0.50	3.00	0.15	0.04	0.01
1.50	2.50	0.63	0.28				

z (m)	r = 1.0 m		r = 2.0 m		r = 3.0 m	
	r / z	σ_z (kPa)	r / z	σ_z (kPa)	r / z	σ_z (kPa)
0.0	-	0.00	-	0.00	-	0.00
0.5	2.000	3.42	4.000	0.16	6.000	0.02
1.0	1.000	8.44	2.000	0.85	3.000	0.15
1.5	0.667	8.46	1.333	1.65	2.000	0.38
2.0	0.500	6.83	1.000	2.11	1.500	0.63
2.5	0.400	5.27	0.800	2.22	1.200	0.82
3.0	0.333	4.08	0.667	2.11	1.000	0.94
3.5	0.286	3.20	0.571	1.92	0.857	0.98
4.0	0.250	2.56	0.500	1.71	0.750	0.97
4.5	0.222	2.09	0.444	1.50	0.667	0.94
5.0	0.200	1.73	0.400	1.32	0.600	0.88
5.5	0.182	1.45	0.364	1.16	0.545	0.82
6.0	0.167	1.24	0.333	1.02	0.500	0.76

$$\frac{\partial \sigma_z}{\partial z} = \frac{3Q}{2\pi} \frac{\partial}{\partial z} [\frac{z^3}{(r^2 + z^2)^{5/2}}] = 0.$$

$$3r^2 - 2z^2 = 0 \rightarrow r/z = \sqrt{2/3} \rightarrow \theta = \tan^{-1} \sqrt{2/3} = 39.23°.$$

$z = 1.2247r$, from Equation 5.19:

$$I_q = 0.47746/[(\sqrt{2/3})^2 + 1]^{5/2} = 0.13314.$$

For $r = 1.0$ m, $z = 1.225$ m $\sigma_z = [100.0/(1.225)^2] \times 0.13314 = 8.9$ kPa.

For $r = 2.0$ m, $z = 2.449$ m $\sigma_z = [100.0/(2.449)^2] \times 0.13314 = 2.2$ kPa.

For $r = 3.0$ m, $z = 3.674$ m $\sigma_z = [100.0/(3.674)^2] \times 0.13314 = 1.0$ kPa.

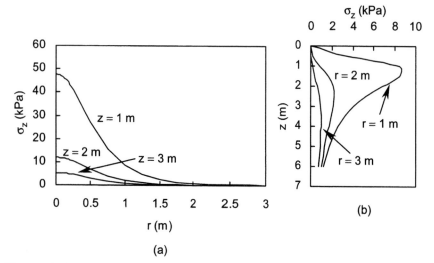

Figure 5.4. Example 5.2.

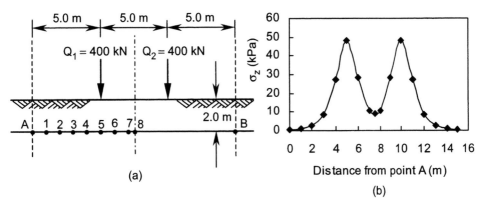

Figure 5.5. Example 5.3.

Example 5.3

Referring to Figure 5.5(a), calculate the distribution of the vertical stress component from point A to point B at a depth of 2 m. Specify the vertical stresses at 1 m intervals.

Solution:

For multiple concentrated loads, superposition is used to evaluate the stresses at any point within the soil. Due to symmetrical loading, we will consider only the left-hand half. The results are presented in the table below and illustrated in Figure 5.5(b).

Points	Q_1			Q_2			Total σ_z (kPa)
	r (m)	r/z	σ_z (kPa)	r (m)	r/z	σ_z (kPa)	
A	5.0	2.50	0.34	10.0	5.00	-	0.3
1	4.0	2.00	0.85	9.0	4.50	-	0.8
2	3.0	1.50	2.51	8.0	4.00	-	2.5
3	2.0	1.00	8.44	7.0	3.50	-	8.4
4	1.0	0.50	27.33	6.0	3.00	0.15	27.5
5	0.0	0.00	47.75	5.0	2.50	0.34	48.1
6	1.0	0.50	27.33	4.0	2.00	0.85	28.2
7	2.0	1.0	8.44	3.0	1.50	2.51	10.9
8	2.5	1.25	4.54	2.5	1.25	4.54	9.1

Example 5.4

Referring to Figure 5.6, calculate the normal and shear stresses at point A with $R = 3$ m on a plane that passes through the point of application of the load and makes an angle of $55°$ below the ground surface. Determine the magnitudes and directions of the principal stresses at this point. Repeat the calculation for another point on this plane with $R = 1.5$ m. Assume $\mu = 0.3$.

Solution:

$z = R\cos\theta = 3.0\cos(90.0° - 55.0°) = 2.457$ m. Using Equation 5.15:

$\sigma_z = (400.0/2\pi \times 2.457^2)(3\cos^5 35.0°) = 11.7$ kPa.

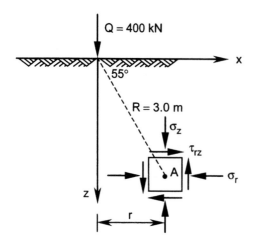

Figure 5.6. Example 5.4.

Substituting $\mu = 0.3$, $z = 2.457$ m, and $\theta = 35°$ into Equations 5.16 and 5.18 we obtain:
$\sigma_r = 4.2$ kPa and $\tau_{rz} = 8.2$ kPa. From Equations 4.1 and 4.2:
$\sigma = (11.7+4.2)/2+(11.7-4.2)/2 \times \cos(-2\times55.0°)+8.2\sin(-2\times55.0°) = -1.0$ kPa.
$\tau = (11.7-4.2)/2\times\sin(-2\times55.0)-8.2\cos(-2\times55.0) = -0.7$ kPa.
$\theta = 1/2 \times \tan^{-1}[2\times8.2/(11.7-4.2)] = 32.7°$.
Using Equations 4.3, 4.4 and 4.5:
$s = (\sigma_1+\sigma_3)/2 = (\sigma_z+\sigma_r)/2 = (11.7+4.2)/2 = 7.9$ kPa,

$t = \sqrt{[(11.7-4.2)/2]^2+(8.2)^2} = 9.0$ kPa,

$\sigma_3 = s-t = 7.9-9.0 = -1.1$ kPa,

$\sigma_1 = s+t = 7.9+9.0 = 16.9$ kPa.

At $R = 1.5$ m:

$z = R\cos\theta = 1.5\cos35.0° = 1.229$ m, $\sigma_z = 46.6$ kPa, $\sigma_r = 16.6$ kPa, $\tau_{rz} = 32.6$ kPa.

It can be seen that the stress components have increased by the ratio of $(3.0 / 1.5)^2 = 4$, therefore the direction of the principal stresses remains constant:
$s = 31.6$ kPa, $t = 35.9$ kPa.

$\sigma_3 = 31.6-35.9 = -4.3$ kPa, $\sigma_1 = 31.6+35.9 = 67.5$ kPa.

5.3.3 *Loaded areas using superposition of concentrated loads*

Stresses at a specified depth beneath a loaded area is found by application of that developed previously for a concentrated load. Mathematically this is possible by integration of the stress equations within the loaded area. Alternatively, the loaded area may be sub-divided into many small areas (Figure 5.7) and the distributed load q within each small area replaced by a concentrated load acting at the centroid of the small area. Superposition of the effects of the finite number of concentrated loads at the point of interest yields the desired stress components.

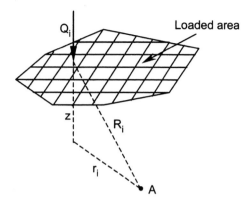

Figure 5.7. Superposition of concentrated loads.

Example 5.5

The footing shown in Figure 5.8(a) is subjected to a uniform load of 300 kPa. Calculate the vertical stress component at 2 m below point A, using the concept of superposition of concentrated loads.

Solution:

Since the footing is symmetrical, the effect of one-half can be calculated and then doubled. The right-hand half is divided into 10 squares of 1 m by 1 m each with 300 kN concentrated load (Figure 5.8(b)). Distances (r) between the centres of each square and point A are calculated to establish r/z ratios. The results are shown in the table below.

Point	r (m)	r/z	I_q	σ_z (kPa)	Point	r (m)	r/z	I_q	σ_z (kPa)
1	3.535	1.7677	0.0138	1.04	6	1.581	0.7906	0.1418	10.64
2	3.808	1.9039	0.0104	0.78	7	2.121	1.0607	0.0725	5.44
3	2.549	1.2747	0.0428	3.21	8	2.915	1.4577	0.0276	2.07
4	2.915	1.4577	0.0276	2.07	9	0.707	0.3535	0.3557	26.68
5	3.535	1.7677	0.0138	1.04	10	1.581	0.7905	0.1418	10.64

Total σ_z due to right-hand half = 63.6 kPa, total σ_z = 63.6 ×2 = 127.2 kPa.

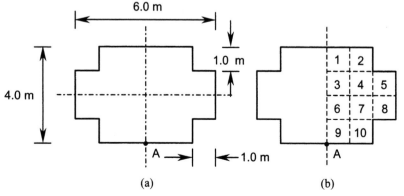

(a)   (b)

Figure 5.8. Example 5.5.

5.3.4   Stresses under a uniformly loaded circular area: influence charts

Consider a circular area on the horizontal ground surface with a uniform vertical loading of q (Figure 5.9(a)). Vertical stress at a specified depth z may be calculated mathematically by integration of Equations 5.11 or 5.15 within the circular loaded area, and the difficulty in this process depends upon the plan position of the point of interest. The complete solution for stresses, strains and deflections (or settlements) of the ground surface is given by Ahlvin & Ulery (1962). If the point is located on a vertical axis passing through the centre of the circle, then the integration process for the vertical stress component is straightforward and yields:

$$\sigma_z = \frac{3q}{2\pi z^2} \int_0^{2\pi} d\alpha \int_0^{R_C} \frac{r\,dr}{[(r/z)^2 + 1]^{5/2}} \tag{5.20}$$

It is assumed that an elemental area dA defined by $r \times dr \times d\alpha$ (Figure 5.9(a)) is subjected to a concentrated load of $q \times dA$. A similar integration may be carried out for the stress component σ_r. The results of the both integrations are presented in Equations 5.21 and 5.22. Due to symmetry, the shear stresses on this axis are zero.

$$\sigma_z = q \left\{ 1 - \frac{1}{[(R_C/z)^2 + 1]^{3/2}} \right\} \tag{5.21}$$

$$\sigma_r = \sigma_t = \frac{q}{2} \left\{ 1 + 2\mu - \frac{2(1+\mu)}{[(R_C/z)^2 + 1]^{1/2}} + \frac{1}{[(R_C/z)^2 + 1]^{3/2}} \right\} \tag{5.22}$$

where R_C is the radius of the circular loaded area. Figure 5.9(b) shows an element of area $abcd$ defined by R_1 and R_2 and a central angle of $2\pi / n$. The integer n is selected as 4, 8, 12, 16, 20 etc to ensure symmetry about the centre. It is assumed that this element is within a circular area that has a very long radius approaching infinity. The vertical stress component calculated from Equation 5.21 due to the loaded circle approaches the applied uniform vertical load q. The vertical stress component produced by area $abcd$ at depth z on the central axis, is obtained by an integration similar to that shown in Equation 5.21:

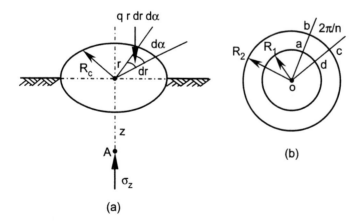

(a)

(b)

Figure 5.9. (a) Vertical stress under uniformly loaded circular area, (b) the basics of influence chart.

$$\sigma_z(abcd) = \frac{q}{n} \{ \frac{1}{[(R_1/z)^2 + 1]^{3/2}} - \frac{1}{[(R_2/z)^2 + 1]^{3/2}} \}$$ (5.23)

Let us assume the area between two circles of radii R_1 and R_2 produces a fraction m of the contact pressure q (at depth z). The vertical stress due to the element *abcd* is therefore:

$$\sigma_z(abcd) = \frac{mq}{n}$$ (5.24)

Equating Equations 5.23 and 5.24 we obtain:

$$\frac{1}{[(R_1/z)^2 + 1]^{3/2}} - \frac{1}{[(R_2/z)^2 + 1]^{3/2}} = m$$ (5.25)

Consider the case where $m = 0.1$ and $R_1 = 0.0$. From Equation 5.25 R_2 equals $0.26975z$. Substituting this value of R_2 as the internal radius R_1 in Equation 5.25 we find the radius of another external circle as $0.40050z$. It is seen that a circular loaded area of radius $0.26975z$ produces a vertical stress of $0.1q$ at depth z on the central axis, while a circular area with radius of $0.40050z$ produces $0.2q$; the difference between these is $0.1q$. Similar calculations can be carried out to produce the results presented in Table 5.1. If z is assumed to have a unit length then Table 5.1 may be presented graphically by selecting a reasonable scale for z and a value for $n = 20$ as shown in Figure 5.10. The total number of elements is $n/m = 200$ and each element produces a vertical stress equal to: $(m/n)q = 0.005q$, where $I_q = m/n = 0.005$ is called the influence factor.

The resulting graph, called *Newmark's influence chart* (Newmark, 1942), is used to estimate the vertical stress component under a loaded irregular area. Initially a plan of the footing is drawn assuming the depth z of the point of interest is equal to the scale of the chart. The plan is placed on Newmark's chart so that the point (under which the vertical stress component is required) is located at the central point of the chart. The vertical stress component due to a unit contact pressure is equal to the number of elements within the plan multiplied by the influence factor, or

$$\sigma_z = (\text{number of elements covered}) \times I_q \times q$$ (5.26)

Table 5.1. Construction of Newmark's influence chart with $m = 0.10$.

R/z	0.26975	0.40050	0.51811	0.63696	0.76642
σ_z/q	0.1	0.2	0.3	0.4	0.5
R/z	0.91761	1.10970	1.38709	1.90829	∞
σ_z/q	0.6	0.7	0.8	0.9	1.0

Example 5.6

Re-do Example 5.5 using the influence chart shown in Figure 5.10.

Solution:

A plan is drawn by assuming that the depth under point A is equal to the length of the scale in Figure 5.10. The re-scaled figure is adjusted on the influence chart as shown in Figure 5.11. The best estimate for the number of elements in one-half of the area is 42.5 Therefore:

$\sigma_z = 0.005 \times 300.0 \times 42.5 \times 2 = 127.5$ kPa.

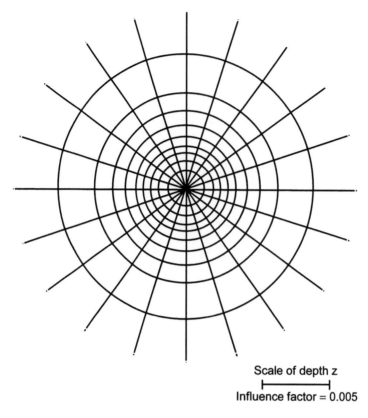

Scale of depth z

Influence factor = 0.005

Figure 5.10. Newmark's influence chart.

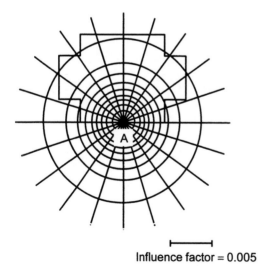

Influence factor = 0.005

Figure 5.11. Example 5.6.

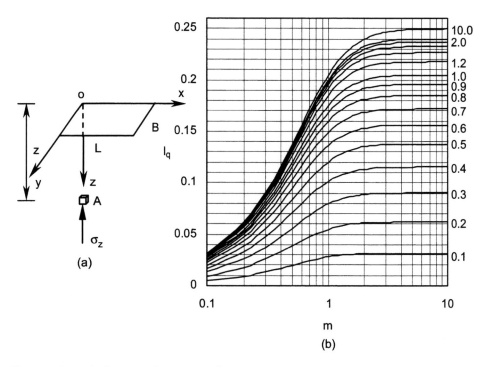

Figure 5.12. Vertical stress under a corner of a uniformly loaded rectangular area.

5.3.5 *Vertical stress under a corner of a uniformly loaded rectangular area*

Fadum (1948) introduced influence factors (based on Newmark's mathematical approach) for the computation of the vertical stress component at a specified depth under a corner of a rectangular loaded area. The mathematical work is based on the integration of Equation 5.11 over a rectangular area. Two dimensionless parameters m and n are defined as:

$$m = \frac{L}{z}, \; n = \frac{B}{z} \tag{5.27}$$

where L and B are the plan dimensions of the loaded rectangle, as shown in Figure 5.12(a). The vertical stress component is computed from:

$$\sigma_z = q \times I_q \tag{5.28}$$

where I_q is the influence factor, a dimensionless parameter defined by:

$$I_q = \frac{1}{4\pi}\left(\frac{2mn\sqrt{m^2+n^2+1}}{m^2+n^2+1+m^2n^2}\frac{m^2+n^2+2}{m^2+n^2+1} + \tan^{-1}\frac{2mn\sqrt{m^2+n^2+1}}{m^2+n^2+1-m^2n^2}\right) \tag{5.29}$$

For negative values of $\tan^{-1}$, π should be added to the calculated angle measured in radians. Figure 5.12(b) shows the graphical presentation of Equation 5.29 where m and n are interchangeable dimensionless parameters defined by Equations 5.27. The practical significance of this method is in its application to loaded areas of irregular geometry. The point at which the vertical stress component is required may be located within or outside the plan projection of the loaded area. Superposition may be applied if the loaded area is divided into a number of rectangles having one common corner. If the point of interest is outside the loaded area then a few of the rectangles that share the common corner may cover areas that are not loaded. In this case it is convenient to assume that the rectangles that cover the unloaded areas are subjected to a negative contact pressure.

Example 5.7

The uniform contact pressure under a rectangular footing of 6 m by 5 m is 200 kPa. Compute the vertical stress component under points A and B (Figure 5.13) at a depth of 2 m.

Solution:

The loaded area is divided into the 4 rectangles shown. Note that all rectangles share a common corner of either A or B. Results of the calculations are tabulated below.

Rectangle	L (m)	B (m)	m	n	I_q	σ_z (kPa)
Aebh	4.0	2.0	2.0	1.0	0.1999	39.98
Ahcf	2.0	1.0	1.0	0.5	0.1202	24.04
Afdg	4.0	1.0	2.0	0.5	0.1349	26.98
Agae	4.0	4.0	2.0	2.0	0.2325	46.50

Total vertical stress component at A: 137.5 kPa.

Rectangle	L (m)	B (m)	m	n	I_q	σ_z (kPa)
Bebj	7.0	2.0	3.5	1.0	0.2039	40.78
Bfcj	2.0	2.0	1.0	1.0	−0.1752	−35.04
Bidf	4.0	2.0	2.0	1.0	−0.1999	−39.98
Biae	7.0	4.0	3.5	2.0	0.2387	47.74

Total vertical stress component at B: 13.5 kPa.

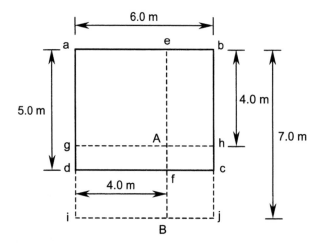

Figure 5.13. Example 5.7.

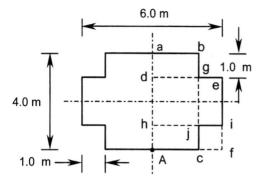

Figure 5.14. Example 5.8.

Example 5.8

Re-work Example 5.5 using Equation 5.29 or Figure 5.12(b).

Solution:

Since the footing is symmetrical about Aa (Figure 5.14), computations are carried out on one-half of the figure, the results of which are tabulated below.

Rectangle	L (m)	B (m)	m	n	I_q	σ_z (kPa)
$Aabc$	4.0	2.0	2.0	1.0	0.1999	59.97
$Adef$	3.0	3.0	1.5	1.5	0.2167	65.01
$Adgc$	3.0	2.0	1.5	1.0	−0.1936	−58.08
$Ahif$	3.0	1.0	1.5	0.5	−0.1313	−39.39
$Ahjc$	2.0	1.0	1.0	0.5	0.1202	36.06

Total σ_z at $A = 2 (59.97 + 65.01 - 58.08 - 39.39 + 36.06) = 127.1$ kPa.

5.3.6   *Concentrated line load of finite length*

Figure 5.15 shows a vertical line load of finite length L applied to a horizontal ground surface whose intensity q is measured by force per unit length (e.g. kN/m). For the arrangement shown, the vertical stress component is defined by:

$$\sigma_z = \frac{q}{z}(I_{q,y} + I_{q,L-y}) \tag{5.30}$$

where $I_{q,y}$ and $I_{q,L-y}$ are influence factors given by:

$$I_{q,y}, I_{q,L-y} = \frac{1}{2\pi(m^2 + 1)^2}[\frac{3n}{\sqrt{m^2 + n^2 + 1}} - (\frac{n}{\sqrt{m^2 + n^2 + 1}})^3] \tag{5.31}$$

in which the dimensionless parameters m and n are:

$$m = \frac{x}{z}, n = \frac{y}{z} \text{ (for } I_{qy}), n = \frac{L - y}{z} \text{ (for } I_{q,L-y}) \tag{5.32}$$

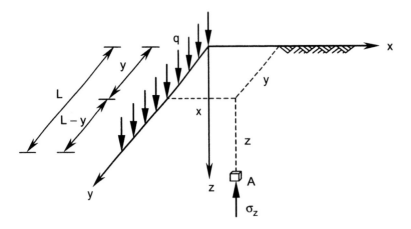

Figure 5.15. Vertical stress due to a line load of finite length.

For values of y greater than L the principle of superposition can be applied:

$$\sigma_z = \frac{q}{z}(I_{q,y} - I_{q,y-L})$$ (5.33)

Example 5.9

A 6 m long line load of intensity 100 kN/m is applied to the ground surface. Compute the vertical stress component at points A, B, C, and D located on a line at a depth of 2 m and 2 m horizontally from the line load. $y_A = 0$, $y_B = 3$ m, $y_C = 6$ m, $y_D = 9$ m.

Solution:

Calculations have been carried out using Equations 5.30 to 5.33, and results are tabulated below.

Point	for $I_{q,y}$	n for (a) $I_{q,L-y}$ (b) $I_{q,y-L}$	m	$I_{q,y}$	(a) $I_{q,L-y}$ (b) $I_{q,y-L}$		σ_z (kPa)
A	0.00	3.00 (a)	1.00	0.00000	0.07852	(a)	3.9
B	1.50	1.50 (a)	1.00	0.07152	0.07152	(a)	7.1
C	3.00	0.00 (a)	1.00	0.07852	0.00000	(a)	3.9
D	4.50	1.00 (b)	1.00	0.07933	0.07152	(b)	0.4

Note: due to symmetry, the vertical stress at point C equals the vertical stress at point A.

Example 5.10

Re-work Example 5.5 using the concepts of a line load of finite length and superposition. Assume that the plan view of the loaded area is divided into horizontal strips 1 m wide and that the line load is applied on the centre line of each strip.

Solution:

The equivalent line loads are idealized by four horizontal strips (Figure 5.8) as follows:

Strip 1 (only right half) includes squares 1 and 2; strip 2 includes squares 3, 4, and 5; strip 3 includes squares 6, 7, and 8; and strip 4 includes squares 9 and 10. The results of computations are tabulated below. A sample calculation for strip 3 is as follows:

For $I_{q,y}$: $m = x / z = 1.5 / 2.0 = 0.75$, $n = y / z = 3.0 / 2.0 = 1.5$.

For $I_{q,L - y}$: $n = 3.0 / 2.0 = 1.5$.

$I_{q,y} = I_{q,L - y} = 0.1207$, $q = 300.0$ kPa $\times 1.0$ m (width of the strip) $= 300.0$ kN/m.

$\sigma_z = (q / z)(I_{q,y} + I_{q,L-y}) = (300.0 / 2.0)(0.1207 + 0.1207) = 36.21$ kPa.

Strip no.	n for: $I_{q, y}$	for: $I_{q, y - L}$	m	$I_{q, y}$	$I_{q, y - L}$	σ_z (kPa)
1	1.00	1.00	1.75	0.0120	0.0120	3.60
2	1.50	1.50	1.25	0.0420	0.0420	12.60
3	1.50	1.50	0.75	0.1207	0.1207	36.21
4	1.00	1.00	0.25	0.2469	0.2469	74.07
Total $\sigma_z = 3.60 + 12.60 + 36.21 + 74.07 = 126.5$ kPa.						

5.3.7 Concentrated line load of infinite length

The analysis of a line load of infinite length is normally considered in two-dimensional space, and stresses are defined in accordance with the sign convention of Figure 5.16:

$$\sigma_x = \frac{2q}{\pi z} \cos^2 \theta \sin^2 \theta = \frac{2q\, x^2 z}{\pi(x^2 + z^2)^2} \tag{5.34}$$

$$\sigma_z = \frac{2q}{\pi z} \cos^4 \theta = \frac{2q\, z^3}{\pi(x^2 + z^2)^2} \tag{5.35}$$

$$\tau_{xz} = \frac{2q}{\pi z} \cos^3 \theta \sin \theta = \frac{2q\, xz^2}{\pi(x^2 + z^2)^2} \tag{5.36}$$

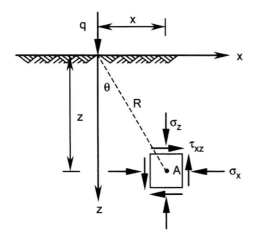

Figure 5.16. Stresses at a point due to a line load of infinite length.

Example 5.11

Determine the stress components, principal stresses and their directions under a line load of infinite length at points with x/z ratio of 0.6.

Solution:

Substituting $x = 0.6 z$ into Equations 5.34, 5.35 and 5.36 we obtain:

$\sigma_x = 0.124\ q/z$, $\sigma_z = 0.344\ q/z$, $\tau_{xz} = 0.207\ q/z$.

The position of the centre and the radius of the Mohr's circle are given by:

$s = (\sigma_z + \sigma_x)/2 = (0.344 + 0.124)/2(q/z) = 0.234q/z$,

$$t = \sqrt{[(\sigma_z - \sigma_x)/2]^2 + \tau_{zx}^2} = \sqrt{[(0.344 - 0.124)/2]^2 + 0.207^2}\ q/z = 0.234q/z,$$

$\sigma_3 = s - t = 0, \sigma_1 = s + t = 0.468q/z$.

The direction of the major principal stress plane is calculated from Equation 4.6. Substituting Equations 5.34, 5.35 and 5.36 into Equation 4.6:

$$\tan 2\theta_1 = \frac{2\tau_{xz}}{\sigma_z - \sigma_x} = \frac{2xz}{z^2 - x^2} = \frac{2 \times 0.6z^2}{z^2 - (0.6z)^2} = 1.875,\ \text{from which}\ \tan\theta_1 = 0.6 = \tan\theta.$$

This means that at any point on a plane with $x/z = $ constant, the angle between the major principal plane and the x-axis is θ.

5.3.8 Uniformly loaded infinite strip

Stress components under a uniformly loaded strip of infinite length are calculated by integrating an infinite line load across the width of the strip. Such a loaded strip is shown in Figure 5.17. Any point under the strip is represented either by its coordinates in the xz plane whose origin is at the centre of its width, or by the angles α and β. Equations 5.37 to 5.42 give the stress components on any plane that is perpendicular to both the

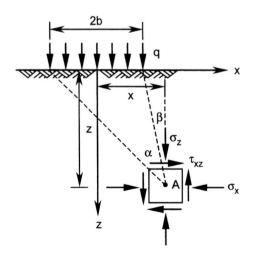

Figure 5.17. Stress components under uniformly loaded infinite strip.

horizontal ground surface and the infinite dimension of the strip. As for the line load, none of the stress components depend on the elastic properties of the material. The vertical stress component may also be approximated using the Newmark's influence chart (Figure 5.12) or Equations 5.28 and 5.29 where either m or n approaches infinity. The stresses are given by:

$$\sigma_x = \frac{q}{\pi}[\tan^{-1}\frac{x+b}{z} - \tan^{-1}\frac{x-b}{z} - \frac{2bz(z^2 - x^2 + b^2)}{(z^2 + x^2 - b^2)^2 + 4b^2z^2}] \tag{5.37}$$

$$\sigma_z = \frac{q}{\pi}[\tan^{-1}\frac{x+b}{z} - \tan^{-1}\frac{x-b}{z} + \frac{2bz(z^2 - x^2 + b^2)}{(z^2 + x^2 - b^2)^2 + 4b^2z^2}] \tag{5.38}$$

$$\tau_{xz} = \frac{q}{\pi}[\frac{4bxz^2}{(z^2 + x^2 - b^2)^2 + 4b^2z^2}] \tag{5.39}$$

$$\sigma_x = \frac{q}{\pi}[\alpha - \sin\alpha\cos(\alpha + 2\beta)] \tag{5.40}$$

$$\sigma_z = \frac{q}{\pi}[\alpha + \sin\alpha\cos(\alpha + 2\beta)] \tag{5.41}$$

$$\tau_{xz} = \frac{q}{\pi}[\sin\alpha\sin(\alpha + 2\beta)] \tag{5.42}$$

Angles α and β are defined by:

$$\alpha = \tan^{-1}\frac{x+b}{z} - \tan^{-1}\frac{x-b}{z}, \quad \beta = \tan^{-1}\frac{x-b}{z} \tag{5.43}$$

Curves representing a constant magnitude of a stress component (isobar) may be constructed by equating each stress component to a specified constant value. This is of particular interest, as the volume of soil subjected to deformation (as a result of the stress distribution) may be evaluated by drawing an isobar representing a fraction of the applied load. The construction of an isobar is fully explained in Example 5.13.

Example 5.12

An infinite strip of width 20 m is subjected to a uniform load of 100 kPa. Compute the vertical stress component at a depth of 8 m at points A ($x = 0$), B ($x = 10$ m) and C ($x = 18$ m).

Solution:

At point A: $\alpha_A = \tan^{-1}(10.0/8.0) - \tan^{-1}(-10.0/8)/z = 51.34° + 51.34° = 102.68°$,
$\beta_A = \tan^{-1}(-10.0/8.0) = -51.34°$,
$\sigma_{zA} = 100.0[102.68(\pi/180.0) + \sin 102.68° \cos(102.68° + 2\times-51.34°)]/\pi = 88.1$ kPa.
At point B: $\alpha_B = \tan^{-1}(20.0/8.0) = 68.20°$, $\beta_B = 0.0$,
$\sigma_{zB} = 100.0[68.20(\pi/180.0) + \sin 68.20° \cos(68.20° + 2\times0.0°)]/\pi = 48.9$ kPa.
At point C: $\alpha_C = \tan^{-1}(28.0/8.0) - \tan^{-1}(8.0/8.0) = 74.05° - 45.0° = 29.05°$,
$\beta_C = \tan^{-1}(8.0/8.0) = 45.0°$,
$\sigma_{zC} = 100.0[29.05(\pi/180.0) + \sin 29.05° \cos(29.05° + 2\times45.0°)]/\pi = 8.6$ kPa.

Example 5.13

Under a uniformly loaded strip construct an isobar for $\sigma_z = 0.2q$.

Solution:

Step procedure:

(a) obtain expressions for x and z in terms of the angles α and β,
(b) select a numerical value for α and substitute this into Equation 5.41 and find β,
(c) calculate x and z for the corresponding angles,
(d) repeat steps (b) and (c). From the geometry of Figure 5.17:
$\tan(\alpha + \beta) = (x + b)/z$, $\tan\beta = (x - b)/z$, solving for x and z:
$x = b[\tan(\alpha + \beta) + \tan\beta]/[\tan(\alpha + \beta) - \tan\beta]$, $z = (x + b)/\tan(\alpha + \beta)$.
Sample calculation for $\alpha = 30°$: $\alpha = \pi/6 = 0.523598$ radian,
$\sigma_z = q[\alpha + \sin\alpha\cos(\alpha + 2\beta)]/\pi = 0.2q$, $\alpha + \sin\alpha\cos(\alpha + 2\beta) = 0.2 \times \pi = 0.628318$,
$0.523598 + \sin 30.0° \cos(30.0° + 2\beta) = 0.628318$, $\cos(30.0° + 2\beta) = 0.209439 \rightarrow \beta = 23.95°$.
$x = b[\tan(30.0° + 23.95°) + \tan 23.95°]/[\tan(30.0° + 23.95°) - \tan 23.95°] = 1.9556b$,
$x/2b = 0.978$. $z = (1.9556b + b)/\tan(30.0° + 23.95°) = 2.1513b \rightarrow z/2b = 1.076$.
Results for the right half are tabulated below and the full isobar is shown in Figure 5.18.

$\alpha°$	$\beta°$	$x/2b$	$z/2b$	$\alpha°$	$\beta°$	$x/2b$	$z/2b$
60.00	29.46	0.505	0.009	24.00	17.50	1.054	1.756
50.00	29.30	0.618	0.211	22.00	13.64	1.011	2.107
38.00	27.62	0.811	0.594	20.00	7.63	0.844	2.567
30.00	23.95	0.978	1.076	19.00	2.65	0.632	2.852
28.00	22.35	1.017	1.257	18.50	-1.39	0.427	3.011
26.00	20.27	1.046	1.479	18.15	-9.075	0.000	3.130

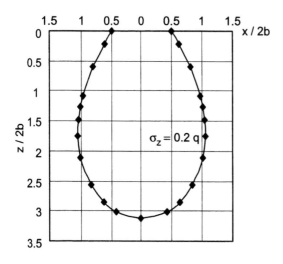

Figure 5.18. Example 5.13.

5.3.9   *Linearly loaded infinite strip*

Figure 5.19 shows a vertical distributed load applied to an infinitely long strip and linearly increasing across its width $2b$. Using the sign convention shown, the stress components may be obtained by integration of the load across the width of the strip. As before, all the stress components are found to be independent of the elastic properties of the material.

$$\sigma_x = \frac{q}{2\pi}(\frac{x}{b}\alpha - \frac{z}{b}\ln\frac{R_1^2}{R_2^2} + \sin 2\beta) \qquad (5.44)$$

$$\sigma_z = \frac{q}{2\pi}(\frac{x}{b}\alpha - \sin 2\beta) \qquad (5.45)$$

$$\tau_{xz} = \frac{q}{2\pi}(1 + \cos 2\beta - \frac{z}{b}\alpha) \qquad (5.46)$$

Angles α and β are defined by:

$$\alpha = \tan^{-1}\frac{x}{z} - \tan^{-1}\frac{x-2b}{z}, \ \beta = \tan^{-1}\frac{x-2b}{z} \qquad (5.47)$$

Example 5.14

A load on an infinitely long strip increases linearly from zero to a maximum of 100 kPa across its width of 8 m. Calculate the vertical stress components at a depth of 8 m at points:
A $(x = 0)$, B $(x = 8$ m$)$ and C $(x = 18$ m$)$.

Solution:

At point A:
$\alpha_A = \tan^{-1}(0.0/8.0) - \tan^{-1}[(0.0-8.0)/8.0] = 45.0°$,
$\beta_A = \tan^{-1}[(0.0-8.0)/8.0] = -45.0°$,

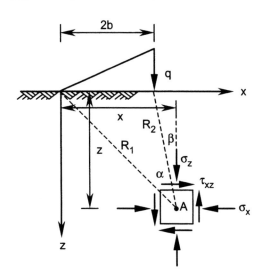

Figure 5.19. Stress components under a linearly loaded infinitely long strip.

$$\sigma_{zA} = \frac{100.0}{2\pi}[0.0 - \sin(2\times -45.0°)] = 15.9 \text{ kPa. At point } B:$$

$$\alpha_B = \tan^{-1}(8.0/8.0) - \tan^{-1}[(8.0-8.0)/8.0] = 45.0°, \quad \beta_B = \tan^{-1}[(8.0-8.0)/8.0] = 0.0°,$$

$$\sigma_{zB} = \frac{100.0}{2\pi}(\frac{8.0}{4.0}\times 45.0°\frac{\pi}{180°} - 0.0) = 25.0 \text{ kPa.}$$

At point C: $\alpha_C = \tan^{-1}(18.0/8.0) - \tan^{-1}[(18.0-8.0)/8.0] = 14.70°$,

$\beta_C = \tan^{-1}[(18.0-8.0)/8.0] = 51.34°$,

$$\sigma_{zC} = \frac{100.0}{2\pi}[\frac{18.0}{4.0}\times 14.7°\frac{\pi}{180°} - \sin(2\times 51.34°)] = 2.8 \text{ kPa.}$$

5.3.10 *Embankment loading of infinite length*

An earth embankment may be idealized as a combination of the uniformly and linearly loaded infinite strips shown in Figure 5.20. The magnitude of the uniform load is γh, where γ is the unit weight of the earth embankment material of height h. The vertical stress component at point A beneath the corner of the uniform load is the sum of Equations 5.41 and 5.45:

$$\sigma_z = \frac{q}{\pi}[\alpha_2 + \sin\alpha_2 \cos\alpha_2 + \frac{\alpha_1(a+b)}{a} - 0.5\sin 2\alpha_2],$$

$$\sigma_z = \frac{q}{\pi}[\alpha_2 + \frac{\alpha_1(a+b)}{a}] \tag{5.48}$$

The vertical stress component may be represented by the dimensionless parameters $m = a/z$ and $n = b/z$, obtained by substituting the magnitudes of $\tan\alpha_1$ and $\tan\alpha_2$ (calculated from the geometry of the embankment) into Equation 5.48:

$$\sigma_z = qI_q, \quad I_q = \frac{1}{\pi}(\frac{m+n}{m}\tan^{-1}\frac{m}{1+n^2+mn} + \tan^{-1}n) \tag{5.49}$$

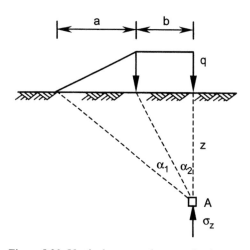

Figure 5.20. Vertical stress under an embankment.

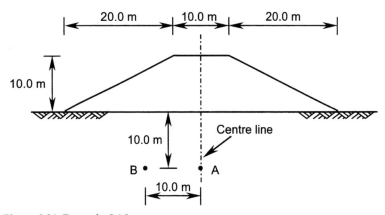

Figure 5.21. Example 5.15.

Depending on the position of the point at which the vertical stress component is required, superposition may be used.

Example 5.15

For the embankment shown in Figure 5.21, calculate the vertical stress component at points *A* and *B*. Assume the unit weight of the embankment material is 20 kN/m³.

Solution:

Vertical stress component at point *A*: Due to symmetry we need only to calculate the vertical stress component due to the left half and double the result. For the left half:
$m = a/z = 20.0 / 10.0 = 2.0$, $n = b/z = 5.0 / 10.0 = 0.5$.
Using Equation 5.49:

$$I_{qA} = \frac{1}{\pi}(\frac{2.0+0.5}{2.0}\tan^{-1}\frac{2}{1+0.5^2+2\times0.5}+\tan^{-1}0.5) = 0.4367,$$

$\sigma_{zA} = 2\times(20.0\times10.0)\times0.4367 = 174.7$ kPa.

Vertical stress component at point *B*:
The section of the earth embankment is divided into the three areas *cde*, *dfhi*, and *efg* (Figure 5.22(a)), where the area *efg* represents a negative linear loading. All three areas are analysed using Equation 5.49 and the results are superimposed to give the vertical stress component at point *B*, as shown in Figure 5.22(b). From the geometry of the earth embankment we find:
$cd = 15.0$ m, $de = 0.5 \times 15.0 = 7.5$ m, $ef = 10.0 - 7.5 = 2.5$ m, $fg = 5.0$ m, $fh = 15.0$ m.
The results of the computations are summarised in the table below.

Area	$q = \gamma \times h$ (kPa)	m	n	I_q	σ_z (kPa)
cde	$20.0 \times 7.5 = 150$	1.5	0.0	0.3128	46.92
dfhi	$20.0 \times 10.0 = 200$	2.0	1.5	0.4853	97.06
efg	$20.0 \times -2.5 = -50$	0.5	0.0	0.1476	-7.38

$\sigma_{zB} = 46.92 + 97.06 - 7.38 = 136.6$ kPa.

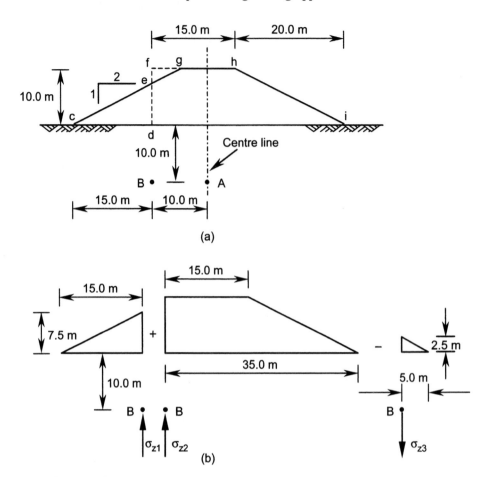

Figure 5.22. Example 5.15.

5.3.11 *Stress distribution due to internal loading*

An approximation for the vertical stress component due to a vertical force in the interior of a semi-infinite elastic soil was proposed by Geddes (1966 and 1969). The approximation is based on the Mindlin (1936) solution and follows a Boussinesq based approach. One important application is in the evaluation of the vertical stresses beneath a pile due either to the surface reaction in the form of an applied shear stress or the end reaction in the form of a single concentrated load. Figure 5.23 shows the arrangement adopted for the solution. The vertical stress component at a point A with a horizontal distance r from the load Q and a depth $z > L$ is:

$$\sigma_z = \frac{Q}{L^2} I_q \tag{5.50}$$

where L is the depth of vertical load Q and I_q is given by Equation 5.51.
The coordinates of point A are expressed in the dimensionless parameters of $m = z/L$, and $n = r/L$.

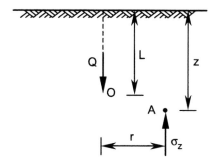

Figure 5.23. Concentrated load applied in the interior of a semi-infinite elastic soil.

$$I_q = \frac{1}{8\pi(1-\mu)}\{\frac{(1-2\mu)(m-1)}{[n^2+(m-1)^2]^{3/2}} - \frac{(1-2\mu)(m-1)}{[n^2+(m+1)^2]^{3/2}} + \frac{3(m-1)^3}{[n^2+(m-1)^2]^{5/2}}$$

$$+\frac{3(3-4\mu)m(m+1)^2 - 3(m+1)(5m-1)}{[n^2+(m+1)^2]^{5/2}} + \frac{30m(m+1)^3}{[n^2+(m+1)^2]^{7/2}}\} \tag{5.51}$$

where μ is Poisson's ratio of the elastic soil. If the applied load Q is transferred to the soil through the surface area of the pile and the end reaction is ignored, the vertical stress component is calculated from Equations 5.52 (uniform shear force of magnitude Q/L) or 5.53 (linear shear force starting from zero at the top of the pile and increasing to a maximum value of $2Q/L$ at the bottom of the pile).

$$I_q = \frac{1}{8\pi(1-\mu)}\{\frac{2(2-\mu)}{[n^2+(m-1)^2]^{1/2}} - \frac{2(2-\mu)+2(1-2\mu)\frac{m}{n}(\frac{m}{n}+\frac{1}{n})}{[n^2+(m+1)^2]^{1/2}}$$

$$+\frac{2(1-2\mu)(\frac{m}{n})^2}{(m^2+n^2)^{1/2}} - \frac{n^2}{[n^2+(m-1)^2]^{3/2}} - \frac{4m^2 - 4m^2(1+\mu)(\frac{m}{n})^2}{(m^2+n^2)^{3/2}}$$

$$-\frac{4m(1+\mu)(m+1)(\frac{m}{n}+\frac{1}{n})^2 - (4m^2+n^2)}{[n^2+(m+1)^2]^{3/2}}$$

$$-\frac{6m^2(\frac{m^4-n^4}{n^2})}{(m^2+n^2)^{5/2}} - \frac{6m[mn^2 - \frac{1}{n^2}(m+1)^5]}{[n^2+(m+1)^2]^{5/2}}\} \tag{5.52}$$

$$I_q = \frac{1}{4\pi(1-\mu)}\{\frac{2(2-\mu)}{[n^2+(m-1)^2]^{1/2}} - \frac{2(2-\mu)(4m+1)-2(1-2\mu)(\frac{m}{n})^2(m+1)}{[n^2+(m+1)^2]^{1/2}}$$

$$-\frac{2(1-2\mu)\frac{m^3}{n^2} - 8(2-\mu)m}{(m^2+n^2)^{1/2}} - \frac{mn^2+(m-1)^3}{[n^2+(m-1)^2]^{3/2}}$$

$$-\frac{4\mu n^2 m + 4m^3 - 15n^2 m - 2(5+2\mu)(\frac{m}{n})^2(m+1)^3 + (m+1)^3}{[n^2+(m+1)^2]^{3/2}}$$

$$-\frac{2(7-2\mu)mn^2 - 6m^3 + 2(5+2\mu)(\frac{m}{n})^2 m^3}{(m^2+n^2)^{3/2}} - \frac{6mn^2(n^2-m^2)+12(\frac{m}{n})^2(m+1)^5]}{[n^2+(m+1)^2]^{5/2}}$$

$$+\frac{12(\frac{m}{n})^2 m^5 + 6mn^2(n^2-m^2)}{(m^2+n^2)^{5/2}}$$

$$+2(2-\mu)\ln(\frac{[n^2+(m-1)^2]^{1/2}+m-1}{(m^2+n^2)^{1/2}+m}\frac{[n^2+(m+1)^2]^{1/2}+m+1}{(m^2+n^2)^{1/2}+m})\} \qquad (5.53)$$

Equations 5.51, 5.52 and 5.53 are presented in Table 5.2 (Bowles, 1996).

Example 5.16

Referring to Figure 5.24, compute the vertical stress components at a depth of 2 m under each pile: (a) the load is carried by end bearing only, (b) one-half of the load is carried by end bearing and the other half by skin friction. $\mu = 0.3$.

Solution:

(a) $z = 2.0 + 8.0 = 10.0$ m, $m = z/L = 10.0/8.0 = 1.25$ for all piles.
The vertical stress component under pile 1 is calculated from Equation 5.51 by superimposing the effects of three loads according to: $\sigma_z = (Q/L^2)(I_{q1} + I_{q2} + I_{q3})$, where I_{q2} and I_{q3} are the corresponding effects of pile 2 and pile 3 at the point of interest under pile 1.
For $m = 1.25$, $n = 0.0$, from Table 5.2(a) and interpolating between 1.2 and 1.3:
$I_{q1} = (4.9099 + 2.2222)/2 = 3.5660$. For $n = 4.0/8.0 = 0.5$, $I_{q2} = 0.1495$,
for $n = 8.0/8.0 = 1.0$, $I_{q3} = 0.0506$.
$\sigma_{z1} = \sigma_{z3} = (1000.0/8.0^2)(3.5660 + 0.1495 + 0.0506) = 58.8$ kPa.
Pile 2: $I_{q1} = I_{q3} = 0.1495$, $I_{q2} = 3.5660$.
$\sigma_{z2} = (1000.0/8.0^2)(0.1495 + 3.5660 + 0.1495) = 60.4$ kPa.
(b) Depth $= 10.0$ m, piles 1 and 3: $m = 1.25$, $n = 0.0$, $I_{q1} = (0.9699 + 0.6430)/2 = 0.8064$.
Interpolating between 1.2 and 1.3 from Table 5.2(b):
For $n = 4.0/8.0 = 0.5$, $I_{q2} = (0.2292 + 0.2207)/2 = 0.2249$.
For $n = 8.0/8.0 = 1.0$, $I_{q3} = (0.0760 + 0.0782)/2 = 0.0771$.

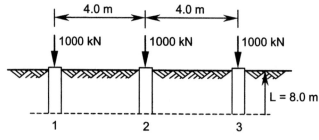

Figure 5.24. Example 5.16.

Table 5.2(a). Vertical stress influence factors: vertical load applied in the interior of the soil.

$\mu = 0.2$

$n \to$ $m \downarrow$	0.00	0.1	0.20	0.30	0.40	0.50	0.75	1.00	1.50	2.00
1.0	∞	0.0960	0.0936	0.0897	0.0846	0.0785	0.0614	0.0448	0.0208	0.0089
1.1	17.9689	1.7753	0.6182	0.2238	0.1332	0.0999	0.0659	0.0467	0.0222	0.0099
1.2	4.5510	2.7458	1.0005	0.3987	0.2056	0.1325	0.0724	0.0490	0.0236	0.0110
1.3	2.0609	1.6287	0.9233	0.4798	0.2672	0.1681	0.0811	0.0520	0.0249	0.0119
1.4	1.1858	1.0382	0.7330	0.4652	0.2926	0.1930	0.0905	0.0555	0.0263	0.0129
1.5	0.7782	0.7153	0.5682	0.4114	0.2875	0.2025	0.0985	0.0592	0.0277	0.0138
1.6	0.5548	0.5238	0.4457	0.3518	0.2664	0.1997	0.1038	0.0625	0.0290	0.0147
1.7	0.4188	0.4018	0.3569	0.2984	0.2399	0.1893	0.1061	0.0651	0.0303	0.0156
1.8	0.3294	0.3193	0.2918	0.2539	0.2133	0.1775	0.1057	0.0668	0.0315	0.0164
1.9	0.2673	0.2609	0.2431	0.2177	0.1890	0.1606	0.1033	0.0675	0.0325	0.0172
2.0	0.2222	0.2180	0.2060	0.1883	0.1676	0.1462	0.0995	0.0673	0.0334	0.0179

$\mu = 0.3$

$n \to$ $m \downarrow$	0.00	0.1	0.20	0.30	0.40	0.50	0.75	1.00	1.50	2.00
1.0	∞	0.1013	0.0986	0.0944	0.0889	0.0824	0.0641	0.0463	0.0209	0.0087
1.1	19.3926	3.9054	0.5978	0.2123	0.1287	0.0986	0.0668	0.0475	0.0222	0.0097
1.2	4.9099	2.9275	1.0358	0.4001	0.2027	0.1303	0.0722	0.0493	0.0235	0.0106
1.3	2.2222	1.7467	0.9757	0.4970	0.2717	0.1687	0.0808	0.0519	0.0247	0.0116
1.4	1.2777	1.1152	0.7905	0.4891	0.3032	0.1974	0.0908	0.0555	0.0260	0.0125
1.5	0.8377	0.7686	0.6070	0.4356	0.3012	0.2098	0.0999	0.0594	0.0274	0.0134
1.6	0.5968	0.5626	0.4768	0.3738	0.2809	0.2086	0.1063	0.0631	0.0288	0.0143
1.7	0.4500	0.4312	0.3819	0.3177	0.2538	0.1988	0.1094	0.0661	0.0302	0.0152
1.8	0.3536	0.3424	0.3122	0.2706	0.2262	0.1849	0.1096	0.0682	0.0315	0.0161
1.9	0.2866	0.2795	0.2600	0.2321	0.2006	0.1697	0.1076	0.0693	0.0326	0.0169
2.0	0.2380	0.2333	0.2201	0.2007	0.1780	0.1547	0.1039	0.0694	0.0336	0.0177

$\mu = 0.4$

$n \to$ $m \downarrow$	0.00	0.1	0.20	0.30	0.40	0.50	0.75	1.00	1.50	2.00
1.0	∞	0.1083	0.1054	0.1008	0.0947	0.0876	0.0676	0.0483	0.0212	0.0083
1.1	21.2910	4.0788	0.5699	0.1970	0.1228	0.0970	0.0680	0.0486	0.0223	0.0093
1.2	5.3884	3.1699	1.0829	0.4020	0.1989	0.1274	0.0720	0.0496	0.0233	0.0102
1.3	2.4373	1.9040	1.0455	0.5200	0.2776	0.1695	0.0804	0.0519	0.0244	0.0111
1.4	1.4002	1.2179	0.8438	0.5208	0.3173	0.2032	0.0913	0.0554	0.0256	0.0120
1.5	0.9172	0.8395	0.6587	0.4678	0.3194	0.2196	0.1017	0.0596	0.0270	0.0129
1.6	0.6527	0.6143	0.5181	0.4033	0.3001	0.2205	0.1095	0.0638	0.0284	0.0138
1.7	0.4915	0.4705	0.4152	0.3435	0.2724	0.2116	0.1138	0.0675	0.0300	0.0147
1.8	0.3858	0.3732	0.3393	0.2929	0.2433	0.1976	0.1148	0.0701	0.0314	0.0156
1.9	0.3123	0.3044	0.2825	0.2512	0.2161	0.1818	0.1133	0.0717	0.0328	0.0166
2.0	0.2590	0.2537	0.2390	0.2173	0.1919	0.1659	0.1098	0.0722	0.0340	0.0174

Table 5.2(b). Vertical stress influence factors: uniform shear force applied in the interior of the soil.

$\mu = 0.2$

$n \rightarrow$ $m \downarrow$	0.00	0.02	0.04	0.06	0.08	0.1	0.15	0.20	0.50	1.0	2.0
1.0	∞	6.4703	3.2374	2.1592	1.6202	1.2962	0.8630	0.6445	0.2300	0.0690	0.0081
1.1	1.7781	1.7342	1.5944	1.4178	1.2418	1.0850	0.7953	0.6138	0.2283	0.0730	0.0096
1.2	0.9015	0.8789	0.8576	0.8269	0.7882	0.7446	0.6317	0.5307	0.2231	0.0759	0.0111
1.3	0.5968	0.5799	0.5725	0.5629	0.5500	0.5340	0.4867	0.4355	0.2138	0.0779	0.0125
1.4	0.4569	0.4288	0.4241	0.4201	0.4142	0.4068	0.3838	0.3562	0.2010	0.0789	0.0139
1.5	0.3482	0.3359	0.3334	0.3313	0.3282	0.3242	0.3113	0.2952	0.1862	0.0790	0.0152
1.6	0.2922	0.2626	0.2716	0.2707	0.2689	0.2666	0.2589	0.2487	0.1708	0.0784	0.0165
1.7	0.2518	0.2304	0.2287	0.2274	0.2261	0.2247	0.2195	0.2127	0.1559	0.0770	0.0175
1.8	0.1772	0.1953	0.1949	0.1942	0.1936	0.1925	0.1891	0.1844	0.1420	0.0750	0.0185
1.9	0.1648	0.1702	0.1698	0.1678	0.1682	0.1675	0.1650	0.1616	0.1295	0.0727	0.0193
2.0	0.1461	0.1482	0.1486	0.1480	0.1478	0.1473	0.1455	0.1429	0.1180	0.0700	0.0201

$\mu = 0.3$

$n \rightarrow$ $m \downarrow$	0.00	0.02	0.04	0.06	0.08	0.10	0.15	0.20	0.50	1.00	2.00
1.0	∞	6.8419	3.4044	2.2673	1.6983	1.3567	0.8998	0.6695	0.2346	0.0686	0.0076
1.1	1.9219	1.8611	1.7072	1.5134	1.3211	1.1503	0.8368	0.6419	0.2335	0.0728	0.0091
1.2	0.9699	0.9403	0.9166	0.8825	0.8400	0.7922	0.6688	0.5588	0.2292	0.0760	0.0105
1.3	0.6430	0.6188	0.6099	0.5992	0.5850	0.5675	0.5157	0.4597	0.2207	0.0782	0.0120
1.4	0.4867	0.4558	0.4507	0.4461	0.4396	0.4316	0.4063	0.3761	0.2082	0.0796	0.0134
1.5	0.3766	0.3561	0.3533	0.3510	0.3476	0.3432	0.3291	0.3115	0.1834	0.0800	0.0148
1.6	0.3339	0.2895	0.2878	0.2863	0.2843	0.2817	0.2732	0.2621	0.1777	0.0796	0.0160
1.7	0.2664	0.2438	0.2414	0.2399	0.2384	0.2369	0.2313	0.2239	0.1623	0.0784	0.0172
1.8	0.2025	0.2065	0.2054	0.2044	0.2038	0.2026	0.1989	0.1938	0.1479	0.0766	0.0182
1.9	0.1847	0.1794	0.1785	0.1777	0.1768	0.1760	0.1733	0.1696	0.1347	0.0744	0.0191
2.0	0.1634	0.1565	0.1561	0.1556	0.1551	0.1545	0.1525	0.1498	0.1229	0.0718	0.0199

$\mu = 0.4$

$n \rightarrow$ $m \downarrow$	0.00	0.02	0.04	0.06	0.08	0.10	0.15	0.20	0.50	1.00	2.00
1.0	∞	7.2744	3.6270	2.4110	1.8026	1.4373	0.9488	0.7029	0.2407	0.0681	0.0069
1.1	2.0931	2.0296	1.8574	1.6409	1.4266	1.2372	0.8921	0.6794	0.2404	0.0725	0.0083
1.2	1.0486	1.0209	0.9947	0.9567	0.9091	0.8556	0.7181	0.5964	0.2373	0.0760	0.0098
1.3	0.6922	0.6694	0.6598	0.6476	0.6318	0.6122	0.5543	0.4921	0.2298	0.0787	0.0113
1.4	0.5347	0.4922	0.4860	0.4807	0.4735	0.4645	0.4362	0.4026	0.2178	0.0805	0.0128
1.5	0.4020	0.3823	0.3798	0.3771	0.3734	0.3684	0.3527	0.3332	0.2029	0.0813	0.0142
1.6	0.3440	0.3096	0.3083	0.3068	0.3045	0.3017	0.2922	0.2800	0.1868	0.0812	0.0155
1.7	0.2943	0.2606	0.2580	0.2564	0.2549	0.2531	0.2469	0.2387	0.1708	0.0803	0.0167
1.8	0.2114	0.2207	0.2189	0.2181	0.2174	0.2161	0.2119	0.2063	0.1558	0.0787	0.0178
1.9	0.1782	0.1907	0.1904	0.1890	0.1881	0.1873	0.1843	0.1802	0.1419	0.0766	0.0188
2.0	0.1741	0.1660	0.1658	0.1652	0.1648	0.1642	0.1620	0.1590	0.1294	0.0741	0.0196

Table 5.2(c). Vertical stress influence factors: linear shear force applied in the interior of the soil.

$\mu = 0.2$

$n \rightarrow$ $m \downarrow$	0.00	0.02	0.04	0.06	0.08	0.10	0.15	0.20	0.50	1.0	2.0
1.0	∞	11.5315	5.3127	3.3023	2.3263	1.7582	1.0372	0.7033	0.1963	0.0618	0.0082
1.1	2.8427	2.7518	2.4908	2.1596	1.8329	1.5469	1.0359	0.7346	0.2074	0.0656	0.0096
1.2	1.2853	1.2541	1.2158	1.1620	1.0930	1.0162	0.8211	0.6529	0.2141	0.0689	0.0110
1.3	0.7673	0.7753	0.7585	0.7420	0.7195	0.6928	0.6142	0.5312	0.2139	0.0717	0.0123
1.4	0.5937	0.5450	0.5343	0.5267	0.5181	0.5063	0.4693	0.4261	0.2068	0.0737	0.0136
1.5	0.4485	0.4051	0.4059	0.4006	0.3960	0.3901	0.3704	0.3460	0.1947	0.0750	0.0148
1.6	0.3635	0.3201	0.2326	0.3183	0.3154	0.3123	0.3008	0.2861	0.1803	0.0754	0.0160
1.7	0.3204	0.2583	0.2635	0.2618	0.2595	0.2574	0.2503	0.2408	0.1652	0.0750	0.0170
1.8	0.2533	0.2222	0.2239	0.2206	0.2181	0.2166	0.2122	0.2059	0.1506	0.0739	0.0180
1.9	0.2382	0.1761	0.1855	0.1880	0.1878	0.1853	0.1827	0.1782	0.1371	0.0722	0.0188
2.0	0.1776	0.1643	0.1648	0.1630	0.1631	0.1614	0.1591	0.1561	0.1248	0.0700	0.0196

$\mu = 0.3$

$n \rightarrow$ $m \downarrow$	0.00	0.02	0.04	0.06	0.08	0.10	0.15	0.20	0.50	1.0	2.0
1.0	∞	12.1310	5.5765	3.4591	2.4320	1.8346	1.0774	0.7276	0.1997	0.0616	0.0777
1.1	3.0612	2.9620	2.6751	2.3119	1.9547	1.6433	1.0908	0.7680	0.2115	0.0654	0.0090
1.2	1.3821	1.3465	1.3052	1.2465	1.1706	1.0864	0.8730	0.6899	0.2198	0.0689	0.0104
1.3	0.8262	0.8035	0.8130	0.7949	0.7705	0.7411	0.6548	0.5639	0.2212	0.0720	0.0117
1.4	0.6194	0.5827	0.5722	0.5630	0.5540	0.5410	0.5005	0.4530	0.2150	0.0744	0.0130
1.5	0.5189	0.4337	0.4332	0.4281	0.4227	0.4163	0.3946	0.3679	0.2033	0.0760	0.0143
1.6	0.3841	0.3415	0.3449	0.3395	0.3361	0.3327	0.3202	0.3039	0.1887	0.0768	0.0155
1.7	0.3332	0.2764	0.2810	0.2782	0.2764	0.2739	0.2660	0.2556	0.1732	0.0767	0.0166
1.8	0.2837	0.2268	0.2381	0.2347	0.2319	0.2300	0.2253	0.2183	0.1580	0.0758	0.0176
1.9	0.2654	0.1873	0.1963	0.1991	0.1988	0.1965	0.1937	0.1887	0.1439	0.0742	0.0186
2.0	0.1872	0.1730	0.1744	0.1732	0.1725	0.1714	0.1684	0.1651	0.1310	0.0721	0.0194

$\mu = 0.4$

$n \rightarrow$ $m \downarrow$	0.00	0.02	0.04	0.06	0.08	0.10	0.15	0.20	0.50	1.0	2.0
1.0	∞	12.9304	5.9282	3.6683	2.5729	1.9365	1.1311	0.7600	0.2042	0.0614	0.0069
1.1	3.3525	3.2423	2.9209	2.5144	2.1171	1.7719	1.1641	0.8125	0.2170	0.0652	0.0083
1.2	1.5030	1.4712	1.4255	1.3588	1.2742	1.1800	0.9422	0.7394	0.2274	0.0689	0.0096
1.3	0.8965	0.9066	0.8862	0.8649	0.8383	0.8056	0.7089	0.6076	0.2308	0.0723	0.0109
1.4	0.6753	0.6350	0.6222	0.6120	0.6018	0.5874	0.5419	0.4890	0.2260	0.0752	0.0123
1.5	0.5629	0.4718	0.4712	0.4641	0.4584	0.4511	0.4270	0.3971	0.2147	0.0773	0.0136
1.6	0.4198	0.3701	0.3730	0.3672	0.3642	0.3600	0.3461	0.3278	0.1999	0.0786	0.0149
1.7	0.3752	0.2840	0.3039	0.3011	0.2984	0.2956	0.2870	0.2754	0.1838	0.0788	0.0161
1.8	0.3158	0.2496	0.2575	0.2530	0.2497	0.2479	0.2427	0.2349	0.1680	0.0782	0.0172
1.9	0.2851	0.2022	0.2122	0.2155	0.2142	0.2113	0.2083	0.2028	0.1530	0.0769	0.0182
2.0	0.2012	0.1929	0.1878	0.1854	0.1850	0.1837	0.1807	0.1771	0.1393	0.0749	0.0191

$\sigma_{z1} = \sigma_{z3} = 500.0(0.8064 + 0.2249 + 0.0771)/8.0^2 + 58.8\,(\text{point load})/2 = 38.1\,\text{kPa}.$
Pile 2: $I_{q2} = 0.8064$, $I_{q1} = I_{q3} = 0.2249.$
$\sigma_{z2} = 500.0(0.2249 + 0.8064 + 0.2249)/8.0^2 + 60.4/2 = 40.0\,\text{kPa}.$

5.4 ELASTIC SETTLEMENT OF FOOTINGS

5.4.1 *Distribution of contact pressure and settlement*

Elastic settlement of the soil under a loaded area may be calculated by integrating Equation 5.6 (vertical strain) along a vertical plane. The boundary conditions for the integration depend on the rigidity of the footing. In a rigid footing the settlement under the loaded area is uniform. While the vertical stress component is independent of the elastic properties of the soil, the elastic settlement, however, depends on the magnitudes of E_s and μ. Available analytical solutions have to be applied with care as E_s may change with depth or with confining pressure. Analytical formulae and discussions for the elastic treatment of heterogeneous soils are suggested by Poulos & Davis (1974), and Gibson (1974). In general, a linear increase in E_s with depth is recommended. In most clay soils it is reasonable to assume a constant E_s, while in sands the elastic formulation has little use as the E_s changes along the loaded area as well as with depth. Figure 5.25 shows the distribution of contact pressure and settlement under both rigid and flexible uniformly loaded footings constructed on clay and sand layers. The assumption of ideal elastic properties for the soil implies that settlement occurs also outside of the loaded area and approaches zero in Boussinesq type behaviour. Most of the available formulations define the settlements at the corner of a rectangular loaded area as well as at the centre and edges of a circular loaded area. In the case of a rectangular loaded area, superposition can be used to estimate the settlements within or outside the loaded area. Bowles (1996) suggested that elastic solutions are reliable and the differences between measured and calculated settlements are mostly due to the errors involved in the determination of the elastic properties of the soil. An alternative approach is to divide the soil beneath the loaded area into horizontal layers of finite thickness and estimate an average vertical stress, an average vertical strain and the vertical deformation for each layer. This method has been found to yield reliable solutions when compared with the traditional elastic numerical analysis (Powrie, 1997).

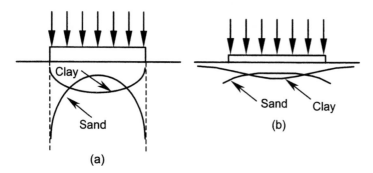

Figure 5.25. Distribution of contact pressure and settlement:
(a) contact pressure under a rigid footing, (b) settlement under a flexible footing.

5.4.2 *Elastic settlement under flexible and rigid footings*

Based on the theory of elasticity (Timoshenko & Goodier, 1982), the elastic settlement S_e of a flexible footing, either rectangular of dimensions $L \times B$ ($L > B$) or circular of diameter B, is given by:

$$S_e = qB \frac{1-\mu^2}{E_s} I_s \qquad (5.54)$$

where I_s is an influence factor depending on the shape and L/B ratio. In this derivation, the Modulus of Elasticity of the soil is assumed to be constant with depth and the soil is infinitely deep. The influence factor for a flexible rectangular footing is based on the original analytical solution developed by Steinbrenner (1934) for a corner of a loaded rectangular area.

$$I_1 = \frac{1}{\pi}[\frac{L}{B}\ln(\frac{1+\sqrt{(L/B)^2+1}}{L/B})+\ln(\frac{L}{B}+\sqrt{(L/B)^2+1})] \qquad (5.55)$$

The centre of a flexible rectangular footing is the common corner for 4 rectangles of $L' = L/2$ and $B' = B/2$. Noting that $L'/B' = L/B$, the settlement at the centre is:

$$S_c = 4 \times qB' \frac{1-\mu^2}{E_s} I_1 = qB \frac{1-\mu^2}{E_s}(2I_1), \text{ and therefore:}$$

$$I_s \text{ (centre)} = 2\,I_1 \qquad (5.56)$$

Thus the elastic settlement at the centre of a flexible rectangular footing is twice that of its corner. It can be shown that the average influence factor for a flexible rectangular footing is approximately 0.848 that of the corresponding influence factor at the centre:
$I_s = 0.848 \times 2I_1$ and therefore:

$$I_s \text{ (average)} = 1.696\,I_1 \qquad (5.57)$$

For a rigid rectangular footing the influence factor is approximately 0.926 of the average influence factor of the flexible footing.

Hence, I_s (rigid) $= 0.926 \times I_s$ (average) $= 0.926 \times 1.696\,I_1$, or:

$$I_s \text{ (rigid)} = 1.570 I_1 = (\pi/2)I_1 \qquad (5.58)$$

The variation of I_s with the L/B ratio is shown in Figure 5.26. For a flexible circular footing the influence factors at the centre and edge of the footing are 1.0 and $2/\pi \approx 0.64$ respectively, and the average influence factor is 0.848. For a rigid circular footing the influence factor is $\pi/4 \approx 0.79$ which is 0.926 of the average value for the flexible footing. Indeed, the factor 0.926 is the ratio of the settlement under a rigid circular footing with non-uniform contact pressure distribution to the settlement corresponding to the uniform contact pressure under the same vertical load. This means that for a rigid footing, regardless of the distribution of the contact pressure, the maximum difference in the settlement is approximately 7.4%. The influence factor of 0.79 for a rigid circular footing is calculated on the basis of a non-uniform contact pressure distribution. Note that the influence factors for circular loading are based on analytical solutions. It is evident that the ratio between rigid and elastic influence factors in a circular footing is equally applicable to a

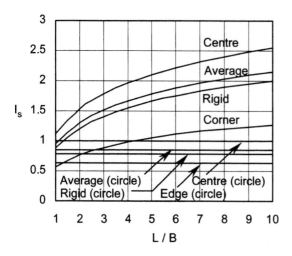

Figure 5.26. Steinbrenner's influence factors for circular and rectangular footings.

rectangular footing. Based on an analysis carried out by Jumikis (1969), the influence factors within and outside a flexible circular footing are presented in Table 5.3. The parameter a represents the distance of the point of interest from the centre of the circular footing whilst R_c is the radius of the circle.

Table 5.3. Influence factors for a flexible circular footing.

a/R_c	I_s	a/R_c	I_s	a/R_c	I_s
0.00	1.000	1.75	0.307	3.50	0.146
0.25	0.984	2.00	0.258	3.75	0.134
0.50	0.934	2.25	0.234	4.00	0.125
0.75	0.840	2.50	0.204	4.25	0.118
1.00	0.637	2.75	0.191	4.50	0.112
1.25	0.444	3.00	0.167	4.75	0.105
1.50	0.360	3.25	0.157	5.00	0.099

Example 5.17

A 4 m by 8 m flexible rectangular footing exerts a uniform contact pressure of 200 kPa on the ground surface. Calculate the elastic settlement at its centre, corners and the centre points of its long and short sides. $E_s = 20000$ kPa, $\mu = 0.35$.

Solution:

$L/B = 2.0$, $I_s = 1.53$ for the centre point and $I_s = 0.76$ for the corners. From Equation 5.54:

S_e (corner) $= 200 \times 4.0 \times [1 - (0.35)^2] \times 0.76 / 20000 = 0.0267 \, \text{m} \approx 27 \, \text{mm}$.

S_e (centre) $= 200 \times 4.0 \times [1 - (0.35)^2] \times 1.53 / 20000 = 0.0537 \, \text{m} \approx 54 \, \text{mm}$.

To find the settlement at the centre point of the longer side the original footing is subdivided into two rectangles (in this example two squares) sharing the centre point at their corners. The total settlement at this point is equal to twice the settlement of the corner of one rectangle:

$L/B = 4.0/4.0 = 1.0$, $I_s = 0.56$,

S_e (centre of long side) $= 2 \times 200 \times 4.0 \times [1 - (0.35)^2] \times 0.56/20000 = 0.0393$ m ≈ 39 mm.

For the centre point of the short side $L/B = 8.0/2.0 = 4$, $I_s = 0.98$,

S_e (centre of short side) $= 2 \times 200 \times 2.0 \times [1 - (0.35)^2] \times 0.98/20000 = 0.0344$ m ≈ 34 mm.

Example 5.18

Find the elastic settlement at point A of the flexible footing shown in Figure 5.14. $E_s = 15000$ kPa, $\mu = 0.3$.

Solution:

The right half of the footing is divided into 5 rectangles all sharing point A as a common corner. The results of calculations are tabulated below.

Rectangle	L (m)	B (m)	L/B	I_s	S_e (m)
Aabc	4.0	2.0	2.0	0.76	0.0277
Adef	3.0	3.0	1.0	0.56	0.0306
Adgc	3.0	2.0	1.5	0.68	−0.0247
Ahif	3.0	1.0	3.0	0.89	−0.0162
Ahjc	2.0	1.0	2.0	0.76	0.0138

$S_e = 2 \, (0.0277 + 0.0306 - 0.0247 - 0.0162 + 0.0138) = 0.0624$ m

5.4.3 Settlement of a soil layer of finite thickness

If an incompressible layer (for which $E_s \to \infty$) is located at depth H from the ground surface, the surface settlement is reduced. In this case, the I_s values for a flexible rectangular footing are given by:

$$I_s(\text{corner}) = I_1 = I_2 + \frac{1 - 2\mu}{1 - \mu} I_3, \, I_s(\text{centre}) = 2I_1, \, I_s(\text{average}) = 1.696I_1 \quad (5.59)$$

where I_2 and I_3 are Steinbrenner's influence factors given by:

$$I_2 = \frac{1}{\pi} \left\{ \frac{L}{B} \ln \left[\frac{(1 + \sqrt{(L/B)^2 + 1})\sqrt{(L/B)^2 + (H/B)^2}}{L/B(1 + \sqrt{(L/B)^2 + (H/B)^2 + 1})} \right] + \right.$$

$$\left. \ln \left[\frac{(L/B + \sqrt{(L/B)^2 + 1})\sqrt{1 + (H/B)^2}}{L/B + \sqrt{(L/B)^2 + (H/B)^2 + 1}} \right] \right\} \quad (5.60)$$

$$I_3 = \frac{(H/B)}{2\pi} \tan^{-1} \left[\frac{L/B}{(H/B)\sqrt{(L/B)^2 + (H/B)^2 + 1}} \right] \quad (5.61)$$

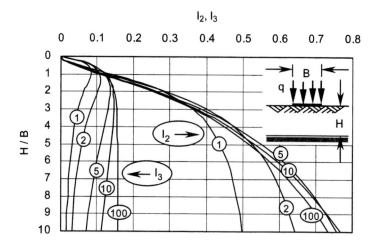

Figure 5.27. Steinbrenner's influence factors for the elastic settlement of a layer of finite thickness.

Equation 5.59 (I_s (centre)) may be used for the centre of a flexible circular footing by converting the circle into an equivalent square. The influence factor at the edge of the circle is $2 / \pi$ times I_s (centre). For rigid rectangular and circular footings the influence factor is 0.926 of the average, or 0.79 of the central values. Figure 5.27 shows the variations of I_2 and I_3 with H / B for selected values of L / B (circled numbers).

Example 5.19

Re-work Example 5.17 with a rigid layer located at 12 m depth from the ground surface and compare the results.

Solution:

$H / B = 12.0 / 4.0 = 3.0, L / B = 2.0$.
Substituting these values into Equations 5.60 and 5.61:
$I_2 = 0.402, I_3 = 0.084$, thus from Equations 5.59:
I_s (corner) $= 0.402 + (1 - 2 \times 0.35) / (1 - 0.35) \times 0.084 = 0.4408$.
I_s (centre) $= 2 \times 0.4408 = 0.8816$.

S_e (corner) $= 200 \times 4.0 \times [1 - (0.35)^2] \times 0.4408 / 20000 = 0.0155 \, \text{m} \approx 16 \, \text{mm}$.

S_e (centre) $= 200 \times 4.0 \times [1 - (0.35)^2] \times 0.8816 / 20000 = 0.0309 \, \text{m} \approx 31 \, \text{mm}$.

For the centre point of the long side:
$H / B = 3.0, L / B = 1.0, I_2 = 0.363, I_3 = 0.048$,
$I_s = 0.363 + (1 - 2 \times 0.35) / (1 - 0.35) \times 0.048 = 0.3851$.

S_e (centre of long side) $= 2 \times 200 \times 4.0 \times [1 - (0.35)^2] \times 0.3851 / 20000 = 0.0270 \, \text{m} \approx 27 \, \text{mm}$.
For the centre point of the short side: $H / B = 6.0, L / B = 4.0, I_2 = 0.609, I_3 = 0.087$,
$I_s = 0.609 + (1 - 2 \times 0.35) / (1 - 0.35) \times 0.087 = 0.6491$.

S_e (centre of short side) $= 2 \times 200 \times 2.0 \times [1 - (0.35)^2] \times 0.6491 / 20000 = 0.0228 \, \text{m} \approx 23 \, \text{mm}$.
The settlements significantly reduced by the presence of the incompressible layer.

5.4.4   *Reduction of settlement due to footing embedment depth*

Fox (1948) investigated the effect of the embedment of a footing (D) on the average elastic settlement of a uniformly loaded rectangular area and introduced the concept of a depth factor. Here, the average settlement for the footing at ground surface level is multiplied by the depth factor I_F which is always less than unity. Plots of the depth factor are given in Figure 5.28. From Figure 5.28(b) it can be seen that the depth factor decreases with a reduction in Poisson's ratio. Table 5.4 shows the variation of I_F for $\mu = 0.5$ for various values of L/B and D/B. For a circular footing an equivalent square area may be used. Fox's depth factor may also be applied to consolidation settlement as recommended by Tomilinson (1995).

Table 5.4. Depth factors I_F for $\mu = 0.5$.

D/B	$L/B=1.0$	$L/B=2.0$	$L/B=3.0$	$L/B=4.0$	$L/B=5.0$	$L/B=10.0$
0.00	1.0000	1.0000	1.0000	1.0000	1.0000	1.0000
0.25	0.9424	0.9650	0.9723	0.9760	0.9783	0.9883
0.50	0.8501	0.8993	0.9184	0.9284	0.9348	0.9489
0.75	0.7760	0.8372	0.8651	0.8808	0.8909	0.9139
1.00	0.7234	0.7864	0.8196	0.8394	0.8525	0.8829
1.50	0.6587	0.7150	0.7510	0.7750	0.7919	0.8334
2	0.6220	0.6698	0.7039	0.7288	0.7473	0.7957
3	0.5828	0.6180	0.6459	0.6684	0.6868	0.7412
4	0.5625	0.5899	0.6126	0.6319	0.6485	0.7030
5	0.5502	0.5724	0.5913	0.6079	0.6226	0.6746
6	0.5419	0.5606	0.5767	0.5911	0.6040	0.6526
8	0.5315	0.5456	0.5580	0.5692	0.5796	0.6210
10	0.5252	0.5366	0.5466	0.5557	0.5643	0.5998
50	0.5050	0.5073	0.5094	0.5113	0.5131	0.5212
100	0.5025	0.5037	0.5047	0.5056	0.5065	0.5106

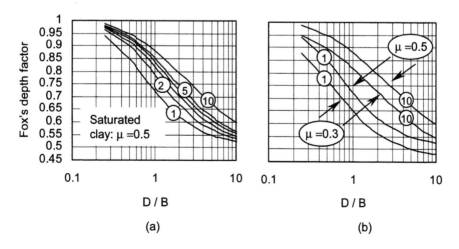

Figure 5.28. Depth factor: (a) saturated clay, $\mu = 0.5$, (b) effect of Poisson's ratio.

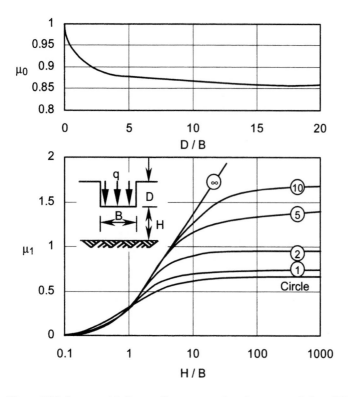

Figure 5.29. Improved influence factors μ_0 and μ_1 for saturated clays (Christian & Carrier, 1978).

5.4.5 *Alternative influence factors for saturated clays*

For shallow and deep flexible footings in saturated clay with $\mu = 0.5$, the chart proposed by Janbu et al. (1956) finds significant use in design. The settlement is calculated from:

$$S_e = \frac{\mu_0 \mu_1 qB}{E_s} \tag{5.62}$$

where μ_0 and μ_1 are influence factors for depth and layer thickness respectively. Improvements to the μ_0 and μ_1 influence factors (Figure 5.29) were proposed by Christian & Carrier (1978) based on the work carried out by Burland (1970) and Giroud (1972).

Example 5.20

A flexible 2 m × 2 m footing is located at a depth of 4 m and exerts a uniform contact pressure of 400 kPa. If an incompressible stratum is at 10 m below the ground surface, calculate the average elastic settlement of the footing. $E_s = 20000$ kPa, $\mu = 0.5$.

Solution:

$H / B = 10.0 / 2.0 = 5.0$ and $L / B = 2.0 / 2.0 = 1.0$, $I_2 = 0.437$, $I_3 = 0.031$,

$$I_s(\text{corner}) = 0.437 + \frac{1 - 2 \times 0.5}{1 - 0.5} \times 0.031 = 0.4370.$$

I_s (average) $= 1.696 \times I_s$ (corner) $= 1.696 \times 0.4370 = 0.7411$.

For $D / B = 4.0 / 2.0 = 2.0$, the depth factor from Table 5.4 is: 0.622,

$$S_e \text{(average)} = 400 \times 2.0 \frac{1-0.5^2}{20000} \times 0.7411 \times 0.622 = 0.0138 \text{ m} \approx 14 \text{ mm}.$$

Using Janbu's improved chart (Figure 5.29) for $H / B = 3.0$ and $L / B = 1.0$: $\mu_0 \approx 0.9$, $\mu_1 \approx 0.58$. From Equation 5.62:

$$S_e = \frac{0.9 \times 0.58 \times 400 \times 2.0}{20000} = 0.0208 \text{ m} \approx 21 \text{ mm}.$$

5.4.6   *Estimation of settlements in sands*

Due to the variation in the compressibility characteristics of sands, Equation 5.54 is not applicable to the estimation of their elastic settlement. This non-uniformity exists under identical footings in the same layer of sand as well as in laboratory samples. This non-uniformity becomes reasonably acceptable for normally consolidated sands. However, as discussed in Chapter 4, the stress level in the sand should be more than 700 kPa in order to approach to the relevant *NCL* (Atkinson & Bransby, 1978), and recompression in the moderate stresses is a horizontal line indicating no volume change. The following methods used to estimate settlements in sands are based on in-situ tests and the theory of elasticity.

Estimation of settlement using plate load tests. Terzaghi & Peck (1948) reported an extrapolation equation that related the settlement of a loaded circular or square plate of 0.305 m ($S_{0.305}$) to the settlement of a footing (S_e) of width B with the same contact pressure.

$$S_e = S_{0.305} \left(\frac{2B}{0.305 + B} \right)^2 \tag{5.63}$$

The unreliability of this extrapolation equation is discussed by Bjerrum & Eggestad (1963), Sutherland (1975), Clayton et al. (1995) and Terzaghi et al. (1996). However, recent findings in cemented residual soils show that extrapolation can be carried out to relate the settlement of a loaded plate to the settlement of an actual footing. Consoli et al. (1998) have shown that, by expressing the load test results in terms of dimensionless variables (e.g. normalised applied stress versus ratio of settlement to diameter / width), clear behavioural patterns emerge.

Estimation of settlements using the Burland and Burbidge method. Based on the results of Standard Penetration Test (*SPT*; Chapter 10) as well as observations and measurements of numerous field cases, Burland & Burbidge (1985) suggested a semi-empirical solution for the settlement of a normally consolidated sand. The surface settlement (S_e), which is equal to the integration of vertical strain ε_z, is estimated using the average value of the coefficient of compressibility m_v within the strain influence depth of Z_I under a surface loading of q:

$$S_e = \int_0^{Z_I} \varepsilon_z dz = Z_I m_v q \tag{5.64}$$

The parameter m_v is defined by: $m_v = \varepsilon_z \text{(average)} / \sigma_z \text{(average)}$.

Empirical expressions for Z_I (in m) and m_v (in MPa^{-1}) are:

$$Z_I = B^{0.75} \qquad (5.65)$$

$$m_v = \frac{1.7}{\overline{N}^{1.4}} \qquad (5.66)$$

where B is the width of the footing in m and $\overline{N}$ is the average *SPT* number within the influence depth Z_I. The value of $\overline{N}$ is corrected for saturated sands and gravelly sands:

$$N' = 15 + 0.5(\overline{N} - 15) \qquad \text{saturated sand} \qquad (5.67)$$

$$N' = 1.25\overline{N} \qquad \text{gravelly sand} \qquad (5.68)$$

For a normally consolidated sand the general equation for its settlement is:

$$S_e = B^{0.75}\frac{1.7}{\overline{N}^{1.4}}q \qquad (5.69)$$

where S_e is the settlement in mm and q is the contact pressure in kPa. Note that Equation 5.69 does not imply a linear relationship between the settlement and the width of the footing, as any change in B will alter Z_I which in turn will change the average *SPT* number $\overline{N}$. For overconsolidated soils, and for q greater than the preconsolidation pressure p'_c, the settlement is the sum of the recompression and normally consolidated stages, and a different value for m_v at the recompression side has to be evaluated. Assuming the compressibility index at the recompression stage is $1/3$ of the normally consolidated stage, then:

$$S_e = B^{0.75}\frac{1.7}{\overline{N}^{1.4}}(q - 2p'_c/3) \qquad \text{for } q > p'_c \qquad (5.70)$$

If the applied contact pressure is less than p'_c this becomes:

$$S_e = B^{0.75}\frac{1.7}{\overline{N}^{1.4}}(q/3) \qquad \text{for } q < p'_c \qquad (5.71)$$

Equations 5.69 and 5.70 are also applicable to footings located at some depth below the ground surface if the preconstruction vertical stress is substituted for p'_c. For rectangular footings with $L/B > 1.0$, the settlement is increased by a shape factor given by:

$$I_s = (\frac{1.25L/B}{L/B + 0.25})^2 \qquad (5.72)$$

The maximum value of this factor occurs in a strip footing when $L/B \to \infty$ and is equal to 1.56. If the incompressible layer is located within the influence depth Z_I, the calculated settlement is multiplied by a correction factor defined by:

$$I_f = \frac{H(2 - H/Z_I)}{Z_I} \qquad (5.73)$$

where H is the depth of the incompressible layer from the footing level.

Example 5.21

A footing on sand has plan dimensions of 2 m × 6 m and exerts a loading of $q = 300$ kPa. The values of N from *SPT* tests are tabulated below. If the water table is 1 m below the ground surface, calculate the settlement of the footing for the following cases: (a) the base of the footing is on the ground surface, (b) the base is 1 m below the ground surface. For both cases find the uniform contact pressure that causes 25 mm average settlement.

Depth (m)	0.0	0.5	1.0	1.5	2.0	2.5	3.0
N	11	12	14	19	19	23	27

Solution:

The corrected N values (for $N > 15$ and in the saturated zone) are tabulated below:

Depth (m)	0.0	0.5	1.0	1.5	2.0	2.5	3.0
N	11	12	14	17	17	19	21

From Equation 5.65: $Z_I = B^{0.75} = 2.0^{0.75} = 1.68$ m; for case (a) the average value of $\overline{N} = 14$, thus from Equations 5.69 and 5.72:

$S_{(2.0\times2.0)} = 1.68\times1.7\times300.0/14^{1.4} = 21.3$ mm,

$I_s = [1.25\times3.0/(3.0+0.25)]^2 = 1.33, S_{(2.0\times6.0)} = S_{(2.0\times2.0)}\times I_s = 21.3\times1.33 = 28.3$ mm.

For case (b) the average value of $\overline{N} = 17$, and

$S_{(2.0\times2.0)} = 1.68\times1.7\times300.0/17^{1.4} = 16.2$ mm,

$S_{(2.0\times6.0)} = S_{(2.0\times2.0)}\times I_s = 16.2\times1.33 = 21.5$ mm.

For 25 mm settlement: $S_{(2.0\times6.0)} = 1.33\times1.68\times1.7\times q/14^{1.4} = 25$ mm,

$q = 264.8$ kPa for case (a).

$S_{(2.0\times6.0)} = 1.33\times1.68\times1.7\times q/17^{1.4} = 25$ mm, $q = 347.5$ kPa for case (b).

Estimation of settlements using Schmertmann's method. An alternative method of estimating settlement is based on a numerical approximation of Equation 5.64 in which the soil beneath the loaded area is divided into a finite number of layers:

$$S_e = \sum_{i=1}^{i=n} (\varepsilon_z z_l)_i \tag{5.74}$$

where n is the number of the layers within an influence depth of Z_I, $(z_l)_i$ is the thickness of a typical layer and $(\varepsilon_z)_i$ is the average vertical strain of the layer. Substituting $(\varepsilon_z)_i$ from Hooke's law (Equation 5.6) into Equation 5.74 we obtain:

$$S_e = \sum_{i=1}^{i=n} \left[\frac{\sigma_z - \mu(\sigma_x + \sigma_y)}{E_s} z_l \right]_i \tag{5.75}$$

Schmertmann (1970) and Schmertmann et al. (1978) proposed solutions for sands in which there was no need to estimate the stress components in the vicinity of the influence depth. The Modulus of Elasticity must be determined for each layer by means of in-situ testing, preferably using the Cone Penetration Test (*CPT*; Chapter 10). Introducing a strain influence factor I_z to replace the stress term in Equation 5.75, the settlement increment is defined by:

$$I_z = [\sigma_z - \mu(\sigma_x + \sigma_y)]/q \rightarrow \Delta S_e = q\times I_z/E_s \tag{5.76}$$

Equation 5.75 may therefore be written as:

$$S_e = q \sum_0^{Z_I} \frac{I_z}{E_s} z_1 \tag{5.77}$$

The influence depth Z_I is defined by:

$$Z_I = 2B(1 + \log\frac{L}{B}) \tag{5.78}$$

In a circular loaded area, the strain influence factor I_z under its centre can be calculated mathematically in terms of depth z or relative depth z / B (B being the diameter) by integrating ε_z over the influence depth assumed under the footing. The variation of I_z with depth has a Boussinesq pattern and, after increasing to a maximum, decreases to a negligible value in the specified influence depth. Mayne & Poulos (1999) presented the strain influence factors for a circular loaded area for both rigid and flexible footings. Papadopoulos (1992) introduced a different model for the distribution of applied stress, which takes into account the density of the soil; strains can then be integrated to yield the settlement of the footing. However, depending on the shape of the footing, the mathematical integration of ε_z is either difficult or, in some cases, impossible. Strain influence factors, given by Schmertmann (1970) and Schmertmann et al. (1978), that idealize the Boussinesq pattern by straight lines (e.g. line *ABC* in Figure 5.30(a)) are based on extensive in-situ tests, and are shown in Figures 5.30. Equation 5.77 is multiplied by depth and time factors defined by:

$$C_1 = 1 - 0.5\frac{p'_o}{q} \geq 0.5 \tag{5.79}$$

$$C_2 = 1 + 0.2\log\frac{t}{0.1} \tag{5.80}$$

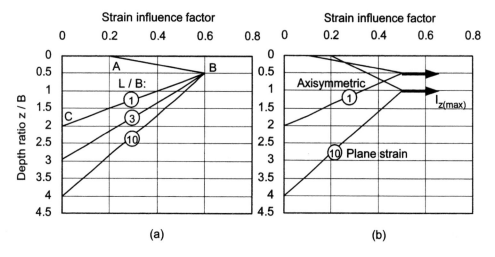

Figure 5.30. (a) Strain influence factor (Schmertmann, 1970), (b) improved strain influence factor (Schmertmann, et al., 1978).

where p'_o is the effective overburden pressure at the footing level and t is the elapsed time in years. Correction of the settlement for the effects of time may not be necessary (Holtz, 1991; Terzaghi et al., 1996) as observed field results are probably due to consolidation of clay or silt lenses within the sand. The term q represents the net contact pressure at the foundation level. If the footing is at some depth below the ground surface, q has to be replaced by $q - p'_o$. In the improved strain influence factor diagram the maximum value of I_z is given by:

$$I_{z(\text{max})} = 0.5 + 0.1 \sqrt{\frac{q - p'_o}{p'_{(B/2 \text{ or } B)}}} \tag{5.81}$$

where $p'_{(B/2 \text{ or } B)}$ is the effective overburden pressure at the depth of $B / 2$ or B from the footing level depending on the L / B ratio. For $L / B = 1$ (axisymmetric) and $L / B > 10$ (plane strain), the effective overburden pressures correspond to $B / 2$ and B respectively.

Example 5.22

A flexible rectangular footing of 2 m × 4 m applies a uniform load of 100 kPa to a horizontal ground surface. The underlying stratum is divided into 6 layers as shown in the table below. Estimate the elastic settlement under the centre of the footing for the following cases: (a) the soil has a constant Modulus of Elasticity of $E_s = 15000$ kPa, (b) the soil is heterogeneous and its Modulus of Elasticity changes linearly from 15000 kPa at the ground surface to 30000 kPa at a depth of 10 m.

Layer (downwards)	1	2	3	4	5	6
Thickness (m)	0.50	1.00	1.00	1.67	1.67	3.00

Solution:

The influence factor for the vertical stress σ_z at the centre of each layer is 4 times that for the influence factor calculated for the corner of $L' = L / 2 = 4.0 / 2 = 2.0$ m and $B' = B / 2 = 2.0 / 2 = 1.0$ m. The average vertical strain ε_z is computed from Equation 5.6 by ignoring the lateral normal stresses. For case (a) $\varepsilon_z = \sigma_z / 15000$.

For case (b) $E_s = E_o + nz$, where $E_o = 15000$ kPa and $n = (30000 - 15000) / 10.0 = 1500$, thus:

$E_s = 15000 + 1500z$ and $\varepsilon_z = \sigma_z / (15000 + 1500z)$. In both cases:

$\Delta s =$ the settlement of each layer $= \varepsilon_z \times z_l$, where z_l is the thickness of each layer.

The results of computations for both cases are tabulated below in which the symbols C and L represent constant and linear E_s respectively.

Layer	Depth (m)	m	n	$4 \times I_q$	σ_z (kPa)	$\varepsilon_z (C)$ $\times 10^{-3}$	$\varepsilon_z (L)$ $\times 10^{-3}$	$\Delta s (C)$ (mm)	$\Delta s (L)$ (mm)
1	0.25	8.000	4.000	0.99343	99.34	6.62	6.46	3.31	3.23
2	1.00	2.000	1.000	0.79976	79.98	5.33	4.85	5.33	4.85
3	2.00	1.000	0.500	0.48070	48.07	3.20	2.67	3.20	2.67
4	3.33	0.600	0.300	0.25174	25.17	1.68	1.26	2.80	2.10
5	5.00	0.400	0.200	0.13119	13.12	0.87	0.58	1.45	0.97
6	7.34	0.272	0.136	0.06562	6.56	0.44	0.25	1.32	0.75
							Total:	17.4	14.6

Example 5.23

A square footing 3 m × 3 m is located at a depth of 2 m and exerts a uniform contact pressure of 200 kPa (includes the weight of the soil above the footing level) on an underlying deep deposit of sand. Calculate S_e using Schmertmann's modified strain influence factor if the variation in E_s with depth z given by: E_s (kPa) = 5000 + 3000 z (m). ρ = 1.7 Mg/m^3.

Solution:

$p'_o = 1.7 \times 9.81 \times 2.0 = 33.3$ kPa, $q - p'_o = 200.0 - 33.3 = 166.7$ kPa.

From Equation 5.79: C_1 = correction factor for depth = $1 - 0.5 (33.3 / 166.7) = 0.9$.

Calculating the influence depth from the footing level (Equation 5.78):

$Z_1 = 2 \times 3.0 \times [1 + \log(3.0/3.0)] = 6.0$ m. From Equation 5.81:

$$I_{z(max)} = 0.5 + 0.1\sqrt{166.7/[33.3 + (3.0/2) \times 1.7 \times 9.81]} = 0.669.$$

For $L / B = 1.0$, the maximum strain influence factor occurs at the depth of $3.0 / 2 = 1.5$ m from the footing level or $1.5 + 2 = 3.5$ m from the ground surface. The equation of the line representing the strain influence factor from the foundation level to the depth of 3.5 m is:

$(I_z - 0.1)/(z - 2.0) = (0.669 - 0.1)/(3.5 - 2.0)$, or $I_z = 0.1 + 0.3793(z - 2.0)$.

The strain influence factor for the range z = 3.5 m to z = 6.0 + 2.0 = 8.0 m is:

$(I_z - 0.0)/(z - 8.0) = (0.669 - 0.0)/(3.5 - 8.0)$, or $I_z = 0.1487(8.0 - z)$.

The two equations above are plotted in Figure 5.31. The influence depth is divided into 6 layers, each 1 m thick. Results of the calculations are tabulated below.

Layer	D (m)	I_z	E_s (kPa)	$I_z \times z_1 / E_s$ (mm/kPa)	Layer	D (m)	I_z	E_s (kPa)	$I_z \times z_1 / E_s$ (mm/kPa)
1	2.5	0.2896	12500	0.0232	4	5.5	0.3717	21500	0.0173
2	3.5	0.6690	15500	0.0432	5	6.5	0.2230	24500	0.0091
3	4.5	0.5204	18500	0.0281	6	7.5	0.0743	27500	0.0027
								Total	0.1236

From Equation 5.77, $S_e = 0.9 \times 166.7 \times 0.1236 = 18.5$ mm.

5.4.7 Estimation of the elastic properties of soil

The Modulus of Elasticity E_s can be estimated from the slope of the experimental stress-strain relationship obtained from an appropriate triaxial compression test. The tangent modulus is the slope of the stress-strain curve at the early stage of a test whilst the secant modulus is the slope of a line from the origin to a point at 50% of the peak strength (Figure 5.32). Poisson's ratio can be obtained using Equation 5.8, as at different stages of the triaxial compression test both axial deformation and volume change are recorded. Samples must accurately represent the field condition in terms of physical properties such as density and moisture content. Any disturbance to the specimen will decrease its Modulus of Elasticity (Bowles, 1996). However, a sample with moisture content drier than that in the field may give a higher value for E_s. In granular materials where the preparation of an undisturbed sample is not possible, care has to be taken to duplicate the void ratio during the preparation of a disturbed sample.

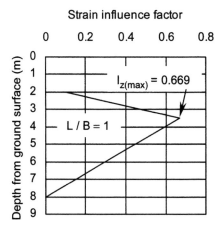

Figure 5.31. Example 5.23.

It is vital that any disturbed or undisturbed sample be prepared in accordance with the relevant code. In granular materials the confining pressure (σ_3) should be chosen to match field conditions. An increase in the confining pressure increases the Modulus of Elasticity of the specimen. The drainage condition in the test must also simulate the field situation, especially in clay soils. It is always good practice to determine E_s in both the fully drained and undrained conditions. Based on laboratory and field experiments, D'Appolonia et al. (1971) reported that the Modulus of Elasticity in clay varies with plasticity index, sensitivity, and overconsolidation ratio, and may be formulated as a linear function of undrained shear strength. Furthermore, the Modulus of Elasticity of clay from different places around the world varies from the very small value of 100 kPa up to about 13000 kPa. Table 5.5 shows the variation of the ratio of undrained modulus to undrained shear strength in terms of overconsolidation ratio and plasticity index (Duncan & Buchignani, 1976).

In sands the magnitudes of the elastic parameters depend upon the particle size as well as the initial void ratio (Harr, 1966 and 1977). The Modulus of Elasticity reduces with decreasing of size of particles. Lower values of initial void ratio yield a high value of 45200 kPa for E_s in coarse and medium coarse sand (Table 5.6).

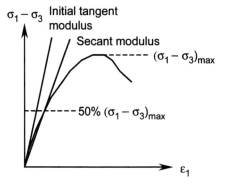

Figure 5.32. Determination of Modulus of Elasticity.

Table 5.5. Variation of E_{su}/τ_f in clay soils (Duncan & Buchignani, 1976).

	E_{su}/τ_f $(\tau_f = c_u + \sigma \tan\phi_u)$		
OCR	$PI < 30$	$30 < PI < 50$	$PI > 50$
<3	600	300	125
3-5	400	200	75
>5	150	75	50

Table 5.6. Modulus of Elasticity and Poisson's ratio in sandy soils (Harr, 1966).

Soil	Poisson's ratio: μ	E_s (kPa)		
		$0.41 < e < 0.5$	$0.51 < e < 0.6$	$0.61 < e < 0.7$
Coarse and medium coarse sand	0.15-0.20	45200	39300	32400
Fine-grained sand	0.25	36600	27600	23500
Sandy silt	0.30-0.35	13800	11700	10000

Recent improvements in the correlation of *SPT* and *CPT* results to the soil strength properties, the Modulus of Elasticity and compressibility index have created useful practical relationships. Bowles (1996) suggested the following empirical relationships between *SPT* results and E_s:

$$E_s = 500\,(N + 15) \qquad \text{for normally consolidated sands} \qquad (5.82)$$

$$E_{s\,(over.)} = E_{s\,(norm.)}\sqrt{OCR} \qquad \text{for overconsolidated sands} \qquad (5.83)$$

$$E_s = 250\,(N + 15) \qquad \text{for saturated sands} \qquad (5.84)$$

$$E_s = 600\,(N + 6) \qquad \text{for gravelly sands, } N \le 15 \qquad (5.85)$$

$$E_s = 600\,(N + 6) + 2000 \qquad \text{for gravelly sands, } N > 15 \qquad (5.86)$$

$$E_s = 320\,(N + 15) \qquad \text{for clayey sands} \qquad (5.87)$$

$$E_s = 300\,(N + 15) \qquad \text{for silts} \qquad (5.88)$$

Using *CPT* results, Meyerhof & Fellenius (1985) proposed the following relationship in which the term q_c represents the tip resistance that has the same units as E_s.

$$E_s = K\,q_c \qquad (5.89)$$

where: $K = 1.5$ for silts and sands, $K = 2.0$ for compacted sand, $K = 3$ for dense sand, and $K = 4$ for sand and gravel. Schmertmann et al. (1978) suggested the use of the following for axisymmetric and plane strain loadings:

$$E_s = 2.5q_c \qquad \text{for axisymmetric} \qquad (5.90)$$

$$E_s = 3.5q_c \qquad \text{for plane strain} \qquad (5.91)$$

5.5 SOIL-FOOTING INTERACTION MODELS

5.5.1 *The concept of rigid footing*

Figure 5.33(a) shows a rigid footing subjected to parallel vertical forces and moments that are applied about the axes located in the plane of footing. This loading system is equivalent to a single resultant force P (Figure 5.33(b)) that is perpendicular to the plane of footing. It is more convenient to express the contact pressure in terms of the coordinate system $Gxyz$ where G represents the centroid of the footing, z is perpendicular to the footing and parallel to the contact pressure, and xGy is the plane of the footing. To simplify the problem the resultant force P is moved from its application point O to point G, which requires the addition of the two external moments, M_x and M_y, to maintain equilibrium:

$$M_x = -P\,y_O,\ M_y = P\,x_O \tag{5.92}$$

where x_O and y_O are the coordinates of point O (assumed positive in Equation 5.92) in the xGy coordinate system, and the right-hand rule sign convention for moments is adopted. The contact pressure under the rigid footing is assumed to be linear and independent of the type of soil:

$$q = Ax + By + C \tag{5.93}$$

where A, B and C are coefficients depending on the geometry of the footing and applied forces, and x, y are the coordinates of any point within the contact area. These coefficients are found by solving the three equilibrium equations of:

$\sum F_z = 0$, $\sum M_x = 0$ and $\sum M_y = 0$:

$$A = -\frac{I_{xy}M_x + I_x M_y}{(I_{xy})^2 - I_x I_y},\ B = \frac{I_{xy}M_y + I_y M_x}{(I_{xy})^2 - I_x I_y},\ C = \frac{P}{S} \tag{5.94}$$

where S is the (plan) contact area of the footing, I_x and I_y are the second moments of area of the footing about the x and y axes, and I_{xy} is the product second moment of area in the xGy plane. If the application point of the resultant P coincides with the centriod of the footing, then $M_x = M_y = 0$ and the soil pressure is uniform and given by $q = P / S$. If the footing is symmetrical about the Gx or Gy axes, then $I_{xy} = 0$ and:

$$q = \frac{M_y}{I_y}x - \frac{M_x}{I_x}y + \frac{P}{S} \tag{5.95}$$

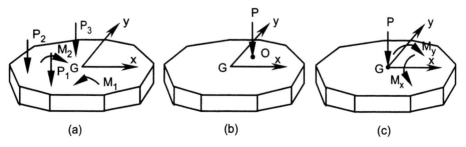

Figure 5.33. Rigid footing subjected to vertical forces and moments.

A negative contact pressure implies tensile stresses under the footing. The position of the zero pressure line (where the contact pressure is zero) can be determined by setting Equations 5.93 or 5.95 to zero. If this line is located outside of the soil-footing contact area then the contact pressure under the footing is compressive. Figure 5.34(a) shows a plan of a rectangular rigid footing $(L \times B)$ in which bc is the zero pressure line (*ZPL*). The triangle *abc* represents the tensile stress applied to the soil beneath the footing; the rest of the contact area is subjected to compressive stress. The vertical force equilibrium equation may be written as:

$$P = F_c - F_t$$

where F_c is the soil reaction under the compressed area and F_t is the tensile force corresponding to area *abc*. The soil cannot sustain tensile stress and therefore F_t approaches zero. In order to maintain equilibrium, the position of the zero pressure line has to be adjusted by reducing F_c to satisfy $P = F_c$ and moving the application point of the new F_c to point O (application point of resultant force). It may be shown that a linear equation of the following form satisfies the equilibrium requirements.

$$q = \frac{Mh}{I} \tag{5.96}$$

where q is the contact pressure at any point in the compressive area under the footing, M is the moment of the resultant about the readjusted zero pressure line (*R-ZPL*), h is the distance of the point from the new zero pressure line and I is the second moment of area of the compressive area about the readjusted zero pressure line. The position of the new zero pressure line is usually found by iteration (Peck et al., 1974) until the above equilibrium conditions are satisfied. Note that a part of the footing (area *afhj* in Figure 5.34(a)) will be ineffective in resisting the applied loading.

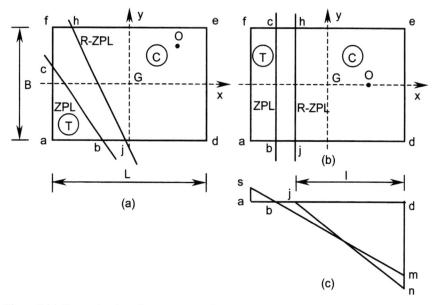

Figure 5.34. Determination of zero pressure line.

For rectangular footings, Irles & Irles (1994) suggested an explicit solution for the position of the zero pressure line. If the application point of the resultant is on the x-axis, as shown in Figure 5.34(b), the zero pressure line will be parallel to the y-axis. The new position of the zero pressure line (point j in Figure 5.34(c)) is found by drawing the line jn to satisfy equilibrium in z direction:

$(1/2)\times jd\times dn\times B = P$.

The resultant P must pass through the centroid of triangle jdn:

$jd = 3(L/2-x_O)$.

Thus the maximum soil pressure dn is:

$$q_{max} = 2P/[3B(L/2-x_O)] \tag{5.97}$$

For a rigid rectangular footing $(L\times B)$ with the symmetric loading (about the x-axis), we can use Equation 5.95 to obtain the contact pressures at the two edges of the rectangle (Figure 5.35).

$$q_{max} = \frac{P}{LB}(1+6e/L) \rightarrow x=L/2, \quad q_{min} = \frac{P}{LB}(1-6e/L) \rightarrow x=-L/2 \tag{5.98}$$

where x_O is replaced by e to give the familiar expressions used in soil mechanics-geotechnical engineering textbooks. For the purpose of the calculation of shear forces and bending moments along the footing, the contact pressure at every point on the centre line is multiplied by B to give the soil reaction (in kN/m) along the assumed one-dimensional beam.

Example 5.24

For the rigid rectangular footing shown in Figure 5.36(a), determine the contact pressure distribution and the position of the R-ZPL. Find the length of the footing required to achieve a uniform contact pressure if its length can be increased only beyond the right-hand column.

Solution:

Take moments about the left edge to find the position of the resultant of the system:
$0.2 \times 500.0 + 5.2 \times 1100.0 + 580 - (500.0 + 1100) x' = 0.0, x' = 4.0$ m.
The resultant is a vertical load of 1600 kN that acts 4 m from the left edge of the footing. It is clear that the integration of the contact pressure is also equal to 1600 kN and acts at the same point but in reverse direction.

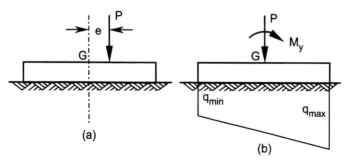

Figure 5.35. Linear contact pressure under a rectangular rigid footing.

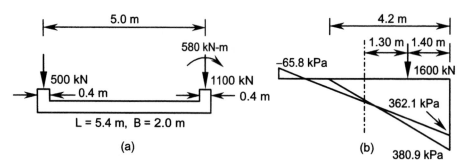

Figure 5.36. Example 5.24.

$e = x' - L/2 = 4.0 - 5.4/2 = 1.3$ m. Using Equations 5.98:

$$q_{max} = \frac{1600.0}{5.4 \times 2.0}(1 + 6 \times \frac{1.3}{5.4}) = 362.1 \, \text{kPa (right edge)}.$$

$$q_{min} = \frac{1600.0}{5.4 \times 2.0}(1 - 6\frac{1.3}{5.4}) = -65.8 \, \text{kPa (left edge)}.$$

The soil may not sustain the tensile stress at the left edge, therefore the zero pressure line moves towards the right and q_{max} increases to satisfy equilibrium:
Effective length of the footing or the distance of zero pressure line from the right edge:
$= 3(5.4/2 - 1.3) = 4.2$ m.

The maximum soil pressure from Equation 5.97 is:

$$q_{max} = \frac{2 \times 1600}{3 \times 2.0(2.7 - 1.3)} = 380.9 \, \text{kPa, (Figure 5.36(b))}.$$

For a uniform soil pressure, the centroid of the footing and the point of application of the resultant must coincide: $L = 2 \times x' = 2 \times 4.0 = 8.0$ m, $q = 1600.0 / (8.0 \times 2.0) = 100.0$ kPa.

5.5.2   Beam on an elastic foundation

The concept of a beam on an elastic foundation provides a general solution in the evaluation of the contact pressure and settlement under a flexible footing, as it takes into account the elastic properties of both the footing and the soil. In this regard the Winkler model (after Winkler in 1867) has given a satisfactory solution for soil-footing interaction as well as for retaining structures such as sheet piles. An early model of Winkler foundation assumed the soil under the footing to be a bed of parallel and discontinuous springs as shown in Figure 5.37. The modulus of subgrade reaction is defined as:

$$k_s = \frac{q}{S_e} \tag{5.99}$$

where q is the contact pressure under a loading plate and S_e is the corresponding settlement (Figure 5.38(a)). This modulus is measured in kN/m^3. A typical stress-settlement behaviour is shown in Figure 5.38(b). The value of k_s depends on the magnitude of q and S_e and decreases as both q and S_e increase; thus the spring model is not linear. The relationship between q and S_e depends also on the size of the contact area, and the results

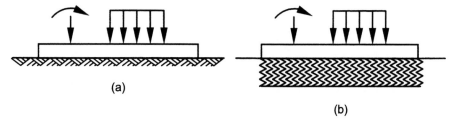

Figure 5.37. Winkler model.

obtained from a 0.3 m square loading plate cannot be readily related to a loading plate of 1 m by 1 m or to real footings. The relationship between k_s and Modulus of Elasticity E_s has been extensively investigated, and the proposal by Vesić (1961a and 1961b) is often used in design:

$$k_s' = 0.65 \times 12 \sqrt{\frac{E_s B^4}{E_f I_f}} \frac{E_s}{1-\mu^2} \tag{5.100}$$

where E_s, E_f are the Moduli of Elasticity of soil and footing respectively, B is the width of the footing, I_f is the second moment of area of the footing section, μ is Poisson's ratio for the soil, and $k_s' = k_s B$. A simplified version of Equation 5.100 (based on Equation 5.54) is also used:

$$k_s = \frac{E_s}{B(1-\mu^2)} \tag{5.101}$$

The Winkler model can be simplified by replacing the bed of springs with a finite number of springs of stiffness K where:

$$K = \text{contact area corresponding to a single spring} \times k_s = B \times a \times k_s = k_s' a \tag{5.102}$$

where a is the horizontal distance between the parallel springs. Consider now the beam shown in Figure 5.39. Each guided spring support allows (only) for vertical deformation of the beam (z_i) equal to R_i / K, where R_i is the reactive force at each support.

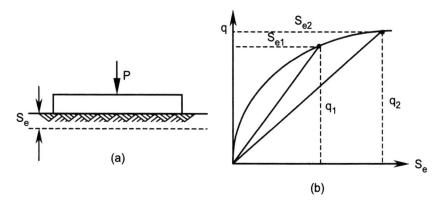

Figure 5.38. Determination of modulus of subgrade reaction by plate loading test.

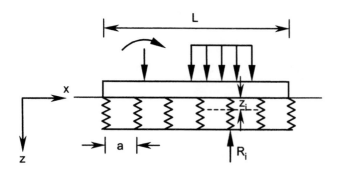

Figure 5.39. Finite beam on parallel springs.

$$R_i = k'_s a z_i = k_s B a z_i = K z_i \tag{5.103}$$

The problem can be solved using traditional methods as well numerical methods such as finite differences and finite elements. In the latter method, the segments connecting the springs may be selected as simple beam elements. In this simplified Winkler model, the modulus of subgrade reaction k_s is assumed to be constant along the length of the beam. Terzaghi (1955) suggested that the error arising from this assumption may be neglected in most practical problems. Ulrich (1991 and 1994) suggested a method based on the theory of elasticity that utilises a spring constant that varies along the length of the beam. The ACI (1988) proposed a design procedure that uses a Boussinesq-type vertical stress distribution as the basis for the variation of k_s. A practical design procedure for plane strain conditions was proposed by Liao (1995) who used finite elements as a replacement for the mathematical treatment of the theory of elasticity. He obtained correction factors for k_s that depend upon the position of the nodal spring and the depth of the incompressible stratum.

A closed form solution for a beam of simple geometry and loading resting on a bed of springs has been derived by Hetenyi (1946). Whilst the solution for an applied concentrated moment and vertical load on infinite and finite beams can be found in many geotechnical engineering textbooks, there has been no practical development of the equations for distributed loads and complicated geometry. Moreover, the flexural stiffness of the beam may vary along its length, and in this case the derivation of a closed form solution is almost impossible. The beam shown in Figure 5.39 is indeterminate to the $n - 2^{th}$ degree, where n is the number of springs. A traditional approach to the solution of this beam establishes supplementary equations based on the geometry of its deformations. The governing differential equation for the deflection of the beam is:

$$\frac{d^2 z}{dx^2} = \frac{-M(x)}{EI} \tag{5.104}$$

where $M(x)$ is the bending moment at distance x from the origin (say left edge) and EI is the flexural stiffness of the beam. Note that by accepting this differential equation, (vertical) deflections due to shear are ignored (an Euler beam rather than a Timoshenko beam). A practical way of handling a beam on an elastic foundation is to use finite elements with constant spring stiffness. An alternative method is to approximate the second derivative in

Equation 5.104 with finite differences (Equation 3.52), an approach demonstrated in the following example.

Example 5.25

Predict the soil settlement and reaction at selected points for the finite beam shown in Figure 5.40. The thickness of the footing is 0.7 m and its width is 1 m. The beam is divided into 5 equal segments of 2 m each. $E_f = 22 \times 10^6$ kPa, $E_s = 10440$ kPa, $\mu = 0.3$.

Solution:

Using Equation 5.100 to calculate k'_s:

$$k'_s = 0.65 \sqrt[12]{\frac{10440 \times 1.0^4}{22 \times 10^6 (0.7)^3 \times 1.0/12}} \frac{10440}{1 - 0.3^2} = 5300 \text{ kPa,}$$

$K = k'_s \times a = 5300 \times 2.0 = 10600$ kN/m. For the edges: $K = k'_s \times (a/2) = 5300$ kN/m.
Although Bowles (1996) recommended that the end stiffness be doubled, the model expressed by Equation 5.103 will be followed. Thus, $R_1 = 5300 \ z_1$, $R_2 = 10600 \ z_2$, $R_3 = 10600 \ z_3$. Because of the symmetrical loading, the required parameters at points 1, 2 and 3 are equal to the corresponding values at points 6, 5, and 4. To solve the problem three linear equations involving the three unknowns of z_1, z_2 and z_3 (settlements at points 1, 2 and 3) must be developed. Applying the finite difference equations at points 2 and 3, we have:

$$(\frac{d^2 z}{dx^2})_i = \frac{z_{i+1} + z_{i-1} - 2z_i}{a^2} = \frac{-M_i}{EI},$$

$$(\frac{d^2 z}{dx^2})_2 = \frac{z_3 + z_1 - 2z_2}{2.0^2} = \frac{-M_2}{EI},$$

$$(\frac{d^2 z}{dx^2})_3 = \frac{z_4 + z_2 - 2z_3}{2.0^2} = \frac{z_2 - z_3}{2.0^2} = \frac{-M_3}{EI}.$$

Note $z_3 = z_4$.

$$EI = 22.0 \times 10^6 \times \frac{1.0 \times 0.7^3}{12} = 628833.33 \text{ kN.m}^2.$$

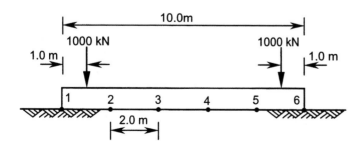

Figure 5.40. Example 5.25.

Calculate bending moments at points 2 and 3:

$M_2 = R_1 \times 2.0 - 1000.0 \times 1.0 = 10600\, z_1 - 1000.0,$

$M_3 = R_1 \times 4.0 + R_2 \times 2.0 - 1000 \times 3.0 = 21200\, z_1 + 21200\, z_2 - 3000.0,$

Substituting M_2 and M_3 into the finite difference equations:

$$\frac{z_3 + z_1 - 2z_2}{4.0} = \frac{-(10600z_1 - 1000.0)}{628833.33},$$

$$\frac{z_2 - z_3}{4.0} = \frac{-(21200z_1 + 21200z_2 - 3000.0)}{628833.33}.$$

The third equation is obtained from (static) equilibrium of the vertical forces:

$2R_1 + 2R_2 + 2R_3 = 1000.0 + 1000.0 = 2000.0,$

$R_1 + R_2 + R_3 = 1000.0$ or:

$5300\, z_1 + 10600\, z_2 + 10600\, z_3 = 1000.0.$

Simplifying and rearranging the equations:

$671233.33\, z_1 - 1257666.66\, z_2 + 628833.33\, z_3 = 4000.0$

$84800\, z_1 + 713633.33\, z_2 - 628833.33\, z_3 = 12000.0$

$5300\, z_1 + 10600\, z_2 + 10600\, z_3 = 1000.0$

Solving for z_1, z_2, and z_3:

$z_1 = 48.92 \times 10^{-3}$ m $= 48.9$ mm,

$z_2 = 38.58 \times 10^{-3}$ m $= 38.6$ mm,

$z_3 = 31.30 \times 10^{-3}$ m $= 31.3$ mm,

$R_1 = 5300 \times 48.92 \times 10^{-3} = 259.3$ kN $\uparrow$

$R_2 = 10600 \times 38.58 \times 10^{-3} = 408.9$ kN $\uparrow$

$R_3 = 10600 \times 31.30 \times 10^{-3} = 331.8$ kN $\uparrow$

5.6 PROBLEMS

5.1 A cylinder of diameter 150 mm and height 300 mm is filled with sand. The surface of the sand is subjected to a vertical stress of 300 kPa causing 4 mm settlement under the loading plate. Calculate the lateral stress on the wall of the cylinder and Modulus of Elasticity of the sand. Poisson's ratio for the sand is 0.2.

Answers: 75 kPa, 20300 kPa

5.2 Referring to Figure 5.5(a), calculate the distribution of vertical stress along a vertical plane passing through one of the two forces. Specify the values of stress at depths of 1 m, 2 m, 3 m, 4 m and 5 m.

Answers: 191.0 kPa, 48.1 kPa, 22.0 kPa, 13.1 kPa, 9.0 kPa

5.3 Figure 5.41(a) shows a plan view of a footing that applies a uniform load of 300 kPa on a horizontal ground surface. Calculate the vertical stress component at point *A* at a depth of 2 m using:

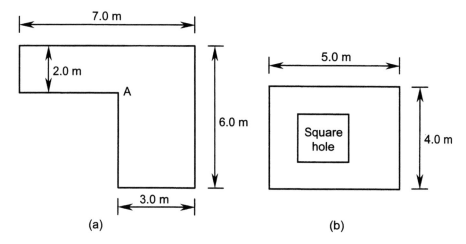

Figure 5.41. Problems 5.3 & 5.5.

(a) the principle of superposition of vertical forces by dividing the area into elemental 1 m squares, (b) Newmark's influence chart of Figure 5.10.

Answers: 186.5 kPa, 184.5 kPa

5.4   Re-work Problem 5.3 using Fadum's chart.

Answer: 185.1 kPa

5.5   Figure 5.41(b) shows a plan view of a rectangular footing with a 2 m × 2 m square hole (through its entire thickness). The hole is located at 1 m from the left edge and is equally positioned between the top and lower edges of the footing. If the uniform bearing pressure under the footing is 200 kPa, use Fadum's chart to compute the vertical stress component at a point 2 m below the centre of the square hole.

Answer: 80.3 kPa

5.6   Compute σ_z at point A (Figure 5.42(a)) due to the two line loads shown.

Answers: $\sigma_x = 1.5$ kPa, $\sigma_z = 5.8$ kPa, $\tau_{xz} = 2.0$ kPa

5.7   Compute the magnitudes of the major and minor principal stresses within a soil that is subjected to an infinitely long line load applied at the ground surface.

Answers: $\sigma_3 = 0.0$, $\sigma_1 = (2q / \pi z) \cos^2\theta$

5.8   Under the centre line of an infinite strip footing, calculate the depth at which the vertical stress component is 10% of the applied load.

Answer: $z = 6.34 B$

5.9   A strip loading of infinite length is shown in Figure 5.42(b). Compute the vertical stress component at points A, B, C, and D.

Answers: 81.8 kPa, 48.0 kPa, 8.4 kPa, and 1.7 kPa

5.10 An earth embankment is 2.5 m high and has a slope of 2 horizontal to 1 vertical. If the base of the embankment is 20 m, find the vertical stress component at a point on the centre line at a depth of 7 m. The unit weight of the embankment material is 18 kN/m^3.

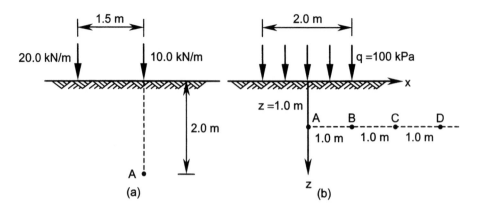

Figure 5.42. Problems 5.6 & 5.9.

Answer: 37.2 kPa

5.11 A vertical pile carrying a load of 1500 kN has been driven 18 m into the ground. Calculate the vertical stress component at a point 19.8 m below the ground surface and 3 m from the centre line of the pile for the following cases:

(a) the entire load is transmitted to the soil through the base of the pile,

(b) the base carries one-half of the load and the rest is carried by skin friction.

$\mu = 0.3$.

Answers: 7.9 kPa, 5.7 kPa

5.12 Two flexible square footings 2 m × 2 m are constructed 4 m apart (centre to centre) on the ground surface. The uniform contact pressure under one footing is 200 kPa and is 400 kPa under the other. Calculate the elastic settlement at the centre of each footing and at the midpoint of the line connecting the two centres.

$E_s = 10000$ kPa, $\mu = 0.35$.

Answers: 51.0 mm ($q = 200$ kPa), 84.5 mm ($q = 400$ kPa), 34.7 mm

5.13 A flexible circular footing 4 m in diameter exerts a pressure 300 kPa at the ground surface. The soil is a saturated clay with $E_s = 8000$ kPa, and an incompressible stratum is located 12 m below the ground surface. Calculate the settlement under the centre of the footing using:

(a) a semi-numerical method by dividing the depth into 6 layers of 2 m thickness,

(b) Steinbrenner's influence factors.

Answers: 100 mm, 77 mm

5.14 A flexible rectangular footing for which $B = 2$ m and $L = 4$ m exerts 300 kPa, 2 m below the ground surface. The Modulus of Elasticity for the soil is 8000 kPa, and $\mu = 0.5$. An incompressible stratum is located at 10 m below the ground surface. Calculate the average settlement of the footing using:

(a) Steinbrenner's influence factors and Fox's correction factor for depth,

(b) the improved chart of Janbu et al.

Answers: 39.3 mm, 55.0 mm

5.15 A footing has plan dimensions of 3 m × 5 m and exerts a uniform net contact pressure of 200 kPa on the underlying saturated sand. The base of the footing is at the ground surface, and the average *SPT* number at a depth of 3 m is 19. Calculate the settlement of the footing using Burland and Burbidge method.

Answer: 17.3 mm

5.16 A rectangular footing of width 2 m and length of 20 m is located at the ground surface. It exerts a net uniform contact pressure of 180 kPa to the underlying deep deposit of sand.

Calculate the settlement of the footing using Schmertmann's modified strain influence factor diagrams. The average results of *CPT* tests are shown in the table below.

Depth (m)	1.0	2.0	3.0	4.0	5.0	6.0	7.0
q_c (kPa)	1600	1600	2000	1800	2400	2400	2600

For calculation of E_s use the plane strain condition expressed by Equation 5.91.

Answer: 85.0 mm

5.17 Calculate the contact pressures under the corners of a rigid square footing using the following data: $P = 600$ kN, $M_x = -60$ kN.m, $M_y = 100$ kN.m, $L = B = 2$ m.

Answers: 120.0 kPa, 270.0 kPa, 180.0 kPa, 30.0 kPa

5.18 Calculate the contact pressures for the rigid footing shown in Figure 5.43(a) at points *a*, *b* and *c*.

Answers: 51.3 kPa, 278.4 kPa, 391.9 kPa

5.19 The following data apply to a square footing located at the ground surface:
$L = B = 1.8$ m, $P = 270$ kN, $M_x = -160$ kN.m, $M_y = 160$ kN.m.

Determine:
(a) the contact pressures at the corners of the footing,
(b) the equation of the zero pressure line, its readjusted position (by formulation) and the maximum contact pressure (assume readjustment occurs parallel to the original zero pressure line),
(c) the size of the footing to limit the maximum contact pressure to 280 kPa and the position of the new zero pressure line.

Answers:
(a) 83.3, 412.5, 83.3, and −245.9 all in kPa,
(b) $x + y + 0.4556 = 0.0$, $h_1 =$ distance from the maximum contact pressure point = 0.87 m, $q_{max} = 1070$ kPa,
(c) 2.4 m, 1.7 m (from the maximum contact pressure point)

5.20 For a trapezoidal rigid footing shown in Figure 5.43(b) compute the dimensions *B* and *A* that will ensure a uniform contact pressure of 200 kPa.

Answers: $B = 3.92$ m, $A = 2.08$ m

5.21 Re-work Example 5.25 by replacing the two vertical forces by a single vertical force of 2000 kN applied at the centroid of the footing.

Answers: $z_1 = 26.2$ mm, $z_2 = 36.4$ mm, $z_3 = 44.8$ mm,
$R_1 = 139$ kN, $R_2 = 386$ kN, $R_3 = 475$ kN

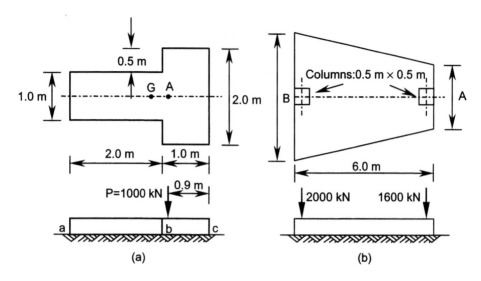

Figure 5.43. Problems 5.18 & 5.20.

5.7 REFERENCES

ACI, 1988. Suggested analysis and design procedures for combined footings and mats. *ACI structural journal* 85(3): 304-324.

Ahlvin, R.G. & Ulery, H.H. 1962. Tabulated values for determining the complete pattern of stresses, strains, and deflections beneath a uniform load on a homogeneous half space. *Highway research board bulletin* (342): 1-13.

Atkinson, J. H. & Bransby, P. L. 1978. *The mechanics of soils*. UK: McGraw-Hill.

Bjerrum, L. & Eggstad, A. 1963. Interpretation of loading test on sand. *Proc. 5th Eurpean conf. on soil mechanics* 1: 199-203. Wiesbaden.

Bowles, J.E. 1996. *Foundation analysis and design*. 5th edition. New York: Mc Graw-Hill.

Burland, J.B. 1970. Discussion. *Proc. conf. on in-situ investigations in soils and rocks*. London: British Geotechnical Society.

Burland, J.B. & Burbidge, M.C. 1985. Settlements of foundations on sands and gravel. *Proc.*, Part 1, 78: 1325-1381. UK: Institution of Civil Engineers.

Christian, J.T. & Carrier, W.D. 1978. Janbu, Bjerrum, and Kjaernsli's chart reinterpreted. *Canadian geotechnical journal* 15(1): 123, 436.

Clayton, C.R.I., Matthews, M.C. & Simons, N.E. 1995. *Site investigation*. 2nd edition. London: Blackwell Science.

Consoli, N.C., Schnaid, F. & Militsky, J. 1998. Interpretation of plate load tests on residual soil site. *Journal GGE, ASCE* 124(9): 857-867.

D'Appolonia, D.J., Poulos, H.G. & Ladd, C.C. 1971. Initial settlements of structures on clay. *Journal SMFEV, ASCE* 97(SM10): 1359-1377.

Duncan, J.M. & Buchignani, A.L. 1976. *An engineering manual for settlement studies*. Berkeley: University of California.

Fadum, R.E. 1948. Influence values for estimating stresses in elastic foundations. *Proc. 2nd intern. conf. SMFE* 3: 77-84. Rotterdam.

Fox, E.N. 1948. The mean elastic settlement of a uniformly loaded area at a depth below the ground surface. *Proc. 2nd intern. conf. SMFE* 1: 129-132. Rotterdam.

Geddes, J.D. 1966. Stresses in foundation soils due to vertical subsurface loading. *Geotechnique* 16(3): 231-255.

Geddes, J.D. 1969. Boussinesq based approximations to the vertical stresses caused by pile type subsurface loadings. *Geotechnique* 19(4): 509-514.

Gibson, R.E. 1974. The analytical method in soil mechanics: 14th Rankine lecture. *Geotechnique* 24(2): 115-140.

Giroud, J.P. 1972. Settlement of rectangular foundation on soil layer. *Journal SMFED, ASCE* 98(SM1): 149-154.

Harr, M.E. 1966. *Foundations of theoretical soil mechanics*. New York: McGraw-Hill.

Harr, M.E. 1977. *Mechanics of particulate media: A probabilistic approach*. New York: McGraw-Hill.

Hetenyi, M. 1946. *Beam on elastic foundations*. Ann Arbor, Michgan: The University of Michigan Press.

Holtz, A.D. 1991. Stress distribution and settlement of shallow foundations. In H.Y. Fang (ed), *Foundation engineering handbook*. New York: Van Nostrad Reinhold.

Irles, R. & Irles, F. 1994. Explicit stresses under rectangular footings. *Journal SMFED, ASCE* 120(2): 444-450.

Janbu, N., Bjerrum, L. & Kjaernsli, B. 1956. Veileduing ved losning, av fundamenteringsopogaver (Soil mechanics applied to some engineering problems). *Noewegian geotechnical institiute* Pub. 16. Oslo.

Jumikis, A.R. 1969. *Theoretical soil mechanics*. New York: Van Nostrand Reinhold.

Liao, S.S.C. 1995. Estimating the coefficient of subgrade reaction for plane strain conditions. *Proc. institution of civil engineering, geotechnical engineering* 113: 166-181.

Mayne, P.W. & Poulos, H.G. 1999. Approximate displacement influence factors for elastic shallow foundations. *Journal GE, ASCE* 129(6): 453-460.

Meyerhof, G.G. & Fellenius, B.H. 1985. *Canadian foundation engineering manual*. 2nd edition. Canada: Canadian Geotechnical Society.

Mindlin, R.D. 1936. Force at a point in the interior of a semi-infinite solid. *Journal of the American institution of physics* 7(5): 195-202.

Newmark, N.M. 1942. Influence charts for computations of stresses in elastic soils. *Enginnering experimental station bulletin* (367). University of Illinois.

Papadopoulos, B.P. 1992. Settlements of shallow foundations on cohesionless soils. *Journal SMFED, ASCE* 118(3): 377-393.

Peck, R.B., Hanson, W.E. & Thornburn, T.H. 1974. *Foundation engineering*. 2nd edition. New York: John Wiley & Sons.

Poulos, H.G. & Davis, E.A. 1974. *Elastic solutions for soil and rock mechanics*. New York: Wiley.

Powrie, W. 1997. *Soil mechanics-concepts and applications*. London: E & FN Spon.

Schmertmann, J.H. 1970. Static cone to compute static settlement over sand. *Journal SMFED, ASCE* 96(SM3): 1011-1043.

Schmertmann, J.H., Hartman, J.P. & Brown, P.R. 1978. Improved strain influence factor diagrams. *Journal GE, ASCE* 104(8): 1131-1135.

Steinbrenner, W. 1934. Tafeln zur Setzungsberechnung. *Die strasse* 1.

Sutherland, S.B. 1975. Granular materials. *Proc. conf. on settlements of structures*: 473-499. BGS. Cambridge. London: Pentech Press.

Terzaghi, K. 1955. Evaluation of coefficients of subgrade reaction. *Geotechnique* 5(4): 297-326.

Terzaghi, K. & Peck, R. B. 1948. *Soil mechanics in engineering practice*. New York: John Wiley & Sons.

Terzaghi, K., Peck, R. B. & Mesri, G. 1996. *Soil mechanics in engineering practice*. 3rd edition. New York: John Wiley & Sons.

Timoshenko, S.P & Goodier, J.N. 1982. *Theory of elasticity*. 3rd edition. 398-409. New York: McGraw-Hill.

Tomilinson, M.J. 1995. *Foundation design and construction.* 6th edition. Harlow, Essex: Longman Scientific & Technical.

Ulrich, E.J., Jr. 1991. Subgrade reaction in mat foundation design. *Concrete international* 13(1): 41-50.

Ulrich, E.J., Jr. 1994. Mat foundation design: A historical perspective. *Proc. ASCE specialty conf. on vertical and horizontal deformations of foundations and embankments.* 107-120. Texas A & M, College Station 1.

Vesić, A.S. 1961a. Beams on elastic subgrade and the Winkler's hypothesis. *Proc. 5th intern. conf. SMFE* 845-850.

Vesić, A.S. 1961b. Bending of beams resting on isotropic elastic solid. *Journal EMD, ASCE* 87(2): 35-53.

Wu, T.H. 1966. *Soil mechanics.* Boston: Allyn & Bacon Inc..

CHAPTER 6

One Dimensional Consolidation

6.1 INTRODUCTION

This chapter introduces methods to predict both the magnitude and rate of one-dimensional consolidation settlement. Consolidation settlement in a fully saturated clay is the result of a gradual reduction in its volume due to the drainage of water through the voids of the soil. The potential that causes this movement of the water is caused by the increase in the pore pressure due to external loading. This increase in pore pressure within the soil is termed the *excess pore pressure* and varies with time and from point to point within the soil mass. In a saturated clay in the undrained condition, the applied load at time $t = 0$ is resisted solely by the pore water. If a drainage condition exists, the initial excess pore pressure dissipates with time leaving the soil, again, in a fully saturated state. The consequent reduction in the volume of the soil is approximately equal to the volume of water entering the free draining boundaries. The increase in the effective stress at any given time equals the decrease in the excess pore pressure (because of the concept behind Equation 2.1). This assumption, along with the continuity conditions of water movement, enables the creation of the necessary equations that express the development of the consolidation process. The settlement due to consolidation is the major controlling factor in the design of footings and the like, and in many cases little error is introduced by assuming consolidation only occurs in the vertical direction. Any immediate settlement that occurs upon application of the load is estimated using elastic theory.

6.2 CONSOLIDATION INDICES AND SETTLEMENT PREDICTION

6.2.1 *Assumptions and basic settlement model*

The magnitude of the theoretical consolidation settlement is dependent on the level of the applied stress (applied by a footing for instance) and the adopted stress-settlement model that is normally based on laboratory tests. Over the period 1925-35, soil mechanics pioneer Terzaghi and his associate Frolich established the conventional consolidation theory that later became one of the most widely used theories in geotechnical engineering (Terzaghi & Peck, 1967). The limitations and shortcomings inherent in this theory (Duncan, 1993) have been overcome by the application of more advanced methods including modifications to the original basic assumptions. In order to develop the equations

necessary to predict the magnitude and rate of one-dimensional consolidation the following assumptions are made:

1. The soil is homogeneous.
2. The soil is fully saturated.
3. Darcy's law is valid.
4. The flow is one-dimensional and the coefficient of permeability is constant during the
consolidation process.
5. The relationship between volume change and effective stress is independent of time but
changes with the level of stress.
6. Strains are small.

Figure 6.1(a) shows behaviour typical of an element of saturated clay undergoing one-
dimensional consolidation in the laboratory. Void ratio is plotted vertically against effective vertical stress on a logarithmic scale. In general, the void ratio decreases with an
increase in effective vertical stress. At small values of effective vertical stress the rate of
decrease of the void ratio is correspondingly low until the effective vertical stress reaches
the *preconsolidation pressure* p'_c. The state of the soil at effective vertical stresses below
p'_c is called *overconsolidated*. For effective vertical stresses greater than p'_c, the e-$\log\sigma'$
curve can be approximated by a straight line, referred to as the *virgin compression line*. In
this region the soil is *normally consolidated*. Under one-dimensional consolidation the
state of stress is either on the virgin compression line or to the left-hand side of this line.
No state of stress can exist on the right-hand side of the virgin compression line. An
unloading-reloading cycle is also shown in Figure 6.1(a). As a result of unloading, minor

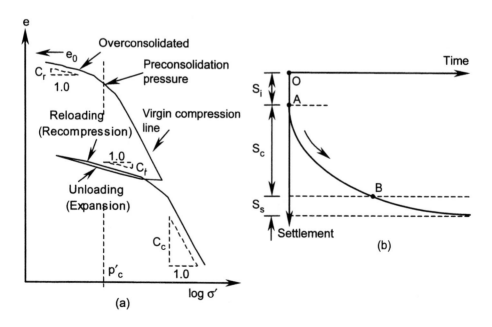

Figure 6.1. (a) Void ratio-$\log\sigma'$ relationship, (b) time settlement relationship.

expansion of the specimen occurs and a fraction of the compression is recovered. The re-loading or recompression segment is slightly above the expansion curve, and approaches the normally consolidated state as the effective vertical stress increases to the preconsoli-dation pressure where it was unloaded. It may be concluded that the overconsolidated segment indicates a recompression stage after an unloading stage in the field. The effec-tive vertical stress at the point where the curve approaches the virgin compression line is the maximum effective vertical stress that has acted on the soil sometime in the past and is termed the preconsolidation pressure (p'_c). The stress history for an overconsolidated segment is better represented by the introduction of a dimensionless parameter called the *overconsolidation ratio (OCR)*. This is the ratio of the preconsolidation pressure to the ex-isting effective vertical stress on the element. In the field the *OCR* may be defined as the ratio of the preconsolidation pressure to the existing overburden pressure.

Figure 6.1(b) shows a typical time-settlement relationship for an element of saturated clay during a vertical load increment. The settlement (or the vertical compressive deformation) of the element is divided into three segments, S_i, S_c, and S_s. The first seg-ment S_i occurs immediately after application of the load and without any change in the amount of water in the element: consequently, this is representative of the elastic settle-ment in one-dimensional conditions. The second segment S_c (A to B on the time-settlement curve) is called *primary consolidation*, or in general, consolidation settlement. As time advances from A to B the excess pore pressure dissipates from its initial value to zero. The third segment S_s represents the settlement under a constant effective vertical stress and is termed *secondary consolidation*. The main reason for this settlement is unknown but is thought to be as a result of colloid-chemical interactions and small resid-ual excess pore pressures. For inorganic soils the amount of secondary consolidation is negligible.

6.2.2   Compression index and coefficient of volume compressibility

The compression index C_c is the absolute value of the gradient of the virgin compression line:

$$C_c = \frac{e_0 - e_1}{\log \sigma'_1 - \log \sigma'_0} = \frac{e_0 - e_1}{\log(\sigma'_1 / \sigma'_0)} \tag{6.1}$$

where the subscripts 0 and 1 represent the initial and final states of void ratios and corre-sponding effective vertical stresses. A similar equation can be written to define the recompression index C_r, which can be obtained from either the overconsolidated segment of the e-logσ' plot or an unloading-reloading cycle. Leonards (1976) suggested that the unloading-reloading cycle must be close to the preconsolidation pressure, and that an average slope must be taken for expansion and recompression assuming that both occur on the same line. By doing this it is assumed that the slopes of all the recompression stages that could happen in the field, or that can be performed in the laboratory, are approximately equal. If a laboratory specimen has been subjected to distortion or com-pression during the preparation process, it is more convenient to establish C_r from an unloading-reloading cycle.

For the secondary consolidation, a compression index is defined by:

$$C_\alpha = \frac{\Delta e}{\Delta \log t} \tag{6.2}$$

where Δe is the change in void ratio in the secondary consolidation stage, and $\Delta \log t$ represents one (log) cycle of time. The secondary consolidation can also be represented by the *coefficient of secondary compression* in the form:

$$C_{\alpha s} = \frac{\Delta H_{\alpha}}{H_0} \qquad (6.3)$$

where ΔH_{α} is the change in the specimen height (in a standard consolidation test) over one (log) cycle of time and H_0 is the initial height of the specimen.

The compressibility of a clay layer undergoing one-dimensional consolidation can also be represented by the *coefficient of volume compressibility* m_v which is the volume change per unit volume per unit increase in effective vertical stress:

$$m_v = \frac{\Delta H / H_0}{\sigma'_1 - \sigma'_0} \qquad (6.4)$$

where ΔH is the consolidation settlement, H_0 is the initial thickness of the layer, and σ'_0 and σ'_1 are the average initial and final states of the effective vertical stress in the layer. The magnitude of m_v is obtained by a standard one-dimensional consolidation test for a stress level represented by $\sigma'_1 - \sigma'_0$. In this case ΔH and H_0 represent the settlement and initial thickness of the soil specimen respectively. In a fully saturated soil undergoing a vertical deformation ΔH, the following can be derived from the phase diagram (Chapter 1):

$$\frac{\Delta e}{\Delta H} = \frac{1 + e_0}{H_0} \qquad (6.5)$$

where e_0 is the initial void ratio and $\Delta e = e_0 - e_1$, in which e_1 is the final void ratio. Thus Equation 6.4 can be written as:

$$m_v = \frac{\Delta H / H_0}{\sigma'_1 - \sigma'_0} = \frac{1}{H_0} \frac{H_0 - H_1}{\sigma'_1 - \sigma'_0} = \frac{1}{1 + e_0} \frac{e_0 - e_1}{\sigma'_1 - \sigma'_0} \qquad (6.6)$$

in which the units of m_v is the inverse of stress (m^2/kN or m^2/MN).

6.2.3   *Discussion on compressibility indices - correlation with basic soil properties*

The magnitude of the compression index C_c in saturated clays varies from 0.1 to 0.5 depending on their plastic characteristics, and increases with increasing plasticity. In organic soils and peat, the compression index may increase to 3, although in some cases (e.g. Mexico City clay) it reaches as high as 10 (Mesri et al., 1975). Although the concept of consolidation is not applicable to sands, a value of C_c of about 0.05 is most likely to be expected for loose sands.

In most engineering projects the compression index C_c may be assumed to be constant within working stress levels. However, in instances where the stress level varies between low and high values (such as airstrips) it is necessary to investigate the variation of C_c with depth and stress level (Balasubramaniam & Brenner, 1981).

In overconsolidated clays the recompression index C_r is small compared to C_c, but in the calculation of settlement it must nevertheless be taken into account. Typical values are within the range 0.015 to 0.035 (Leonards, 1976) and the index decreases with a decrease in plasticity. Disturbed samples exhibit high values of recompression index. However, the unloading and reloading cycles for void ratios less than $0.42e_0$ may give a reasonable value for recompression and compression indices (Schmertmann, 1953). This means that the stress-settlement behaviour, plotted in the form of e-$\log\sigma'$, approaches the field or undisturbed condition as the stress level passes the corresponding effective vertical stress of the void ratio of $0.42e_0$.

Secondary compression is quantified by the parameters C_α and $C_{\alpha s}$ (Equations 6.2 and 6.3). Values of $C_{\alpha s}$ commonly vary from 0.0005 for an overconsolidated clay to 0.1 for organic soils and peat. From the definition of both parameters it would appear that the secondary compression indices are dependent on both time and stress level. However, the ratio of C_α / C_c may be assumed constant in both the compression and recompression stages (Mesri & Godlewski, 1977; Mesri & Castro, 1987). This means that, for the calculation of secondary consolidation settlement, the magnitude of C_α can be assumed to be independent of the time span of interest. Furthermore, it is convenient to assume that the parameter C_α in the field is independent of the thickness of the soil layer and the load increment ratio (the ratio of the applied effective vertical stress to the effective overburden pressure) provided primary consolidation occurs (Raymond & Wahls, 1976).

The statistical treatment of plasticity and compressibility has been the subject of extensive investigation. Numerous linear equations in which the compression index is expressed in terms of the liquid limit and some basic physical properties, such as the field moisture content w_n and initial void ratio e_0, have been proposed (Table 6.1). Whilst these equations are very useful, they are not universal. Rendon-Herrero (1980 and 1983) suggested that a universal relationship between C_c and dry unit weight can be constructed by relating the slope of the zero-air curve (Equation 1.17) to the slope of e-$\log\sigma'$ curve. Usually, a simple linear regression equation with e_0 can explain up to 80% of the variation of C_c and C_r (Balasubramaniam & Brenner, 1981). This percentage increases if the applicability of the equation is narrowed.

Table 6.1. Empirical relationships between compressibility indices and index or physical properties.

Empirical equation	Author
$C_c = 0.007\,(LL - 7)$	Skempton (1944)
$C_c = 0.009\,(LL - 10)$	Terzaghi & Peck (1967)
$C_c = 0.37\,(e_0 + 0.003\,LL + 0.0004\,w_n - 0.34)$	Azzouz et al. (1976)
$C_c = 0.48827\,(\gamma_w / \gamma_d)^{0.19167}$	Rendon - Herrero (1980)
$C_c = 0.009\,w_n + 0.005\,LL$	Koppula (1986)
$C_c = -0.156 + 0.411\,e_0 + 0.00058\,LL$	Al-Khafaji & Andersland (1992)
$C_r = 0.14\,(e_0 + 0.007)$, $C_r = 0.003\,(w_n + 7)$	Azzouz et al. (1976)
$C_r = 0.00566\,w_n - 0.037$, $C_r = 0.00463\,LL - 0.013$	Balasubramaniam & Brenner (1981)
$C_r = 0.00463\,LL\,G_s$	Nagaraj & Srinivasa Murthy (1985)
$C_\alpha / C_c = 0.032$, $0.025 < C_\alpha < 0.1$	Mesri & Godlewski (1977)
$C_\alpha / C_c = 0.06$ to 0.07, Peats, organic soils	Mesri (1986)
$C_\alpha / C_c = 0.015$ to 0.030, Sandy clays	Mesri et al. (1990)

Example 6.1

In a standard one-dimensional consolidation test the vertical load on the sample was increased from 107 kPa to 428 kPa and the consequent reduction in the thickness of the sample was measured at 1.044 mm. The initial thickness of the sample (at 107 kPa) was 19 mm and the void ratio was 0.841. Calculate C_c and m_v assuming that the specimen is in the normally consolidated state. Using the C_c concept, calculate the total load on the sample to achieve a void ratio of 0.71.

Solution:

From Equation 6.5: $\dfrac{\Delta e}{1.044} = \dfrac{1+0.841}{19.0} \rightarrow \Delta e = 0.1011$, thus

$$C_c = \frac{0.1011}{\log(428.0/107.0)} = 0.1679.$$

Using Equation 6.6: $m_v = \dfrac{1}{1+0.841}\dfrac{0.1011}{428.0-107.0} = 1.71\times10^{-4}\ \text{m}^2/\text{kN}.$

From definition of C_c by Equation 6.1: $C_c = \dfrac{0.841-0.71}{\log(\sigma_1'/107.0)} = 0.1679 \rightarrow \sigma_1' = 645.1\,\text{kPa}.$

6.2.4   Determination of preconsolidation pressure

The following procedure to establish the preconsolidation pressure is based on extensive laboratory testing, and was originally proposed by Casagrande (1936):

1. Plot the e-$\log\sigma'$ curve from laboratory results (Figure 6.2(a)).
2. Visually establish the point of maximum curvature M on the recompression line.

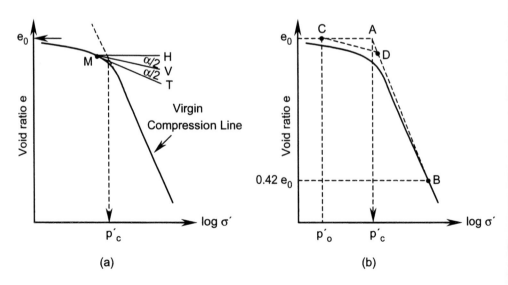

Figure 6.2. (a) Casagrande method to estimate preconsolidation pressure, (b) in-situ e-$\log\sigma'$ curve.

3. From point M draw two lines MH parallel to the $\log\sigma'$-axis and MT tangent to the curve.

4. Draw the line MV that equally subdivides the angle between these two lines and find the intersection point of the virgin compression line with this line.

5. Draw a vertical line through the above intersection point to intersect the e-$\log\sigma'$ curve at the preconsolidation pressure p'_c.

In general, the compressibility parameters obtained in the laboratory are assumed to represent those in the filed. However, it has been found that the slope of the virgin compression line in the field is slightly greater than that in the laboratory. Schmertmann's method for the correction of the laboratory results (Figure 6.2(b)) is as follows:

1. In the e-$\log\sigma'$ plot, establish point A with coordinates of p'_c and e_0 (initial void ratio).

2. Draw a horizontal line from $0.42e_0$ and find the intersection point of this line and the experimental curve (point B). The laboratory and in-situ void ratios are equal at this point.

3. Connect A to B to obtain the in-situ virgin compression line.

4. Calculate the effective overburden pressure on the overconsolidated soil (p'_0) and locate this pressure on the $\log\sigma'$-axis. Draw a vertical line from this point and find the intersection point of this line with a horizontal line drawn from e_0 (point C).

5. From point C draw a line parallel to the average slope of the overconsolidated segment of the experimental curve to obtain the overconsolidated segment in the field. The broken line CDB represents the in-situ e-$\log\sigma'$ relationship.

6.2.5   One-dimensional consolidation test

A diagrammatic section of a consolidation apparatus (oedometer), consisting of a consolidation cell and a sample ring, is shown in Figure 6.3. The cylindrical soil specimen has a diameter in the range 50 to 75 mm and a height of 15 to 20 mm. There are also oedometers with thicker samples and larger diameters (Section 3.3.2), but their use in standard testing is infrequent as it takes much longer any excess pore pressure to dissipate. The specimen is inside a metal ring that is held by the consolidation cell, and the thickness of the ring must be sufficient to ensure zero radial deformation. This ring is either fixed to the base of the cell or is left to float where it is supported by the friction developed on the periphery of the soil specimen. To provide free draining boundaries, porous plates (stones) with a diameter slightly less than the diameter of the ring are positioned on both faces of the specimen. The ring with the soil specimen and porous plates is submerged in the water inside the consolidation cell. Load is applied through a loading plate and the consolidation settlement is measured by means of mechanical or electronic devices to facilitate a plot of the time-settlement (or time-void ratio) relationship. For each increment of load the void ratio is calculated after the settlement of the specimen is completed, which is normally within 24 hours. The time-settlement plot for each load increment is used to obtain the co-efficient of consolidation c_v (Section 6.3.4) which in turn is used to relate time to settlement (or time to degree of consolidation) in the field. It is customary to double the load every 24 hours from an initial value of 6 kPa to 3200 kPa and, for each increment, a value of m_v is computed. However, an alternative loading schedule, possibly including a rebound (unloading-reloading) cycle, may be more representative of field conditions. Swelling that may occur in some clays under low stresses should be avoided and

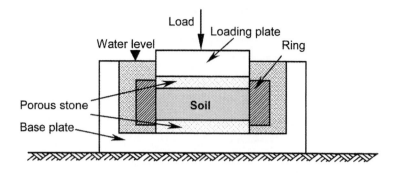

Figure 6.3. Diagrammatic view of a floating oedometer apparatus.

therefore the loading should be increased to the next higher level. The initial effective vertical stress may be set equal to the effective overburden pressure in the field for stiff clays but should be lower for soft clays. Reference is made to standard test methods: ASTM D-2435-96, BS 1377-6: 1990 and AS 1289.6.6.1-1998.

Example 6.2

Data from a laboratory consolidation test is given in the first two rows of the table below:

Test points	1	2	3	4	5	6	7	8
σ' (kPa)	14.0	28.0	56.0	112.0	224.0	448.0	896.0	0.0
Total ΔH (mm)	0.100	0.170	0.495	0.860	1.390	2.130	2.855	1.530
Void ratio	0.714	0.708	0.681	0.650	0.606	0.544	0.483	0.594
m_v (m^2/MN)	-	0.250	0.565	0.329	0.238	0.172	0.088	-

Plot the e-logσ' curve and calculate C_c and m_v. $G_s = 2.70$, $H_0 = 20.5$ mm and $w = 0.22$.

Solution:

The final void ratio after unloading is calculated and substituted into Equation 6.5 to obtain the initial void ratio at zero pressure:

$$e_8 = wG_s = 0.22 \times 2.7 = 0.594, \quad \frac{\Delta e}{1.53} = \frac{1+e_8}{H_8} = \frac{1+0.594}{20.5-1.53} \rightarrow \Delta e = 0.1286 \text{ (positive)},$$

$e_0 = 0.594 + 0.1286 = 0.7226$.

Similarly, we can relate the initial condition to the condition at the end of each load increment. Sample calculation for test point 1:

$(\Delta e / 0.1) = (1+0.7226)/20.5 \rightarrow \Delta e = 0.0084$, (negative),

$e_1 = 0.7226 - 0.0084 = 0.7142$.

The results are summarized in the above table and are shown in Figure 6.4.

For points 5 and 6: $C_c = (0.606 - 0.544)/\log(448.0/224.0) = 0.206$.

For points 6 and 7: $C_c = (0.544 - 0.483)/\log(896.0/448.0) = 0.203$.

Assume an average of 0.204.

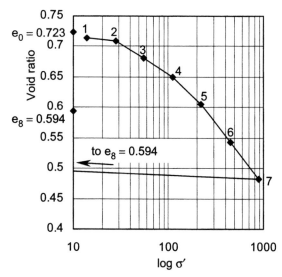

Figure 6.4. Example 6.2.

Sample calculation for m_v for the stress range of 112 kPa to 224 kPa:

$$m_v = \frac{1}{1+0.650}\frac{0.650-0.606}{224.0-112.0} = 2.38\times10^{-4} \text{ m}^2/\text{kN}$$

$m_v = 0.238 \text{ m}^2/\text{MN}$.

6.2.6   Computation of consolidation settlement

Consolidation settlement in the field is calculated based on either the C_c or m_v concept. Consider Equation 6.5 and replace Δe with $e_0 - e_1$, ΔH with S_c and H_0 with H (the initial thickness of a layer).

$$S_c = \frac{e_0 - e_1}{1+e_0}H \qquad (6.7)$$

Replacing $(e_0 - e_1)$ using Equation 6.1, we obtain:

$$S_c = \frac{C_c H}{1+e_0}\log(\sigma_1'/\sigma_0') \qquad (6.8)$$

Similarly, by combining Equations 6.2 and 6.5, the secondary consolidation is given by:

$$S_s = \frac{C_\alpha H}{1+e_0}\Delta \log t \qquad (6.9)$$

where e_0 and H are the initial values at the beginning of the secondary consolidation, which correspond to the final state of primary consolidation. The term $\Delta \log t$ represents the time cycle in the field ($t / t_{primary}$) and is often less than 10 (Terzaghi et al., 1996). This means that $\Delta \log t$ may be taken as unity and is always smaller than 2.

In using Equation 6.8 we note that both the initial and final states are located on the virgin compression line. If the initial effective vertical stress σ'_0 is less than the preconsolidation pressure (overconsolidated state) and the final state is on the virgin compression line (normally consolidated), then the settlement should be calculated in two stages:

$$S_c = \frac{C_r H}{1+e_0} \log(p'_c/\sigma'_0) + \frac{C_c H}{1+e_p} \log(\sigma'_1/p'_c) \tag{6.10}$$

where e_p is the void ratio corresponding to the preconsolidation pressure.

For a layered system or when the compressibility characteristics change with depth, Equation 6.8 becomes:

$$S_c = \sum_{i=1}^{i=n} (S_c)_i \tag{6.11}$$

where n is the number of layers and $(S_c)_i$ represents the consolidation settlement of each layer. Equation 6.11 can also be used for a soil with a constant C_c throughout its depth if a realistic average incremental effective vertical stress for each layer is selected.

From the definition of m_v, we can find the consolidation settlement for a layer of thickness H:

$$S_c = \Delta H = m_v H(\sigma'_1 - \sigma'_0) \tag{6.12}$$

The general form of this equation may be written as:

$$S_c = \sum_{i=1}^{i=n} (\Delta\sigma_1 m_v \Delta z)_i \tag{6.13}$$

where $\Delta\sigma_1$ is the incremental effective vertical stress and Δz is the thickness of each layer.

Example 6.3

The coordinates of two points on the virgin compression line are: $\sigma'_1 = 400$ kPa, $e_1 = 0.80$; $\sigma'_2 = 800$ kPa, $e_2 = 0.75$. In the field, a 3 m normally consolidated layer of this soil is to be subjected to construction works and the average effective vertical stress will increase from 250 kPa to 450 kPa. Determine: (a) the consolidation settlement, (b) the load increment to cause a 25 mm final consolidation settlement and the corresponding void ratio.

Solution:

(a) Calculation of C_c from Equation 6.1: $C_c = \dfrac{0.80 - 0.75}{\log(800.0/400.0)} = 0.166$.

Equation of the virgin compression line:

$C_c = \dfrac{0.80 - e}{\log\sigma'/400.0} = 0.166 \rightarrow e = 0.8 - 0.166(\log\sigma' - \log 400.0)$.

Initial void ratio in the field: $e_0 = 0.8 - 0.166(\log 250.0 - \log 400.0) = 0.834$.

$S_c = [0.166 \times 3000/(1 + 0.834)]\log(450.0/250.0) = 69.3 \approx 69$ mm.

(b) $S_c = [0.166 \times 3000/(1 + 0.834)]\log[(250.0 + \Delta\sigma')/250.0] = 25.0 \rightarrow \Delta\sigma' = 59.0$ kPa.

Thus $\sigma'_1 = 250.0 + 59.0 = 309.0$ kPa and

$e_1 = 0.8 - 0.166(\log 309.0 - \log 400.0) = 0.819$.

6.2.7 *Consolidation settlement due to lowering of the water table*

The drawdown of the water table in soft clays and silts introduces consolidation effects resulting in settlement of the ground surface. This settlement is progressive and its magnitude and rate are dependent on the thickness of the layer, the drainage conditions and the amount of drawdown. The main reason for the consolidation is the decrease in the pore pressure that ultimately increases the effective vertical stress approximately in proportion to the depth of the drawdown. The magnitude of this type of consolidation settlement recorded around the world has varied from 0.2 m up to 8 m (Terzaghi et al., 1996) and may be predicted by conventional methods. However, this prediction in thick layers is unreliable due to uncertainties in the drainage conditions and the variation with depth of the compressibility characteristics that are not detected from somewhat disturbed samples.

Undesirable side effects of dewatering have been fully documented (Powers, 1985) together with procedures to minimize the dangers of the resulting consolidation settlement to the main construction project and nearby facilities. Drawdown of the water table may result from the pumping and removal of water from the ground aquifer and construction site in water supply projects and excavations for footings and the like. In the case of temporary retaining structures such as sheet piles, the increase in effective vertical stresses results in the increase of the effective horizontal stress or effective soil pressure that has to be compared with the reduction of total water thrust due to the lowering of the water table. Zeevaert (1957) and Parsons (1959) suggested that consolidation settlement could be prevented by injecting the water from the excavations into the backfill. However, an environmental study may be appropriate for this process of removal and injection of water because of bacterial concerns. Despite the problems associated with drawdown of the water table, it is an effective technique to preconsolidate the soil in order to minimize the effects of consolidation settlement on a structure, which is constructed after the settlement has occurred.

Parry & Wroth (1981) proposed a simplified analysis to estimate the overconsolidation ratio due to dewatering. The effective vertical stress prior to lowering of the water table is calculated from Equation 2.1: $\sigma_0 = \sigma'_0 + u_0$ where σ_0 is the total vertical stress and u_0 is the pore pressure. Similarly the effective vertical stress after lowering the water table by the amount of ΔH is expressed by:

$$\sigma_1 = \sigma'_1 + (u_0 - \Delta H \gamma_w)$$

where γ_w is the unit weight of water. Considering the change in the unit weight of the zone subjected to dewatering, the amount of increase in the effective vertical stress is:

$$\sigma'_1 - \sigma'_0 = \Delta\sigma' = \Delta H \gamma_w - \Delta H (\gamma_{sat} - \gamma_m) \tag{6.14}$$

where γ_m is the unit weight of the dewatered zone and γ_{sat} is the saturated unit weight. The negative term in Equation 6.14 may be ignored if it is assumed that the unit weight of the dewatered zone remains unchanged as a result of capillary action. If pumping is stopped after the completion of the primary consolidation, the water level will recover and the overconsolidation ratio is defined by:

$$OCR = \frac{\sigma'_1}{\sigma'_0} = 1 + \frac{\Delta H \gamma_w - \Delta H (\gamma_{sat} - \gamma_m)}{\sigma'_0} \tag{6.15}$$

The magnitude of ΔH is controlled by the amount of consolidation settlement predicted for the specified loads exerted by the structure. If the initial state is in overconsolidated range the magnitude of $\Delta H \gamma_w$ must be greater than p'_c in order to increase the *OCR* to a required value. Note that the increase in the effective vertical stress is constant within the thickness.

Example 6.4

A 4.5 m thick normally consolidated layer of clay is overlain by a free draining sand of 3 m thickness. The clay layer is resting on a very thick sand layer. The water table is at the ground surface. Calculate the settlement of the ground surface if the water table is lowered 3 metres by pumping. The compression index of the clay is 0.64 and a point on the field virgin compression line has a void ratio of 0.823 corresponding to an effective vertical stress of 35 kPa.

$\gamma_{sand} = \gamma_{clay} = 19.1 \ \text{kN/m}^3$.

Solution:

Calculate the initial and final effective vertical stresses at the centre of the layer:

$\sigma'_0 = (4.5/2 + 3.0) \times 19.1 - (4.5/2 + 3.0) \times 9.81 = 48.8 \ \text{kPa}$.

$\sigma'_1 = (4.5/2 + 3.0) \times 19.1 - 2.25 \times 9.81 = 78.2 \ \text{kPa}$.

Equation of the virgin compression line is:

$e = 0.823 - 0.64(\log \sigma' - \log 35.0)$, thus:

$e_0 = 0.823 - 0.64(\log 48.8 - \log 35.0) = 0.731$. Using Equation 6.8:

$S_c = 0.64 \times \log(78.2 / 48.8) \times 4500 / (1 + 0.731) = 341 \ \text{mm}$.

Example 6.5

A strip footing is constructed on the ground surface of the soil described in Example 6.4. The clay layer is divided into 3 layers of 1.5 m thickness. The effective vertical stresses at the centre points of each layer (downward) have been increased by 103.2 kPa, 86.5 kPa and 74.4 kPa respectively. Calculate the final consolidation settlement of the footing under the following conditions:
(a) The load is applied after completion of the pumping stage of Example 6.4,
(b) the load is applied after the water table has returned to its original level at the ground surface without affecting the void ratios (no expansion has occurred),
(c) the load is applied before the lowering of the water table.

Solution:

(a) The results of calculations are summarized in the table below:

Layer	σ'_0 (kPa)	σ'_1 (kPa)	e_0	S_c (mm)
1	64.3	167.5	0.654	219
2	78.2	164.7	0.600	179
3	92.1	166.5	0.554	149
Total consolidation settlement (mm)				547

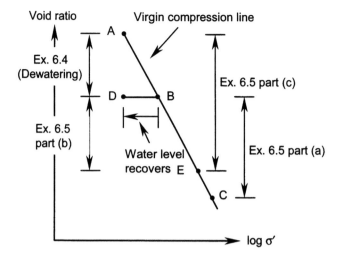

Figure 6.5. Examples 6.4 and 6.5.

The settlements of layers due to dewatering are 140 mm, 114 mm and 97 mm (calculations not shown); note that the sum of the above settlements does not match with the average answer obtained in Example 6.4. Sample calculation for layer 1:

$\sigma'_0 = (0.75 + 3.0) \times 19.1 - 0.75 \times 9.81 = 64.3 \, \text{kPa}$.

$\sigma'_1 = 64.3 + 103.2 = 167.5 \, \text{kPa}$.

$e_0 = 0.823 - 0.64(\log 64.3 - \log 35.0) = 0.654$.

$S_c = 0.64 \times \log(167.5/64.3)(1500.0 - 140.0)/(1 + 0.654) = 219 \text{mm}$.

(b) The unloading and reloading curves are equivalent to a horizontal line, as shown in Figure 6.5. The initial state is represented by point D, and there is no settlement during the recompression from point D to point B. The consolidation settlement is due to the virgin compression from B to E as tabulated below. Sample calculation for layer 1 is as follows:

Layer	σ'_0 (kPa) at point B	e_0 at point B	σ'_1 (kPa) at point E	S_c (mm)
1	64.3	0.654	138.0	175
2	78.2	0.600	135.3	132
3	92.1	0.554	137.1	100
		Total consolidation settlement (mm)		407

$\sigma'_0 = (\text{before lowering or after recovering}) =$
$(0.75 + 3.0) \times 19.1 - (0.75 + 3.0) \times 9.81 = 34.8 \, \text{kPa}$;

similar calculations will yield 48.8 kPa and 62.7 kPa for layers 2 and 3 respectively.

$\sigma'_1 = 34.8 + 103.2 = 138.0 \, \text{kPa}$,

$S_c = 0.64 \times \log(138.0/64.3)(1500.0 - 140.0)/1.654 = 175 \text{ mm}$.

(c) The results of calculations are tabulated below (see also Figure 6.5).
For layer 1: $S_c = 0.64 \times \log(138.0/34.8) \times 1500.0/(1 + 0.825) = 315 \text{ mm}$.

Layer	σ'_0 (kPa) Point A	e_0	σ'_1 (kPa) Point E	S_c (mm)
1	34.8	0.825	138.0	315
2	48.8	0.731	135.3	246
3	62.7	0.661	137.1	196
	Total consolidation settlement (mm)			757

6.3 SOLUTION OF ONE DIMENSIONAL CONSOLIDATION DIFFERENTIAL EQUATION

6.3.1 *Time and excess pore pressure relationships*

The final consolidation settlement theoretically takes an infinite amount of time to occur. Prediction of the magnitude of the consolidation settlement at any given time after the loading requires the solution of the one-dimensional consolidation differential equation constructed on the basis of assumptions described in Section 6.2.1. The continuity equation in one-dimensional flow, where the volume of the soil decreases over time, is:

$$k\frac{\partial^2 h}{\partial z^2}dx \times dy \times dz = \frac{\Delta V}{\Delta t}, \text{ or } k\frac{\partial^2 h}{\partial z^2} = \frac{\Delta V/V}{\Delta t} = \frac{m_v \Delta \sigma'}{\Delta t} = m_v \frac{\partial u_e}{\partial t}.$$

Note that in one-dimensional volume change $\Delta V/V = \Delta H/H$. Furthermore, the change in effective vertical stress is equal to the change in the excess pore pressure ($\Delta \sigma' = \Delta u$). Expressing h in terms of the pore pressure ($u_e = \gamma_w \times h$), the above differential equation becomes:

$$\frac{\partial u_e}{\partial t} = c_v \frac{\partial^2 u_e}{\partial z^2} \tag{6.16}$$

The parameter c_v is called the *coefficient of consolidation* and is defined by:

$$c_v = \frac{k}{m_v \gamma_w} \tag{6.17}$$

and is expressed in mm^2/min or m^2/year. Consider a layer of thickness $2d$ with two free draining boundaries (*FDB*) at the top and the base of the layer (Figure 6.6(a)). Assuming that the origin of the z-axis is located at the top free draining boundary, the following boundary conditions apply:

(a) $t = 0$ and $0 \le z \le 2d$, $u_e = u_i$ (note: u_i equals the applied total vertical stress),

(b) $t > 0$, for $z = 0$ and $z = 2d$, $u_e = 0$,

(c) $t > 0$, for $z = d$, $\partial u_e/\partial z = 0$ (conditions of symmetry and location of the maximum u_e).

The solution for the two-way drainage condition using a Fourier series is:

$$u_e = \sum_{n=1}^{n=\infty} \frac{1}{d}\left(\int_0^{2d} u_i \sin\frac{n\pi}{2d}dz\right)\sin\frac{n\pi}{2d}\exp\left(-\frac{n^2\pi^2 c_v t}{4d^2}\right) \tag{6.18}$$

Assuming a constant value for u_i at $t = 0$ (Figure 6.6(a)), we have:

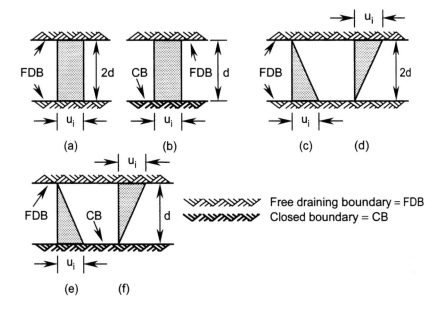

Figure 6.6. Different types of initial excess pore pressure distribution.

$$u_e = \sum_{n=1}^{n=\infty} \frac{2u_i}{n\pi}(1 - \cos n\pi)\left(\sin\frac{n\pi z}{2d}\right)\exp\left(-\frac{n^2\pi^2 c_v t}{4d^2}\right) \qquad (6.19)$$

A dimensionless variable called the *time factor* T_v is defined as:

$$T_v = \frac{c_v t}{d^2} \qquad (6.20)$$

For even values of n, the magnitude of $1 - \cos n\pi$ is zero, while for odd values it is equal to 2. By assuming $n = 2m + 1$ and $M = \pi(2m+1)/2$, Equation 6.19 becomes:

$$u_e = \sum_{m=0}^{m=\infty} \frac{2u_i}{M}\left(\sin\frac{Mz}{d}\right)\exp(-M^2 T_v) \qquad (6.21)$$

6.3.2   *Degree of consolidation*

The degree of consolidation is defined as:

$$U_z = (u_i - u_e)/u_i \qquad (6.22)$$

For a uniform u_i at time $t = 0$, $u_e = u_i$ and the degree of consolidation is zero. As $t \to \infty$, $u_e \to 0$ and degree of consolidation approaches 1. At time t, the degree of consolidation gives the increase in effective vertical stress as a proportion of the initial excess pore pressure, which indicates the progress of consolidation. Substituting u_e from Equation 6.21 into Equation 6.22:

$$U_z = 1 - \sum_{m=0}^{m=\infty} \frac{2}{M}\left(\sin\frac{Mz}{d}\right)\exp(-M^2 T_v) \qquad (6.23)$$

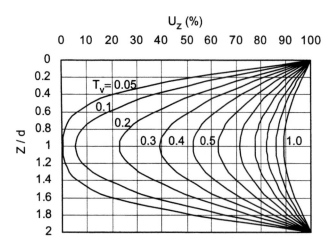

Figure 6.7. Consolidation isochrones.

Equation 6.23 may be presented graphically in a two-dimensional coordinate system as shown in Figure 6.7. The degree of consolidation shown on the abscissa varies from 0 to 1 (0% to 100%). The ordinate shows the relative depth parameter z / d that varies from 0 at the upper surface of the layer to 2 at the lower surface of the layer. This results in a series of curves for various (constant) values of T_v. These curves are termed *Isochrones*. Each isochrone shows the development of consolidation within the soil mass at a specified time.

6.3.3   Characteristics of isochrones

Figure 6.8 shows an isochrone for a value of $T_v = 0.4$. At point B with a depth ratio of 0.5 the degree of consolidation is $AB = 66.44\%$ which means that 66.44% of the initial excess pore pressure created at this point is dissipated. This is also equal to the increase in effective vertical stress at point B. The remaining excess pore pressure is $CB = 33.56\%$ of the initial value. For a two-way drainage system the maximum excess pore pressure is at the centre of the layer. For $T_v = 0.4$ this value is 47.45%. At this point the degree of consolidation is a minimum that is given by $100.0 - 47.45 = 52.55\%$.

For a one-way drainage system, only the half of the isochrone is used. Note that in this case the maximum drainage path d is equal to the depth of the layer and point B represents the midpoint of the layer. The maximum excess pore pressure is on the impermeable boundary and consequently points on this boundary will have the minimum degree of consolidation.

An average degree of consolidation for the layer can be defined by dividing the area between the isochrone and the left-hand axis (side AB) by the total area of $z / d \times 1$. The area within the isochrone, if divided by the total area, gives the remaining average pore pressure (in terms of a ratio or percentage of the initial excess pore pressure). For the example above ($T_v = 0.4$), it can be shown that the area between the isochrone and the z / d axis is 1.3958. Since the total area is 2, the average degree of consolidation for the layer is therefore $1.3958 / 2.0 = 69.8\%$.

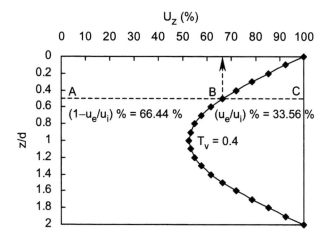

Figure 6.8. Characteristics of an isochrone.

If the U_z and z/d values are multiplied by $u_i/100$ and d respectively to give $u_i - u_e$ and z, then the area between the isochrone (plotted in the u_e-z coordinate system) and z-axis, when multiplied by m_v, will yield the settlement at that specific time (based on Equation 6.12). Similarly, the total area $(z \times u_i)$ multiplied by m_v will equal the final settlement providing that m_v is constant within the layer. Consequently, the area between two isochrones represents the incremental settlement between two given times.

The slope at any point on the isochrone represents the hydraulic gradient related to the steady flow of water. For an isochrone plotted in the u_e-z coordinate system, the hydraulic gradient is equal to $(1/\gamma_w)(\partial u_e/\partial z)$.

The average degree of consolidation is expressed mathematically by:

$$U = \frac{\left(u_i - \dfrac{1}{2d} \displaystyle\int_0^{2d} u_e dz \right)}{u_i}, \text{ or} \tag{6.24}$$

$$U = \frac{u_i(2d) - \displaystyle\int_0^{2d} u_e dz}{u_i(2d)} = \frac{\displaystyle\int_0^{2d} (u_i - u_e) m_v dz}{\displaystyle\int_0^{2d} u_i m_v dz} = \frac{\displaystyle\int_0^{2d} \sigma'_{(t)} m_v dz}{\displaystyle\int_0^{2d} \sigma'_{(t \to \infty)} m_v dz},$$

$$U = \frac{S}{S_c} \tag{6.25}$$

where S is the consolidation settlement of the layer of thickness $2d$ at time t and S_c is the final consolidation settlement. The average degree of consolidation can be expressed in terms of the time factor T_v by substituting the excess pore pressure (Equation 6.21) into Equation 6.24:

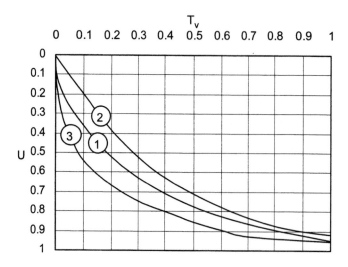

Figure 6.9. Relationship between average degree of consolidation and time factor.

$$U = 1 - \sum_{m=0}^{m=\infty} \frac{2}{M^2} \exp(-M^2 T_v) \tag{6.26}$$

This equation is plotted in Figure 6.9 (curve 1), and is approximated with reasonable accuracy by the following:

$$T_v = \frac{\pi U^2}{4} \rightarrow U \le 0.6, \; T_v = -0.933 \log(1-U) - 0.085 \rightarrow U \ge 0.6 \tag{6.27}$$

For a half-closed layer (for which only one boundary is free draining), Equation 6.26 – with d representing the thickness of the layer – is used if the distribution of u_i is uniform within the layer (Figure 6.6(b)). Equation 6.26 is also valid for the case of a linear distribution of initial excess pore pressure with two free draining boundaries, where at one boundary the initial excess pore pressure is zero and at the other is equal to u_i (Figures 6.6(c) and 6.6(d)). If the distribution of initial pore pressure is linear, from zero at the open boundary to u_i at the closed boundary with $u_{iz} = z \times u_i / d$ (Figure 6.6(e)), then the average degree of consolidation can be obtained by substituting this linear distribution into Equation 6.24:

$$U = 1 - 4 \sum_{m=0}^{m=\infty} \frac{1}{M^3} (\sin M) \exp(-M^2 T_v) \tag{6.28}$$

If the distribution of initial pore pressure is linear, from u_i at the open boundary to zero at the closed boundary (Figure 6.6(f)), then the average degree of consolidation is:

$$U = 1 - 4 \sum_{m=0}^{m=\infty} \frac{1}{M^3} (\sin M)(M \sin M - 1) \exp(-M^2 T_v) \tag{6.29}$$

In practice, sufficient accuracy is achieved if the first seven terms are used, i.e., m increases from 0 to 6. Equations 6.28 and 6.29 are shown in Figure 6.9 (curves 2 and 3 respectively). For a non-linear initial excess pore pressure, the average degree of consolidation may be calculated from:

$$U = \frac{u_i - u_e}{u_i} = 1 - \frac{\int_0^{2d} u_e dz}{\int_0^{2d} u_i dz} \qquad (6.30)$$

Table 6.2 shows the U-T_v relationship for a parabolic distribution of initial excess pore pressure in which both sides of the layer are free draining boundaries, and u_i is the maximum excess pore pressure at the middle of the layer.

Table 6.2. U and T_v values for parabolic initial excess pore pressure.

T_v	0.000	0.048	0.090	0.115	0.207	0.281
U	0.0	0.1	0.2	0.3	0.4	0.5

T_v	0.371	0.488	0.652	0.933	∞
U	0.6	0.7	0.8	0.9	1.0

Example 6.6

A normally consolidated 4 m thick layer of clay lies between two strata of free draining coarse sand. Plot the corresponding isochrone and determine the average degree of consolidation after 6 months. $m_v = 0.24$ m^2/MN, $c_v = 2.8$ m^2/year.

Solution:

From Equation 6.20, $T_v = \dfrac{2.8 \times 0.5 (\text{year})}{2.0^2} = 0.35$.

Use Equation 6.23 to find the degree of consolidation at depths 0.0, 0.5, 1.0, 1.5, 2.0, 2.5, 3.0, 3.5 and 4.0 metres. Due to symmetry, the calculation is carried out for 0.0 to 2.0 metres. The corresponding relative depths (z / d) are 0.0, 0.25, 0.50 and 1.00. For values of $m > 1$, the magnitude of $\exp(-M^2 T_v)$ is negligible, and calculations are carried out only for $m = 0$ and $m = 1$. The results are tabulated below and shown in Figure 6.10.

m	M	$2/M$	$2/M^2$	$\exp(-M^2 T_v)$	$\sin(Mz/d)$			
					z/d			
					0.250	0.500	0.750	1.000
0	0.5π	1.2732	0.8106	0.4216	0.3827	0.7071	0.9239	1.0000
1	1.5π	0.4244	0.0901	0.0004	0.9239	0.7071	-0.3827	-1.0000

z/d	0.250	0.500	0.750	1.000
$u_{(0)}/u_i$	205.445×10^{-3}	379.614×10^{-3}	495.990×10^{-3}	536.855×10^{-3}
$u_{(1)}/u_i$	0.165×10^{-3}	0.126×10^{-3}	-0.068×10^{-3}	-0.178×10^{-3}
Total u_e/u_i	205.610×10^{-3}	379.740×10^{-3}	495.922×10^{-3}	536.677×10^{-3}
$U_z = 1 - u_e/u_i$	0.7944	0.6203	0.5040	0.4633

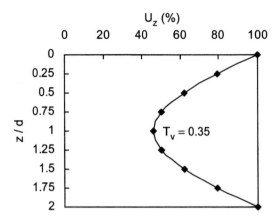

Figure 6.10. Example 6.6.

Using Equation 6.26: $U = 1 - (0.8106 \times 0.4216 + 0.0901 \times 0.0004) = 0.6582 = 65.82\%$.

From Equation 6.27: $T_v = -0.933\log(1-U) - 0.085 = 0.35 \rightarrow U = 0.6582 = 65.82\%$.

6.3.4 Determination of the coefficient of consolidation c_v

The parameter c_v is determined on the assumption that a theoretical relationship accurately represents the experimental relationship. Traditionally, two methods based on different time bases have been used.

Square root time fitting method. In this method, settlements measured during a load increment are plotted against the square root time. Figure 6.11 shows the theoretical consolidation curve (Equations 6.26 or 6.27) in which the variation of degree of consolidation is plotted against square root of time factor. From this plot the following observations can be made. The theoretical curve represents a straight line up to $U = 60\%$. From Equation 6.26 (or 6.27) for $U = 90\%$, $T_v = 0.848$. This is 1.333 times greater than the time factor calculated from the linear part of the curve: $T_v = \pi U^2/4 = \pi \times (0.9)^2/4 = 0.636$ and therefore $0.848 / 0.636 = 1.333$. Consequently, a line drawn from the point of zero consolidation with a slope (measured from the vertical axis) of $\sqrt{1.333} \approx 1.15$ times greater than the slope of linear part of the curve will intersect the curve at $\sqrt{T_v}$ of 90% consolidation. These observations lead to the following procedure for obtaining t_{90} and the calculation of c_v from the experimental curve:

1. Plot the settlement against the square root time.
2. Find the intersection of the linear part of the plot with the vertical axis and hence establish the position of zero consolidation settlement. This point should be below the experimental zero settlement point indicating an immediate settlement just before the commencement of the consolidation process. Note that in the theoretical curve these two points are represented by one point, as the immediate settlement is not included in the theoretical equation.

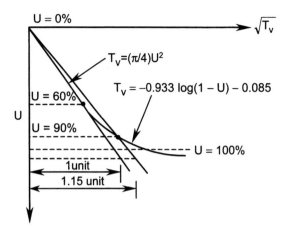

Figure 6.11. Theoretical consolidation in the square root (T_v)-U coordinate system.

3. From the point in (2) above, draw a line with a slope of 1.15 times of the slope of the linear portion of the plot (measured from the vertical axis). The intersection point of this line with the experimental plot represents $\sqrt{t_{90}}$.

4. Substitute t_{90} into Equation 6.20, noting that at 90% consolidation $T_v = 0.848$.

$$c_v = \frac{0.848 d^2}{t_{90}} \tag{6.31}$$

Equation 6.31 can also be arranged in the following form:

$$c_v = \frac{0.1114 H_{av}^2}{t_{90}} \tag{6.32}$$

where t_{90} is in minutes, $H_{av} = 2d$ is the average thickness (mm) of the specimen for the load increment and c_v is in m^2/year.

5. Establish the primary consolidation or 100% level by proportion, increasing the 90% settlement by ten-ninths.

log of time fitting method. In this method, settlements are plotted against log time. Figure 6.12 shows the theoretical consolidation curve, where the variation in the degree of consolidation in terms of time factor is plotted on a semi-logarithmic scale. While the early portion of the theoretical curve is parabolic, the middle portion is linear. The intersection of this line, represented by AB, with the tangent to the third and lower part of the theoretical curve located on the horizontal axis shows the position of 100% consolidation. The foregoing observations lead to a straightforward step-by-step procedure for obtaining t_{50} and calculation of c_v from experimental data:

1. Plot the settlement in terms of log time. Note that zero time corresponding to zero settlement cannot be shown on a logarithmic scale.

2. Select a point on the curve close to the settlement axis and find the corresponding time t for this point. Locate another point such that its time is $4t$. The settlement between these

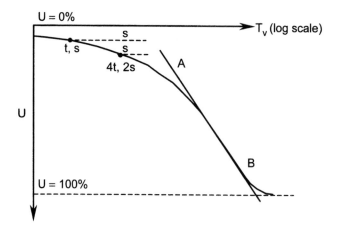

Figure 6.12. Theoretical consolidation in the log (T_v)-U coordinate system.

two points is equal to the settlement between times $t/4$ and t. The point with $t/4$ could be taken as the start time if t is relatively small. Thus, the position of zero settlement can be found. This point should be beneath the experimental zero settlement point indicating an immediate settlement just before the commencement of the consolidation process.

3. Draw a tangent to the lower portion of the plot. This line is either horizontal or slightly inclined thereby showing a secondary consolidation process. Find the intersection point of this line with the linear portion of the plot and hence establish the settlement corresponding to 100% consolidation. Find the midpoint between 0% and 100% to determine t_{50}.

4. Substitute t_{50} into Equation 6.20 noting that at 50% consolidation $T_v = 0.197$.

$$c_v = \frac{0.197 d^2}{t_{50}} \tag{6.33}$$

Equation 6.33 can also be arranged in the following form:

$$c_v = \frac{0.0259 H_{av}^2}{t_{50}} \tag{6.34}$$

where t_{50} is in minutes, H_{av} is the average thickness of the specimen for the load increment in mm and c_v is in m²/year. If the immediate settlement and secondary consolidation do not control the consolidation behaviour, the average thickness and other parameters are calculated based on the initial and final conditions recorded in the test; otherwise, correction factors have to be applied according to the relevant codes.

Example 6.7

Data obtained from a laboratory consolidation test on a clay is given below.
$\sigma'_0 = 100$ kPa, $\sigma'_1 = 200$ kPa, $G_s = 2.72$, $H_0 = $ (initial thickness at 100 kPa) = 18 mm, $w = $ (moisture content at the end of the test) = 0.245. Determine: (a) c_v from the square root time plot in m²/year, (b) c_v from the log time plot in m²/year, (c) the coefficient of permeability, and (d) the compression index C_c. Assume the specimen is normally consolidated.

Solution:

(a) & (b) $H_{av} = 18.0 - 1.03 / 2 = 17.485$ mm, $d = 17.485 / 2 = 8.74$ mm. The square root time plot is shown in Figure 6.13. $\sqrt{t_{90}} = 3.45$ from which $t_{90} = 11.90$ min.

Time (min)	Total ΔH (mm)	Time (min)	Total ΔH (mm)	Time (min)	Total ΔH (mm)	Time (min)	Total ΔH (mm)
0.25	0.160	4.00	0.507	25	0.834	81	0.894
0.50	0.208	6.25	0.611	36	0.850	100	0.905
1.00	0.276	9	0.694	49	0.866	300	0.954
2.25	0.391	16	0.789	64	0.878	1440	1.030

$c_v = 0.848 \times 8.74^2 \times 10^{-6} / [11.90 / (365 \times 1440)] = 2.86 \ \mathrm{m^2/year}$.

The log time plot is shown in Figure 6.14 from which $t_{50} = 2.7$ min.

$c_v = 0.197 \times 8.74^2 \times 10^{-6} / [2.7 / (365 \times 1440)] = 2.93 \ \mathrm{m^2/year}$.

(c) Calculate m_v: e_1 (final) $= w \ G_s = 0.245 \times 2.72 = 0.666$. From Equation 6.5:

$\Delta e / 1.03 = (1 + \Delta e + 0.666) / 18.0 \rightarrow \Delta e = 0.101$, $e_0 = \Delta e + e_1 = 0.101 + 0.666 = 0.767$.

From Equation 6.6: $m_v = [1/(1.0 + 0.767)] \times 0.101 / (200.0 - 100.0) = 5.72 \times 10^{-4} \ \mathrm{m^2/kN}$.

Using Equation 6.17 with an average value of $c_v = 2.9 \ \mathrm{m^2/year}$:

$k = c_v m_v \gamma_w = 2.9 \times 5.72 \times 10^{-4} \times 9.81 / (365 \times 24 \times 60 \times 60) = 5.16 \times 10^{-10} \ \mathrm{m/s}$.

(d) $C_c = 0.101 / \log(200.0 / 100.0) = 0.336$.

Example 6.8

Re-working Example 6.3 (with free draining boundaries), determine: (a) the time in days for 10%, 20%, ... 50%, ..., and 90% consolidation in the field, (b) the time in days for 25 mm of settlement to occur, (c) the settlement at the end of 3 months. In a laboratory consolidation test on this soil, the time required for 50% consolidation of a 20 mm thick specimen was 7.5 min.

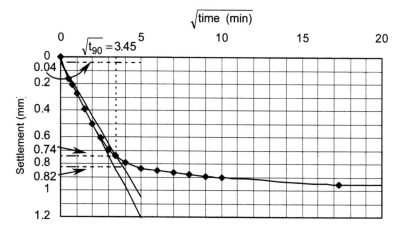

Figure 6.13. Settlement versus square root time: Example 6.7.

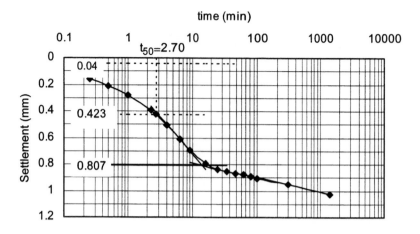

Figure 6.14. Settlement versus log time: Example 6.7.

Solution:

(a) $c_v = 0.197 \times 10^2 \times 10^{-6} / (7.5/1440) = 3.78 \times 10^{-3}\,\text{m}^2/\text{day}$.

The times required for 10%, 20%,... and 90% of consolidation in the field are calculated and tabulated below. Sample calculation for 50% consolidation:

$c_v = 3.78 \times 10^{-3} = 0.197 \times (3.0/2)^2 / t_{50} \rightarrow t_{50} = 117.3$ days.

(b) $U = 25.0$ mm / total consolidation settlement $= 25.0 / 69.3 = 0.3607 = 36.07\%$,

$T_v = \pi(0.3607)^2 / 4 = 0.102$, $c_v = 3.78 \times 10^{-3} = 0.102 \times (3.0/2)^2 / t \rightarrow t = 60.7$ days.

U (%)	10	20	30	40	50	60	70	80	90
T_v	0.008	0.031	0.071	0.126	0.197	0.286	0.403	0.567	0.848
Time (days)	4.8	18.4	42.3	75.0	117.3	170.2	239.9	337.5	504.8

(c) $c_v = 3.78 \times 10^{-3} = T_v \times (3.0/2)^2 / (3 \times 30) \rightarrow T_v = 0.151$,

$T_v = 0.151 = \pi U^2 / 4 \rightarrow U = 0.4385$, $S_c = 0.4385 \times 69.3 = 30.4$ mm.

6.3.5 Vertical drains

To facilitate the horizontal flow of water and thereby accelerate the consolidation process within a layer of saturated soft clay, vertical sand drains or wick drains are sometimes employed. The diameter of a sand drain varies between 160 mm and 610 mm. It is constructed using either an auger or a mandrel (Landau, 1966). In both methods some disturbance to the soil occurs around the perimeter of the drain, making a smear zone with lower permeability characteristics. Wick drains are prefabricated from a geotextile to produce drains 100 mm to 300 mm wide by 4 to 6 mm thick. Whilst different types of prefabricated drains have different characteristics (Rixner et al., 1986; Bergado et al., 1996), they normally consist of a plastic core for vertical draining paths and a filter jacket to prevent the passage of fine particles into the core.

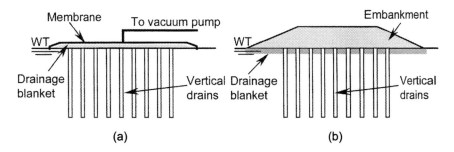

Figure 6.15. Applications of vertical drains.

The characteristics include geometric shape, configuration, drainage capacity and flow resistance, and its likely effect on the permeability of the adjacent soil (see case studies in Pilot, 1981). Figure 6.15(a) shows the application of vertical drains in conjunction with a preloading system in the form of a vacuum pump, as originally suggested by Kjellman (1952). In a deep clay layer this technique may not be sufficient when used alone, and vertical drains are designed either with preloading at the surface or under gradually increasing construction load as shown in Figure 6.15(b). Drains are arranged in triangular or square patterns (Figure 6.16). The influence zone of each drain defines a soil cylinder of diameter D_e that is a function of the distance L between the drains:

$$D_e = 1.128\,L \quad \text{(rectangular pattern)}, \quad D_e = 1.505\,L \quad \text{(triangular pattern)} \tag{6.35}$$

The differential equation of one-dimensional consolidation for radial draining towards a sand or a wick drain is constructed using principles similar to those in Section 6.3.1.

$$c_h \left(\frac{\partial^2 u_e}{\partial r^2} + \frac{1}{r} \frac{\partial u_e}{\partial r} \right) = \frac{\partial u_e}{\partial t} \tag{6.36}$$

where u_e is the excess pore pressure due to radial drainage only, t is time, and r is the distance of any point from the vertical drain. The parameter c_h is called the *coefficient of consolidation in the horizontal direction* and is defined by:

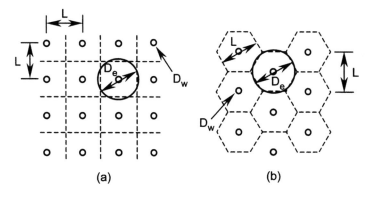

Figure 6.16. Two traditional patterns for vertical drains.

$$c_h = \frac{k_h}{m_v \gamma_w} \tag{6.37}$$

where k_h is the coefficient of permeability in the horizontal direction. The solution of Equation 6.36 was originally obtained by Barron (1948) after assuming uniform surface loading and uniform vertical deformation together with the following: at $t = 0$, $u_e = u_i$; and at $t = t$, $u_e = 0$ in the drain. The degree of consolidation of the layer was found to be:

$$U_h = 1 - \exp\frac{-8T_h}{F(n)} \tag{6.38}$$

where n, $F(n)$, and T_h are defined below in which D_w is the diameter of the drain:

$$n = \frac{D_e}{D_w} \tag{6.39}$$

$$T_h = \frac{c_h t}{D_e^2} \tag{6.40}$$

$$F(n) = \frac{n^2}{n^2 - 1} \ln(n) - \frac{3n^2 - 1}{4n^2} \tag{6.41}$$

For the values of $n > 10$, $F(n)$ may be approximated by:

$$F(n) = \ln(n) - \frac{3}{4} \tag{6.42}$$

Figure 6.17 shows the variation of the degree of consolidation with the time factor and n. For comparison Terzaghi's one-dimensional consolidation (Equations 6.26 or 6.27) is also shown. Hansbo (1979, 1981, and 1987) modified Equation 6.38 to include the effects of the smear zone and drain resistance by replacing $F(n)$ with a general parameter F:

$$F = F(n) + F_s + F_r \tag{6.43}$$

$$F_s = (\frac{k_h}{k_{hs}} - 1)\ln(\frac{D_s}{D_w}), \ F_r = \frac{\pi z(H - z)k_{hs}}{q_w} \tag{6.44}$$

The diameter of the smear zone D_s is taken to be twice the diameter of the mandrel used in the installation of the vertical drain, whilst k_{hs} defines its coefficient of permeability in the horizontal direction. In the absence of experimental data k_{hs} could be assumed to be equal to the coefficient of permeability in the vertical direction. The parameters H and q_w are the length and discharge capacity of the vertical drain respectively, where the discharge capacity is the flow that passes through the drain under a hydraulic gradient of 1.0. If drainage occurs only at one end the length H must be doubled. The parameter F_r represents the drain resistance at any point with a distance z from the drainage end and has the following average values for the two possible drainage conditions at the ends of the drain:

$$F_r = \frac{2\pi H^2 k_{hs}}{3q_w} \ \text{(one end)}, \ F_r = \frac{\pi H^2 k_{hs}}{6q_w} \ \text{(both ends)} \tag{6.45}$$

The contribution of any vertical flow can be incorporated by the inclusion of the corresponding differential equation:

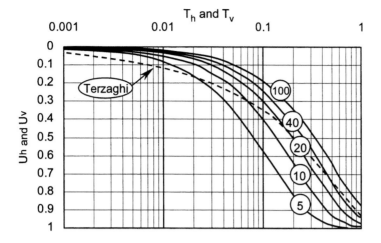

Figure 6.17. Degree of consolidation versus time factor for a clay layer with vertical drains.

$$c_h\left(\frac{\partial^2 u_e}{\partial r^2}+\frac{1}{r}\frac{\partial u_e}{\partial r}\right)+c_v\frac{\partial^2 u_e}{\partial z^2}=\frac{\partial u_e}{\partial t} \tag{6.46}$$

The combined excess pore pressure originated by Carillo (1942) is:

$$u_e=\frac{u_{ev}\times u_{eh}}{u_i} \tag{6.47}$$

where u_{ev} is the excess pore pressure due to vertical drainage only (Terzaghi's solution) and u_{eh} is the excess pore pressure due to radial flow (Barron's solution). Consequently, by applying the definition of average degree of consolidation in Equation 6.47, the combined degree of consolidation can be written in the following form:

$$U=1-(1-U_h)(1-U_v) \tag{6.48}$$

where U_h and U_v are the degrees of consolidation for radial and vertical flow respectively.

Example 6.9

A soft clay layer 10 m thick is resting on an impermeable boundary and is overlain by 3 m of sand, whose upper boundary is the ground surface. The clay layer is in the normally consolidated state and the water table is 2 m below the ground surface. A load of 80 kPa is applied at the ground surface over a large area. If sand drains are constructed in square pattern with $L=3$ m, determine the consolidation settlement after 6 months and the time required for 90% consolidation. γ (sand above water table) = 18 kN/m³, γ_{sat} (sand) = 20.2 kN/m³, γ_{sat} (clay) = 19 kN/m³, c_v = 5.15 m²/year, c_h = 3.20 m²/year, C_c = 0.25, e_0 = 0.8.

Solution:

Initial and final effective vertical stresses at the mid point of layer are:

$\sigma_0' = 19.0\times5.0+20.2\times1.0+18.0\times2.0-9.81\times6.0 = 92.3$ kPa,

$\sigma_1' = 92.3+80.0 = 172.3$ kPa.

$S_c = 0.25\times10.0\times\log(172.3/92.3)/(1+0.8) = 0.377$ m.

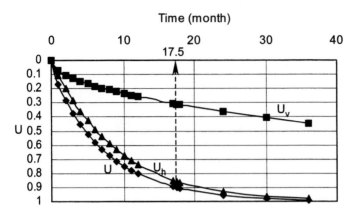

Figure 6.18. Example 6.9.

From Equations 6.35, 6.39 and 6.42 D_e, n and $F(n)$ are calculated:

$D_e = 3.0 \times 1.128 = 3.384$ m, $n = D_e / D_w = 3.384/0.3 = 11.28$,

$F(n) = \ln(11.28) - 3/4 = 1.673$.

Applying Equation 6.20: $T_h = 3.20 \times 0.5/3.384^2 = 0.1397$ and from Equation 6.38:

$U_h = 1 - \exp(-8 \times 0.1397/1.673) = 0.4873 = 48.73\%$.

$T_v = 5.15 \times 0.5/10.0^2 = 0.02575 = \pi U_v^2/4 \rightarrow U_v = 0.1811 = 18.11\%$.

With no contribution from horizontal flow the degree of consolidation after 6 months is 18.11%. The overall degree of consolidation is calculated from Equation 6.48:

$U = 1 - (1 - 0.4873)(1 - 0.1811) = 0.5801 = 58\%$,

S_c (6 months) = 0.377 m × 0.5801 = 0.219 m. The results are tabulated below and shown in Figure 6.18, from which $t_{90} = 17.5$ months.

Time (months)	U_v	U_h	U	Time (months)	U_v	U_h	U
1	0.0730	0.1054	0.1707	10	0.2337	0.6716	0.7483
2	0.1045	0.1996	0.2832	11	0.2452	0.7062	0.7782
3	0.1280	0.2840	0.3756	12	0.2561	0.7372	0.8045
4	0.1478	0.3594	0.4541	17	0.3048	0.8494	0.8953
5	0.1653	0.4269	0.5216	17.5	0.3092	0.8575	0.9016
6	0.1811	0.4873	0.5801	18	0.3136	0.8653	0.9075
7	0.1956	0.5414	0.6311	24	0.3621	0.9309	0.9559
8	0.2091	0.5897	0.6755	30	0.4049	0.9646	0.9789
9	0.2218	0.6329	0.7143	36	0.4435	0.9818	0.9899

6.3.6 Numerical solution using the finite difference method

The finite difference approach has been proved to be a convenient method to model the consolidation theory with reasonable accuracy (Olson et al., 1974). Olson (1998) reported that the incremental small strain approach seemed to work acceptably well and to execute

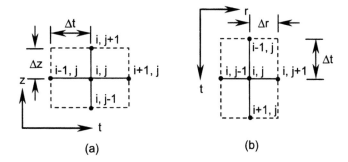

Figure 6.19. Finite difference grid for calculation of u_e: (a) vertical drainage, (b) radial drainage.

faster than the more complicated solutions. The simplified version is based on the finite difference approximation of Equations 6.16 and 6.36 using the depth - time grid shown in Figure 6.19. For vertical drainage the clay layer is divided into equal parts of thickness of Δz. The time interval at which the excess pore pressure is required is Δt. From Taylor's theorem:

$$\frac{\partial u_e}{\partial t} = \frac{1}{\Delta t}(u_{i+1,j} - u_{i,j}) \tag{6.49}$$

$$\frac{\partial^2 u_e}{\partial z^2} = \frac{1}{(\Delta z)^2}(u_{i,j+1} + u_{i,j-1} - 2u_{i,j}) \tag{6.50}$$

Substituting Equations 6.49 and 6.50 into Equation 6.16:

$$u_{i+1,j} = u_{i,j} + \frac{c_v \Delta t}{(\Delta z)^2}(u_{i,j+1} + u_{i,j-1} - 2u_{i,j}) \tag{6.51}$$

As higher order derivatives have been neglected and care must be taken in the selection of Δt and Δz so that the term $c_v \Delta t /(\Delta z)^2$ is approximately $1/6$ but certainly not exceeding $1/2$ (Craig, 1997). This equation applies only within the grid; for the points on the impermeable boundary:

$\frac{\partial u}{\partial z} = 0$, or $\frac{1}{2\Delta z}(u_{i,j-1} - u_{i,j+1}) = 0$ and $u_{i,j-1} = u_{i,j+1}$ and hence,

$$u_{i+1,j} = u_{i,j} + \frac{c_v \Delta t}{(\Delta z)^2}(2u_{i,j+1} - 2u_{i,j}) \tag{6.52}$$

For a layered soil with different permeabilities and consolidation characteristics, Scott (1963) showed that the first and second derivations are given by:

$$\frac{k}{c_v}\frac{\partial u_e}{\partial t} = \frac{1}{2}(\frac{k_T}{c_{vT}} + \frac{k_B}{c_{vB}})\frac{1}{\Delta t}(u_{i+1,j} - u_{i,j}) \tag{6.53}$$

$$k\frac{\partial^2 u_e}{\partial z^2} = \frac{1}{2}(\frac{k_T + k_B}{(\Delta z)^2})(\frac{2k_T}{k_T + k_B}u_{i,j+1} + \frac{2k_B}{k_T + k_B}u_{i,j-1} - 2u_{i,j}) \tag{6.54}$$

where the symbols T and B represent the top and the bottom layers respectively. Substituting these into the primary differential equation of consolidation (Equation 6.16):

$$u_{i+1,j} = u_{i,j} + \frac{k_T + k_B}{k_T + k_B(c_{vT}/c_{vB})} \frac{\Delta t}{(\Delta z)^2} (\frac{2k_T}{k_T + k_B} u_{i,j+1} + \frac{2k_B}{k_T + k_B} u_{i,j-1} - 2u_{i,j})$$

(6.55)

For radial drainage the typical radius-time grid of Figure 6.19(b) is used to obtain:

$$u_{i+1,j} = u_{i,j} + \frac{c_h \Delta t}{(\Delta r)^2} \left[u_{i,j+1} + u_{i,j-1} - 2u_{i,j} + \frac{\Delta r}{2r}(u_{i,j+1} - u_{i,j-1}) \right]$$

(6.56)

Example 6.10

A saturated clay layer of 7.5 m is subjected to the following vertical loading distribution:

Depth (m)	0.00	1.5	3.00	4.5	6.00	7.5
$\Delta\sigma$ (kPa)	200.00	145.45	114.28	94.11	80.00	69.56

The clay layer is resting on an impermeable rock, but the upper boundary has free draining conditions. Using the finite difference method obtain the values of excess pore water pressures at the end of each month during one year of consolidation and calculate the average degree of consolidation after one year. $\Delta z = 1.5$ m, $c_v = 2.6$ m^2/year.

Solution:

It is reasonable to assume that the applied load represents the initial distribution of excess pore pressure throughout the depth of the soil except at the upper boundary where it is always zero.

$c_v \Delta t /(\Delta z)^2 = 2.6 \times (1/12)/1.5^2 = 0.09629$. Therefore, for the points within the grid:

$u_{i+1,j} = u_{i,j} + 0.09629(u_{i,j+1} + u_{i,j-1} - 2u_{i,j})$.

For the points on the impermeable boundary at the depth of 7.5 m:

$u_{i+1,j} = u_{i,j} + 0.09629(2u_{i,j+1} - 2u_{i,j})$.

The excess pore pressures (in kPa) are shown in the table below and plotted in Figure 6.20. It is evident that the excess pore pressure in the vicinity of the closed boundary slightly increases over time, but eventually will decease.

Sample calculation for depth $= 4.5$ m ($j = 3$), and $t = 8$ months ($i = 8$):

$u_e = 95.54$ (corresponding to $i = 7$) $+ 0.09629$ ($101.23 + 84.35 - 2 \times 95.54$) $= 95.0$ kPa,

Calculation of average degree of consolidation:
The average initial excess pore pressure of the layer:
$u_{i\text{-}average} = $ Area of isochrone / 7.5 =

$(0.00 / 2 + 145.45 + 114.28 + 94.11 + 80.00 + 69.56 / 2)$ $1.5 / 7.5 = 93.7$ kPa.
The average excess pore pressure of the layer at the end of 1 year is:

$u_{e\text{-}average} = $ Area of isochrone / 7.5 =

$(0.00 / 2 + 56.68 + 86.99 + 91.79 + 86.55 + 83.45 / 2)$ $1.5 / 7.5 = 72.7$ kPa.
Average degree of consolidation $U = 1 - 72.7 / 93.7 = 0.224 = 22.4\%$.

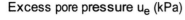

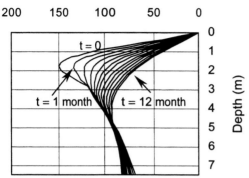

Figure 6.20. Example 6.10.

$i \rightarrow$ $j \downarrow$	0	1	2	3	4	5	6
0	0.00	0.00	0.00	0.00	0.00	0.00	0.00
1	145.45	128.44	114.81	103.73	94.61	87.00	80.58
2	114.28	115.34	114.61	112.77	110.26	107.38	104.33
3	94.11	94.69	95.30	95.77	96.04	96.08	95.91
4	80.00	80.35	80.88	81.53	82.25	82.98	83.69
5	69.56	71.57	73.26	74.73	76.04	77.23	78.34

$i \rightarrow$ $j \downarrow$	7	8	9	10	11	12
0	0.00	0.00	0.00	0.00	0.00	0.00
1	75.11	70.39	66.29	62.69	59.51	56.68
2	101.23	98.17	95.19	92.32	89.59	86.99
3	95.54	95.01	94.34	93.57	92.71	91.79
4	84.35	84.95	85.47	85.91	86.27	86.55
5	79.37	80.33	81.22	82.04	82.78	83.45

6.4 APPLICATION OF PARABOLIC ISOCHRONES

6.4.1 *Uniform initial excess pore pressure in open and half-closed layers*

The general shape of the isochrone shown in Figure 6.8 may be approximated by a parabolic equation. This leads to another U-T_v relationship that is rather different from the curve (1) shown in Figure 6.9. Figure 6.21(a) shows two isochrones in the u_e-z coordinate system. The middle portion of the isochrone corresponding to a low value of T_{v1} is approximately a vertical line, indicating high excess pore pressures and that the process of consolidation has not yet progressed into this part. Identical parabolas may be used to approximate the lower and upper portions of this isochrone. For a specific range of T_v the

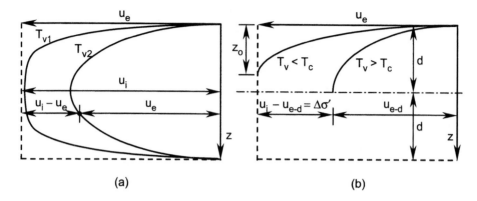

Figure 6.21. (a) Isochrones for low and high values of T_v, (b) parabolic isochrones.

parabolas are tangent to the vertical line drawn from the zero consolidation point (Figure 6.21(b)). The maximum depth of the tangent point (z_o) is equal to half the thickness d of the layer. If T_c is assumed to be the time factor corresponding to $z_o = d$, the isochrone with a time factor of $T_{v2} > T_c$ represents a more developed case and is well approximated by a parabola. Assuming a parabolic equation of the form: $u_e = az^2 + bz + c$ for $T_v < T_c$, then at $z = 0$, $u_e = 0$, and at $z = z_o$ (unknown), $u_e = u_i$. Substituting these conditions in the equation of the parabola and noting that the latter point is the maximum, we have:

$$u_e = -\frac{u_i}{z_o^2}z^2 + \frac{2u_i}{z_o}z \qquad (6.57)$$

From the properties of a parabola the volume change is expressed by:

$\Delta V = S_c \times 1$ unit area $= m_v \times$ area outside of the isochrone $= m_v \times z_o \times u_i / 3$.

The basic concept of consolidation states that the rates of volume change and flow are equal:

$$m_v\frac{dz_o}{dt}\frac{u_i}{3} = 1 \times k \times i = k\frac{\partial h}{\partial z} = \frac{k}{\gamma_w}\partial(-\frac{u_i}{z_o^2}z^2 + \frac{2u_i}{z_o}z)/\partial z, \text{ at } z = 0:$$

$$m_v\frac{dz_o}{dt}\frac{u_i}{3} = \frac{2ku_i}{z_o\gamma_w}, \text{ thus: } z_o\frac{dz_o}{dt} = \frac{6k}{m_v\gamma_w}.$$

Integrating the equation above between $t = 0$, $z = 0$ and $t = t$, $z = z_o$:

$z_o = \sqrt{12kt / m_v\gamma_w}$, using Equation 6.17 for k:

$$z_o = \sqrt{12c_vt} \qquad (6.58)$$

Replacing z_o in Equation 6.57:

$$u_e = u_i(\frac{-z^2}{12c_vt} + \frac{2z}{\sqrt{12c_vt}}) \qquad (6.59)$$

Equation 6.59 may be represented in U-T_v coordinate system using the main definition for the average degree of consolidation:

$$U = (\frac{z_0 u_i}{3})/(d \times u_i) = \frac{\sqrt{12 c_v t}}{3d} \rightarrow 2\sqrt{\frac{T_v}{3}}$$

(6.60)

This equation is valid until z_0 becomes equal to d. At this depth:

$z_0 = \sqrt{12 c_v t} = d$, or $z_0 = \sqrt{12 T_v d^2} = d$, thus $T_v = 1/12$ and $t = d^2/12 c_v$.

For the values of $T_v > 1/12$ the equation of parabola is:

$$u_e = -\frac{u_{e-d}}{d^2} z^2 + \frac{2 u_{e-d}}{d} z$$

(6.61)

where u_{e-d} is the excess pore pressure at the centre of the layer as shown in Figure 6.21(b). Equating the rate of change of volume with the flow rate and integrating:

$$u_{e-d} = u_i \exp(\frac{1}{4} - \frac{3 c_v t}{d^2}).$$

Replacing u_{e-d} in the equation of the parabola (Equation 6.61):

$$u_e = u_i(-\frac{z^2}{d^2} + \frac{2z}{d}) \exp(\frac{1}{4} - \frac{3 c_v t}{d^2})$$

(6.62)

In the U-T_v coordinate system:

$$U = 1 - \frac{2}{3} \exp(\frac{1}{4} - 3T_v)$$

(6.63)

Equations 6.60 (for $T_v < 1/12$) and 6.63 (for $T_v > 1/12$) are shown in Figure 6.22. If the linear portion of the curve is extended downwards it will intersect the $U = 1$ axis at $T_v = 0.75$.

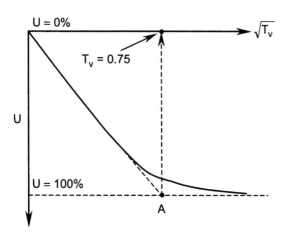

Figure 6.22. Reference point to calculate c_v.

This could be a reference point to calculate c_v from the square root time-settlement curve produced from experimental results. If the time corresponding to $T_v = 0.75$ is t_1 then:

$$c_v = \frac{3d^2}{4t_1} \tag{6.64}$$

It may be shown that the average degree of consolidation estimated by parabolic isochrones is slightly higher than the traditional method, and results in a higher value of effective stress (higher settlement) and a lower value of excess pore pressure.

Example 6.11

Resolve Example 6.6 using the parabolic isochrones.

Solution:

Calculations are carried out using Equation 6.62 for $T_v = 0.35$ (only for z / d from 0 to 1.0) and are tabulated below. The results are plotted for the whole thickness in Figure 6.23. Average U from Equation 6.63:

$$U = 1 - (2/3)\exp(1/4 - 3 \times 0.35) = 0.700 = 70.0\%.$$

z / d	u_e / u_i	$U_z = 1 - u_e / u_i$	$U_z \%$ (parabola)	$U_z \%$ (Exact)
0.00	0.0000	1.0000	100	100
0.25	0.1966	0.8034	80.34	79.43
0.50	0.3370	0.6630	66.30	62.02
0.75	0.4212	0.5788	57.88	50.40
1.00	0.4493	0.5507	55.07	46.33

Example 6.12

Using the following consolidation test data determine c_v from the root time plot in m^2/year. The initial thickness of the sample is 20 mm.

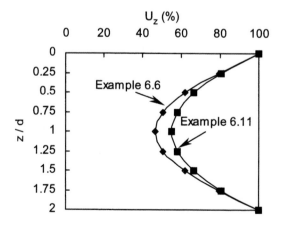

Figure 6.23. Example 6.11.

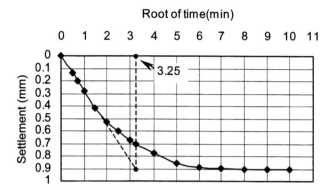

Figure 6.24. Example 6.12.

Time (min)	Total ΔH (mm)	Time (min)	Total ΔH (mm)	Time (min)	Total ΔH (mm)
0.25	0.138	6.25	0.524	49	0.894
0.50	0.196	9	0.597	64	0.900
1.00	0.277	16	0.775	81	0.900
2.25	0.375	25	0.855	100	0.900
4.00	0.415	36	0.890		

Solution:

The given data are plotted in Figure 6.24, from which: $t_1 = (3.25)^2 = 10.56$ min.

$2d$ (average) $= [20.0 + (20.0 - 0.9)] / 2 = 19.55$, $d = 9.77$ mm.

$c_v = [0.75 (9.77)^2 / 10.56] \times 60 \times 24 \times 365 \times 10^{-6} = 3.56$ m^2/year.

6.4.2   *Linear initial excess pore pressure in a half-closed layer*

The solution of one-dimensional consolidation using parabolic isochrones with linear initial excess pore pressure is limited to the case where the maximum initial excess pore pressure is applied at the closed boundary while at the free draining boundary its value is zero. The reverse conditions cannot be approximated by parabolic isochrones due to the increase of excess pore pressure on the closed boundary in low values of T_v. If the excess pore pressure on the free draining boundary is not zero and has a value less than the excess pore pressure on the closed boundary then a superposition method can be used. The initial linear excess pore pressure can be divided into two loading of uniform and linear. The magnitude of the uniform pressure is equal to its minimum value on the free draining boundary. Its linear part starts from zero at the free draining boundary and linearly increases to its maximum at the closed boundary. The magnitude of this maximum is equal to the difference between boundary values. Basic assumptions for the parabolic isochrones are shown in Figure 6.25. For T_v values smaller than a specific value T_c each isochrone is tangent to the initial excess pore pressure line (point T) and perpendicular to the closed boundary at B. For a value of $T_v > T_c$ the isochrone is located under the initial excess

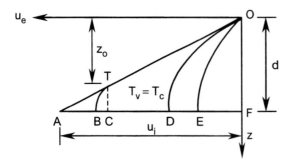

Figure 6.25. Approximation of isochrone with parabola for linear initial excess pore pressure.

pore pressure line and is perpendicular to the closed boundary at point E. To establish the isochrone for $T_v = T_c$ we assume $u_e = az^2 + bz + c$. At $z = 0$, $u_e = 0$, and the slope of the curve is u_i / d. Hence, the equation of parabola is:

$$u_e = -\frac{u_i}{2d^2}z^2 + \frac{u_i}{d}z \tag{6.65}$$

At $z = d$, $u_{e-d} = u_i / 2$ therefore: $S_c = m_v(d \times u_i / 2 - 2/3d \times u_{e-d}) = du_i m_v / 6$.
Equating S_c to the integration of the rate of flow:

$$\frac{du_i m_v}{6} = \int_0^t \frac{ku_i}{\gamma_w d}dt = \frac{ku_i}{\gamma_w d}t \rightarrow t = t_1 = \frac{d^2}{6c_v} \text{ or } T_v = \frac{1}{6}.$$

For the values of $T_v < 1/6$ the typical equation of a parabolic isochrone is constructed based on the geometry shown in Figure 6.25 using z_o as the depth for the point T:

$$u_e = \frac{-u_i}{2d(d-z_o)}(z-z_o)^2 + \frac{u_i}{d}(z-z_o) + \frac{u_i z_o}{d} \tag{6.66}$$

In order to relate z_o to time we first find the excess pore pressure at the closed boundary. Replacing $z = d$ in Equation 6.66:

$$u_{e-d} = FB = u_i(d + z_o)/2d \tag{6.67}$$

The settlement at this particular time is $S_c = m_v$ (area of TAC – area of TBC). Using the geometry of the figure we obtain:

$$S_c = \frac{m_v u_i(d-z_o)^2}{6d} \tag{6.68}$$

Using a unit area of soil, then volume change and settlement are numerically equal and the rate of volume change is equal to the rate of flow. In other words, settlement is equal to the integration of the rate of flow:

$$S_c = \frac{m_v u_i(d-z_o)^2}{6d} = \int_0^t \frac{ku_i}{\gamma_w d}dt = \frac{ku_i}{\gamma_w d}t, \text{ this yields a relationship between } z_o \text{ and time:}$$

$$(d - z_o)^2 = 6c_v t \tag{6.69}$$

Combining Equations 6.68 and 6.69 we get:

$$S_c = \frac{m_v u_i (d - z_o)^2}{6d} = \frac{m_v u_i \times 6 c_v t}{6d} = \frac{m_v u_i c_v t}{d} \tag{6.70}$$

The final settlement of the layer is:

S_c (final) $= m_v u_i d / 2$; from the basic definition of average degree of consolidation:

$U = S_c$ (at specific time) / final settlement, thus:

$$U = 2 c_v t / d^2 = 2 T_v \qquad \text{(for } T_v < 1/6 \text{)} \tag{6.71}$$

Using a similar procedure for the values of $T_v > 1/6$ the following equations are obtained.

$$u_e = \frac{-u_{e-d}}{d^2} z^2 + \frac{2 u_{e-d}}{d} z \tag{6.72}$$

$$u_{e-d} = \frac{u_i}{2} \exp(\frac{1}{2} - 3 T_v) \tag{6.73}$$

$$S_c = m_v (\frac{u_i d}{2} - \frac{2 u_{e-d} d}{3}) \text{ (at any specified time)} \tag{6.74}$$

$$U = 1 - \frac{2}{3} \exp(\frac{1}{2} - 3 T_v) \tag{6.75}$$

Equations 6.60 and 6.63 are represented by curve $P1$ in Figure 6.26 that approximates the exact solution shown by curve (1) in Figure 6.9. Similarly, Equations 6.71 and 6.75 are represented by curve $P2$ which approximates the exact solution shown by curve (2).

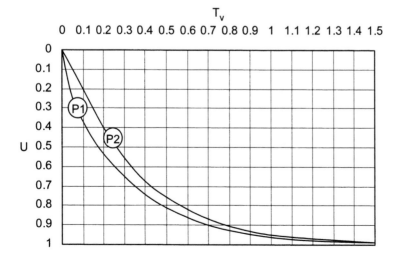

Figure 6.26. Relationship between U and T_v using parabolic isochrones.

Example 6.13

A slurry material comprised of fly ash and water is deposited in a large lagoon with an impermeable surface. The material thickness after deposition is 4 m. If the density of the slurry is 1.25 Mg/m^3, $m_v = 7.5$ m^2/MN, and $c_v = 5.5$ m^2/day, determine: (a) the time required for 90% consolidation, the corresponding settlement and the excess pore pressure at the surface of the impermeable layer, (b) the equations and plots of isochrones after 8 and 16 hours, the average degrees of consolidation and the relevant surface settlements, (c) the equation of excess pore pressure on the surface of the impermeable boundary in terms of time and a plot of the results.

Solution:

(a) The initial excess pore pressure is linear and varies from zero at the surface of the slurry to u_i at the surface of the lagoon:

$u_i = (\gamma - \gamma_w) \times 4.0 = (1.25 - 1.0) \times 9.81 \times 4.0 = 9.8$ kPa.

For $U = 90\%$ the value of T_v is calculated from Equation 6.75:

$U = 1 - (2/3)\exp(1/2 - 3T_v) = 0.9 \rightarrow T_v = 0.80$,

$T_v = c_v t / d^2 = 5.5 \times t / 16.0 = 0.8 \rightarrow t = 2.3$ days.

$S_c(\text{final}) = m_v u_i d / 2 = 7.5 \times 10^{-3} \times 9.8 \times 4.0 / 2 = 0.147$ m,

$S_c(90\%) = 0.147 \times 0.9 = 0.132$ m.

From Equation 6.73: $u_{e-d} = (9.8/2)\exp(0.5 - 3 \times 0.80) = 0.7$ kPa.

(b) $T_v = (5.5 \times 8.0/24)/16.0 = 0.1146 < 1/6$. Calculate z_o from Equation 6.69:

$(4.0 - z_o)^2 = 6c_v t = 6 \times 5.5 \times 8/24 \rightarrow 4.0 - z_o = 3.31$ m, $z_o = 0.69$ m.

Substitute z_o in Equation 6.66 to find the excess pore pressure parabola:

$u_e = -0.37z^2 + 2.96z - 0.176$.

Similarly the equation of isochrone at $t = 16$ hours is: $u_e = -0.254z^2 + 2.031z$.

Note that in this case $T_v = 0.2292 > 1/6$. The results are plotted in Figure 6.27(a).

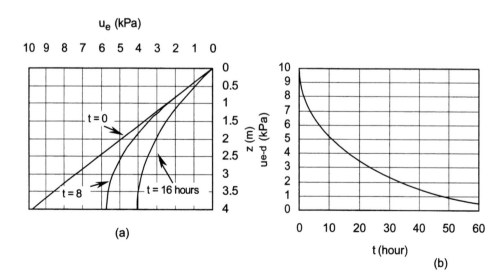

Figure 6.27. Example 6.13.

For $t = 8$ hours: $T_v = 0.1146 < 1/6$. From Equation 6.71:

$U = 2\,T_v = 2 \times 0.1146 = 0.2292 = 22.92\%$, $S_c = 0.2292 \times 0.147 = 0.03369$ m $= 33.7$ mm.

For $t = 16$ hours: $T_v = 0.2292 > 1/6$. From Equation 6.75:

$U = 1 - 2/3 \times \exp(1/2 - 3 \times 0.2292) = 0.4474 = 44.74\%$,

$S_c = 0.4474 \times 0.147 \times 1000 = 65.8$ mm.

(c) $c_v = 5.5$ m^2/day $= 5.5 / 24 = 0.2292$ m^2/hour.

For $T_v < 1/6$ or $t < T_v d^2 / c_v = (1/6) \times 16.0 / 0.2292 = 11.63$ hours.

Substitute z_o from Equation 6.69 into Equation 6.67 to obtain:

$u_{e-d} = (u_i / 2d)(2d - \sqrt{6c_v t}) = (9.8/8.0)(8.0 - \sqrt{6 \times 0.2292t}) = (9.8/8.0)(8.0 - 1.173\sqrt{t})$.

For $T_v > 1/6$ or $t > 11.63$ hours use Equation 6.73, but substitute T_v in terms of t:

$u_{e-d} = (9.8/2)\exp(0.5 - 3 \times 0.2292t/16.0) = (9.8/2)\exp(0.5 - 0.043t)$.

The results are plotted in Figure 6.27(b).

6.5 LIMITATIONS OF ONE DIMENSIONAL CONSOLIDATION THEORY

6.5.1 *Correction of primary consolidation settlement*

In the derivation of the U-T_v relationships it was assumed that the increase in the external load is instantaneous. However, construction work is normally a gradual process and may be idealized as shown in Figure 6.28, where t_c is the time when the applied vertical stress reaches its full amount $\Delta\sigma$. When using a numerical method, any type of non-linear time dependent load can generally be incorporated in the solution. Terzaghi's method for the correction of the time-settlement relationship (curve (*b*), Figure 6.28) for an instantaneous loading is based on the following assumptions. The corrected settlement at $t \geq t_c$ is equal to the settlement on the uncorrected curve corresponding to $t / 2$. For $t < t_c$ further correction is needed by considering the load ratio at the time of interest. The load ratio is the ratio of the load at the time t to the final load $\Delta\sigma$. The settlement corresponding to $t / 2$ is multiplied by this ratio. The curve (c) in Figure 6.28 is the corrected time – settlement relationship. This method implies that for a specified U the amount of time is twice of the time required for instantaneous loading.

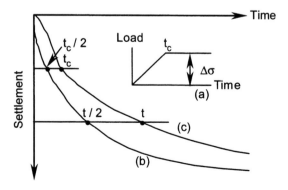

Figure 6.28. Terzaghi's correction for time dependent loading.

Example 6.14

At the centre of a 2 m thick saturated clay layer the effective vertical stress is increased from 60 kPa to 160 kPa within a period of 6 months. Plot the variation of consolidation settlement against time up to 48 months. The layer is half-closed and has the following properties: $C_c = 0.15$, $c_v = 1.5$ m^2/year, $e_0 = 0.7$.

Solution:

S_c (final) $= 0.15 \times 2000 \times \log(160.0/60.0)/(1+0.7) = 75.2$ mm.

Results of the computations are tabulated below and plotted in Figure 6.29. Sample calculations for 3, 6, and 12 months are as follows.

For $t = 3$ months:

$T_v = c_v t / d^2 = 1.5 \times (3.0/12.0)/2.0^2 = 0.0937 = \pi U^2 / 4 \rightarrow U = 0.345$,

$S_c = 0.345 \times 75.2 = 25.9$ mm. Corrected T_v, U and S_c:

$T_v = 1.5 \times (1.5/12.0)/2.0^2 = 0.0469 = \pi U^2 / 4 \rightarrow U = 0.244$,

U (corrected) $= 0.244$ (3.0 / 6.0) $= 0.122$, S_c (corrected) $= 0.122 \times 75.2 = 9.2$ mm.

For $t = 6$ months:

$T_v = 1.5 \times (6.0/12.0)/2.0^2 = 0.1875 = \pi U^2 / 4 \rightarrow U = 0.488$,

$S_c = 0.488 \times 75.2 = 36.7$ mm. Corrected T_v, U and S_c:

For 6.0 / 2 = 3.0 months we have already calculated T_v and U: $T_v = 0.0937$, $U = 0.345$,
U (corrected) $= 0.345$ (6.0 / 6.0) $= 0.345$, S_c (corrected) $= 0.345 \times 75.2 = 25.9$ mm.

For $t = 12$ months:

$T_v = 1.5 \times (12.0/12.0)/2.0^2 = 0.375 = -0.933\log(1-U) - 0.085 \rightarrow U = 0.679$,

$S_c = 0.679 \times 75.2 = 51.1$ mm. Corrected T_v, U and S_c:

For 12.0 / 2 = 6.0 months we have already calculated T_v and U: $T_v = 0.1875$, $U = 0.488$, no further correction required, thus $S_c = 0.488 \times 75.2 = 36.7$ mm.

Time (Month)	U Uncorrected	U Corrected	S_c (mm) Uncorrected	S_c (mm) Corrected
0.5	0.141	0.008	10.6	0.6
1	0.199	0.023	15.0	1.7
2	0.282	0.066	21.2	5.0
3	0.345	0.122	25.9	9.2
4	0.399	0.188	30.0	14.1
5	0.446	0.263	33.5	19.8
6	0.488	0.345	36.7	25.9
9	0.598	0.423	45.0	31.8
12	0.679	0.488	51.1	36.7
18	0.798	0.598	60.0	45.0
24	0.873	0.679	65.6	51.1
30	0.920	0.745	69.2	56.0
36	0.949	0.798	71.4	60.0
42	0.968	0.839	72.8	63.1
48	0.979	0.873	73.6	65.6

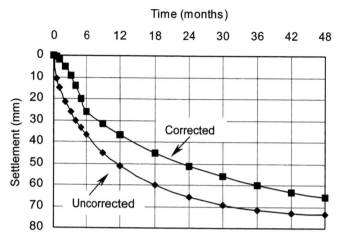

Figure 6.29. Example 6.14.

6.5.2   Sources of inaccuracy and three-dimensional consolidation

The main sources of inaccuracy in the prediction of both the magnitude and rate of consolidation settlement compared with in-situ measurements are (Duncan, 1993):

1. Inaccuracies in the measurement of consolidation characteristics such as preconsolidation pressure p'_c, compression index C_c, coefficient of volume compressibility m_v, and the coefficient of consolidation c_v in the laboratory conditions.
2. Difficulties in evaluating the drainage conditions in the field.
3. Shortcomings in the conventional consolidation theory.

Inaccuracies in the magnitude of p'_c, even with undisturbed specimens, could occur due to adoption of the 24 hour-based e-$\log\sigma'$ as being representative of field behaviour. Secondary consolidation may alter the overconsolidation ratio from 0 to a large value under the same effective vertical stress as suggested by Bjerrum (1967), introducing time-related consolidation curves. Point A in Figure 6.30 represents the state of stress on the virgin compression line constructed in laboratory conditions. As time passes beyond the 24 hour mark the state of stress drops vertically to point B on a consolidation curve t, indicating a secondary consolidation settlement. A point in the field with the same state of stress as B will, if it is tested in the laboratory, pass an apparent preconsolidation pressure σ'_m corresponding to an overconsolidation ratio of σ'_m / σ'_0. The overconsolidated soil can also be created by the changes in the ground water level, as discussed earlier in this chapter.

The final primary consolidation is calculated either using C_c or m_v. Accuracy of the prediction depends on the selection of C_c and m_v as well as the method of obtaining the incremental effective vertical stresses due to the surface loading. The parameter involved in the calculation of degree of consolidation (rate of settlement) is c_v, which is determined in the laboratory. Inaccuracies arise from both the method used in the laboratory as well as the evaluation of the drainage length.

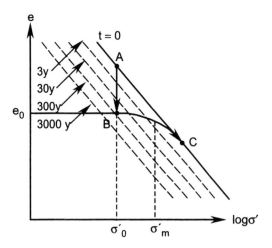

Figure 6.30. The concept of apparent overconsolidation (after Bjerrum, 1967).

Most of the standard codes recommend both root time and log time methods. However, the experimental results obtained in the laboratory do not perfectly satisfy the theoretical equations and the c_v obtained from the methods above become different within a small range. In some clays this difference may give rise to a high value, as reported by Duncan (1993) for San Francisco Bay mud (highly plastic organic clayey silt), where the root time method gave 50% more than the value determined by the log time method.

It must be noted that the magnitude of c_v is dependent on the stress level and changes during the dissipation of the excess pore pressure, considering its definition by Equation 6.17 ($c_v = k/m_v\gamma_w$). As a result of the progress of consolidation the coefficient of permeability k decreases. Furthermore, the decrease in volume results in an increase in strength and resistance to volume change and consequently, m_v decreases. In general the value of c_v decreases as the effective vertical stress approaches the preconsolidation pressure. However, in normally consolidated clays, c_v may increase with the increase of effective vertical stress along the virgin compression line. Sometimes reductions in k and m_v occur such that c_v remains constant during the consolidation progress.

Computations are simplified if the drainage length (or the average thickness of the specimen) is calculated using identical equations for the field and the laboratory. Olson & Ladd (1979) suggest the use of following relationship for a constant average thickness:

$$H = H_0 - \frac{S_T}{2N_D} \tag{6.76}$$

where H_0 is the initial thickness, S_T is the ultimate settlement and N_D is the number of free draining boundaries.

One of the major basic assumptions of the conventional consolidation theory is the assumption of a linear elastic soil, which is inherent in the differential equation of one-dimensional consolidation. The elastic properties E_s and μ are replaced by m_v in the form of proportionality of volume change and effective vertical stress. From the principles of elastic theory it can be shown that, for zero lateral strain, the magnitude of m_v is given by:

$$m_v = \frac{(1+\mu)(1-2\mu)}{(1-\mu)E_s} \tag{6.77}$$

However, soil is not a linear elastic material and the inclusion of non-linear behaviour in the main differential equation is only possible by means of numerical analysis (Mesri & Rokhsar, 1974). Note that the e-$\log\sigma'$ curve is non-linear, but this is only used to calculate the final settlement and is not incorporated in the time-settlement solution.

Conventional consolidation theory is applicable for small strains and cannot be applied to hydraulic fills and other slurry materials where the large strains control the consolidation progress. For large strain consolidation, most of the methods developed for the estimation of final settlements are based on the mathematical treatment of data obtained from the field measurement at the early and advanced stages of consolidation (Tan et al., 1991).

It must be noted that most consolidation settlements in the field have a two or three-dimensional nature. To better represent field conditions, attempts have been made to correct settlements predicted using one-dimensional theory. Skempton & Bjerrum (1957) proposed the following equation for the computation of final primary consolidation settlement:

$$S_c = \sum_{i=1}^{i=n} (m_v \Delta u \Delta z)_i \tag{6.78}$$

where Δz is the thickness of a finite layer and Δu is the initial average excess pore pressure (final increment in effective vertical stress) in the layer and is calculated from Equation 4.27 by assuming $B = 1$:

$$\Delta u = \Delta\sigma_3 + A(\Delta\sigma_1 - \Delta\sigma_3),$$

where $\Delta\sigma_1$ and $\Delta\sigma_3$ represent the incremental stresses (total) in the vertical and lateral directions respectively. Substituting Δu in Equation 6.78 we obtain the three-dimensional consolidation settlement:

$$S_c(3-D) = \sum_{i=1}^{i=n} [(A\Delta\sigma_1 + (1-A)\Delta\sigma_3)m_v\Delta z]_i \tag{6.79}$$

The combination of Equation 6.79 and 6.13 (one-dimensional consolidation settlement S_c) may be written in the following form:

$$S_c(3-D) = \lambda S_{c1} \tag{6.80}$$

where: $\lambda = A + \alpha(1-A)$ and, $\alpha = \sum_{i=1}^{i=n}(\Delta\sigma_3 m_v\Delta z)_i \Big/ \sum_{i=1}^{i=n}(\Delta\sigma_1 m_v\Delta z)_i$.

The parameter λ varies with pore pressure and α, the latter being dependent on the vertical and lateral stress distribution within the consolidated layer. The variation of λ with the pore pressure coefficient A for both circular and strip footings has been presented by Scott (1963). Equation 6.80 can be applied to a circular footing with reasonable accuracy assuming that a uniform load transmitted from the circular footing will create triaxial stress conditions in the soil underneath. For a strip footing, Scott (1963) modified Equation 6.80 to the form:

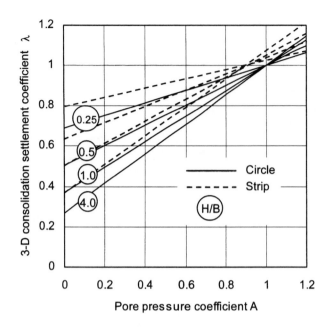

Figure 6.31. Settlement coefficients for 3-dimensional consolidation.

$$\lambda = N + \alpha(1-N) \rightarrow N = (\sqrt{3}/2)(A-1/3)+0.5 \tag{6.81}$$

The coefficient λ varies from 0.25 to 1.2 for a range of pore pressure coefficient A between 0 and 1.2 within a practical range of thickness of clay layer H to the width of footing B ratios (Figure 6.31). Balasubramaniam & Brenner (1981) challenged this method and indicated that Equation 6.79 is strictly valid only for triaxial stress conditions, and that a stress path method (Lambe & Marr, 1979) can better take into account the effects of lateral deformation on vertical settlement.

6.6 PROBLEMS

6.1 Data obtained from a laboratory consolidation test are tabulated as follows:

σ' (kPa)	20	50	100	200	400	800
Total ΔH (mm)	0.23	0.87	1.90	3.62	5.55	7.25

$G_s = 2.70$, $H_0 =$ (initial thickness at zero pressure) = 22.5 mm,
$w =$ (moisture content at the beginning of the test) = 0.78.
Plot the e-logσ' curve and calculate C_c.

Answer: 0.832

6.2 An open layer of clay 4 m thick is subjected to loading that increases the average effective vertical stress from 185 kPa to 310 kPa. Determine (a) the total settlement, (b) the settlement at the end of one year, (c) the time in days for 50% consolidation, (d) the time in days for 25 mm of settlement to occur.
$m_v = 0.00025$ m^2/kN, $c_v = 0.75$ m^2/year.

Answers: 125 mm, 61 mm, 383.5 days, 61 days

6.3 Data obtained from a laboratory consolidation test are shown in table below:

Time (min)	0.25	1	4	9	16	25	36	81	1440
Total ΔH (mm)	0.622	1.244	2.468	3.400	3.838	3.970	4.000	4.051	4.100

$\sigma'_0 = 100$ kPa, $\sigma'_1 = 200$ kPa, $H_0 = 23.6$ mm. Determine: (a) c_v from the root time plot in m^2/year, (b) c_v from the log time plot in m^2/year, (c) k in m/s.

Answers: (root time) 4.9 m^2/year, (log time) 5.1 m^2/year, 2.7×10^{-9} m/s

6.4 In a one-dimensional consolidation test the time required for 50% consolidation has been measured at 154 seconds (through the observation and measurement of pore pressure). The settlement of the sample at the end of the test was 2.5 mm. $e_0 = 0.65$, $\sigma'_0 = 60$ kPa, $\sigma'_1 = 120$ kPa, $H_0 = 20$ mm. Determine (a) the time required for 90% consolidation, (b) the coefficient of permeability (m/s), (c) the compression index.

Answers: 663 s, 2.3×10^{-9} m/s, 0.685

6.5 For a 4 m layer of the clay of Problem 6.4, how long would it take to reach 50% degree of consolidation under the same drainage, physical and stress conditions? What will be the settlement of the clay layer at this stage?

Answers: 81.1 days, 250 mm

6.6 A surface load of 60 kPa is applied on the ground surface over a large area. The soil profile consists of a sand layer 2 m thick, the top of which is the ground surface, overlying a 4 m thick layer of clay. An impermeable boundary is located at the base of the clay layer. The water table is 1 m below the ground surface. If the preconsolidation pressure for a sample of soil from the mid-point of the clay layer is 60 kPa, calculate the consolidation settlement of the clay layer. The properties of the soil section are: sand: $\rho_{dry} = 1.6$ Mg/m^3, $\rho_{sat} = 1.9$ Mg/m^3,
clay: $\rho_{sat} = 1.65$ Mg/m^3, $e_0 = 1.5$, $C_c = 0.6$, $C_r = 0.1$.

Answer: 236 mm

6.7 A soil profile consists of a sand layer 2 m thick, whose top is the ground surface, and a clay layer 3 m thick with an impermeable boundary located at its base. The water table is at the ground surface. A widespread load of 100 kPa is applied at the ground surface. Construct isochrones corresponding to 10%, 50%, and 90% consolidation, indicate the amount of excess pore pressures on the impermeable boundary and determine the amount of settlement after 2 years. Assume that the soil is in a normally consolidate state. The properties of the soil section are:
sand: $\gamma_{sat} = 20$ kN/m^3,
clay: $\gamma_{sat} = 16$ kN/m^3, $e_0 = 1.3$, $C_c = 0.5$, $c_v = 6.5$ m^2/year.

Answers: 100.0 kPa, 77.8 kPa, 15.7 kPa, 0.41 m

6.8 If the water level in Problem 6.7 is lowered to the surface of the clay layer; calculate the consolidation settlement of the clay layer after 6 months if (a) the drawdown is instantaneous, (b) the drawdown takes 2 months. Assume there is no surface load and take the unit weight of the sand layer 17.5 kN/m^3 after the draw down has taken place.

Answers: 75 mm, 54 mm

6.9 A stratum of clay is 5 m thick and is overlain by 3 m of sand, the top of which is the ground surface that is subjected to a widespread load of 200 kPa. The water table is 1.5 m below the ground surface, and the pore pressure at the impermeable boundary was measured to be 242.5 kPa after 18 months. If the settlement of the ground surface was 230 mm, determine the field values of c_v and C_c, the final consolidation settlement, and the settlement and pore pressure at the base of the clay layer after 3 years, using the concept of parabolic isochrones. The properties of the soil section are: sand: $\rho_{dry} = 1.75$ Mg/m^3, $\rho_{sat} = 2$ Mg/m^3, clay: $\rho_{sat} = 1.95$ Mg/m^3, $e_0 = 0.8$.

Answers: $c_v = 2.0$ m^2/year, $C_c = 0.332$, 0.57 m, 0.33 m, 188.8 kPa

6.10 In the soil profile of Problem 6.9, vertical drains of diameter 0.3 m are constructed in a square pattern. It is required that 95% of the combined consolidation be achieved after 1.5 years. Calculate the required distance between the vertical drains. $c_h = 4$ m^2/year.

Answer: 3.0 m

6.11 In Problem 6.7, calculate the consolidation settlements at 3 months, 6 months and 2 years if the load is increased linearly to 100 kPa over 6 months.
Answers: 71 mm, 200 mm, 360 mm

6.7 REFERENCES

Al-Khafaji, A.W. & Andersland, O.B. 1992. Equations for compression index approximations. *Journal GED, ASCE* 118(1): 148-153.

ASTM D-2435. 1996. *Standard test method for one-dimensional consolidation properties of soils.* PA, West Conshohocken: American Society for Testing and Materials.

Australian Standard AS 1289.6.6.1. 1998. *Methods of testing for engineering purposes, method 6.6.1: soil strength and consolidation tests-determination of one-dimensional consolidation properties of a soil-standard method.* Australia, NSW: Standard Association of Australia.

Azzouz, A.S., Krizek, R.J. & Corotis, R.B. 1976. Regression analysis of soil compressibility. *Soils and foundations* 16(2): 19-29.

Balasubramaniam, A.S. & Brenner, R.P. 1981. Consolidation and settlement of soft clay. In E.W. Brand & R.P. Brenner (eds), *Soft clay engineering: Developments in geotechnical engineering 20.* New York: Elsevier Scientific Publishing Company.

Barron, R.A. 1948. Consolidation of fine-grained soil by drain wells. *Transactions, ASCE* 113: 718-742.

Bergado, D.T., Anderson, L.R., Miura, A.S. & Balasubramaniam, A.S. 1996. *Soft ground improvement in lowland and other environments.* USA: ASCE Press.

Bjerrum, L. 1967. Engineering geology of Norwegian normally consolidated marine clays as related to settlements of buildings. *Geotechnique* (17): 81-118.

British Standard 1377-6. 1990. *Methods of test for soils for civil engineering purposes. Consolidation and permeability tests in hydraulic cells and with pore pressure measurement.* London: British Standards Institution.

Carillo, N. 1942. Simple two-and three-dimensional cases in the theory of consolidation of soils. *Journal of mathematics and physics* 21(1): 1-5.

Casagrande, A. 1936. The determination of preconsolidation load and its practical significance, discussion 34; *proc. 1st intern. conf. SMFE* 3: 60-64. Cambridge.

Craig, R.F. 1997. *Soil mechanics.* 6th edition. London: E & FN SPON.

Duncan, J.M. 1993. Limitations of conventional analysis of consolidation settlement. *Journal GE, ASCE* 119(9): 1333-1359.

Hansbo, S. 1979. Consolidation of clay by band shaped prefabricated drains. *Ground engineering* 12(5): 16-25.

Hansbo, S. 1981. Consolidation of fine-grained soils by prefabricated drains. *Proc. 10th intern. conf. SMFE* 3: 12-22. Stockholm.

Hansbo, S. 1987. Design aspects of consolidation of vertical drains and lime column installation. *Proc. 9th Southeast Asian geotechnical con.* 2: 8-12. Bangkok, Thailand.

Kjellman, W. 1952. Consolidation of clay by means of atmosphere pressure. *Proc. soil stabilization conf.*: 258-263. Mass..

Koppula, S.D. 1986. Discussion: consolidation parameters derived from index tests. *Geotechnique* 36(2): 68-73.

Lambe, T.W. & Marr, W.A. 1979. Stress path method. *Journal GE, ASCE* 105(GT6): 727-738.

Landau, R.E. 1966. Method of installation as a factor in sand drain stabilization design. *Highway research board* HRR (133): 75-96.

Leonards, G.A. 1976. Estimating consolidation settlements of shallow foundations on overconsolidated clays. *Special report* 163, *transportation research board*: 13-16.

Mesri, G. 1986. Discussion: postconstruction settlement of an expressway built on peat by precompression. *Canadian geotechnical journal* (23)3: 403-407.

Mesri, G. & Castro, A. 1987. The C_α / C_c concept and k_o during secondary compression. *Journal GE, ASCE* 112(3): 230-247.

Mesri, G., Feng, T.W. & Benak, J.M. 1990. Postdensification penetration resistance of clean sands. *Journal GED, ASCE* 116(GT7): 1095-1115.

Mesri, G. & Godlewski, P.M. 1977. Time and stress compressibility interrelationship. *Journal GE, ASCE* 103(5): 417-430.

Mesri, G. & Rokhsar, A. 1974. Theory of consolidation for clays. *Journal GED, ASCE* 100(GT8): 889-904.

Mesri, G., Rokhsar, A. & Bohor, B.F. 1975. Composition and compressibility of typical samples of Mexico City clay. *Geotechnique* (25): 527-554.

Nagaraj, T.S. & Srinivasa Murthy, B.R. 1985. Prediction of the preconsolidation pressure and recompression index of soils. *Geotechnical testing journal, ASTM* (4): 199-202.

Olson, R.E. & Ladd, C.C. 1979. One-dimensional consolidation problems. *Journal GED, ASCE* 105(GT1): 11-30.

Olson, R.E., Daniel, D.E. & Liu, T.K. 1974. Finite difference analysis for sand drain problems. *Analysis and design in geotechnical engineering, ASCE* 1: 85-110.

Olson, R.E. 1998. Settlement of embankments on soft clays. *Journal GGE, ASCE* 124(8): 659-669.

Parry, R.H.G. & Wroth, C.P. 1981. Shear stress-strain properties of soft clay. In E.W. Brand & R.P. Brenner (eds), *Soft clay engineering: Developments in geotechnical engineering 20.* New York: Elsevier Scientific Publishing Company.

Parsons, J.D. 1959. Foundation installation requiring recharging of ground water. *Journal of construction division, ASCE* 85(CO2): 1-21.

Pilot, G. 1981. Methods of improving the engineering properties of soft clay. In E.W. Brand & R.P. Brenner (eds), *Soft clay engineering: Developments in geotechnical engineering 20*. New York: Elsevier Scientific Publishing Company.

Powers, J.D. 1985. Dewatering-avoiding its unwanted side effects. *Underground technology research council, ASCE*. New York.

Raymond, G.P. & Wahls, H.E. 1976. Estimating 1-dimensional consolidation, including secondary compression of clay loaded from overconsolidated to normally consolidated state. *Special report 163, transportation research board*: 17-23.

Rendon-Herrero, O. 1980. Universal compression index equation. *Journal GED, ASCE* 106(GT11): 1179-1200.

Rendon-Herrero, O. 1983. Closure: universal compression index equation. *Journal GED, ASCE* 109(GT5): 755-761.

Rixner, J.J., Kraemer, S.R. & Smith, A.D. 1986. Prefabricated vertical drains. *Engineering guidelines, federal highway administration* 1(FHWA-RD-86 / 168). Washington DC.

Schmertmann, J.H. 1953. Estimating the true consolidation behavior of clay from laboratory test results. *Journal SMFE, ASCE* 79 (311): 26.

Scott, R.E. 1963. *Principles of soil mechanics*. Reading, Massachusetts: Addison-Wesley.

Skempton, A.W. 1944. Notes on the compressibility of clays. *Quart. journal of geological society* (100): 119-135. London.

Skempton, A.W. & Bjerrum, L. 1957. A contribution to the settlement analysis of foundations on clay. *Geotechnique* (7): 168-178.

Tan, T., Inoue, T. & Lee, S. 1991. Hyperbolic method for consolidation analysis. *Journal GED, ASCE* 117(11): 1723-1737.

Terzaghi, K. & Peck, R. B. 1967. *Soil mechanics in engineering practice*. New York: John Wiley.

Terzaghi, K., Peck, R. B. & Mesri, G. 1996. *Soil mechanics in engineering practice*. 3rd edition. New York: John Wiley & Sons.

Zeevaert, L. 1957. Foundation design and behaviour of tower Latino Americana in Mexico City. *Geotechnique* 7(3): 115-133.

CHAPTER 7

Application of Limit Analysis to Stability Problems in Soil Mechanics

7.1 INTRODUCTION

The main objective of this chapter is to investigate the stability of a soil structure using the lower and upper bound theorems of plasticity. These theorems are used to predict collapse loads where analytical solutions either do not exist or are inconsistent with the governing equations of mechanics (Bottero et al., 1980). They are also used when the deformations of the soil structure are negligible. A lower bound solution provides a safe limit load, whereas an upper bound solution estimates an unsafe limit load under which the failure of material has taken place already. In a lower bound solution only the equilibrium and yield criterion are satisfied, whilst in an upper bound solution, only the compatibility and the yield criterion are considered. These solutions, obtained either manually or numerically, bracket the exact solution within usually acceptable accuracy (Aysen & Sloan, 1991a). These bounds also provide useful guidance on the accuracy of finite element solutions (Naylor & Pande, 1981). In a simplified upper bound solution (in two dimensions) the continuum is converted to a mechanism consisting of rigid blocks sliding on their contact areas. For a virtual displacement, the external work done by the external forces is equated to the internal work done by the internal forces to obtain the unknown collapse load as an upper bound to the true collapse load. Although the actual behaviour is not the same as the assumed mechanisms, in many cases the load obtained for an optimised mechanism is very close to the true load.

The stress strain model for the soil may be idealized as an elastic-plastic or a rigid-perfectly plastic material. In the latter case, the collapse load is the same as for an elastic-plastic material being independent from the Modulus of Elasticity or stress path. Therefore, in this chapter, a rigid-perfectly plastic model is assumed. Figure 7.1 shows the stress-strain relationships for a rigid plastic material. There is no elastic deformation as the Modulus of Elasticity is infinite. On the horizontal plateau the material is in the plastic state and the work done by an increment of plastic strain is dissipated. Thus the plastic strain increments cannot be recovered during unloading. The term *plastic flow* describes the deformation behaviour of the material on this plateau. In a plastic analysis the plastic strains at failure cannot be determined. However, using incremental plasticity, relative rates of strains may be evaluated. In a rigid-perfectly plastic material the yield point is represented by a yield function F in the stress space. The yield function represents a fixed surface called the yield surface. In the case of work hardening and work softening materials the yield function is not fixed and changes with the development of the plastic strains. The rate of plastic strain is given by the *flow rule* according:

259

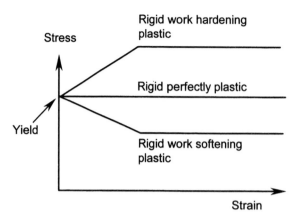

Figure 7.1. Stress-strain behaviour of rigid plastic material.

$$\dot{\varepsilon}^P = d\lambda \frac{\partial G}{\partial \sigma} \qquad (7.1)$$

where G is a plastic potential function (Chen & Saleeb, 1982 and 1986) and $d\lambda$ represents a positive proportionality parameter (plastic multiplier). When $G = F$, Equation 7.1 represents the associated flow rule, otherwise it is a non-associated flow rule. For a rigid-perfectly plastic material $G = F$. Furthermore, Equation 7.1 implies that the incremental plastic strains are normal to the yield surface (Figure 7.2) and that the yield surface must be convex. The yield function for a two-dimensional Mohr-Coulomb failure criterion (Equation 4.11) is expressed by:

$$F(\sigma'_x, \sigma'_z, \tau_{xz}) = (\sigma'_z - \sigma'_x)^2 + (2\tau_{xz})^2 - [2c'\cos\phi' + (\sigma'_z + \sigma'_x)\sin\phi']^2 \qquad (7.2)$$

In the undrained conditions Equation 7.2 will form the Tresca yield function given by:

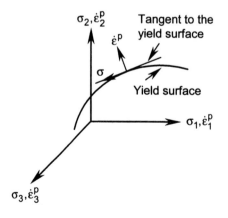

Figure 7.2. Yield surface and plastic strain rates.

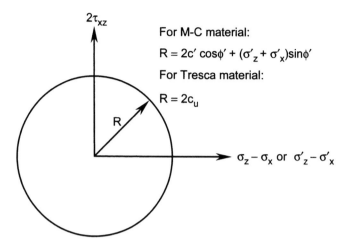

For M-C material:

$R = 2c' \cos\phi' + (\sigma'_z + \sigma'_x)\sin\phi'$

For Tresca material:

$R = 2c_u$

$2\tau_{xz}$

$\sigma_z - \sigma_x$ or $\sigma'_z - \sigma'_x$

R

Figure 7.3. Tresca and Mohr-Coulomb yield criteria.

$$F(\sigma_x, \sigma_z, \tau_{xz}) = (\sigma_z - \sigma_x)^2 + (2\tau_{xz})^2 - (2c_u)^2 \tag{7.3}$$

The yield surface is defined by $F = 0$. In terms of the quantities $(\sigma'_z - \sigma'_x)$ or $(\sigma_z - \sigma_x)$ and $2\tau_{xz}$, each of these functions represents a circle of radius $2c' \cos\phi' + (\sigma'_z - \sigma'_x)\sin\phi'$ or $2c_u$, as shown in Figure 7.3. Both yield functions can be generalized and expressed in three-dimensional stress space. In this chapter we will consider only plane strain loading. For appropriate three-dimensional yield surfaces reference may be made to Chen & Baladi (1985).

Lower and upper bound solutions may be obtained using finite elements and linear programming (Turgeman & Pastor, 1982; Aysen, 1987; Sloan, 1988 and 1989). In a lower bound solution, the need for linear programming results from discretisation of the yield criterion into linear segments and the mathematical expression of a safe load, which creates inequality constraints rather than equalities. Equality constraints are due to the equilibrium and stress boundary conditions. In an upper bound solution inequality constraints are needed to define that the rate of work dissipated within the material during plastic deformation or plastic flow is positive, and that the material cannot generate energy. Equality constraints arise from satisfying the yield criterion and boundary displacement conditions. In both cases the collapse load is expressed as a linear function of the relevant (Cartesian) stresses, which is treated as the objective function of the linear programming process.

7.2   LOWER BOUND SOLUTION

7.2.1   *Statically admissible stress field*

The lower bound theorem of classical plasticity states that any statically admissible stress field will provide a lower bound on the limit load. This lower bound limit load is a safe load and the failure of the material will not occur under this load. The necessary conditions required for a lower bound solution are (Chen, 1975):

1. The stress field must satisfy equilibrium everywhere within the domain of the problem.
2. The stress field must satisfy the specified stress boundary conditions.
3. The stress field must nowhere violate the yield criterion.

A statically admissible stress field may not necessarily represent the actual stresses of the true solution and include stress discontinuities. There are various ways to construct an admissible stress field. An analytical solution relies on reasonable simplifying assumptions and gives a conservative lower bound load (Mulhaus, 1985). Alternatively, the powerful finite element method can be used to predict the collapse loads in complex stability problems (Naylor & Pande, 1981). A numerical solution due to numerous equality and inequality constraints needs to be supported by a reliable algorithm to locate the optimal solution on a workstation or microcomputer.

An alternative method is to construct the stress field by commencing from a specified boundary and moving from one zone to another zone that is separated by stress discontinuities in the hope that the stress field will satisfy the numerous boundary conditions of the problem. This method needs a lot of practice and ingenuity and has been applied to a variety of tunnel problems (Davis et al., 1980). It is, however, only really useful in demonstrating the applications of lower bound solutions rather than in the search for an optimal answer. The third condition of the lower bound theorem, where the yield criterion is not violated, is expressed by the following inequality:

$$F \leq 0 \tag{7.4}$$

where F is the yield function. This means the state of stress at each point must lie inside the yield surface, and creates non-linear constraints in terms of the unknown stress components.

7.2.2   Stress discontinuity

Consider the state of stress at point C the corner of the strip footing shown in Figure 7.4. The two states of stress immediately to the right and left of this point show a rotation of

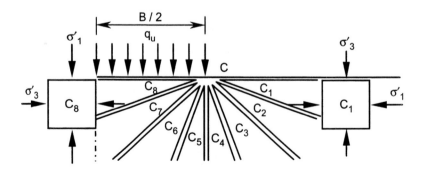

Figure 7.4. Stress discontinuities under the corner of a strip footing.

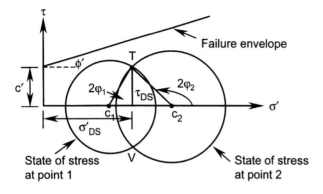

Figure 7.5. States of stress at a point on a discontinuity.

90° in the principal stress directions. The magnitudes of the principal stresses are also different due to the different stress boundary conditions. In order to facilitate the construction of an analytical or a numerical stress field about this point, the adjacent area is divided into a number of fan type zones. Thus, the point C is represented by 8 points, each with a different state of stress. Due to the geometry of the problem it is often necessary to insert stress discontinuities within the deforming mass in order to satisfy the boundary conditions (Parry, 1995). The states of stress within the zones are connected to each other through stress discontinuities, each having a zero (physical) thickness. Equilibrium requires that the normal and shear stresses along each discontinuity, calculated from the neighbouring zones, be equal. However, in the theory of plasticity only continuity in the direction normal to the stress discontinuity is required. Figure 7.5 shows the geometrical expression of equilibrium at a stress discontinuity. The coordinates of the intersection points of the Mohr's circles (point T or V) are the normal and shear stresses at the stress discontinuity. If the state of stress at point 1 is known, then there will be an infinite number of stress states at point 2 (or infinite number of circles centred on the σ-axis and passing through points T and V) to satisfy equilibrium of the zones as well as the stress discontinuity. However, if it assumed that circle 2 satisfies the condition $F = 0$, then there will be only two solutions. If the condition $F = 0$ is imposed on both circles then, with the specified state of stress at point 1, there will be only one possible state for point 2. In this case the following equations can be derived from the geometry of Figure 7.6 (Atkinson, 1993; Parry, 1995):

$$\frac{Oc_2}{Oc_1} = \frac{\dfrac{\sigma'_{z2} + \sigma'_{x2}}{2} + c' \cot \phi'}{\dfrac{\sigma'_{z1} + \sigma'_{x1}}{2} + c' \cot \phi'} = \frac{\sin(\omega + \lambda)}{\sin(\omega - \lambda)} \qquad c' > 0 \qquad (7.5)$$

$$\frac{Oc_2}{Oc_1} = \frac{\sigma'_{z2} + \sigma'_{x2}}{\sigma'_{z1} + \sigma'_{x1}} = \frac{\sin(\omega + \lambda)}{\sin(\omega - \lambda)} \qquad c' = 0 \qquad (7.6)$$

$$\sin \omega = \frac{\sin \lambda}{\sin \phi'} \qquad (7.7)$$

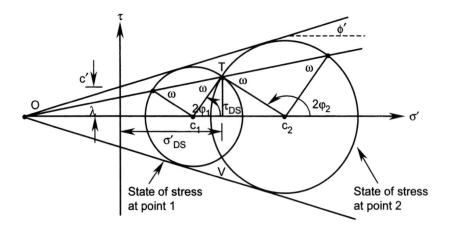

Figure 7.6. Stress discontinuity satisfying Mohr-Coulomb failure criterion.

where ω and λ are defined in Figure 7.6 and the stresses are in accordance with the sign convention of Section 4.2.1 or Figure 4.1. Rotation of the principal stress directions due to the stress discontinuity is defined by ψ that is given by:

$$\psi = \frac{2\varphi_2 - 2\varphi_1}{2} = \frac{180.0° - 2\omega}{2} = 90° - \omega \tag{7.8}$$

For c_u, $\phi_u = 0$ soil all the Mohr's circles have identical radii and:

$$\varphi_1 + \varphi_2 = 90° \tag{7.9}$$

Example 7.1

A stress discontinuity in a two-dimensional stress field has an angle of 60° with the x-axis. The state of stress at point 1 at one side of the stress discontinuity is: $\sigma'_x = 100$ kPa, $\sigma'_z = 200$ kPa and $\tau_{xz} = 50$ kPa. Determine: (a) the principal stresses and their direction at point 1; check if the state of stress satisfies the failure criterion $(F = 0)$, (b) the magnitudes of the shear and normal stresses along the stress discontinuity, (c) the principal stresses and their direction at point 2 on the other side of the stress discontinuity that has the same co-ordinates as point 1 and satisfies the Mohr-Coulomb failure criterion. (Note that if the condition of $F < 0$ applies at point 1, there will be two sets of answers for point 2. Consider the case that has the higher principal stresses – this is not a rule but just a choice for this example.) (d) The Cartesian stresses at point 2. The effective shear strength parameters are $c' = 0$, $\phi' = 30°$.

Solution:

(a) From Equation 7.2:

$$F = (200.0 - 100.0)^2 + (2 \times 50.0)^2 - [(200.0 + 100.0)\sin 30.0°]^2 = -2500.0 < 0,$$

which means point 1 has not failed.
Using Equations 4.3 and 4.4:

$s' = (200.0 + 100.0)/2 = 150.0$ kPa, t or $t' = \sqrt{[(200.0 - 100.0)/2]^2 + 50.0^2} = 70.7$ kPa.

$\sigma'_3 = 150.0 - 70.7 = 79.3$ kPa, $\sigma'_1 = 150.0 + 70.7 = 220.7$ kPa.

The direction of the major principal plane is obtained from Equation 4.6:

$\theta = 1/2 \times \tan^{-1}[2 \times 50.0/(200.0 - 100.0)] = 22.5°$.

(b) Calculate the normal and shear stresses at the stress discontinuity:

From Equations 4.1 and 4.2:

$\sigma' = (200.0 + 100.0)/2 + (200.0 - 100.0)/2 \times \cos(2 \times 60.0°) + 50.0 \sin(2 \times 60.0°)$,

$\sigma' = 168.3$ kPa.

$\tau = (200.0 - 100.0)/2 \times \sin(2 \times 60.0°) - 50.0 \cos(2 \times 60.0°) = 68.3$ kPa.

(c) Relationship between σ'_1 and σ'_3 at point 2, which is at the state of failure, is according to Equation 4.13: $\sigma'_1 = \sigma'_3 \tan^2(45.0° + 30.0°/2) = 3\sigma'_3$.

Substituting the calculated values of the normal and shear stresses along the stress discontinuity into Equations 4.7 and 4.8 (stress transformation equations using principal stresses), and noting that $\sigma'_1 = 3\sigma'_3$, two equations with two unknowns σ'_3 and φ are obtained, where φ is the angle of stress discontinuity with the σ'_1 plane:

$\sigma' = (\sigma'_1 + \sigma'_3)/2 + (\sigma'_1 - \sigma'_3)/2 \times \cos 2\varphi = 2\sigma'_3 + \sigma'_3 \cos 2\varphi = 168.3$ kPa.

$\tau = (\sigma'_1 - \sigma'_3)/2 \times \sin 2\varphi = \sigma'_3 \sin 2\varphi = 68.3$ kPa.

The solution of the two equations above yields: $3\sigma'^2_3 - 673.2\sigma'_3 + 32990.4 = 0$.

Considering the higher value: $\sigma'_3 = 152.1$ kPa, and $\sigma'_1 = 3\sigma'_3 = 456.3$ kPa, $2\varphi = 153.3°$. Taking the sign convention into account it is evident that the σ_1 plane makes an angle of: $153.3°/2 - 60.0° = 16.65°$ with the x-axis.

(d) The φ values for the x and z axes are $16.65°$ and $16.65° + 90° = 106.65°$ respectively. From Equations 4.1 and 4.2:

$\sigma'_z = (456.3 + 152.1)/2 + (456.3 - 152.1)/2 \times \cos(2 \times 16.65°) = 431.3$ kPa.

$\sigma'_x = (456.3 + 152.1)/2 + (456.3 - 152.1)/2 \times \cos(2 \times 106.65°) = 177.1$ kPa.

$\tau_{zx} = (456.3 - 152.1)/2 \times \sin(2 \times 16.65°) = 83.5$ kPa, $\tau_{xz} = -83.5$ kPa.

Example 7.2

Two states of stress on two sides of a stress discontinuity are defined by: $s'_1 = 100$ kPa, and $s'_2 = 300$ kPa. Both states satisfy the failure criterion. Determine: (a) the magnitudes of the principal stresses of both states, (b) the rotation of the principal stresses due to stress discontinuity, (c) the shear and normal stresses along the stress discontinuities T and V (Figure 7.6).

$c' = 20$ kPa, $\phi' = 30°$.

Solution:

(a) Referring to Figure 7.6 calculate the radius of each Mohr's circle:

$R_1 = Oc_1 \times \sin \phi' = (c' \cot \phi' + s'_1) \times \sin \phi' = (20.0 \times \cot 30.0° + 100.0) \sin 30.0°$,

$R_1 = 67.3$ kPa.

$R_2 = Oc_2 \times \sin \phi' = (20.0 \times \cot 30.0° + 300.0) \sin 30.0° = 167.3$ kPa.

$\sigma'_{11} = 100.0 + 67.3 = 167.3$ kPa, $\sigma'_{31} = 100.0 - 67.3 = 32.7$ kPa.

$\sigma'_{12} = 300.0 + 167.3 = 467.3$ kPa, $\sigma'_{32} = 300.0 - 167.3 = 132.7$ kPa $\leq \sigma'_{11} = 167.3$ kPa.

Thus the presence of a stress discontinuity is confirmed.

(b) Using Equation 7.7: $\sin\omega = \sin\lambda/\sin\phi' = \sin\lambda/\sin 30.0° = 2\sin\lambda$,

$$\cos\omega = \sqrt{1-\sin^2\omega} = \sqrt{1-4\sin^2\lambda} = \sqrt{1-4(1-\cos^2\lambda)} = \sqrt{4\cos^2\lambda-3}.$$

From Equation 7.5:

$$\frac{300.0+20.0\times\cot30.0°}{100.0+20.0\times\cot30.0°} = 2.4854 = \frac{\sin(\omega+\lambda)}{\sin(\omega-\lambda)} = \frac{\sin\omega\cos\lambda+\cos\omega\sin\lambda}{\sin\omega\cos\lambda-\cos\omega\sin\lambda}.$$

Replacing $\sin\omega$ and $\cos\omega$ in the equation above we obtain an equation in terms of $\cos\lambda$ only, from which: $\lambda = 16.8°$.

Thus $\sin\omega = 2.0 \times \sin\lambda = 2.0 \times \sin 16.8° = 0.578$, $\omega = 35.3°$.

Using the geometry of Figure 7.6; for the stress discontinuity T:

$\phi_1 = (\omega+\lambda)/2 = (35.3°+16.8°)/2 = 26.05°$,

$\phi_2 = [(\omega+\lambda)+(180.0°-2\omega)]/2 = [(35.3°+16.8°)+(180.0°-2\times35.3°)]/2 = 80.75°$.

The angle between σ'_{32} and $\sigma'_{31} = 80.75° - 26.05° = 54.7°$.

(c) For the stress discontinuity T:

$\sigma' = (167.3+32.7)/2+(167.3-32.7)/2\times\cos(2\times26.05°) = 141.3$ kPa.

$\tau = (167.3-32.7)/2\times\sin(2\times26.05°) = 53.1$ kPa.

For the stress discontinuity V:

$\sigma' = 141.3$ kPa, $\tau = -53.1$ kPa.

7.2.3 Ultimate bearing capacity of a strip footing in Tresca material

The ultimate bearing capacity of a strip footing (Chapter 10) is calculated using Terzaghi's equation:

$$q_u = c'N_c + qN_q + 0.5\gamma BN_\gamma \tag{7.10}$$

in which c' is the effective cohesion, q is the surcharge at the footing level ($q = \gamma D$ when this level is lowered to the depth D), B is the width of the footing, N_c, N_q and N_γ are Terzaghi's bearing capacity factors that are functions of the effective internal friction angle ϕ'.

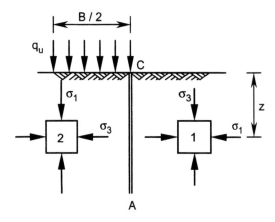

Figure 7.7. Statically admissible stress field under a strip footing.

The bearing capacity factors may be estimated using the lower bound theorem and by means of the fan-type stress discontinuities all of which pass through the corner of the strip footing. A classic example of a statically admissible stress field comprising the two zones 1 and 2 is shown in Figure 7.7. The directions of the principal stresses on both sides of the vertical stress discontinuity CA comply with the stress boundary conditions. In undrained conditions the following values for the principal stresses satisfy all of the conditions of a lower bound solution:

zone 1: $\sigma_3 = \gamma z, \sigma_1 = \sigma_3 + 2c_u = \gamma z + 2c_u$, zone 2: $\sigma_3 = \gamma z + 2c_u, \sigma_1 = \sigma_3 + 2c_u = \gamma z + 4c_u$.

where γ is the unit weight of the soil and z is the depth of the point. In the stress zone 2 and under the strip footing $z = 0$ and $\sigma_3 = 2c_u, q_u = \sigma_1 = 4c_u$.

It can be seen that the ultimate bearing capacity of a strip footing in undrained conditions is independent of the magnitude of the unit weight: $N_\gamma = 0$ and $N_c = 4$. The value of N_c is well below the theoretical value of $N_c = (2 + \pi) = 5.14$ and therefore the number of stress discontinuities has to be increased. It is convenient to assume equal rotation of the principal stress directions across each stress discontinuity. Thus the angles between the stress discontinuities and the state of stress in each zone are calculated in a manner to ensure equal rotation of the principal stress axes. It is also convenient to neglect the weight of the material, as this will reduce the computations and lead to a direct estimation of N_c.

Example 7.3

Figure 7.8 shows a strip footing on c_u, $\phi_u = 0$ with 3 stress discontinuities making 4 stress zones. It is required to calculate the ultimate bearing capacity (q_u) of the footing and the stress field within each zone. Assume that the footing is smooth and that the material is weightless. The stress field within each zone is constant and $F = 0$, and therefore: $\sigma_1 - \sigma_3 = 2c_u$.

Solution:

The state of stress in zone 1 is: $\sigma_3 = 0, \sigma_1 = 2c_u$.

The rotation of the principal stress directions across each discontinuity is: $90.0° / 3 = 30.0°$.

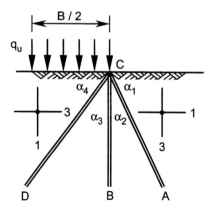

Figure 7.8. Example 7.3.

The angles of the major principal planes in zones 1 and 2 with the stress discontinuity *CA* are ϕ_1 and ϕ_2 respectively. Thus: $\psi = \phi_2 - \phi_1 = 30.0°$ and from Equation 7.9: $\phi_2 + \phi_1 = 90°$. Solving for ϕ_2 and ϕ_1: $\phi_1 = 30.0°$, $\phi_2 = 60.0°$. The normal and shear stresses on the stress discontinuity *CA* are calculated from the state of stress in zone 1:

$$\sigma_{CA} = (\sigma_1 + \sigma_3)/2 + (\sigma_1 - \sigma_3)/2 \times \cos 2\phi_1,$$

$$\sigma_{CA} = (2c_u + 0)/2 + (2c_u - 0)/2 \times \cos(2 \times 30.0°) = 1.5c_u.$$

$$\tau_{CA} = (\sigma_1 - \sigma_3)/2 \times \sin 2\phi_1 = (2c_u - 0)/2 \times \sin(2 \times 30.0°) = 0.866c_u.$$

To calculate the principal stresses in zone 2, the normal and shear stresses on the stress discontinuity *CA* are expressed in terms of the principal stresses in zone 2:

$$\sigma_{CA} = 1.5c_u = (\sigma_1 + \sigma_3)/2 + (\sigma_1 - \sigma_3)/2 \times \cos 2\phi_2.$$

$$\tau_{CA} = 0.866c_u = (\sigma_1 - \sigma_3)/2 \times \sin 2\phi_2.$$

Noting that $\sigma_1 - \sigma_3 = 2c_u$, and $\phi_2 = 60.0°$, the stress state at zone 2 is given by: $\sigma_3 = c_u, \sigma_1 = 3c_u$.

Performing a similar calculation and relating the states of stress between zones 2 and 3 and between zones 3 and 4 we find:

zone 3: $\alpha_2 = 30.0°$, $\sigma_3 = 2c_u, \sigma_1 = 4c_u$,

zone 4: $\alpha_3 = 30.0°$, $\sigma_3 = 3c_u, \sigma_1 = q_u = 5c_u$, and $\alpha_4 = 60.0°$, where q_u is less than 5.14 c_u.

General equation for N_c in undrained conditions. By considering the geometry of Mohr's circles of Figure 7.9, the following general equation for q_u can be obtained:

$$q_u = 2(c_u + nc_u \sin \psi) = 2c_u(1 + n \sin \psi),$$

$$N_c = 2(1 + n \sin \psi) \tag{7.11}$$

where *n* is the total number of stress discontinuities, and ψ is the rotation of the principal stress directions per one stress discontinuity. For example for $n = 10$, $\psi = 90° / 10 = 9°$: $N_c = 2(1 + 10 \sin 9.0°) = 5.13$.

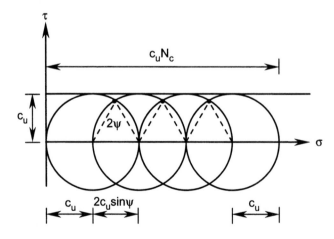

Figure 7.9. General solution for N_c in undrained conditions and Example 7.3.

7.2.4   *Ultimate bearing capacity of a strip footing in a Mohr-Coulomb material*

For $c' = 0$ and for the single stress discontinuity of Figure 7.7, with the surcharge q applied on the ground surface in zone 1, stresses in both zones are obtained using Equation 4.13 after considering the continuity of normal stress on the stress discontinuity:

zone 1: $\sigma_3' = q, \sigma_1' = q \tan^2(45° + \phi'/2)$,

zone 2: $\sigma_3' = q \tan^2(45° + \phi'/2), \sigma_1' = q_u = q \tan^4(45° + \phi'/2)$, therefore:

$$N_q = \tan^4(45° + \phi'/2).$$

The states of stress for two stress discontinuities are summarized in Figures 7.10 and 7.11. From the geometry of Figure 7.11 it can be seen that the rotations of the principal stresses due to each discontinuity are identical and equal to $90° - \omega$. The total rotation is $90°$, and therefore the rotation due to each discontinuity is $90° / 2 = 45°$.

Consequently: $90° - \omega = 45°, \omega = 45°$.

The angle λ is calculated from Equation 7.7:

$\sin \omega = \sin 45° = \sin \lambda / \sin \phi' \rightarrow \lambda = \sin^{-1}(\sin 45° \times \sin \phi')$.

Using the geometry of Figure 7.10 the direction of the major principal stresses can be found as follows:

$$\phi_1 = 90° - \alpha_1, \quad \phi_{21} = 135° - \alpha_1, \quad \phi_{22} = 135° - \alpha_1 - \alpha_2, \quad \phi_3 = \alpha_3.$$

As all stress circles are tangent to the failure envelope; the directions above can also be defined from Figure 7.11:

$$\phi_1 = 22.5° + 0.5\lambda, \quad \phi_{21} = 67.5° + 0.5\lambda, \quad \phi_{22} = 22.5° + 0.5\lambda, \quad \phi_3 = 67.5° + 0.5\lambda.$$

By equating the two sets of angles given above, the unknown values of α_1, α_2 and α_3 are found to be:

$$\alpha_1 = 90° - 0.5(\omega + \lambda) = 67.5° - 0.5\lambda, \quad \alpha_2 = \omega = 45°, \alpha_3 = 90° - 0.5(\omega - \lambda) = 67.5° + 0.5\lambda.$$

Applying Equation 7.6 for successive pairs of stress states of (1, 2) and (2, 3) generates the equations necessary to calculate the principal stresses in stress zone 3 which is the requirement of the problem. Note that the sum of the Cartesian normal stresses to be used in Equation 7.6 is equal to the sum of the principal stresses. The states of stress in the three zones above are:

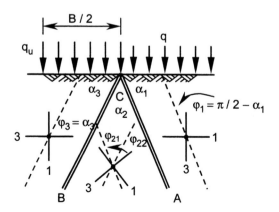

Figure 7.10. Strip footing with two stress discontinuities.

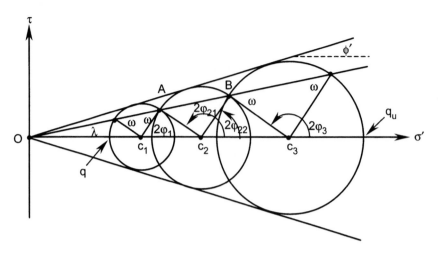

Figure 7.11. States of stress under a smooth strip footing with two stress discontinuities.

zone 1: $\sigma_3' = q, \sigma_1' = \sigma_3' \times \tan^2(45° + \phi'/2) = q \tan^2(45° + \phi'/2)$,

zone 2: $\sigma_3' = q \sin(\omega+\lambda)/\sin(\omega-\lambda), \sigma_1' = q \sin(\omega+\lambda)/\sin(\omega-\lambda) \tan^2(45° + \phi'/2)$,

zone 3: $\sigma_3' = q[\sin(\omega+\lambda)/\sin(\omega-\lambda)]^2, \sigma_1' = q[\sin(\omega+\lambda)/\sin(\omega-\lambda)]^2 \tan^2(45° + \phi'/2)$

Thus, for two stress discontinuities the parameter N_q is given by:

$N_q = [\sin(\omega+\lambda)/\sin(\omega-\lambda)]^2 \tan^2(45° + \phi'/2)$.

For n stress discontinuities with $n + 1$ stress zones, the bearing capacity factor N_q is given by Equation 7.12:

$$N_q = [\sin(\omega+\lambda)/\sin(\omega-\lambda)]^n \tan^2(45° + \phi'/2) \tag{7.12}$$

When n approaches infinity the parameter N_q approaches the value obtained from the analytical method (Chapter 10, Equations 10.2):

$N_q = \exp(\pi \tan \phi') \tan^2(45° + \phi'/2)$.

In a c', ϕ' soil, N_c can be calculated from Figure 7.11 (with n approaching infinity) by moving the vertical axis τ to the right until it becomes tangent to the first circle representing $\sigma'_3 = 0$. This is equivalent to reducing the σ'_1 of the last circle or q_u by the amount: $q = c' \times \cot\phi'$. Consequently, an identical expression to Equations 10.2 is obtained:

$N_c = \cot \phi'(N_q - 1)$.

With the number of stress discontinuities approaching infinity, the magnitudes of α_1 and α_{n+1} are calculated using the method above with α_3 replacing α_{n+1}:
For $n \to \infty$, $\omega = 90°$, and $\lambda = \phi'$. Thus: $\alpha_1 = 45° - \phi'/2, \alpha_{n+1} = 45° + \phi'/2$.

Example 7.4

A strip footing of width 2 m is located 1.5 m below the ground surface. The soil properties are: $c' = 60$ kPa, $\phi' = 23°$, $\gamma = 18.5$ kN/m³. Calculate the ultimate load per metre of strip footing by ignoring the weight of the soil beneath the footing level.

Solution:

$q = 18.5 \text{ kN/m}^3 \times 1.5 \text{ m} = 27.7 \text{ kPa}.$

$N_q = \exp(\pi \tan \phi') \tan^2(45.0° + \phi'/2) = \exp(\pi \times \tan 23.0°) \tan^2(45.0° + 23.0°/2) = 8.7.$

$N_c = \cot \phi'(N_q - 1) = \cot 23.0°(8.7 - 1) = 18.1.$

$q_u = c'N_c + qN_q = 60.0 \times 18.1 + 27.7 \times 8.7 = 1327.0 \text{ kPa}.$

Q_u (per metre of strip) $= 1327.0 \times 2.0 \times 1.0 = 2654.0 \text{ kN}.$

7.2.5 Stability of a vertical cut in Tresca and Mohr-Coulomb materials

Figure 7.12 illustrates the plane strain vertical cut under consideration. The soil is assumed to be either a Tresca or a Mohr-Coulomb material. For fixed values of H, failure of
the cut may be achieved by increasing either the vertical boundary loading q or the unit
weight γ. For the isotropic conditions assumed, the stability of the plane strain cut is a
function of the dimensionless parameters $\gamma H /(c_u$ or $c')$ and $q /(c_u$ or $c')$ called the stability number and the load parameter respectively. The stability number is determined assuming the load parameter is zero; whilst the load parameter is calculated assuming the
stability number is zero (e.g. weightless material). The results may be combined to illustrate the real situation. A statically admissible stress field with two stress discontinuities is
shown in Figure 7.12. In the stress zone 1 the normal stress in the vertical direction is approximated by $\gamma z + q$, while the horizontal normal stress is zero. Both stresses represent
the major and minor principal stresses, as there are no shear stresses on the ground surface
and the vertical face of the wall. Fan type stress discontinuities in this zone cannot be constructed, as there is no rotation of the principal stresses between the horizontal and vertical
boundaries. The stress field in stress zone 2 is hydrostatic in nature because the vertical
and lateral normal stresses are assumed to be equal.

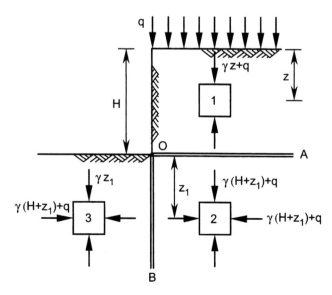

Figure 7.12. Statically admissible stress field for a vertical cut with two stress discontinuities.

In stress zone 3 the vertical normal stress is approximated by γz_1 and the principal stresses have rotated 90° when compared to those in stress zone 1. The stress field (either effective or total) in the three zones above may be summarized as follows:

zone 1: $\sigma_3 = 0, \sigma_1 = \gamma z + q$,

zone 2: $\sigma_3 = \gamma(H + z_1) + q, \sigma_1 = \gamma(H + z_1) + q$,

zone 3: $\sigma_3 = \gamma z_1, \sigma_1 = \gamma(H + z_1) + q$.

The above stress field satisfies equilibrium in the three stress zones and at the two stress discontinuities *OA* and *OB*. Next, the stress zone 1 is assumed to be on the verge of failure. The major principal stress varies linearly with depth and becomes a maximum at depth *H*. For the failure of the toe in a Tresca material we need to enforce the condition: $\sigma_1 - \sigma_3 = 2c_u$. Hence $\gamma H + q - 0 = 2c_u$:

$$\gamma H / c_u = 2 \rightarrow q = 0 \tag{7.13}$$

$$q / c_u = 2 \rightarrow \gamma H / c_u = 0 \tag{7.14}$$

For the stability number $\gamma H / c_u$ (with the load parameter equal to zero), Heyman (1973) presented an analytical solution with $2\sqrt{2} = 2.83$ and $1 + 2\sqrt{2} = 3.83$ as the lower and upper bounds respectively. For a Mohr-Coulomb material with $c' = 0$, it is impossible to construct a vertical slope as, on the vertical face of the slope, $\sigma'_3 = \sigma'_1 = 0$. For a c', ϕ' material, failure occurs only at the toe of the slope. The relationship between σ'_1 and σ'_3 at the toe is given by Equation 4.13 (Chapter 4). Replacing σ'_3 and σ'_1 by the corresponding values defined above, two identical equations in terms of the stability number and the load parameter are obtained.

$$\gamma H / c' = 2 \tan(45° + \phi' / 2) \rightarrow q = 0 \tag{7.15}$$

$$q / c' = 2 \tan(45° + \phi' / 2) \rightarrow \gamma H / c' = 0 \tag{7.16}$$

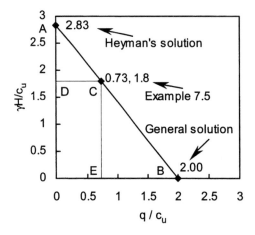

Figure 7.13. Stability chart for a vertical cut using the lower bound theorem.

The highest lower bounds found for the stability number and load parameter may be related by a linear relationship, as shown by the line *AB* in Figure 7.13 (Bottero et al., 1980). This allows the estimation of a lower bound for one of the parameters when the magnitude of the other parameter is known.

Example 7.5

Calculate the bearing capacity at the upper ground surface in a vertical cut 10 m high. The soil properties are: $c_u = 100$ kPa, $\phi_u = 0$, $\gamma = 18$ kN/m^3.

Solution:

$\gamma H / c_u = 18.0 \times 10.0 / 100.0 = 1.8$ (point *D* in Figure 7.13); substituting this value into the equation of the line *AB*:

$\gamma H / c_u = 2\sqrt{2} - \sqrt{2}q / 100.0 = 1.8 \rightarrow q = 72.7$ kPa (point *E* in Figure 7.13) and:

$N_c = 72.7 / 100.0 = 0.73$.

7.2.6 *Lateral earth pressure in Tresca and Mohr-Coulomb materials*

A vertical section of a smooth retaining wall is illustrated in Figure 7.14. The failure of the backfill soil may occur by either of two mechanisms depending on the direction of the wall displacement. If the wall moves towards the outside and away from the soil, the resulting failure is called active. A passive failure occurs if the wall is pushed towards the backfill until the limiting displacement is achieved. In a smooth retaining wall the shear stress immediately behind the wall cannot be mobilized and therefore the vertical and horizontal represent the principal stress directions. The normal stress in the vertical is approximated by $\gamma z + q$ where q is the vertical boundary loading as shown. In the active state this stress represents the major principal stress, while in the passive state it is the minor principal stress. Unlike the vertical cut where on the vertical boundary only the toe point is at the verge of failure, all the points immediately behind the wall are assumed to satisfy the yield criterion ($F = 0$). As the direction of the principal stresses does not change along the wall, there is no need for stress discontinuity to be included. In the active state the lateral soil pressure p_{ah} is equal to the minor principal stress σ'_3 while $\sigma'_1 = \gamma z + q$.

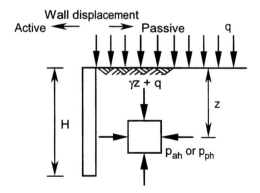

Figure 7.14. Idealized section of a smooth retaining wall subjected to active or passive failure.

In the passive state the lateral soil pressure p_{ph} is equal to the major principal stress σ'_1 and $\sigma'_3 = \gamma z + q$. Using Equation 4.13 the active and passive soil pressures are given by the following equations. For undrained conditions:

$$p_{ah} = \sigma_3 = \sigma_1 - 2c_u = \gamma z + q - 2c_u \qquad (7.17)$$

This linear equation is integrated along the wall to give the total active thrust on the wall:

$$P_{ah} = \gamma H^2 / 2 + (q - 2c_u)H \qquad (7.18)$$

The distribution of the passive pressure on the wall and the total passive thrust are:

$$P_{ph} = \sigma_1 = \sigma_3 + 2c_u = \gamma z + q + 2c_u \qquad (7.19)$$

$$P_{ph} = \gamma H^2 / 2 + (q + 2c_u)H \qquad (7.20)$$

For a c', ϕ' soil the active and passive pressures and the total thrusts are given by:

$$p_{ah} = (\gamma z + q)\tan^2(45° - \phi'/2) - 2c'\tan(45° - \phi'/2) \qquad (7.21)$$

$$P_{ah} = (\gamma H^2 / 2)\tan^2(45° - \phi'/2) + H[q\tan(45° - \phi'/2) - 2c']\tan(45° - \phi'/2)$$
$$(7.22)$$

$$p_{ph} = (\gamma z + q)\tan^2(45° + \phi'/2) + 2c'\tan(45° + \phi'/2) \qquad (7.23)$$

$$P_{ph} = (\gamma H^2 / 2)\tan^2(45° + \phi'/2) + H[q\tan(45° + \phi'/2) + 2c']\tan(45° + \phi'/2)$$
$$(7.24)$$

The lateral soil pressure is linear indicating an increase in active or passive soil pressure with depth. In the active state there is a possibility that tensile stress may develop behind the retaining wall to a depth of z_0, which is estimated by setting the active pressure equation to zero. As the soil cannot sustain tensile stress, tension cracks develop to the depth z_0. In a rough wall shear stress will develop along the back face of the wall causing rotation of the principal stresses. The shear stress is either of a cohesive or frictional nature and causes a reduction in the lateral soil pressure.

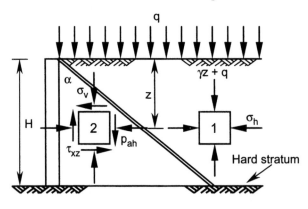

Figure 7.15. Active failure of a rough wall.

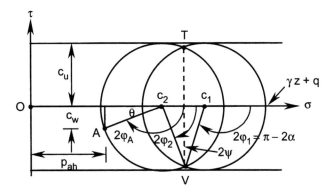

Figure 7.16. Stress circles of active failure for c_u, $\phi_u = 0$ soil behind a rough vertical retaining wall.

To allow for the rotation of the principal stresses, a stress discontinuity is introduced which makes an unknown angle α with the back face of the wall. In an active state and in the undrained conditions, the major principal stress in zone 1 (Figure 7.15) is vertical and its magnitude is $\gamma z + q$. The minor principal stress in this zone is $\sigma_h = \gamma z + q - 2c_u$. In the stress zone 2, the normal stress in the vertical σ_v and in the horizontal p_{ah} and the angle α are all unknown. However, the shear stress τ_{xz} is known and is equal to the cohesion c_w developed on the wall and the soil contact surface. Assuming that the stress field in zone 2 meets the requirements for failure, the geometry of the stress circles shown in Figure 7.16 is used to find all unknown parameters. The rotation of the principal stresses due to the stress discontinuity is $\psi = \phi_1 - \phi_2$. Using Equation 7.9, ψ, ϕ_1 and ϕ_2 can be expressed in terms of the angle α.

$\phi_1 + \phi_2 = \pi/2, \pi/2 - \alpha + \phi_2 = \pi/2 \rightarrow \phi_2 = \alpha$. Thus:

$$\psi = \phi_1 - \phi_2 = \pi/2 - \alpha - \alpha = \pi/2 - 2\alpha \tag{7.25}$$

The angles of the major principal planes of zones 1 and 2 with the back face of the wall are $\pi/2$ and ϕ_A (Figure 7.16) respectively. Hence:

$$\psi = \pi/2 - \phi_A = \pi/2 - (\pi - \theta)/2 = \theta/2 \tag{7.26}$$

where θ is defined by: $\sin \theta = c_w / c_u$. Equating Equations 7.25 by 7.26:

$$\alpha = \pi/4 - \theta/4 \tag{7.27}$$

From the geometry of Figure 7.16 the active lateral pressure p_{ah} is:

$p_{ah} = (\gamma z + q) - c_u - c_1 c_2 - c_u \cos \theta$, or:

$$p_{ah} = (\gamma z + q) - c_u (1 + 2\sin \theta/2 + \cos \theta) \tag{7.28}$$

Integrating the equation above along the wall, the total active thrust is expressed by:

$$P_{ah} = \gamma H^2 /2 + \left[q - c_u (1 + 2\sin \theta/2 + \cos \theta) \right] H \tag{7.29}$$

If the number of stress discontinuities is increased from 1 to n, then the term $2\sin(\theta/2)$ in Equation 7.29 may be replaced by $2n\sin(\theta/2n)$. However, the improvement (decrease) in

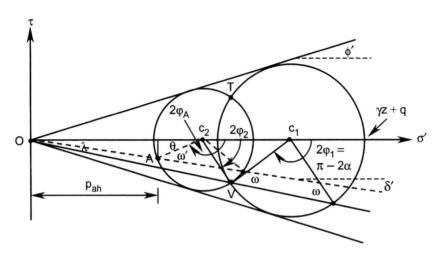

Figure 7.17. Stress circles of active failure for $c' = 0$, ϕ' soil behind a rough vertical retaining wall.

the active pressure is not significant. The passive pressure can be calculated in a similar manner:

$$P_{ph} = (\gamma z + q) + c_u(1 + 2\sin\theta/2 + \cos\theta) \tag{7.30}$$

The total passive thrust is:

$$P_{ph} = \gamma H^2/2 + [q + c_u(1 + 2\sin\theta/2 + \cos\theta)]H \tag{7.31}$$

The stress circles for a $c' = 0$, ϕ' material are shown in Figure 7.17. The shear stress on an element of soil immediately behind the retaining wall is: $\tau_{xz} = p_{ah}\tan\delta'$, where δ' is the friction angle mobilized on the interface. The active pressure is expressed by:

$$p_{ah} = Oc_2 - R_2\cos\theta = Oc_2 - Oc_2\sin\phi'\cos\theta = Oc_2(1 - \sin\phi'\cos\theta),$$

where the angle θ is defined in Figure 7.17. To relate Oc_2 to Oc_1 we use Equation 7.6:

$$p_{ah} = Oc_1 \times \frac{\sin(\omega - \lambda)(1 - \sin\phi'\cos\theta)}{\sin(\omega + \lambda)}, \text{ but}$$

$$Oc_1 = \gamma z + q - R_1 = \gamma z + q - Oc_1\sin\phi' \rightarrow Oc_1 = (\gamma z + q)/(1 + \sin\phi'), \text{ and hence}$$

$$p_{ah} = (\gamma z + q)\frac{\sin(\omega - \lambda)(1 - \sin\phi'\cos\theta)}{\sin(\omega + \lambda)(1 + \sin\phi')} = (\gamma z + q)k_{ah} \tag{7.32}$$

The term k_{ah} (soil pressure coefficient in the horizontal for an active failure) is given by:

$$k_{ah} = \frac{\sin(\omega - \lambda)(1 - \sin\phi'\cos\theta)}{\sin(\omega + \lambda)(1 + \sin\phi')} \tag{7.33}$$

Using a similar approach, the soil pressure coefficient for the passive case becomes:

$$k_{ph} = \frac{\sin(\omega + \lambda)(1 + \sin\phi'\cos\theta)}{\sin(\omega - \lambda)(1 - \sin\phi')} \tag{7.34}$$

The total active or passive thrusts are the integral of the soil pressure along the wall:

$$P_{ah} = (\gamma H^2/2 + qH)k_{ah} \qquad (7.35)$$

$$P_{ph} = (\gamma H^2/2 + qH)k_{ph} \qquad (7.36)$$

The angle θ can be expressed in terms of δ' and ϕ' by using Equation 7.7. Introducing the angle ω' defined in Figure 7.17, then:

$$\sin\omega' = \sin(\theta + \delta') = \sin\delta'/\sin\phi' \qquad (7.37)$$

For the passive case this equation has the following form:

$$\sin\omega' = \sin(\theta - \delta') = \sin\delta'/\sin\phi' \qquad (7.38)$$

Note that the angle $\theta/2$ represents the rotation of the principal stresses ψ (Equation 7.26) due to the stress discontinuity. At the same time, the rotation of the principal stresses is equal to $\psi = 90° - \omega$ (Equation 7.8). Thus, by equating the two values above, the angle ω can be calculated. It can be shown that increasing of number of stress discontinuities will not significantly improve the solution.

Example 7.6

Calculate the total horizontal thrust on a retaining wall of height 10 m in both the active and passive cases. The soil properties are: $c' = 0$, $\phi' = 30°$, $\delta' = 20°$, and $\gamma = 18$ kN/m^3.

Solution:

For the active state use Equation 7.37:
$\sin(\theta + \delta') = \sin(\theta + 20.0°) = \sin\delta'/\sin\phi' = \sin 20.0°/\sin 30.0° = 0.684 \rightarrow \theta = 23.16°$,
$\theta/2 = $ rotation of the principal stresses $= 23.16°/2 = 11.58°$,
$\psi = \theta/2 = 23.16°/2 = 90.0° - \omega \rightarrow \omega = 78.42°$,
$\sin 78.42° = \sin\lambda/\sin 30.0° \rightarrow \lambda = 29.33°$,
$\alpha = (\omega - \lambda)/2 = (78.42° - 29.33°)/2 = 24.54°$. From Equation 7.33:

$$k_{ah} = \frac{\sin(78.42° - 29.33°)(1 - \sin 30.0° \times \cos 23.16°)}{\sin(78.42° + 29.33°)(1 + \sin 30.0°)} = 0.2858.$$

This value is slightly greater than 0.2794, which is based on the analytical approach using the limit equilibrium method (Chapter 8).
$P_{ah} = (\gamma H^2/2 + qH)k_{ah} = (18.0 \times 10.0^2/2)0.2858 = 257.2$ kN.

For the passive state use Equation 7.38:
$\sin(\theta - \delta') = \sin(\theta - 20.0°) = \sin\delta'/\sin\phi' = \sin 20.0°/\sin 30.0° = 0.684 \rightarrow \theta = 63.16°$,
$\theta/2 = $ rotation of the principal stresses $= 63.16°/2 = 31.58°$,
$\psi = \theta/2 = 63.16°/2 = 90.0° - \omega \rightarrow \omega = 58.42°$,
$\sin 58.42° = \sin\lambda/\sin 30.0° \rightarrow \lambda = 25.21°$,
$\alpha = (\omega + \lambda)/2 = (58.42° + 25.21°)/2 = 41.81°$,

$$k_{ph} = \frac{\sin(58.42° + 25.21°)(1 + \sin 30.0° \times \cos 63.16°)}{\sin(58.42° - 25.21°)(1 - \sin 30.0°)} = 4.4483.$$

This is much smaller than the value of 5.7372 obtained form the limit equilibrium method (Chapter 8), and therefore the number of discontinuities must be increased.

$$P_{ph} = (\gamma H^2 / 2 + qH)k_{ph} = (18.0 \times 10.0^2 / 2)4.4483 = 4003.4 \text{ kN}.$$

7.3   UPPER BOUND SOLUTION

7.3.1   *Kinematically admissible velocity field*

The upper bound theorem of limit analysis defines a velocity field (or plastic strain rate field), which is said to be kinematically admissible if the following requirements are met:

1. The velocity field must satisfy strain compatibility.
2. The yield function expressed by Equations 7.2 or 7.3 must be satisfied.
3. The stress-strain relationship of Equation 7.1 must be satisfied within the field.
4. The boundary velocities must be satisfied.

There is no requirement for the equilibrium condition, and the velocity field may contain velocity discontinuities. A velocity discontinuity is needed to distinguish between parts of the material that may remain rigid while the other parts are on the verge of failure or have already failed. This could occur at both the boundaries of the domain of the problem as well as in its interior. By applying the virtual work theorem, which equates the external work done by boundary and gravity loading to the internal work dissipated by plastic straining within the material and by sliding along the velocity discontinuities, an upper bound limit for the collapse of the rigid body can be found. This means that a stress field associated with the kinematic velocity field is created to evaluate the stresses in the interior of the domain and also at its boundaries. From this the unknown collapse load that may rise from the boundary stresses (e.g. the ultimate bearing capacity of a strip footing) or from the gravity related parameter (e.g. $\gamma H / c'$ in the stability of slopes) can be found. Since the stress field does not necessarily satisfy the equilibrium conditions, the upper bound load is an unsafe load and mathematically defines an external approach to the yield surface. The term upper bound does not mean that the collapse load is greater than the lower bound collapse load as this depends on the mechanism of the collapse. An upper bound solution is an external approach to the yield surface while a lower bound solution implies an internal approach. For example, in the case of a strip footing, the upper bound value for the ultimate bearing capacity is higher than the lower bound value. In the active failure of a retaining structure, the lower bound value is greater than the upper bound value. This simply implies that if there is a smaller resistance from the retaining wall against the applied pressure from the soil, then the structure will fail. As was mentioned earlier, the exact solution remains in between the upper and lower bound solutions.

7.3.2   *Velocity discontinuity and the concept of two-dimensional collapse mechanisms*

In a rigid plastic body some parts may be deforming continuously while other parts may not be deforming at all, and this gives rise to the strain rate or *velocity discontinuity*

(Naylor & Pande, 1981). The velocity normal to discontinuity must be continuous otherwise a gap will develop and parts of body will penetrate into each other. A velocity discontinuity is also referred to as a *slip surface* (Atkinson, 1993) if a rigid body is modelled by a mechanism of rigid blocks sliding on their contact surfaces with constant velocity. Although the external work is dissipated by general plastic yielding and sliding along the discontinuities, in a collapse mechanism it is solely dissipated along the velocity discontinuities. This approach is commonly applied in plane strain conditions and the dissipated work is minimized by optimising the geometry of the mechanism. Figure 7.18(a) is an example of a rigid block mechanism used to investigate the stability of a plane strain slope. The mechanism satisfies the velocity compatibility condition and is geometrically defined by relevant angles or lengths. Each velocity vector makes an angle of ϕ' with the direction of the discontinuity. This does not satisfy the continuity of a velocity discontinuity, as defined in the theory of plasticity, as in a collapse mechanism a jump in normal and tangential velocity may occur. For a Tresca material with $\phi_u = 0$ there is no jump in the normal velocity and the velocity vector is parallel to the direction of the discontinuity. For a Mohr-Coulomb material, the ϕ' angle assumed between the velocity vector and the discontinuity facilitates the flow rule conditions.

In order to establish a relationship between the magnitudes of the velocities and thereby satisfy the compatibility of the displacements, a diagram of velocities is drawn (Figure 7.18(b)). This diagram is called a *displacement or velocity diagram*. To construct this diagram for the mechanism shown in Figure 7.18(a), three vectors parallel to the relevant velocities are drawn. The end points of these vectors are connected to each other to represent the relative velocities on the discontinuities. Assuming that one of these velocities is equal to a specified value (e.g. $v_1 = 1$ unit) we can calculate the other velocities from the geometry of the displacement diagram. The work done by the external loading includes the boundary loading q and the weight of each block. The algebraic value of the work is equal to the magnitude of the force multiplied by the component of the velocity in the direction of the force. If this component and the load are at the same direction then the work may be assumed positive, otherwise it is negative. The sign of the dissipated work is always positive and is due solely to cohesion. The resultant of the normal stress σ' and the shear stress component $\sigma' \tan\phi'$ (R in Figure 7.19) is perpendicular to the velocity – as a result the internal work due to this stress is zero.

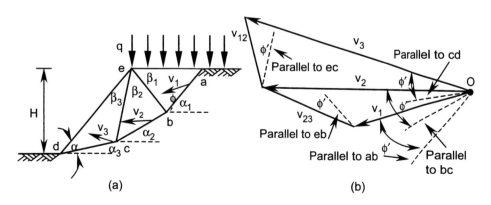

Figure 7.18. (a) Mechanism of rigid blocks, (b) displacement diagram.

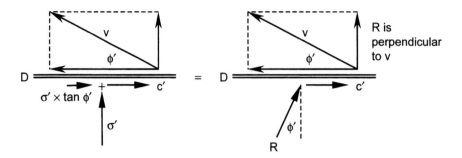

Figure 7.19. The concept of dissipated work along a velocity discontinuity or a slip surface.

Referring to Figure 7.19 and noting that the incremental displacement is inclined to the slip surface at an angle ϕ', the internal work dissipated along the velocity discontinuity of length l may be expressed as:

$$E_i = \int_0^l c'v\cos\phi'\,dl \tag{7.39}$$

Equation 7.39 must be applied for each discontinuity and algebraically combined to represent the overall internal work dissipated in the collapse mechanism.

7.3.3   Application of collapse mechanisms to ascertain bearing capacity

Fan type velocity discontinuities radiate from the corner of a strip footing and satisfy compatibility of the displacements. The following example illustrates the evaluation of an upper bound value for the bearing capacity factor N_c of a Tresca material.

Example 7.7

For the collapse mechanism given in Figure 7.20(a), calculate the upper bound value for the ultimate bearing capacity of the strip footing if the soil is undrained.

Solution:

From the geometry of the problem: $ab = be = ec = cd = \sqrt{2}B/2 = 0.707B$, $bc = B$,
$w_1 = w_2 = w_3 = ab\times eb/2\times\gamma = (0.707B\times0.707B/2)\times\gamma = 0.25B^2\gamma$.
From the displacement diagram: $v_1 = v_3 = 1.0$, $v_2 = \sqrt{2} = 1.414$, $v_{12} = v_{23} = 1.0$.
Calculate the external work due to the weight of each block and q_u:
$E_{e1} = w_1\times v_1\times\cos45.0° = 0.177B^2\gamma$, $E_{e2} = w_2\times v_2\times\cos90.0° = 0.0$,
$E_{e3} = w_3\times v_3\times-\cos45.0° = -0.177B^2\gamma$, $E_{e4} = q_u\times B\times v_1\times\cos45.0° = 0.707Bq_u$.
Total external work $= 0.707Bq_u$.
Internal work along each velocity discontinuity:
$E_{i1}= ab\times c_u\times v_1\times\cos\phi_u = 0.707Bc_u$, $E_{i2} = bc\times c_u\times v_2\times\cos\phi_u = 1.414Bc_u$,
$E_{i3} = cd\times c_u\times v_3\times\cos\phi_u = 0.707Bc_u$, $E_{i4} = be\times c_u\times v_{12}\times\cos\phi_u = 0.707Bc_u$,
$E_{i5} = ce\times c_u\times v_{23}\times\cos\phi_u = 0.707Bc_u$, total internal work $= 4.243Bc_u$.
Equating the external and internal work: $0.707Bq_u = 4.243Bc_u$, $q_u = 6.0c_u$.
Thus, the upper bound for $N_c = 6.0 < 5.14$.

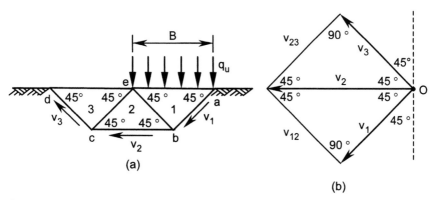

Figure 7.20. Example 7.7.

7.3.4 *Application of two-dimensional collapse mechanisms to the stability of slopes*

For the mechanism shown in Figure 7.18(a), a simple computer program may be written to calculate the stability number or the load parameter. The number of fan angles and their magnitudes β_1, β_2, and β_3 along with parameters α_1, α_2 and α_3 can be given as data, while the final values may be found through an optimisation procedure. In general, a fan of 3 to 4 triangular blocks is adequate to yield reasonable values for $\gamma H / c'$ or q / c'. For a vertical slope (or vertical cut), one variable mechanism consisting of one triangle yields a reasonable solution.

Table 7.1(a). Load parameter q / c' for a vertical cut with $\gamma H / c' = 0$.

ϕ'(deg.)	LB FEM	UB FEM	FEM	FDM	UB Mech.	Slice method
0.0	1.93	2.00	2.00	2.00	2.00	2.00
10.0	2.29	2.40	2.40	2.38	2.38	2.38
20.0	2.71	2.93	2.90	2.85	2.85	2.85
30.0	3.24	3.67	3.50	3.46	3.46	3.46
40.0	3.90	4.97	4.50	4.34	4.29	4.30
50.0	4.77	7.70	5.20	5.00	5.49	5.50

Table 7.1(b). Stability number $\gamma H / c'$ for a vertical cut with $q / c' = 0$.

ϕ'(deg.)	LB FEM	UB FEM	FEM	FDM	UB Mech.	Slice method
0.0	3.60	4.00	3.60	3.48	4.00	3.83
10.0	4.35	5.10	4.37	4.54	4.67	4.66
20.0	5.16	6.50	5.90	6.46	5.71	5.64
30.0	6.12	8.55	7.37	7.65	6.92	6.87
40.0	7.24	13.16	9.56	10.00	8.57	8.53
50.0	8.62	26.14	18.00	-	10.98	10.89

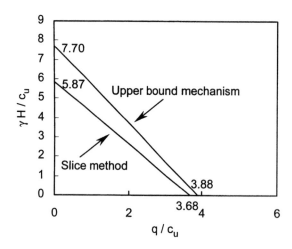

Figure 7.21. Stability bounds for a 45° slope in undrained conditions.

Aysen & Loadwick (1995) compared the mechanism solution with the traditional slice method based on the limit equilibrium (Chapter 9) and available upper and lower bound solutions based on the finite element method (Aysen & Sloan, 1992). Furthermore, a displacement finite element method (Carter & Balaam, 1990) and an explicit finite difference method (Cundall, 1980 and 1987) were also used to evaluate the mechanism results. Tables 7.1(a) and 7.1(b) present a detailed comparison for a vertical slope. In general, the difference between the lower bound and mechanism solutions is very small and at low values of ϕ' they are almost identical. At high values of ϕ', the upper bound finite element solution does not yield reliable answers. Results obtained from the displacement type finite element method and finite difference method using the *fast Lagrangian analysis of continuum (FLAC)* are reliable and fall between the two bounds. Figure 7.21 shows the stability chart for a 45° slope with $\phi_u = 0$ where the two extreme points of each solution have been connected by a straight line.

Example 7.8

For a 45° plane strain slope, compute the upper bound value of q for the collapse mechanism shown in Figure 7.22(a).
The soil properties are: $c_u = 100$ kPa, $\phi_u = 0$, and $\gamma = 18$ kN/m^3.
Solution:
From the geometry of the problem: $h_1 = 17.32$ m, $h_2 = 12.68$ m, $l = 40.0$ m, $\alpha_1 = 17.6°$, $\delta = 27.4°$, $ab = bd = 20.0$ m, $bc = 41.96$ m.
The weight of each block is calculated to be:
$w_1 = 3117.6$ kN, $w_2 = 7376.54$ kN.
From the displacement diagram of Figure 7.22(b), $v_1 = 1$, $v_2 = 0.887$ and $v_{12} = 0.690$.
Components of the external work:
$E_{e1} = w_1 \times v_1 \cos30.0° = 3117.60 \times 1.0 \times \cos30.0° = 2699.92$,

$E_{e2} = w_2 \times v_2 \cos72.4° = 7376.54 \times 0.887 \times \cos72.4° = 1978.40$,

$E_{e3} = 20.0 \, q \times v_1 \times \cos30.0° = 17.32q$. Hence, the total external work $= 4678.32 + 17.32q$.

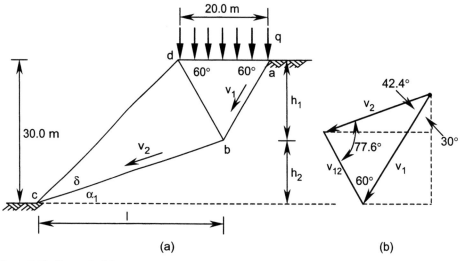

Figure 7.22. Example 7.8.

$E_{i1} = ab \times c_u \times v_1 \cos\phi_u = 20.0 \times 100.0 \times 1.0 = 2000,$
$E_{i2} = bc \times c_u \times v_2 \cos\phi_u = 41.96 \times 100.0 \times 0.887 = 3721.85,$
$E_{i3} = bd \times c_u \times v_{12} \cos\phi_u = 20.0 \times 100.0 \times 0.690 = 1380.00.$
Total internal work = 7101.85. Equating the external work and the internal work:
$4678.32 + 17.32q = 7101.85$, $q = 140$ kPa, $q / c_u = 140.0 / 100.0 = 1.4$.
Note that the given angles in the geometry of the mechanism are not optimised values.

7.3.5   *Application of two-dimensional collapse mechanisms to retaining walls*

Figure 7.23(a) illustrates a simple upper bound mechanism for the active (solid lines) or passive failure of a retaining wall. There may be friction at the interface of the retaining wall and the backfill, as well as cohesion. It is convenient to assume that a hard stratum occurs below the base of the retaining wall that is not subject to failure. Whilst the back face of the retaining wall may or may not be vertical, the upper bound analysis applied to the conditions shown in Figure 7.23(a) is equally applicable for non-vertical retaining wall. For the active failure case, the external work due to the weight of the block, the load q, and the horizontal and vertical components of P_a are:
$E_{e1} = H(H \tan\alpha / 2)\gamma \times v_a \cos(\phi' + \alpha)$, $E_{e2} = H \tan\alpha \times q \times v_a \cos(\phi' + \alpha)$,
$E_{e3} = -P_a \cos\delta' v_a \sin(\phi' + \alpha)$, $E_{e4} = -P_a \sin\delta' v_a \cos(\phi' + \alpha)$.
The internal work is due to the cohesion c' along the sliding surface of length $H / \cos\alpha$:
$E_i = (H / \cos\alpha)c' \times v_a \cos\phi'$.
Equating the total external work to the total internal work:

$$P_a = \frac{1}{2}\gamma H^2 \frac{\tan\alpha \cos(\phi' + \alpha)}{\sin(\delta' + \phi' + \alpha)} + qH\frac{\tan\alpha \cos(\phi' + \alpha)}{\sin(\delta' + \phi' + \alpha)} - c'H\frac{\cos\phi'}{\cos\alpha\sin(\delta' + \phi' + \alpha)}$$

$$(7.40)$$

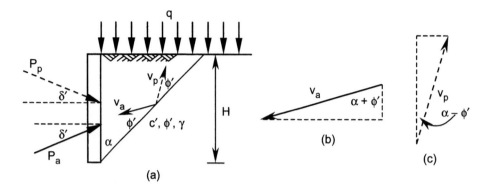

Figure 7.23. Application of a single variable mechanism to a retaining wall.

In undrained conditions with $\delta' = 0$, Equation 7.40 simplifies to:

$$P_a = \frac{\gamma H^2}{2} + qH - \frac{c_u H}{\sin \alpha \cos \alpha} \tag{7.41}$$

Setting the derivative of P_a (in terms of α) to zero, $\alpha = 45°$ and:

$$P_a = \frac{\gamma H^2}{2} + (q - 2c_u)H,$$

which is identical to Equation 7.18 derived from the lower bound theorem. Similarly for the passive case, an equation identical to Equation 7.20 can be obtained.

For the case when the effective cohesion is zero, $E_i = 0$ and the total thrust is given by:

$$P_a = \frac{\gamma H^2}{2} \frac{\tan \alpha \cos(\phi' + \alpha)}{\sin(\delta' + \phi' + \alpha)} \tag{7.42}$$

A soil pressure coefficient and a lateral component can be defined as follows:

$$k_a = \frac{\tan \alpha \cos(\phi' + \alpha)}{\sin(\delta' + \phi' + \alpha)}, \quad k_{ah} = k_a \cos \delta' \tag{7.43}$$

For passive failure of the backfill the corresponding values of k_p and k_{ph} are:

$$k_p = \frac{\tan \alpha \cos(\alpha - \phi')}{\sin(\alpha - \delta' - \phi')}, \quad k_{ph} = k_p \cos \delta' \tag{7.44}$$

Example 7.9

Re-work Example 7.6 using the single variable mechanism of Figure 7.23(a).

Solution:

The optimum value of angle α is calculated by trial and error using Equations 7.43 and 7.44, which is 34° and 72° for the active and passive states respectively. (Refer to the table below).

$\alpha°$	30	33	<u>34</u>	35	36	40		
k_{ah}	0.2754	0.2791	<u>0.2794</u>	0.2791	0.2784	0.2697		
$\alpha°$	51	60	65	70	71	<u>72</u>	73	74	75
k_{ph}	62.0744	8.1172	6.3779	5.7826	5.7473	<u>5.7373</u>	5.7530	5.7957	5.8677

$$P_{ah} = (18.0 \times 10.0^2 / 2) \times 0.2794 = 251.5 \, \text{kN} < 257.2 \, \text{kN},$$

$$(LB), P_{av} = P_{ah} \times \tan 20.0° = 91.5 \, \text{kN}, \ P_a = \sqrt{251.5^2 + 91.5^2} = 267.6 \, \text{kN}.$$

$$P_{ph} = (18.0 \times 10.0^2 / 2) \times 5.7373 = 5163.6 > 4003.4 \, \text{kN} \ (UB),$$

$$P_{pv} = P_{ph} \times \tan 20.0° = 1879.4 \, \text{kN}, \ P_p = \sqrt{5163.6^2 + 1879.4^2} = 5495.0 \, \text{kN}.$$

7.3.6   *Application of two-dimensional collapse mechanisms to shallow tunnels*

Collapse mechanisms have been applied to the stability of shallow tunnels in conjunction with centrifugal tests on soft clay performed at Cambridge University (Mair, 1979). Davis et al. (1980) obtained reasonable upper bound solutions for circular tunnels (Figures 7.24(a) - 7.24(d)) and tunnel headings (Figure 7.26(a)) using 1 to 3 sliding blocks. The mechanisms included local collapse (not shown) as well as the active total collapse of the section. Stability ratios were formulated in terms of the geometry of the sliding blocks. Sloan & Aysen (1992) used a seven variable mechanism (Figure 7.24(e)) with cohesion increasing linearly with depth. An application of this mechanism for uniform strength is shown in Table 7.2. Britto & Kusakabe (1985) applied the upper bound mechanisms to

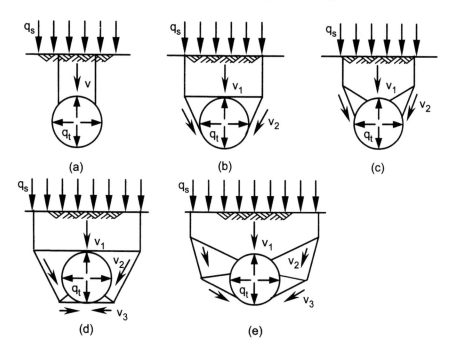

Figure 7.24. Collapse mechanisms for a plane strain shallow circular tunnel in undrained conditions.

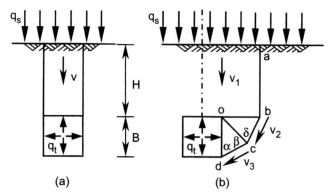

Figure 7.25. Collapse mechanisms for a shallow square tunnel in undrained conditions (Aysen & Sloan, 1991a).

Table 7.2. $(q_s - q_t) / c_u$ for circular tunnel using the collapse mechanism of Figure 7.24(e).

H/D	$\gamma D / c_u = 0$	$\gamma D / c_u = 1$	$\gamma D / c_u = 2$	$\gamma D / c_u = 3$
1.0	2.549	1.366	0.151	− 1.121
2.0	3.676	1.410	− 0.907	− 3.281
3.0	4.505	1.179	− 2.197	− 5.627
4.0	5.174	0.807	− 3.611	− 8.076
5.0	5.745	0.345	− 5.103	− 10.595
6.0	6.246	− 0.179	− 6.651	− 13.163
7.0	6.697	− 0.750	− 8.241	− 15.770
8.0	7.108	− 1.358	− 9.865	− 18.404
9.0	7.486	− 1.995	− 11.516	− 21.062
10.0	7.838	− 2.657	− 13.189	− 23.740

axisymmetric problems with an emphasis on simplicity and the ease with which calculations could be carried out. The ability of the collapse mechanisms in tunnels to capture the results from the numerical based methods is reported by Aysen & Sloan (1991a). The stability ratio may be represented by the dimensionless parameter $q_s - q_t / c'$ or c_u where q_s and q_t are the vertical boundary load and the tunnel pressure (reaction of the tunnel lining or an actual hydrostatic pressure) respectively. Figures 7.25 and 7.26 (Aysen & Sloan, 1991c) illustrate the application of collapse mechanisms to square tunnels and tunnel headings.

Example 7.10

Using the collapse mechanism of Figure 7.25(b) and the following data, compute the upper bound value for $(q_s - q_t) / c_u$ in a square tunnel in undrained conditions (c_u, $\phi_u = 0.0$). $\alpha = 60°$, $\beta = 75°$, $\delta = 75°$, $H / B = 5$, $\gamma B / c_u = 3$, where γ is the unit weight of the soil.

Solution:

The geometry is symmetric, and therefore only one-half of the mechanism is considered.

Relevant dimensions are: $dc = 0.732B$, $bc = 0.732B$, $oc = 0.896B$, $ob = B$, and $ab = 5B$. The weights of the blocks: $w_1 = 7.5B^2\gamma$, $w_2 = w_3 = 0.317B^2\gamma$.

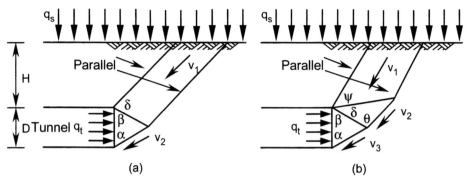

Figure 7.26. Collapse mechanism for a plane strain shallow tunnel heading in undrained conditions: (a) Davis et al. (1980), (b) Aysen & Sloan (1991c).

From the displacement diagram shown in Figure 7.27 and assuming $v_1 = 1$:
$v_2 = v_3 = 1.155$, $v_{12} = 0.577$, $v_{23} = 0.598$.
$E_{e1} = w_1 \times v_1 = 7.5B^2\gamma \times 1.0 = 7.5B^2\gamma$,
$E_{e2} = w_2 \times v_2 \times \cos30.0° = 0.317B^2\gamma \times 1.155 \times \cos30.0° = 0.317\ B^2\gamma$,
$E_{e3} = w_3 \times v_3 \times \cos60.0° = 0.317B^2\gamma \times 1.155 \times \cos60.0° = 0.183\ B^2\gamma$.
Calculate the external work due to the applied stresses:
$E_{e4} = q_s (0.5B + B) \times v_1 = 1.5Bq_s$,
$E_{e5} = - q_t \times B \times v_3 \sin60.0° = - q_t \times B \times 1.155 \times \sin60.0° = - Bq_t$,
$E_{e6} = - q_t \times B / 2 \times v_1 = - B / 2\ q_t$.
Total external work $= 8B^2\gamma + 1.5B\ (q_s - q_t)$.
Calculate internal work along each discontinuity:
$E_{i1} = ab \times c_u \times v_1 \cos\phi_u = 5B \times c_u \times 1.0 = 5Bc_u$,
$E_{i2} = bc \times c_u \times v_2 \cos\phi_u = 0.732B \times c_u \times 1.155 = 0.845Bc_u$,
$E_{i3} = cd \times c_u \times v_3 \cos\phi_u = 0.732B \times c_u \times 1.155 = 0.845Bc_u$,
$E_{i4} = ob \times c_u \times v_{12} \cos\phi_u = B \times c_u \times 0.577 = 0.577Bc_u$,
$E_{i5} = oc \times c_u \times v_{23} \cos\phi_u = 0.896B \times c_u \times 0.598 = 0.536Bc_u$.
Total internal work $= 7.803B\ c_u$. Equating the external work to the internal work:
$8B^2\gamma + 1.5B\ (q_s - q_t) = 7.803B\ c_u \rightarrow (q_s - q_t)\ /\ c_u = - 10.8$,
From the *FEM* solution using Table 7.4, $(q_s - q_t)\ /\ c_u = - 11.2$.

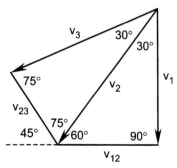

Figure 7.27. Example 7.10.

Example 7.11

A vertical section of a plane strain circular tunnel with a 4 variable upper bound mechanism is shown in Figure 7.28(a). The magnitudes of the variables are selected as follows: $\alpha = 15°$, $\beta = 60°$, $\delta = \theta = 75°$. For the undrained condition find the ratio of q_s / c_u if $\gamma D / c_u = 1.0$. What will be the tunnel pressure q_t if the applied boundary stress $q_s = 4\ c_u$?

Solution:

As the section is symmetrical about OE, only half of the section is considered.
$s_1 =$ area of $EDCAF =$ area $EDGO -$ area $OACG -$ area OAF,

$$s_1 = OE \times OG - [OK \times AK/2 + (AK + CG)/2 \times (OG - OK)] - 15.0° \times \pi OA^2 / 360.0°.$$

Solving the geometry:
$OE = 7.0$ m, $AB = 1.0$ m, $CJ = 1.866$ m, $OC = 2.732$ m, $CG = 1.932$ m,
$OG = 1.932$ m, $ED = OG = 1.932$ m, $AK = 0.966$ m, $OK = 0.259$ m, thus: $s_1 = 10.844$ m^2.
$S(ABC) = S_2 = AB \times OB \cos 30.0°/2 + AB \times CJ/2 - 60.0° \times \pi OB^2 / 360.0° = 0.842$ m^2.
The length of the velocity discontinuities are: $CA = CB = 1.932$ m, $CD = 5.068$ m.
The velocity of each block is calculated from the displacement diagram of Figure 7.28(c):
$v_1 = 1.0$, $v_2 = 1.732$, $v_{12} = 1.0$.

Calculation of components of the external virtual work:
$E_{e1} = w_1 \times v_1 = s_1 \times \gamma \times v_1 = 10.844 \times \gamma \times 1.0 = 10.844\gamma$,
$E_{e2} = w_2 \times v_2 \times \cos 30.0° = s_2 \times \gamma \times v_2 \times \cos 30.0° = 0.842 \times \gamma \times 1.732 \times \cos 30.0° = 1.263\gamma$,
$E_{e3} = q_s \times ED \times v_1 = 1.932\ q_s$, total external work $= 12.107\gamma + 1.932\ q_s$.

Calculation of components of the internal virtual work:
$E_{i1} = DC \times c_u \times v_1 \cos\phi_u = 5.068 \times c_u \times 1.0 = 5.068\ c_u$,
$E_{i2} = CA \times c_u \times v_{12} \cos\phi_u = 1.932 \times c_u \times 1.0 = 1.932\ c_u$,
$E_{i3} = CB \times c_u \times v_2 \cos\phi_u = 1.932 \times c_u \times 1.732 = 3.346\ c_u$, total internal work $= 10.346\ c_u$.
Equating the external work to the internal work: $12.107\gamma + 1.932\ q_s = 10.346\ c_u$.
Noting that: $\gamma D / c_u = 1.0$, $\gamma \times 2.0 / c_u = 1.0 \rightarrow \gamma = 0.5\ c_u$, thus:

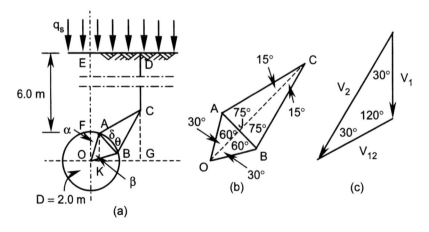

Figure 7.28. Example 7.11.

$12.107 \times 0.5\, c_u + 1.932\, q_s = 10.346\, c_u,\ q_s = 2.22\, c_u.$

$(q_s - q_t)\, /\, c_u = 2.22,$

$(4c_u - q_t)\, /\, c_u = 2.22,\ q_t = 1.78\, c_u.$

7.4 FINITE ELEMENT FORMULATION OF THE BOUND THEOREMS

7.4.1 *General remarks*

The two-dimensional lower and upper bound numerical solutions are based on a finite element formulation of the plastic limit theorems and lead to major linear programming problems. Lysmer (1970) appears to be the first solve the lower bound theorem using a linear programming procedure. A full description of both methods may be found in Anderheggen & Knopfel (1972), Bottero et al. (1980), Aysen (1987), Sloan (1989), Aysen & Sloan (1991a, 1991b and 1992). Both techniques assume a perfectly plastic soil model with a linear Mohr-Coulomb failure criterion. The solution to the lower bound linear programming problem defines a statically admissible stress field whilst the upper bound solution defines a kinematically admissible velocity field. In the finite element formulation of the lower bound solution, a set of equality and inequality constraints is constructed to satisfy the requirements of equilibrium, stress boundary conditions and yield criterion. The collapse load is formulated in terms of unknown stresses and maximized as an objective function to yield the corresponding collapse load. The finite element formulation of the upper bound theorem includes equality and inequality constraints to satisfy the yield criterion, the compatibility of the induced velocity discontinuities and the velocity boundary conditions. The objective function represents the dissipated work and is minimized subject to the prescribed constraints. Both techniques have been applied to variety of geomechanical problems. It has been shown that in most cases the solutions obtained bracket the exact collapse load within 15% or less. In some cases (e.g. plane strain slope), especially those with high values of ϕ', the upper bound collapse load is very much greater than the lower bound load, which makes the solution unacceptable.

7.4.2 *Finite element formulation of the lower bound theorem*

A typical triangular lower bound element is shown in Figure 7.29(a). Stress discontinuities occur at all edges that are shared by adjacent elements. Thus, unlike more familiar type of finite elements, each node is unique to a particular element and several nodes may share the same coordinates.

The stress components within each element vary linearly according to:

$$\sigma_x = \sum_{i=1}^{i=3} N_i \sigma_{xi}, \quad \sigma_z = \sum_{i=1}^{i=3} N_i \sigma_{zi}, \quad \tau_{xz} = \sum_{i=1}^{i=3} N_i \tau_{xzi} \tag{7.45}$$

where N_i are linear shape functions and σ_{xi}, σ_{zi} and τ_{xzi} are the nodal stresses. In order to satisfy the equilibrium condition, the derivatives of the stress components given in Equation 7.45 are substituted into Equations 5.1. Consequently, the nodal stresses corresponding to the three nodes of each element are subject to two equilibrium constraints.

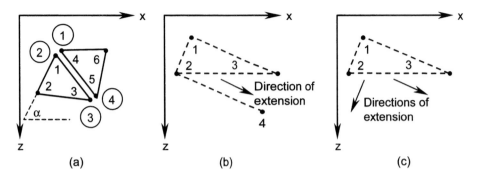

Figure 7.29. Elements for a lower bound numerical analysis.

Pastor (1978) developed the concept of the extension elements shown in Figures 7.29(b) and 7.29(c) which enables statically admissible stress fields to be obtained for semi-infinite domains. For each rectangular extension element, three additional linear equalities are necessary to extend the linear stress distribution to the fourth node. Each equation implies that the stress component at the centre of each rectangle is half the sum of the stresses corresponding to the nodes at the diagonal of the rectangle.

A typical stress discontinuity is shown in Figure 7.29(a), where the nodes in each pair of (1, 2) and (3, 4) have identical coordinates. Equilibrium of a discontinuity is enforced by ensuring that all pairs of nodes on opposite sides of the discontinuity have equal and opposite shear and normal stresses. This gives rise to four equality constraints. These equations are constructed by using the stress transformation of Equations 4.1 and 4.2, which relate the normal and shear stress on an arbitrary plane to the Cartesian stresses. If the normal and shear stresses are replaced by their nodal Cartesian stresses, the equality constraints are obtained in terms of the nodal stresses.

To satisfy the boundary stress conditions, equality constraints are forced on the Cartesian stresses for the boundary nodes. Referring to Figure 7.29(a), it is assumed that the side of the triangular element defined by nodes 1 and 2 are located on the boundary and that the normal and shear stresses along the boundary are known. Substituting these known values into the stress transformation Equations 4.1 and 4.2, four equality constraints may be constructed.

For a lower bound solution the stresses at each node must lie within the yield surface such that Equation 7.4 is satisfied. Since, in the numerical approach to the limit analysis, linear constraints are preferred, then the non-linear function expressed by Equation 7.4 is linearised by an inscribed polygon with p equal sides (Figure 7.30). From the geometry of Figure 7.30 it may be shown that Equation 7.4 is equivalent to the following linear p inequalities:

$$F_k = a_k \sigma_x + b_k \sigma_z + c_k \tau_{xz} + d \le 0 \tag{7.46}$$

where $k = 1, 2, 3..., p$. For the Tresca material:

$$a_k = \cos(2\pi k / p), \quad b_k = -\cos(2\pi k / p), \quad c_k = 2\sin(2\pi k / p),$$

$$d = -2c_u \cos(\pi / p) \tag{7.47}$$

For a Mohr-Coulomb material:

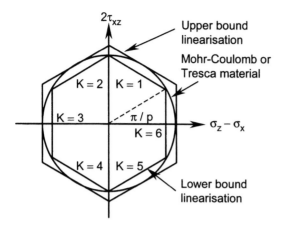

Figure 7.30. Internal and external linear approximations to the yield criterion.

$$a_k = \cos(2\pi k / p) - \sin\phi' \cos(\pi / p), \quad b_k = -\cos(2\pi k / p) - \sin\phi' \cos(\pi / p),$$

$$c_k = 2\sin(2\pi k / p), \quad d = -2c' \cos\phi' \cos(\pi / p) \tag{7.48}$$

When the right-hand term of Equation 7.46 equals zero it represents the equation of the corresponding side of the inscribed polygon in the stress coordinate system of Figure 7.30. Thus, the inequality expressed in Equation 7.46 simply implies that the state of stress is always at the left-hand side of the line while increasing k in the anti-clockwise direction. This ensures that the state of stress is always inside the linearised yield surface, thereby meeting the requirement of the lower bound theorem. Equation 7.46 needs to be enforced at each node of each triangular element. This ensures that the linearised yield criterion is satisfied throughout each triangle (Pastor, 1978; Aysen & Sloan, 1991a) and leads to a set of inequality constraints on the nodal stresses. For a four-node rectangular extension element (Figure 7.29(b)) the yield criterion is satisfied within the extension zone if the following inequalities are enforced:

$$F_{k1} \leq 0, F_{k2} \leq 0, F_{k3} \leq F_{k1} \text{ or } F_{k3} - F_{k1} \leq 0 \tag{7.49}$$

Note that the first two constraints are already included within the set of inequalities expressed by Equation 7.46. For the triangular extension element of Figure 7.29(c) the following inequalities are imposed on the nodal points:

$$F_{k1} \leq 0, F_{k2} \leq F_{k1} \text{ or } F_{k2} - F_{k1} \leq 0, F_{k3} \leq F_{k1} \text{ or } F_{k3} - F_{k1} \leq 0 \tag{7.50}$$

The first constraint in Equation 7.50 may be included in the assembly of inequalities defined in Equation 7.46, while the two remaining have the same form of constraints described by Equation 7.49. It is now required to define the objective function Q that represents the collapse load applied over a specified length of the boundary (as an example); this is given by:

$$Q = l(\sigma_{n1} + \sigma_{n2}) / 2 \tag{7.51}$$

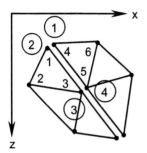

Figure 7.31. Elements for an upper bound numerical analysis.

where l is the length of the edge and σ_{n1}, σ_{n2} are the unknown normal stresses at its two ends. Using the stress transformation equations, the objective function can be expressed in terms of the nodal stresses at its two ends. The elemental constraints and objective function now are assembled by the usual rules of finite elements to give the overall linear programming problem.

7.4.3   Finite element formulation of the upper bound theorem

Figure 7.31 shows the triangular elements that are used to model the velocity field under plane strain loading. Each node has two velocities and each element has p plastic multiplier rates (Equation 7.1) where p is the number of sides in the linearised yield polygon. The velocities vary linearly throughout each element according to:

$$u = \sum_{i=1}^{i=3} N_i u_i, \quad v = \sum_{i=1}^{i=3} N_i v_i \qquad (7.52)$$

where N_i are linear shape functions and u_i, v_i are the nodal velocities in the x and z directions respectively. Whilst the linear triangular element, or the constant strain triangle, is considered to be an unsophisticated element in the general finite element formulation, in the numerical application of the upper bound theorem it has been used extensively to solve quite difficult problems. The shape functions N_i can be formulated in terms of global coordinates and the corresponding equations can be found in many texts (e.g. Zienkiewicz, 1970) and will not be discussed here. Combining Equation 7.1 (plane strain deformation of a rigid-perfectly plastic soil) and Equation 5.3 (general definition of a plane strain deformation), we obtain:

$$\dot{\varepsilon}_x^p = -\frac{\partial u}{\partial x} = d\lambda \frac{\partial F}{\partial \sigma_x}, \quad \dot{\varepsilon}_z^p = -\frac{\partial v}{\partial z} = d\lambda \frac{\partial F}{\partial \sigma_z}, \quad \dot{\gamma}_{xz}^p = -\left(\frac{\partial v}{\partial x} + \frac{\partial u}{\partial z}\right) = d\lambda \frac{\partial F}{\partial \tau_{xz}} \qquad (7.53)$$

where F represents the yield criterion and $d\lambda$ is a plastic multiplier; compressive strains are taken as positive. In order to remove the stress terms from Equation 7.53 and thereby provide a linear relationship between the unknown velocities and plastic multiplier rates, an external approximation to the yield criterion is needed. A six-sided approximation is shown in Figure 7.30. The linearised yield condition may be written as:

$$F_k = a_k \sigma_x + b_k \sigma_z + c_k \tau_{xz} + d = 0 \tag{7.54}$$

where $k = 1, 2, 3 \ldots, p$.
For a Tresca material:

$$a_k = \cos(2\pi k / p), \ b_k = -\cos(2\pi k / p), \ c_k = 2\sin(2\pi k / p), \ d = -2c_u \tag{7.55}$$

For a Mohr-Coulomb material:

$$a_k = \cos(2\pi k / p) - \sin\phi', \ b_k = -\cos(2\pi k / p) - \sin\phi',$$

$$c_k = 2\sin(2\pi k / p), \ d = -2c'\cos\phi' \tag{7.56}$$

Differentiating Equations 7.52 and 7.54 and substituting into Equation 7.53 furnishes three flow rule equations in which the stress terms are removed.

$$\sum_{i=1}^{i=3} \frac{\partial N_i}{\partial x} u_i + \sum_{k=1}^{k=p} d\lambda_k a_k = 0, \ \sum_{i=1}^{i=3} \frac{\partial N_i}{\partial z} v_i + \sum_{k=1}^{k=p} d\lambda_k b_k = 0,$$

$$\sum_{i=1}^{i=3} \frac{\partial N_i}{\partial x} v_i + \sum_{i=1}^{i=3} \frac{\partial N_i}{\partial z} u_i + \sum_{k=1}^{k=p} d\lambda_k c_k = 0 \tag{7.57}$$

The geometry of a velocity discontinuity element and the configuration of the nodes can be described similar to the stress discontinuity and is shown in Figure 7.31. For each velocity discontinuity two types of constraints are necessary to enforce the following conditions: (a) the work dissipated by relative sliding along the discontinuity (internal work) is positive which generates a set of inequalities to ensure the sign convention for the discontinuity, (b) the flow rule must be satisfied within the discontinuity for the plastic deformation. For a Tresca material the jump in the normal velocity $u_n = 0$ (zero dilation). For a Mohr-Coulomb material the flow rule in the discontinuity is satisfied if an equality constraint is enforced to ensure that the angle between the discontinuity and the direction of the velocity jump is equal to ϕ'. The final type of equality constraints to be imposed on the nodal velocities arise from the boundary conditions. The objective function is the linear expression for the dissipated work, which has two terms. Its first term expresses the sum of the work dissipated within the velocity discontinuities. In a Tresca material and for one velocity discontinuity, this term is:

$$E_{i1} = \int_l c_u |u_t| dl \tag{7.58}$$

where l is the length of a discontinuity and $|u_t|$ is the absolute value of the tangential velocity jump. This equation is integrated to express E_{i1} in terms of the nodal velocities. The second term expresses the work dissipated by the plastic stresses for triangular elements. In a Tresca material and for one triangle, it can be shown that this term is:

$$E_{i2} = \int_A (\sigma_x \dot{\varepsilon}_x + \sigma_z \dot{\varepsilon}_z + \tau_{xz} \dot{\gamma}_{xz}) dA = 2c_u A \sum_{k=1}^{k=p} d\lambda_k \tag{7.59}$$

where A is the area of the triangular element.

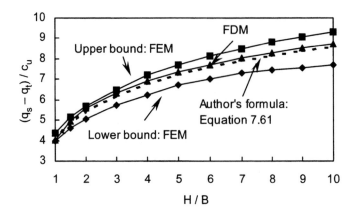

Figure 7.32. Lower and upper bounds for a tunnel heading for $\gamma B / c_u = 0$, compared with *FDM*.

For a Mohr-Coulomb material, an equation similar to Equation 7.59 can be obtained. Once the various coefficients are assembled, the dissipated work is minimized and a kinematically admissible velocity field can be found. The stress components at each node are obtained by substituting the calculated velocities and plastic multipliers in the set of Equations 7.53. We note in passing that the set of equality and inequality constraints may be solved very efficiently by applying the steepest edge active set algorithm, which fully exploits the extreme sparsity of the constraints. As a result, very large analyses may be conducted on a workstation or microcomputer.

7.4.4 Applications to slopes and tunnels

A summary of stability bounds obtained from the above two numerical methods for slopes are shown in Tables 7.1(a) and 7.1(b). Applications for circular and square tunnels, and the tunnel heading, are shown in Tables 7.3, 7.4 and 7.5 respectively. For the plane strain heading a finite difference method developed by Cundall (1980 and 1987) was used by the author to verify the effectiveness of the bound solutions. The results of these comparisons are shown in Table 7.5 and Figure 7.32. For the active failure of the tunnel heading (where the heading collapses towards inside the tunnel) and for $\gamma B / c_u > 0$, it can be shown that the upper bound value of the load parameter $(q_s - q_t) / c_u$ is:

$$\frac{q_s - q_t}{c_u} = [\frac{q_s - q_t}{c_u}]_{\gamma B / c_u = 0} - \frac{\gamma B}{c_u}(\frac{H}{B} + \frac{1}{2}) \tag{7.60}$$

The following formula for undrained conditions is suggested by the author (shown in Figure 7.32 for $\gamma B / c_u = 0$):

$$\frac{q_s - q_t}{c_u} = 4.0 + 2\ln\frac{H}{B} - \frac{\gamma B}{c_u}(\frac{H}{B} + \frac{1}{2}) \tag{7.61}$$

Whilst the results from this equation may be slightly higher or lower than the exact solution, they are always bounded by the lower and upper bounds with reasonable accuracy.

Table 7.3. Undrained stability of a shallow circular tunnel (Aysen & Sloan, 1991b).

H/D	$\gamma D/c_u = 0$		$\gamma D/c_u = 1$		$\gamma D/c_u = 2$		$\gamma D/c_u = 3$	
	LB	UB	LB	UB	LB	UB	LB	UB
1	2.27	2.55^d	1.08	1.23^d	-0.16	-0.16^d	-1.60	-1.60^d
2	3.25	3.48^d	0.97	1.03^d	-1.40	-1.35^d	-3.87	-3.77^d
3	3.78	4.29^d	0.47	0.88^d	-2.95	-2.53^d	-6.49	-5.96^d
4	4.30	5.03^d	-0.08	0.62^d	-4.57	-3.80^d	-9.11	-8.23^d
5	4.65	5.50^d	-0.74	0.28^d	-6.20	-5.14^d	-11.80	-10.57^d
6	4.48	5.90	-1.40	-0.35	-7.90	-6.65^d	-14.50	-13.10^d
7	5.30	6.30	-2.10	-1.05	-9.65	-8.40	-17.48	-15.75
8	5.48	6.65	-2.95	-1.75	-11.50	-10.15	-20.40	-18.55
9	5.75	6.90	-3.80	-2.50	-13.46	-11.90	-23.17	-21.30
10	5.92	7.15	-4.71	-3.22	-15.40	-13.60	-26.10	-24.06

Note: The numbers with superscripts d indicate a collapse mechanism (improved version of Figure 7.24(d)) which gives best upper bound.

Table 7.4. Undrained stability of a shallow square tunnel (Aysen & Sloan, 1991a).

H/B	$\gamma B/c_u = 0$		$\gamma B/c_u = 1$		$\gamma B/c_u = 2$		$\gamma B/c_u = 3$	
	LB	UB	LB	UB	LB	UB	LB	UB
0.3	0.522	0.600^a	0.261	0.300^a	-0.050	0.000^a	-0.401	-0.300^a
0.4	0.712	0.800^a	0.344	0.400^a	-0.055	0.000^a	-0.590	-0.400^a
0.5	0.915	1.00^a	0.477	0.500^a	-0.072	0.000^a	-0.760	-0.500^a
0.6	1.12	1.20^a	0.538	0.600^a	-0.124	0.000^a	-0.985	-0.656^b
0.7	1.32	1.40^a	0.631	0.700^a	-0.181	0.000^a	-1.21	-0.874^b
0.8	1.52	1.60^a	0.704	0.800^a	-0.261	0.000^a	-1.41	-1.10^b
0.9	1.70	1.80^a	0.758	0.900^a	-0.343	-0.115^b	-1.62	-1.32^b
1.0	1.88	2.00^a	0.810	1.00^a	-0.440	-0.234^b	-1.83	-1.54^b
1.5	2.50	2.77^b	0.750	0.960^b	-1.09	-0.854^b	-3.01	-2.68^b
2.0	2.89	3.15^b	0.600	0.825^b	-1.78	-1.50^b	-4.22	-3.84^b
3.0	3.45	3.82^b	0.140	0.478^b	-3.24	-2.87^b	-6.73	-6.23^b
4.0	3.90	4.37	-0.410	0.039	-4.83	-4.31^b	-9.38	-8.68^b
5.0	4.23	4.79	-1.06	-0.554	-6.51	-5.88	-12.10	-11.21
6.0	4.55	5.19	-1.78	-1.15	-8.25	-7.48	-14.87	-13.81
7.0	4.81	5.57	-2.57	-1.77	-10.05	-9.10	-17.65	-16.43
8.0	5.03	5.79	-3.36	-2.54	-11.87	-10.87	-20.50	-19.21
9.0	5.28	6.05	-4.10	-3.29	-13.79	-12.62	-23.57	-21.95
10.0	5.47	6.29	-5.00	-4.05	-15.79	-14.38	-26.60	-24.71

Note: The numbers with superscripts a, b (corresponding to Figures 7.25(a) and 7.25(b)) indicate the mechanism which gives best upper bound.

Table 7.5. Undrained stability of a plane strain heading (Aysen & Sloan, 1991c).

H/B	$\gamma B/c_u = 0$			$\gamma B/c_u = 1$		$\gamma B/c_u = 2$		$\gamma B/c_u = 3$	
	LB	FDM	UB	LB	UB	LB	UB	LB	UB
1.0	4.00	4.05	4.39	2.46	2.89	0.85	1.39	− 0.74	− 0.11
1.5	4.60	4.97	5.13	2.56	3.13	0.49	1.13	− 1.76	− 0.87
2.0	5.05	5.57	5.68	2.40	3.18	− 0.20	0.68	− 2.84	− 1.82
3.0	5.75	6.35	6.50	2.20	3.00	− 1.40	− 0.50	− 5.03	− 4.00
4.0	6.25	6.92	7.21	1.71	2.71	− 2.86	− 1.79	− 7.49	− 6.29
5.0	6.70	7.37	7.70	1.15	2.20	− 4.48	− 3.30	− 10.11	− 8.80
6.0	7.02	7.70	8.12	0.41	1.62	− 6.10	− 4.88	− 12.90	− 11.38
7.0	7.33	8.03	8.49	− 0.32	0.00	− 7.90	− 6.51	− 15.53	− 14.01
8.0	7.46	8.28	8.83	− 1.21	0.33	− 9.80	− 8.17	− 18.43	− 16.67
9.0	7.57	8.53	9.08	− 2.16	− 0.42	− 11.75	− 9.92	− 21.39	− 19.42
10.0	7.70	8.71	9.32	− 3.16	− 1.18	− 13.75	− 11.68	− 24.39	− 22.18

7.5 LIMIT EQUILIBRIUM METHOD AND CONCLUDING REMARKS

7.5.1 *The concept of the limit equilibrium method*

The theory of limit equilibrium uses the combined features of the bound theorems and calculates the collapse loads that are neither lower bound nor upper bound. Its application to soil mechanics problems, particularly in the areas of lateral earth pressures, stability of slopes and bearing capacity problems – has been very successful (Chapters 8 to 10). In a limit equilibrium method a collapse mechanism is selected to represent the failure of the material. Each block of the mechanism is assumed to be a rigid-perfectly plastic material. It is assumed that on the slip or sliding surfaces the mobilized shear stress τ_m, satisfies the following equation:

$$F = \frac{\tau_f}{\tau_m} \tag{7.62}$$

where τ_f is the shear strength of the soil and F is the factor of safety. The factor of safety is assumed constant along all sliding surfaces. If the loading is increased and the shear stresses on the sliding surfaces are forced to approach τ_f (defined by an appropriate failure criterion), then the material will fail by following the pattern of the collapse mechanism. The factor of safety for an assumed mechanism is calculated by applying the equilibrium conditions to the mechanism. An iterative procedure is carried out to establish the optimised values of the variables of the mechanism and subsequently the factor of safety. For $F = 1$ the collapse mechanism will yield a collapse load that is usually between the lower and upper bounds.

If the number of unknown forces is larger than the number of available equilibrium equations, then some simplifying (but reasonable) assumptions have to be made. Using the equilibrium equations, the collapse load, or any stability related parameter for $F \geq 1$, may be formulated in terms of the shear strength characteristics for an optimised mechanism. If the formulation is not possible, then a trial and error method has to be employed to achieve the optimal mechanism.

7.5.2   Concluding remarks

The lower and upper bound theorems of classical plasticity provide powerful methods for the evaluation of various stability problems in geotechnical engineering. The efficiency of the both techniques can be verified by careful analysis using numerical methods (e.g. displacement type finite elements and explicit finite difference method).

In a lower bound solution a statically admissible stress field is constructed so that it satisfies the equilibrium, stress boundary conditions and the yield criterion. The rigid-perfectly plastic material is assumed to obey either the Tresca or Mohr-Coulomb yield criteria. Points representing the state of stress are located inside the convex yield surface and therefore the material does not fail.

In an upper bound solution, a kinematically admissible velocity field is constructed so that it satisfies the compatibility, velocity boundary conditions and yield criteria but not necessarily equilibrium. The points representing the state of stress are located on the yield surface and thus the material fails under the computed collapse load.

In both methods manual computations can be used to estimate the corresponding collapse load. The exact solution lies between the two bounds. In the lower bound solution the selection of the stress field surrounded by the stress discontinuities needs ingenuity and practice. In the upper bound solution a convenient collapse mechanism yields results that are reasonably close to the exact solution. Finite element formulation of both methods leads to a large linear programming problem and needs a suitable algorithm to obtain a feasible solution.

7.6   PROBLEMS

7.1   A stress discontinuity makes an angle of 60° with the x-axis. On the right-hand side of the discontinuity: $\sigma_x = 100$ kPa, $\tau_{xz} = 50$ kPa. Determine:
(a) the magnitude of σ_z to satisfy the failure criterion,
(b) the normal and shear stresses on the discontinuity, and
(c) the state of the stresses at points on the left-hand side of the stress discontinuity to satisfy the failure criterion. $c' = 20$ kPa, $\phi' = 30°$.

Answers: σ_z (at R) = 327.2 kPa, $\sigma = 200.1$ kPa, $\tau = 123.4$ kPa, σ_z (at L) = 521.8 kPa, σ_x (at L) = 164.8 kPa, τ_{xz} (at L) = $-$ 62.3 kPa

7.2   Resolve Example 7.3 by taking into account the weight of the material. Assume the vertical stress at every point increases by γz, ($\gamma =$ unit weight, $z =$ depth).

Answer: $5c_u$

7.3   Calculate the bearing capacity factor N_c for undrained conditions for a smooth strip footing if the ground surface outside of the foundation makes an angle of 15° above or below the horizontal foundation level.

Answers: 5.66, 4.62

7.4 For the plane strain slope shown in Figure 7.22(a), calculate the lower and upper bound values for q / c_u. For the lower bound solution select a reasonable value for the number of the stress discontinuities passing through point d and apply the concept used in Equation 7.11. For the upper bound solution use the mechanism shown in Figure 7.22(a). Assume the stability number $\gamma H / c_u = 0$.

Answers: 3.57 (lower bound), 4.10 (upper bound)

7.5 A plane strain vertical cut is subjected to a vertical uniform load q at the upper ground surface. If the lower bound for the stability number $\gamma H / c'$ (with $q / c' = 0$) is 6.12, (this is obtained from a finite element analysis) calculate a lower bound for the load parameter q / c' for the given soil parameters.
$c' = 20$ kPa, $\phi' = 30°$, $\gamma = 19$ kN/m^3, and $H = 5$ m.

Answer: 0.774

7.6 For a 45° plane strain slope in a c_u, $\phi_u = 0$ soil, compute the upper bound values for the stability number $\gamma H / c_u$ using the collapse mechanism shown in Figure 7.18(a) for:
(a) a single block with $\alpha_1 = 22.5°$, and
(b) two blocks with $\beta_1 = 75°$, $\alpha_1 = 45°$, $\alpha_2 = 15°$.

Answers: 9.64, 7.78

7.7 For a 45° plane strain slope in a c_u, $\phi_u = 0$ soil, calculate the load parameter q / c_u using the collapse mechanism of Figure 7.18(a) together with the following data:
$\gamma H / c_u = 0$, $\beta_1 = 60°$, $\alpha_1 = 60°$, $\alpha_2 = 10°$.

Answer: 3.89

7.8 A 10 m height of saturated clay is supported by a rough retaining wall. The properties of the soil are: $c_u = 60$ kPa, $\phi_u = 0$, and $\gamma = 18$ kN/m^3. The vertical boundary load is $q = 50$ kPa and $c_w = 30$ kPa. Calculate the lower bound value of the horizontal active thrust and the position of its point of application.

Answers: 210.0 kN, 1.61 m

7.9 A rough retaining wall of height 6 m retains a sandy soil for which $c' = 0$ and $\phi' = 35°$, and $\gamma = 19$ kN/m^3. A uniform boundary load of $q = 50$ kPa acts on the upper ground surface. Assuming $\delta' = 35°$, calculate the lower and upper bounds for the active thrust (the resultant of the horizontal and vertical components) on the wall. For the upper bound solution assume a single variable mechanism similar to Figure 7.23(a).

Answers: 176.9 kN (lower bound), 160.3 kN (upper bound)

7.10 Using the single variable mechanism of Figure 7.23(a), find the upper bound value for the active thrust using the following data:
$H = 5$ m, $\alpha = 34°$, $q = 80$ kPa, $\delta' = 20°$, $c' = 10$ kPa, $\phi' = 25°$, and $\gamma = 18$ kN/m^3.

Answer: 165.5 kN

7.11 For the upper bound mechanism of Figure 7.26(a) compute an upper bound value for q_t. The given data are:
$\alpha = \beta = 60°$, $\delta = 90°$, $H = 10$ m, $D = 2$ m, $c_u = 40$ kPa, $\phi_u = 0$, $\gamma = 20$ kN/m^3.

Answer: 334.3 kPa (tensile)

7.7 REFERENCES

Anderheggen, E. & Knopfel, H. 1972. Finite element limit analysis using linear programming. *Intern. journal of solids and structures* (8): 1413-143.

Atkinson, J.H. 1993. *An introduction to the mechanics of soils and foundations.* London: McGraw-Hill.

Aysen, A. 1987. Lower bound solution for soil mechanics problems using finite element method, *Proc. 2nd national conf. on soil mechanics and foundation engineering* 1: 121-136. University of Bogazici, Istanbul, Turkey.

Aysen, A. & Loadwick, F. 1995. Stability of slopes in cohesive frictional soil using upper bound collapse mechanisms and numerical methods. *Proc. 14th Australasian conf. on the mechanics of structures and materials* 1: 55-59.

Aysen, A. & Sloan, S.W. 1991a. Undrained stability of shallow square tunnel. *Journal GE, ASCE* 117(8): 1152-1173.

Aysen, A. & Sloan, S. W. 1991b. Stability of a circular tunnel in a cohesive frictional soil. *Proc. 6th intern. conf. in Australia on finite element methods.* University of Sydney. 1: 68-76.

Aysen, A. & Sloan, S.W. 1991c. *Undrained stability of a plane strain heading.* Research Report No 059.02.1991, ISBN 0 7259 07134, NSW, Australia: The University of Newcastle.

Aysen, A. & Sloan, S.W. 1992. Stability of slopes in cohesive frictional soil. *Proc. 6th Australia-New Zealand conf. on geomechanics*: Geotechnical risk-identification, evaluation and solutions: 414-419. New Zealand: New Zealand Geomechanics Society.

Bottero, A., Negre, R., Pastor, J. & Turgeman, S. 1980. Finite element method and limit analysis: Theory for soil mechanics problems. *Computer methods in applied mechanics and engineering* (22): 131-149.

Britto, A.M. & Kusakabe, O. 1985. Upper bound mechanisms for undrained axisymmetric problems. *Proc. 5th intern. conf. on numerical methods in geomechanics.* Nagoya. 1691-1698.

Carter, J.P. & Balaam, N.P. 1990. Program AFENA, a general finite element algorithm. *Centre for geotechnical research*, University of Sydney, Australia.

Chen, W.F. 1975. *Limit analysis and soil Plasticity.* Amsterdam: Elsevier.

Chen, W.F. & Baladi, G.Y. 1985. *Soil plasticity: theory and implementation.* Amsterdam: Elsevier.

Chen, W.F. & Saleeb, A.F. 1982. *Constitutive equations for engineering materials, Vol.1- Elasticity and modeling.* New York: Wiley-Interscience.

Chen, W.F. & Saleeb, A.F. 1986. *Constitutive equations for engineering materials, Vol.2- Plasticity and modeling.* New York: Wiley-Interscience.

Cundall, P.A. 1980. *UDEC - A generalized distinct element program for modelling jointed rock.* Peter Cundall Associates. Report PCAR-1-80, US Army, European Research Office.

Cundall, P.A. 1987. Distinct element models of rock and soil structure. In E. T. Brown (ed), *Analytical and computational methods in engineering rock mechanics*: 129-139. London: George Allen & Unwin.

Davis, E.H., Gunn, M.J., Mair, R.J. & Seneviratne, H.N. 1980. The stability of shallow tunnels and underground openings in cohesive material. *Geotechnique* 30(4): 397-416.

Heyman, J. 1973. The stability of a vertical cut. *Intern. Journal of mechanical science* (15): 845-854.

Lysmer, J. 1970. Limit analysis of plane problems in soil mechanics. *Journal GE, ASCE* 96(4): 1311-1334.

Mair, R.J. 1979. *Centrifugal modelling of tunnel construction in soft clay.* Ph.D. Thesis, Cambridge University, UK.

Mulhaus, H.B. 1985. Lower bound solutions for circular tunnels in two and three dimensions. *Journal of rock mechanics and rock engineering* 18: 37-52.

Naylor, D.J. & Pande, G.N. 1981. *Finite elements in geotechnical engineering.* Swansea: Pineridge Press.

Parry, R.H.G. 1995. *Mohr circles, stress paths and geotechnics*. London: E & Spon.

Pastor, J. 1978. Limit analysis: numerical determination of complete statical solutions: application to the vertical cut. *Journal de mecanique appliquee, (in French)* (2): 167-196.

Sloan, S.W. 1988. Lower bound limit analysis using finite elements and linear programming. *Intern. journal for numerical and analytical methods in geomechanics* 12(1): 61-77.

Sloan, S.W. 1989. Upper bound limit analysis using finite elements and linear Programming. *Intern. journal for numerical and analytical methods in geomechanics* 13(3): 263-282.

Sloan, S.W. & Aysen, A. 1992. Stability of shallow tunnels in soft ground. In G.T. Houlsby & A.N. Schofield (eds), *Predictive soil mechanics*. London: Thomas Telford.

Turgeman, S. & Pastor, J. 1982. Limit analysis: a linear formulation of the kinematic approach for axisymmetric mechanic problems. *Intern. journal for numerical and analytical methods in geomechanics* 6: 109-128.

Zienkiewicz, O.C. 1970. *The finite element method in engineering science*. London: McGraw-Hill.

CHAPTER 8

Lateral Earth Pressure and Retaining Walls

8.1 INTRODUCTION

A retaining wall is a structure that supports a vertical or near vertical face of soil and is used to facilitate a major change in the ground surface level. The resulting horizontal stress from the soil on the wall is commonly referred to as *lateral earth pressure*. In the evaluation of the magnitude and distribution of this lateral earth pressure, it is assumed that the soil behind the wall (backfill) is on the verge of failure and obeys either the Tresca or Mohr-Coulomb failure criterion. Failure of the backfill occurs by two mechanisms depending on the direction of the wall displacement. If the displacement of the wall is away from the backfill soil the resulting failure is called *active*. A *passive* failure occurs if the wall is displaced towards the backfill until the limiting displacement is achieved. When the wall has no displacement, the lateral earth pressure *at-rest* can be estimated from elastic equilibrium or field measurements. In general there are basic theories used to estimate lateral earth pressures, viz., the Rankine method and the Terzaghi and Peck method. Rankine's theory assumes a long and smooth vertical wall where the lateral pressure increases linearly with depth. This analysis, although convenient, imposes some limitations and consequential errors. For a wall with significant friction, or if there is an irregular ground surface, a Coulomb wedge analysis, in conjunction with the charts produced by Terzaghi & Peck (1967), is applied. In this method the actual lateral pressure distribution is not defined but, based on experimental data, the location of the resultant of these (lateral) pressures is defined. However, for long frictional walls such as sheet piles, Rankine's method is applied using the earth pressure coefficients obtained from a Coulomb wedge analysis. This combination, although theoretically incorrect, appears to give reasonable estimates of the magnitude and distribution of the lateral soil pressure.

8.2 EARTH PRESSURE AT-REST

8.2.1 *The coefficient of soil pressure at-rest*

The magnitude of the lateral pressure depends on the type of the wall, the amount of the wall displacement, the type of (backfill) soil, the drainage conditions behind the wall, and the magnitude of any externally applied surface loading. In the absence of any lateral expansion or compression of the soil, the state of stress within the soil is termed as *at-rest*.

In general the two-dimensional stress field immediately behind a retaining wall is indeterminate and only the effective vertical stress can be defined in simple cases. In a homogeneous half-space the effective vertical stress at depth z is defined by Equation 2.1:

$$\sigma'_z = \gamma z - u.$$

where γ is the unit weight and u is the pore pressure at depth z. For a homogeneous, isotropic and elastic soil with zero lateral strain the lateral stress σ'_x (Figure 8.1) is expressed by:

$$\sigma'_x = \frac{\mu}{1-\mu}\sigma'_z = k_o\sigma'_z \tag{8.1}$$

where μ is Poisson's ratio whose laboratory or field determination is problematic, and k_o is the coefficient of soil pressure for the at-rest condition. Equation 8.1 implies that the axial strains ε_x, and ε_y (y is normal to the plane of the page) are both zero. The magnitude of k_o varies from 0.42 to 0.66 depending on the type of the soil, its degree of compaction and plasticity characteristics, and the degree of its disturbance (Bishop, 1958). For truly normally consolidated soils that exhibit zero cohesion during drained shear, a value for k_o may be calculated from the following generally accepted empirical equation (Jáky, 1948):

$$k_o = 1 - \sin\phi' \tag{8.2}$$

For overconsolidated soils the value of k_o is higher than that given by Equation 8.2. Alpan (1967) suggested the following relationship:

$$\frac{k_{o,OCR}}{k_{o,NC}} = (OCR)^n \tag{8.3}$$

where $k_{o,OCR}$ and $k_{o,NC}$ are the coefficients of earth pressure at-rest for overconsolidated and normally consolidated soil respectively, OCR is the overconsolidation ratio and n is a number depending on the plasticity characteristics of the soil. Wroth & Houlsby (1985) give $n = 0.42$ for soil with low plasticity ($PI < 40\%$) and $n = 0.32$ for soil with high plasticity ($PI > 40\%$). After a statistical analysis of reported data, Mayne & Kulhawy (1982) proposed that $n = \sin\phi'$. Hence,

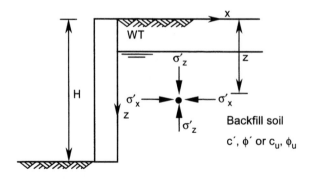

Figure 8.1. State of stress behind a retaining wall.

$$k_o = (1 - \sin \phi')(OCR)^{\sin \phi'} \tag{8.4}$$

Most laboratory methods for the determination of lateral earth pressures are based on tri-axial and odeometer tests. A full discussion on the laboratory measurement of lateral stress in soft clays may be found in Benoit & Lutenegger (1993). In-situ measurements used to determine k_o involve the use of *SPT* and *CPT* type field tests that give rise to empirical correlations but, unfortunately, there is usually a wide scatter in the actual data, especially for soft clays (Kulhawy et al., 1989).

8.2.2   *Earth pressure due to surface loading and compaction*

The additional lateral stress along a wall caused by surface loading can be estimated from a Boussinesq type distribution. In a smooth wall the lateral component of stress at-rest is doubled to take account the presence of the rigid smooth wall. This doubling is equivalent to the application of a mirror of the real load (symmetric about the back face of the wall) that enforces zero lateral strain and zero shear stress on the back face of the wall.

The increase of lateral earth pressure due to the effect of compaction equipment moving across the backfill has been investigated thoroughly and there is (field) evidence that shows a considerable increase in lateral pressures on the wall. This increase, which is regarded as a residual lateral pressure, is concentrated in the first few metres. The state of stress approaches the k_o state at a depth of approximately 9 m (Duncan & Seed, 1986). Following these findings, Williams et al. (1987) developed charts and tables relating the induced lateral stress to the type of the compaction equipment.

8.3   RANKINE'S THEORY FOR ACTIVE AND PASSIVE SOIL PRESSURES

8.3.1   *The concepts of active and passive states and wall displacement*

Rankine's theory of active and passive soil pressures requires a long smooth wall and assumes a linear distribution of lateral earth pressure. These assumptions were challenged by Terzaghi (during 1930s) when he developed a Coulomb wedge analysis. This was a turning point in the computation of the resultant of the lateral earth pressure and the determination of its point of application behind a frictional wall (Peck, 1990). Today, the Coulomb wedge analysis has changed from its original computational-graphical method to a computer-based method. Rankine's theory, along with the lateral earth pressure coefficients obtained from the Coulomb wedge analysis and with numerical modelling of the retaining system, are applied to long retaining structures such as sheet piles.

Consider the retaining wall shown in Figure 8.1 to be subjected to a horizontal displacement away from the retained soil. This displacement causes an expansive strain within the retained soil where the affected volume of the soil depends on the height of the retaining wall. If the height of the wall is infinitely long then the affected volume will progress to infinity to the right of the retaining wall. However, for a finite height of wall, the finite volume of the soil will be subjected to horizontal expansive strains. As a result, the lateral earth pressure acting on the wall will decrease and eventually the soil behind the wall will fail. This means that the state of the stress within the soil is transferred from elastic equilibrium to plastic equilibrium. If there is no surface loading then the magnitude of the effective vertical stress (defined by: $\sigma'_z = \gamma z - u$) remains constant. In the presence of a surface loading q, the effective vertical stress increases in a Boussinesq pattern. At failure, the vertical contribution of the surface loading at every point becomes equal to q. During the progress of the lateral displacement, the diameter of the Mohr's circle of stress increases until at failure, it becomes tangent to the failure envelope. The lateral stress, which is the lateral soil pressure, becomes a minimum at every point along the wall and it is termed as *active earth (or soil) pressure* as it refers to the active state of failure behind the retaining wall. For a long smooth wall, the major principal stress is vertical ($\sigma'_1 = \gamma z - u$). The minor principal stress represents the active soil pressure and acts horizontally. The magnitude of this pressure can be found from the relationship established between σ'_1 and σ'_3 at failure (Equation 4.13 or 4.14).

If the wall moves towards the backfill the lateral pressure will increase, reaching its maximum value called the *passive earth (or soil) pressure*. This represents a passive failure behind the wall for which the effective vertical stress represents the minor principal stress ($\sigma'_3 = \gamma z - u$), and the horizontal major principal stress represents the passive earth pressure. The magnitude of the pressure at any depth can be found by substituting σ'_3 into Equations 4.13 or 4.14. The lateral pressure-displacement ratio (Δ / H) behaviour is shown in Figure 8.2 from which it can be seen that to achieve a passive failure, more displacement of the wall (in comparison with the active failure) is needed. Duncan et al. (1990) summarized the existing laboratory tests on model retaining walls and controlled field experiments that have been carried out from 1934 to 1990. According to this report, the ratio Δ / H required to reduce earth pressure to active values in sands, silty sands and sandy gravels, varies from 0.0003 to 0.008 for rotation and from 0.001 to 0.005 for translation, with the larger values of Δ / H applicable to larger walls. In the reported data two definitions of active failure have been used. The first corresponds to the active thrust required to reach its minimum value whilst the second definition corresponds to the development of a linear lateral earth pressure. In many experiments the first definition is achieved while having a non-linear earth pressure at an elevation higher than $0.33H$. Continuing displacement results in a linear distribution of pressure and the resultant at $0.33H$. Displacement ratios required are between 0.005 to 0.132 for rotation (higher values as the clay percentage increases) and an average value of 0.025 for translation. Changes in the lateral earth pressure with time have been reported by number of investigators. Field observations have shown that the earth pressure changes with time and approaches the at-rest values. For a long-term active state, the average of the active and at-rest conditions may be used in design (Munfakh, 1990).

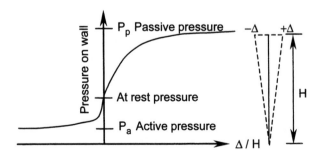

Figure 8.2. Displacement requirements for active and passive states of failure.

8.3.2 *Linear earth pressure distribution: smooth wall and horizontal ground surface*

Figures 8.3(a) and 8.3(b) provide illustrations of the active and passive failure conditions respectively whilst the states of stress are shown in Figure 8.3(c). The inclinations of the potential failure planes from the horizontal ground surface are $45° + \phi'/2$ and $45° - \phi'/2$ for the active and passive states respectively. For the active state, the lateral earth pressure at depth z is $\sigma'_3 = p_a$ and $\sigma'_1 = \sigma'_z = \gamma z - u$.

Substituting these into Equation 4.13 we obtain:

$$p_a = \sigma'_z \frac{1-\sin\phi'}{1+\sin\phi'} - 2c'\sqrt{\frac{1-\sin\phi'}{1+\sin\phi'}} \rightarrow p_a = \sigma'_z k_a - 2c'\sqrt{k_a} \tag{8.5}$$

where k_a is the lateral active earth pressure coefficient given by:

$$k_a = \frac{1-\sin\phi'}{1+\sin\phi'} = \tan^2(45° - \phi'/2) \tag{8.6}$$

If a uniformly distributed vertical load q is applied at the ground surface of the backfill:

$$p_a = (\sigma'_z + q)k_a - 2c'\sqrt{k_a} \tag{8.7}$$

The general form of the lateral earth pressure for the passive state is:

$$p_p = (\sigma'_z + q)k_p + 2c'\sqrt{k_p} \tag{8.8}$$

where k_p is the lateral passive earth pressure coefficient given by:

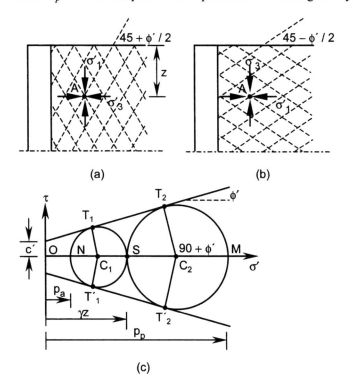

Figure 8.3. Active and passive states of stress at the back of a long smooth retaining wall.

$$k_p = \frac{1+\sin\phi'}{1-\sin\phi'} = \tan^2(45°+\phi'/2) \tag{8.9}$$

Equations 8.7 and 8.8 quantify the linear increase in lateral earth pressure with increasing depth z. Assuming that the same distribution applies for a wall of finite height, then, for a homogeneous soil with no water in the backfill, $\sigma'_z = \gamma z$. The total active or passive thrust is the integral of the lateral earth pressure behind the wall.

$$P_a = \frac{1}{2}\gamma H^2 k_a + qHk_a - 2c'H\sqrt{k_a} \tag{8.10}$$

$$P_p = \frac{1}{2}\gamma H^2 k_p + qHk_p + 2c'H\sqrt{k_p} \tag{8.11}$$

Note that for a purely cohesive saturated clay with undrained shear strength parameters of c_u and $\phi_u = 0$, then $k_a = k_p = 1$. Figure 8.4 shows the distribution of lateral active earth pressure behind a smooth wall. Figure 8.4(a) represents the case where there is no surface loading. At $z = 0$ the soil behind the wall is subjected to a maximum tensile stress of $2c'\sqrt{k_a}$. The tensile stress decreases with depth and becomes zero at depth z_0. Since it is common to ignore the tensile strength of the soil, the cohesion is the minimum shear strength at zero normal stress and the failure criterion does not apply for the tensile normal stress. Consequently cracks will develop down to the depth z_0. Figure 8.4(b) represents the effect of surface loading in the form of uniform lateral pressure equal to qk_a. The combined effects of gravity and surface loading are shown in Figure 8.4(c). Note that the area between the line representing the earth pressure and the vertical axis is equal to the total active thrust. If there is no hydrostatic water pressure, the effective vertical stress corresponding to z_0 is $\sigma'_z = \gamma z_0$. Upon substituting this into Equation 8.7 and setting the active pressure to zero we obtain:

$$z_0 = \frac{2c'}{\gamma\sqrt{k_a}} - \frac{q}{\gamma} = \frac{2c'\tan(45°+\phi'/2)}{\gamma} - \frac{q}{\gamma} \tag{8.12}$$

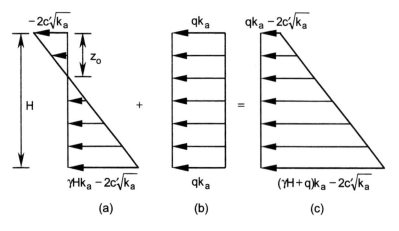

(a)   (b)   (c)

Figure 8.4. Distribution of active lateral earth pressure behind a smooth retaining wall.

The total active thrust may be corrected by ignoring the stress distribution in the tensile zone.

Example 8.1

Calculate the total active thrust per metre length of an 8 m high smooth vertical retaining wall. The properties of the backfill soil are: $c' = 20$ kPa, $\phi' = 25°$, and $\gamma = 17.5$ kN/m³.

Solution:

$k_a = \tan^2(45.0° - \phi'/2) = \tan^2(45.0° - 25.0°/2) = 0.406$.

At $z = 0$, $\sigma'_z = 0$ and:

$p_a = -2c'\sqrt{k_a} = -40.0\sqrt{0.406} = -25.5$ kPa.

At $z = 8.0$, $\sigma'_z = 8.0 \times 17.5 = 140.0$ kPa,

$p_a = 140.0 \times 0.406 - 40.0\sqrt{0.406} = 31.3$ kPa.

The depth at which the active pressure becomes zero is calculated using Equation 8.12:

$z_o = 2 \times 20.0/(17.5 \times \sqrt{0.406}) = 3.59$ m. Ignoring the tensile force:

$P_a = (8.0 - 3.59) \times 31.3 \times 1.0/2 = 69.0$ kN. The point of application of this force is:

$(8.0 - 3.59)/3 = 1.47$ m above the base of the wall (Figure 8.5).

Example 8.2

A smooth vertical retaining wall is 5 m high. Determine the total active thrust for the following cases: (a) there is no hydrostatic water pressure in the backfill: $c' = 0$, $\phi' = 36°$, and $\gamma = 17.5$ kN/m³, (b) the water table is at the ground surface: $c' = 0$, $\phi' = 36°$, $\gamma_{sat} = 19.5$ kN/m³.

Solution:

For both cases: $k_a = (1 - \sin 36.0°)/(1 + \sin 36.0°) = 0.260$.

(a) At $z = 0$, $\sigma'_z = 0$, and $p_a = 0$.

At $z = 5.0$ m, $\sigma'_z = 5.0 \times 17.5 = 87.5$ kPa, and $p_a = 87.5 \times 0.260 = 22.7$ kPa.

$P_a = 22.7 \times 5.0/2 \times 1.0 = 56.8$ kN.

(b) At $z = 0$, $\sigma'_z = 0$, and $p_a = 0$. At $z = 5.0$ m, pore pressure $u = 5.0 \times 9.81 = 49.0$ kPa.

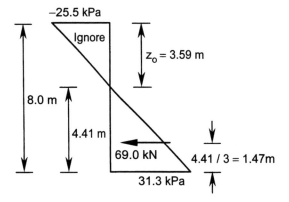

Figure 8.5. Example 8.1.

$\sigma'_z = 5.0 \times 19.5 - 49.0 = 48.5$ kPa, $p_a = 48.5 \times 0.260 = 12.6$ kPa.

$P_a = 12.6 \times 5.0 / 2 \times 1.0 = 31.5$ kN.

P_w = force due to water pressure = $49.0 \times 5.0 / 2 \times 1.0 = 122.5$ kN.

Total thrust on the back of the wall = $P_a + P_w = 3\ 1.5 + 122.5 = 154.0$ kN.

8.3.3   Linear earth pressure distribution for a sloping ground surface in $c' = 0$, ϕ' soil

Figure 8.6(a) illustrates a retaining wall with the sloping ground surface at an angle β to the horizontal. The soil is granular with shear strength parameters $c' = 0$, ϕ' and $\beta < \phi'$. To evaluate the active earth pressure p_a at depth z consider a rhombic element of soil of width of dl. On the vertical sides of the element the stress is p_a and is assumed to be parallel to the ground surface. This implies that the wall has a friction angle of β with the backfill soil. The vertical stress at depth z is equal to the weight of the material in the volume $abcd$ divided by ab:

$$\sigma'_z = (dl \times z \times 1.0 \times \gamma) / (dl / \cos \beta) = \gamma z \cos \beta.$$

Thus the normal and shear stresses on plane ab are:

$$\sigma'_{ab} = \gamma z \cos^2 \beta, \ \tau_{ab} = \gamma z \cos \beta \sin \beta \tag{8.13}$$

The states of stress on plane ab and the vertical plane are represented by points B and A respectively on the active stress circle shown in Figure 8.6(b). The length OB is equal to the vertical stress on plane ab; similarly OA represents p_a. Numerically $OA' = OA$, and therefore from the geometry of Figure 8.6(b), the earth pressure coefficient is given by:

$$k_a = \frac{p_a}{\sigma'_z} = \frac{OA'}{OB} = \frac{OC - A'C}{OC + CB} \rightarrow k_a = \frac{\cos \beta - \sqrt{\cos^2 \beta - \cos^2 \phi'}}{\cos \beta + \sqrt{\cos^2 \beta - \cos^2 \phi'}} \tag{8.14}$$

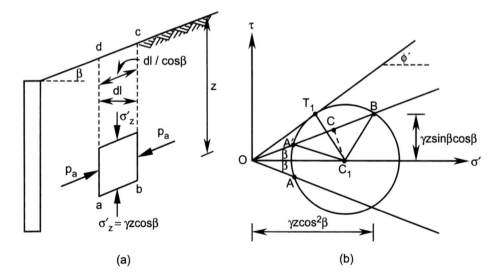

(a)   (b)

Figure 8.6. Active state behind a wall with sloping backfill in $c' = 0$, ϕ' soil.

The active earth pressure at depth z is:

$$p_a = \gamma z k_a \cos\beta \tag{8.15}$$

and the total active thrust that acts at height $H/3$ above the base is given by:

$$P_a = 0.5\gamma H^2 k_a \cos\beta \tag{8.16}$$

Using a similar procedure for the passive state we obtain:

$$k_p = \frac{\cos\beta + \sqrt{\cos^2\beta - \cos^2\phi'}}{\cos\beta - \sqrt{\cos^2\beta - \cos^2\phi'}} \tag{8.17}$$

The passive earth pressure at depth z and the total passive thrust are calculated using equations similar to Equations 8.15 and 8.16 in which k_a is replaced by k_p. In the presence of a surface gravity load of q, γz must be replaced by $\gamma z + q$.

Example 8.3

A retaining wall with a vertical back is 5 m high and retains a sloping soil with $\beta = 20°$. Determine the magnitude of the active thrust and the inclination of the failure planes developed behind the wall. $c' = 0$, $\phi' = 36°$, and $\gamma = 17.5$ kN/m^3.

Solution:

Using Equation 8.14 gives $k_a = 0.326$. The total thrust from Equation 8.16 is:
$$P_a = 0.5 \times 17.5 \times 5.0^2 \times 0.326 \times \cos 20.0° \times 1.0 = 67.0 \text{ kN}.$$

Calculate the state of stress on a rhombic element at an arbitrary depth (say 5 m):
$$\sigma'_z = \gamma z \cos\beta = 17.5 \times 5.0 \times \cos 20.0° = 82.2 \text{ kPa},$$

$$p_a = \gamma z k_a \cos\beta = 82.2 \times 0.326 = 26.8 \text{ kPa}.$$

From the stress circle of Figure 8.7:
$$A'C = A'B/2 = (82.2 - 26.8)/2 = 27.7 \text{ kPa},$$

$$OC = 26.8 + 27.7 = 54.5 \text{ kPa}, \quad OC_1 = OC/\cos\beta = 54.5/\cos 20.0° = 58.0 \text{ kPa},$$

$$C_1 T_1 = OC_1 \times \sin\phi' = 58.0 \times \sin 36.0° = 34.1 \text{ kPa}.$$

$$\cos(\angle CBC_1) = CB/C_1 B = A'C/C_1 T_1 = 27.7/34.1 \rightarrow \angle CBC_1 = 35.6°,$$

$$\angle SC_1 B = \angle CBC_1 + \beta = 35.6° + 20.0° = 55.6°.$$

The angles between the inclined planes of the rhombic element (represented by point B), with the failure plane of T_1 and T_2, are α_1 and α_2 respectively:

$$\alpha_1 = 1/2(\angle SC_1 T_1 - \angle SC_1 B) = 1/2(90.0° + 36.0° - 55.6°) = 35.2°.$$

Thus the angle of failure plane T_1 from the horizontal is:

$35.2° + 20° = 55.2°$ (anticlockwise).

Consequently, the angle of failure plane T_2 from the horizontal is: $89.2° + 20° = 109.2°$ (anticlockwise). The results are presented in Figure 8.7. It may be shown that the failure planes T_1 and T_2 are parallel to the lines AT_1 and AT_2 respectively. Note that these directions are independent of the unit weight and depend entirely on the ϕ' and β values.

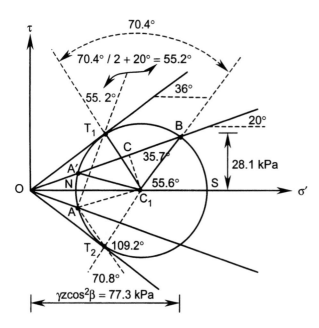

Figure 8.7. Example 8.3.

8.3.4   Stratified backfill

At the boundary of the two soils in a stratified backfill, two lateral earth pressures exist, each calculated using the corresponding earth pressure coefficient. The total vertical stress is always linear with depth but its slope changes with the change in unit weight. The effective vertical stress has a similar pattern if the water table is continuous within the soil. If water is resting on an impervious layer located behind the wall, then for the pervious layer the effective vertical stress is used whilst for the impervious layer the total vertical stress is used.

Example 8.4

A retaining wall 6 m high supports two layers of soil each having a thickness of 3 m. The properties of the layers are:
upper layer: $c' = 0$, $\phi' = 30°$, $\gamma_{dry} = 17.5$ kN/m^3, and $\gamma_{sat} = 19.5$ kN/m^3;
lower layer: $c' = 10$ kPa, $\phi' = 18°$, and $\gamma_{sat} = 19$ kN/m^3.
There is a surface load of 50 kPa and the water table is 1.5 m below the ground surface. Determine the total lateral thrust and its line of action above the base of the wall.

Solution:

For the upper layer: $k_a = (1 - \sin 30.0°)/(1 + \sin 30.0°) = 0.333$.

For the lower layer: $k_a = (1 - \sin 18.0°)/(1 + \sin 18.0°) = 0.528$.

At $z = 0$, $\sigma_z = 50.0$ kPa, $u = 0$, $\sigma'_z = 50.0$ kPa,

$p_a = \sigma'_z k_a = 50.0 \times 0.333 = 16.7$ kPa.

At $z = 1.5$ m, $\sigma_z = 50.0 + 1.5 \times 17.5 = 76.2$ kPa, $u = 0$, $\sigma'_z = 76.2$ kPa,

$p_a = 76.2 \times 0.333 = 25.4$ kPa.

At $z = 3.0$ m, $\sigma_z = 50.0 + 1.5 \times 17.5 + 1.5 \times 19.5 = 105.5$ kPa, $u = 1.5 \times 9.81 = 14.7$ kPa,

$\sigma'_z = 105.5 - 14.7 = 90.8$ kPa, using $k_a = 0.333$:

$p_a = \sigma'_z k_a = 90.8 \times 0.333 = 30.3$ kPa.

Using $k_a = 0.528$:

$$p_a = \sigma'_z k_a - 2c' \times \sqrt{k_a} = 90.8 \times 0.528 - 2 \times 10.0 \sqrt{0.528} = 33.4 \text{ kPa}.$$

At $z = 6.0$ m, $\sigma_z = 50.0 + 1.5 \times 17.5 + 1.5 \times 19.5 + 3.0 \times 19.0 = 162.5$ kPa,

$u = 4.5 \times 9.81 = 44.1$ kPa,

$\sigma'_z = 162.5 - 44.1 = 118.4$ kPa,

$$p_a = 118.4 \times 0.528 - 2 \times 10.0 \sqrt{0.528} = 48.0 \text{ kPa}.$$

Note that due to the linear distribution of the lateral pressure with depth we can find the lateral earth pressure at the bottom of any saturated layer by adding $\gamma' \times h \times k_a$

Force (kN)	Arm above the base (m)	Force × arm (kN.m)
$F_1 = 16.7 \times 1.5 \times 1.0$ (per meter run) $= 25.05$	5.25	131.51
$F_2 = (25.4 - 16.7) \times 1.5 \times 1/2 \times 1.0 = 6.52$	5.00	32.60
$F_3 = 25.4 \times 1.5 \times 1.0 = 38.10$	3.75	142.87
$F_4 = (30.3 - 25.4) \times 1.5 \times 1/2 \times 1.0 = 3.67$	3.5	12.84
$F_5 = 33.4 \times 3.0 \times 1.0 = 100.20$	1.5	150.30
$F_6 = (48.0 - 33.4) \times 3 \times 1/2 \times 1.0 = 21.90$	1.00	21.90
$F_7 = 44.1 \times 4.5 \times 1/2 \times 1.0 = 99.22$	1.5	148.83
Total Thrust: 294.66		Total: 640.85

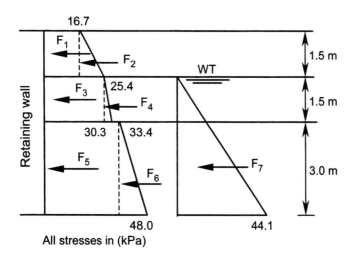

Figure 8.8. Example 8.4.

(or $\gamma \times h \times k_a$ with no water in the system) to the lateral pressure at the top of the layer that has already been calculated. The term γ' represents the *submerged* or *buoyant* unit weight. Computations are summarized in the above table and results are shown in Figure 8.8. Distance of the total horizontal thrust from the base = 640.85 / 294.66 = 2.17 m.

8.4 COULOMB WEDGE ANALYSIS

8.4.1 *Principles of the Coulomb wedge analysis*

The Coulomb wedge analysis is a limit equilibrium method. A wedge shape collapse (or failure) mechanism is assumed which is bounded by the back face of the retaining wall, a horizontal or inclined, loaded or unloaded, ground surface and a linear failure plane (Figure 8.9(a)). The wedge mechanism slides downwards or upwards on the failure plane simulating the active and passive states respectively. The aim of the analysis, which is based on the force equilibrium of the wedge, is to obtain the critical value of the sliding angle α and the corresponding active or passive thrust. The Mohr-Coulomb failure criterion is applied along the sliding surface.

For a granular material the wedge and corresponding force diagram for the active state are shown in Figures 8.9(b) and 8.9(c) respectively. In the case of a granular material with no surface loading, it is possible to formulate the problem (obtaining the values of active and passive thrusts) mathematically without a need for iteration. However, with the inclusion of cohesion on the sliding surface and on the back face of the wall, an irregular ground surface or surface loading, an iterative procedure must be used. Note that the soil reaction R and the active thrust P_a are both located under the line perpendicular to the direction of sliding. This ensures that the senses of the components of these forces along the sliding surfaces (shear resistance) are opposite to the movement of the wedge. If the base of the retaining wall is constructed on a soft soil and the settlement is predicted to be higher than the displacement needed for active failure, then the force P_a may be relocated by making an angle $-\delta'$ with the line perpendicular to the back face of the wall.

A graphical representation of the iteration of the angle α, and locating its critical value and the corresponding maximum active thrust, is illustrated in Figure 8.10. The weight of each trial wedge W is computed from the geometry of the wedge and all such weights are

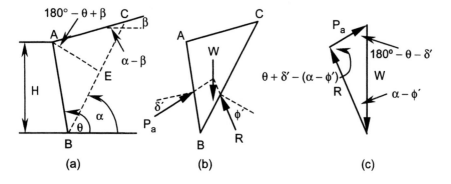

Figure 8.9. Coulomb wedge analysis in $c' = 0$, ϕ' soil (active state).

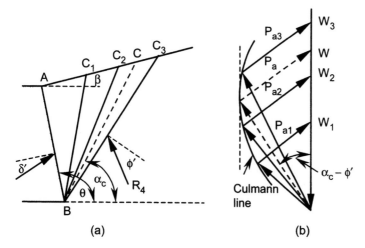

Figure 8.10. Graphical presentation of wedge analysis in $c' = 0$, ϕ' soil (active state).

plotted on a common vertical load line (a process known as Culmann procedure). For each trial wedge in which α is known, the angle between W and P_a is:

$180° - \theta - \delta'$,

and the angle between W and R is: $\alpha - \phi'$.

It follows that two lines representing forces P_a and R can be drawn from both ends of the load line so that they intersect thereby giving the magnitude of P_a for each trial. The maximum value of P_a and the corresponding triangle of forces are found by drawing a vertical tangent to the curve passing from the intersection points described above. If there is a uniform vertical surface loading (q) on the ground surface then the weight of each wedge is increased by $q \times AC \times \cos\beta$.

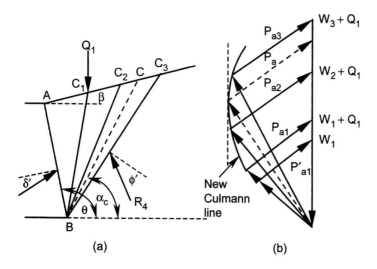

Figure 8.11. Wedge analysis: effect of a line surcharge load in the active state.

In the case of a line surcharge load it is convenient to locate a trial failure plane passing through the point of application of the load as shown in Figure 8.11(a). The corresponding wedge is considered twice, once without the surface load and once with the surface load and there will be a jump on the Culmann line at this point. However the maximum P_a is found in the same way as explained previously. The surface load is included in the weights of the wedges thereafter. The above procedure is easily adapted to include the cohesion resistance developed on the sliding surface and the back face of the wall. For the passive failure where the wedge is pushed upwards, the inclination of forces R and P_p must be selected so as to hold down the wedge and create the shear resistances in the opposite direction of the movement of the wedge. Note that in the passive state we are searching for the smallest force that can move the wedge upwards.

When there is water behind the retaining wall, forces due to the water pressure on the boundaries of the trial wedge (*BA* and *BC*$_1$, *BC*$_2$,...) must be included in the force diagram. The force diagram will yield the effective active or passive thrusts. It can be shown that if the total weight of the wedge is considered along with water pressure on the BC_1, BC_2,...edges and the water force on the *BA* edge is deleted, the force diagram will yield the total thrust.

The effect of wall friction near the base of the wall is to produce a curved failure surface (Figure 8.12). Assuming a linear failure surface introduces errors in the estimation of the active and passive thrusts. In the active state the error is negligible, but the passive thrust will be significantly overestimated depending on the value of the mobilized friction angle on the back face of the wall (δ') and the wall displacement.

In the traditional method of evaluating the passive thrust it is assumed that the failure surface immediately behind the base of the wall is either a logarithmic spiral or a circle. As the distance of the failure surface from the wall is increased it becomes linear and makes an angle of $45° - \phi' / 2$ with the horizontal. The analysis is carried out in two steps and begins by ignoring the cohesion from which a minimum value of the passive thrust is obtained. The point of application of this thrust is assumed to be H / 3 above the base of the wall. In the second step the passive thrust due to cohesion and friction is calculated by neglecting the weight of the material. The point of application of this thrust is assumed to be H / 2 above the base. The failure surfaces from these two steps may not be identical;

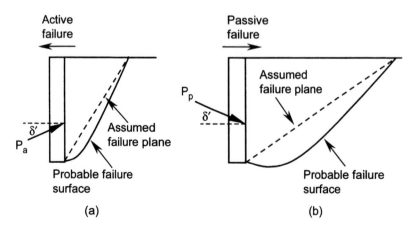

Figure 8.12. Development of the active and passive failure surfaces due to wall friction.

the results are combined to give the total passive thrust located between $H/2$ and $H/3$. A difference of approximately 10% is expected with the results obtained using linear failure plane and a triangular wedge (Terzaghi et al., 1996). The magnitude of δ' depends on the angle of internal friction of the backfill soil and the roughness of the wall surface. For a concrete wall a value between $\phi'/2$ and $2\phi'/3$ is used extensively in design. Values can also be determined in the laboratory using direct shear type tests on composite concrete-soil samples. In this way an estimate can be made of the adhesion developed between the wall material and the backfill soil.

8.4.2 Coulomb wedge analysis for $c' = 0$, ϕ' soil

Referring to Figure 8.9(c), equilibrium of the forces results in:

$$\frac{W}{\sin[(\theta + \delta') - (\alpha - \phi')]} = \frac{P_a}{\sin(\alpha - \phi')},$$

where $W = S_{ABC} \times 1.0 \times \gamma$ is the weight of the trial wedge, and P_a is the active thrust corresponding to the trial angle α. The area of the wedge, calculated from its geometry, is given by:

$$S_{ABC} = \frac{H^2}{2}\left[\frac{\sin(\theta - \alpha)\sin(\theta - \beta)}{\sin^2\theta\sin(\alpha - \beta)}\right] \rightarrow W = \frac{\gamma H^2}{2}\left[\frac{\sin(\theta - \alpha)\sin(\theta - \beta)}{\sin^2\theta\sin(\alpha - \beta)}\right].$$

Substituting W into the force equilibrium equation we obtain:

$$P_a = \frac{\gamma H^2}{2}\left\{\frac{\sin(\theta - \alpha)\sin(\theta - \beta)\sin(\alpha - \phi')}{\sin^2\theta\sin(\alpha - \beta)\sin[(\theta + \delta') - (\alpha - \phi')]}\right\} = \frac{\gamma H^2}{2}k_a\{\alpha, \beta, \theta, \delta', \phi'\}$$

(8.18)

A maximum value for k_a can be found by setting $\partial P_a / \partial \alpha = 0$. Thus, for the active case:

$$k_a = \frac{\sin^2(\theta - \phi')}{\sin^2\theta\sin(\theta + \delta')\left[1 + \sqrt{\dfrac{\sin(\phi' + \delta')\sin(\phi' - \beta)}{\sin(\theta + \delta')\sin(\theta - \beta)}}\right]^2}$$

(8.19)

For a vertical wall ($\theta = 90°$) and a horizontal ground surface ($\beta = 0$):

$$k_a = \left[\frac{\cos\phi'}{\sqrt{\cos\delta'} + \sqrt{\sin(\phi' + \delta')\sin\phi'}}\right]^2$$

(8.20)

For a smooth ($\delta' = 0$) vertical wall with horizontal ground surface, k_a becomes identical to Equation 8.6. Coulomb's theory is also used to establish the passive soil pressure coefficient using the linear failure surface. By considering force equilibrium of the passive wedge shown in Figure 8.13(a), and using the force diagram of Figure 8.13(b), the following expression is determined:

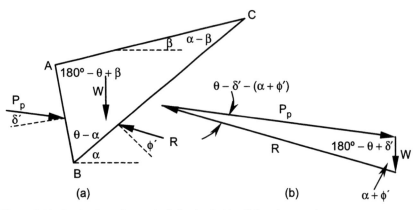

Figure 8.13. Coulomb wedge analysis in $c' = 0$, ϕ' soil (passive state).

$$k_p = \frac{\sin^2(\theta+\phi')}{\sin^2\theta\sin(\theta-\delta')\left[1-\sqrt{\dfrac{\sin(\phi'+\delta')\sin(\phi'+\beta)}{\sin(\theta-\delta')\sin(\theta-\beta)}}\right]^2} \qquad (8.21)$$

For a vertical wall ($\theta = 90°$) and a horizontal ground surface ($\beta = 0$):

$$k_p = \left[\frac{\cos\phi'}{\sqrt{\cos\delta'}-\sqrt{\sin(\phi'+\delta')\sin\phi'}}\right]^2 \qquad (8.22)$$

For a smooth ($\delta' = 0$) vertical wall with a horizontal ground surface, this equation reduces to Equation 8.9.

8.4.3   *Coulomb wedge analysis in c_u, $\phi_u = 0$ soil*

Undrained conditions in both the active and passive states may occur behind a retaining wall and the analysis is representative of a short-term stability. The forces acting on the wedge are shown in Figure 8.14(a), where T_c and T_w are the shear forces developed on the sliding surface and wall-soil interface respectively. In the active state, tension cracks are permitted to extend to a depth z_o that can be evaluated in terms of the undrained cohesion c_u and wall-soil adhesion c_w. The lateral thrust due to the water pressure in the tension cracks (P_w) may be taken into account separately. In the development of an analytical solution we first consider the active case for a vertical wall with cohesion, and a horizontal ground surface. The sliding surface terminates at the bottom of a tension crack and the reaction force R and shear resistance T_c are both mobilized along the length BD. Figure 8.14(b) shows the force diagram excluding P_w. Equilibrium of the vertical and horizontal components of the forces yields:

$$T_c\sin\alpha+T_w+R\cos\alpha-W = 0, T_c\cos\alpha-R\sin\alpha+P_a = 0.$$

Combining these equilibrium equations and replacing T_w, T_c and W with $T_w = c_w(H-z_o)$, $T_c = c_u(H-z_o)/\sin\alpha$, $W = \gamma(H^2-z_o^2)\cot\alpha/2$, and setting $\partial P_a/\partial\alpha = 0$ we obtain:

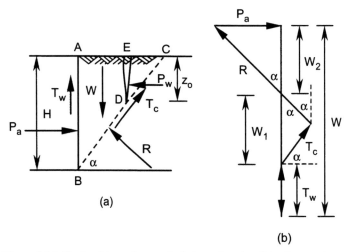

Figure 8.14. Coulomb's wedge analysis in c_u, $\phi_u = 0$ soil (active state).

$$\cot \alpha = \sqrt{1 + c_w / c_u} \tag{8.23}$$

$$P_a = \gamma(H^2 - z_0^2)/2 - 2c_u(H - z_0)\sqrt{1 + c_w/c_u} \qquad \text{for } q = 0 \tag{8.24}$$

If vertical surface loading q exists then:

$$P_a = \gamma(H^2 - z_0^2)/2 + q(H - z_0) - 2c_u(H - z_0)\sqrt{1 + c_w/c_u} \quad \text{for } q > 0 \tag{8.25}$$

Note that the magnitude of z_0 is unknown but may be estimated from Equation 8.12. Equation 8.24 is equivalent to a linear distribution of lateral earth pressure given by:

$$p_a = \gamma z - 2c_u\sqrt{1 + c_w/c_u} \tag{8.26}$$

where $z = 0$ at the ground surface and $z = H$ at the base of the wall. With the assumption of a linear lateral pressure distribution, the depth at which the earth pressure becomes zero is:

$$z_0 = (2c_u/\gamma)\sqrt{1 + c_w/c_u} \tag{8.27}$$

The total active thrust is the integral of Equation 8.26 with a correction for tension zone:

$$P_a = \gamma H(H - z_0)/2 - c_u(H - z_0)\sqrt{1 + c_w/c_u} \tag{8.28}$$

For $q > 0$, and $p_a < 0$ at $z = 0$, the equivalent linear lateral earth pressure, depth of tension crack and total active thrust are given by:

$$p_a = \gamma z + q - 2c_u\sqrt{1 + c_w/c_u} \tag{8.29}$$

$$z_0 = (2c_u/\gamma)\sqrt{1 + c_w/c_u} - q/\gamma \tag{8.30}$$

$$P_a = \gamma H(H - z_0)/2 + q(H - z_0)/2 - c_u(H - z_0)\sqrt{1 + c_w/c_u} \tag{8.31}$$

The foregoing analysis for the active state can be extended to include an inclined wall and sloping ground surface. However, due to the complexity of the mathematics, an iterative method using the variable α is preferred. For the passive state, the direction of the forces T_c and T_w are reversed and, using a similar analysis, we obtain:

$$P_p = \gamma H^2/2 + qH + 2c_u H\sqrt{1 + c_w/c_u} \tag{8.32}$$

The critical magnitude of α is given by Equation 8.23.

Example 8.5

An 8 m high wall retains a soil with the following properties: $c_u = 35$ kPa, $\phi_u = 0$, $\gamma = 19.5$ kN/m^3, and $c_w = 16$ kPa. Determine the magnitude of the active thrust P_a.

Solution:

From Equation 8.27: $z_o = (2\times35.0/19.5)\sqrt{1+16.0/35.0} = 4.33$ m.

Using Equation 8.24 (or 8.28) to compute P_a:

$P_a = 19.5(8.0^2 - 4.33^2)/2 - 2\times35.0(8.0 - 4.33)\sqrt{1+16.0/35.0} = 131.1$ kN.

From Equation 8.23 the corresponding wedge angle α at failure is:

$\cot\alpha = \sqrt{1+16.0/35.0} = 1.207 \rightarrow \alpha = 39.6°$.

The force due to the hydrostatic water pressure in the tension crack is:

$P_w = (1/2)\,9.81 \times 4.33^2 = 92.0$ kN.

Hence, the total horizontal thrust $= P_a + P_w = 131.1 + 92.0 = 223.1$ kN.

8.4.4 Coulomb wedge analysis in c', ϕ' soil

The trial wedge for the active state, together with the corresponding forces and the force diagram, is shown in Figure 8.15. Tension cracks are allowed to extend to the depth z_o that may be estimated from Equations 8.12 or 8.30. As before, these cracks may be filled with water (from rain for example) and so the force due to the hydrostatic water pressure P_w must be taken into account. For the passive failure condition where the wedge is pushed upwards, the inclination of the forces R and P_p must be selected so as to hold down the wedge and create the shear resistances in the direction opposite to the movement of the wedge. Similarly, the direction of the forces T_c and T_w are reversed.

Example 8.6

A vertical retaining wall has a height of 8 m with ground surface sloping upward at an angle of 15°. Determine the active thrust P_a due to trial wedges of soil having failure planes at 45° and 60° to the horizontal. $c' = 10$ kPa, $\phi' = 25°$, $\gamma = 18$ kN/m^3, $c_w = 0$, and $\delta' = 20°$.

Solution:

The forces acting on the trial wedge $ABDE$ are shown in Figure 8.16(a).

From Equation 8.12: $z_o = 2\times10.0\times\tan(45.0° + 25.0°/2)/18.0 = 1.74$ m.

Calculate the area of block ABC from the geometry of triangle ABC:

$AB/\sin(\alpha - 15.0°) = AC/\sin(90.0° - \alpha) \rightarrow AC = AB\cos\alpha/\sin(\alpha - 15.0°)$.

$GC = AC\cos15.0°, GC = AB\cos\alpha\cos15.0°/\sin(\alpha - 15.0°)$.

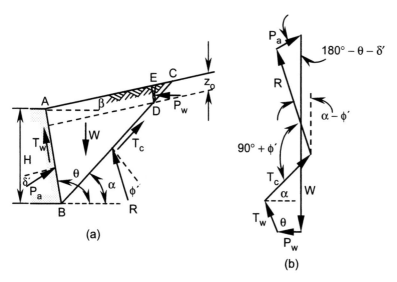

Figure 8.15. Coulomb's wedge analysis in c', ϕ' soil (active state).

$Area_{ABC} = AB \times GC/2 = AB^2 \cos\alpha\cos 15.0°/2\sin(\alpha-15.0°)$, similarly,

$Area_{EDC} = ED^2 \cos\alpha\cos 15.0°/2\sin(\alpha-15.0°)$. Calculate the weight of the wedge W:

$W = (Area_{ABC} - Area_{EDC})\gamma$, $W = 530.0533\cos\alpha/\sin(\alpha-15.0°)$.

For $\alpha = 45°$, $W = 749.6$ kN, and for $\alpha = 60°$, $W = 374.8$ kN.

The lengths BC and DC are calculated from triangles ABC and EDC:

$BC = AB\sin 105.0°/\sin(\alpha-15.0°)$, $DC = ED\sin 105.0°/\sin(\alpha-15.0°)$, thus:

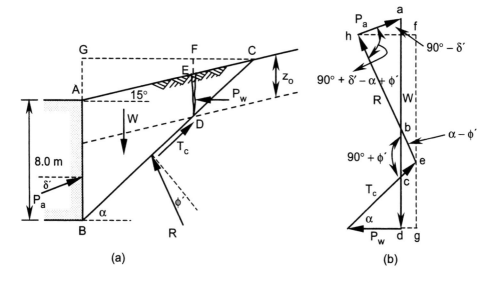

Figure 8.16. Example 8.6.

$T_c = (BC - DC) \times 1.0 \times c' \rightarrow T_c = 60.467 / \sin(\alpha - 15.0°)$.

For $\alpha = 45°$, $T_c = 120.9$ kN, and for $\alpha = 60°$, $T_c = 85.5$ kN.

Calculate: $P_w = 1.74 \times 1.74 \times 9.81 / 2 = 14.8$ kN.

Using the force diagram of Figure 8.16(b):

For $\alpha = 45°$:

$dg = T_c \cos\alpha - 14.8 = 120.9 \times \cos 45.0° - 14.8 = 70.7$ kN.

$ce = dg / \cos\alpha = 70.7 / \cos 45.0° = 100.0$ kN.

In triangle *bec*:

$100.0 / \sin(\alpha - \phi') = bc / \sin(90.0° + \phi') \rightarrow bc = 265.0$ kN.

$ab = W - bc - cd = 749.6 - 265.0 - 14.8 \tan 45.0° = 469.8$ kN.

In triangle *abh*:

$P_a / \sin(\alpha - \phi') = ab / \sin(90.0° + \delta' - \alpha + \phi')$,

$P_a / \sin(45.0° - 25.0°) = 469.8 / \sin(90.0° + 20.0° - 45.0° + 25.0°)$,

$P_a = 160.7$ kN.

For $\alpha = 60°$, $ce = 55.9$ kN, $bc = 88.3$ kN, $ab = 260.9$ kN and $P_a = 154.9$ kN.

8.4.5 The point of application of the active thrust

The precise determination of the point of application of the active thrust remains a difficult task despite significant improvements in the theoretical and experimental techniques in the analysis of retaining structures. This is mostly due to uncertainties in the lateral earth pressure distribution along the wall. The traditional method used to locate the point of application of the active thrust is shown in Figure 8.17. In the absence of any surface line load, the point of application above the base (point *D* in Figure 8.17(a)) is $1/3$ of the height of the wall. If, however, a line load is located within the critical wedge, then the point of application moves upwards so that the length $FD = 1/3 FK$. The lines *EF* and *EK* are parallel to the lines *BB'* and *BC* respectively where *BB'* makes an angle of ϕ' with the horizontal (Figure 8.17(b)). If the line load is located outside the critical wedge (Figure 8.17(c)), the length $FD = 1/3 FB$ where *EF* is parallel to *BB'*.

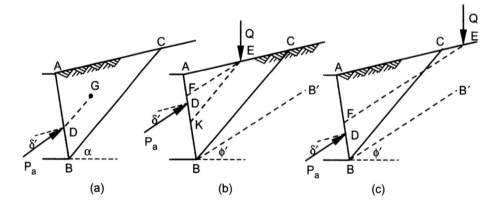

Figure 8.17. Location of the point of application of the active thrust.

8.5 COMMON TYPES OF RETAINING STRUCTURES AND FACTOR OF SAFETY

8.5.1 *Conventional or externally stabilized systems*

Retaining structures may be classified in a variety of ways. O'Rourke & Jones (1990) proposed a classification of retaining structures into two major groups, viz., externally and internally stabilized systems. An externally stabilized system uses a structural wall external to the soil as in the conventional retaining walls shown in Figure 8.18. An internally stabilized system, on the other hand, involves reinforcement installed within the backfill soil.

The gravity type wall of Figure 8.18(a) may be constructed from brick, masonry and plain or reinforced concrete. Both the front and back faces may be vertical, inclined or stepped. When the base and stem are designed and constructed as separate elements, a shear key must be included in the stem to increase the factor of safety against sliding. Similar shear keys are designed to increase the frictional and adhesion forces acting on the interface between the base and the foundation soil. The stability of a gravity wall is maintained mostly by its weight and partly by the passive resistance mobilized at the front of the wall. Assuming the wall remains intact, failure occurs by horizontal displacement or by rotation about the toe. Therefore factors of safety must be determined for both sliding and overturning of the wall. Excessive contact pressure and settlement of the toe may cause a shear failure of the foundation soil. Consequently, it is necessary to evaluate the stability of the wall and soil mass as a whole using the slope stability methods.

Cantilever walls of the type illustrated in Figure 8.18(b) are typically constructed from reinforced concrete. Such a wall is either L-shaped or T-shaped and is considered to be a flexible structure. L-shaped prestressed concrete cantilever walls are also utilized. The projection of the base (the heel portion of the wall) must be positioned inside the backfill. Shear key(s) are constructed on the cut off points and the base of the wall. Stability of the wall is maintained by its own weight and the weight of the retained soil above the heel. Depending on the height of the soil, any mobilized passive resistance at the front of the wall may also be taken into account. Long cantilever retaining walls are often supported along their length by counterforts to increase their rigidity and strength. The thickness of the base is normally governed by an allowable shear stress in the concrete; reinforcement is provided in zones of tensile stress and to control shrinkage cracks.

Figure 8.18(c) represents a gravity type retaining wall called *gabion wall* constructed from cubic (or other shapes) baskets made of metal wire or hard plastic (geogrids) and filled with coarse aggregate. If the soil behind the wall is also reinforced by geogrids, structures as high as 8 m may be built using this technique. A gabion wall may be vertical or inclined depending on the angle of internal friction of the aggregate. If the grading of the aggregate complies with the specification of a filter material then there is no need for the construction of a drainage system. Excessive contact pressure under the base of a gabion wall generates settlement that may lead to a gradual shear distortion and lateral deformation of the wall (O'Rourke, 1987). Contact pressures may be reduced by reinforced solutions. Applications of gabion walls are spreading with the proliferation of dual-purpose products (reinforcing and filtering) on the market (Pálossy et al., 1993).

A general layout of a crib wall is shown in Figure 8.18(d). It is composed of precast concrete elements that are either solid or hollow. Hollow elements may be filled with soil to provide weight and improve stability.

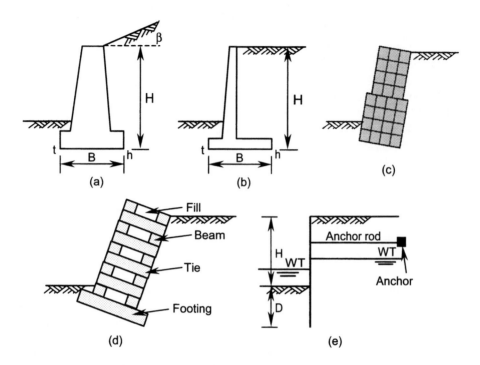

Figure 8.18. Conventional types of retaining structures: (a) gravity wall, (b) cantilever wall, (c) gabion wall, (d) crib wall, (e) cantilever or anchored sheet pile.

Construction of this type of retaining structure is rapid, as it does not require special skills. Similar to a gabion wall, there is no need to provide a drainage system and therefore this form of construction has an economic advantage over other types of retaining structures. Recently, the longitudinal beam elements at the rear of crib walls have been omitted to improve economy.

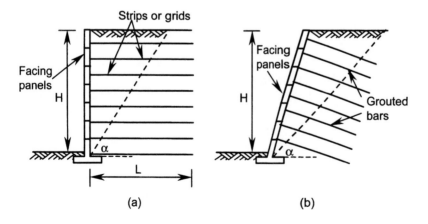

Figure 8.19. Internally stabilized earth retaining structures: (a) reinforced soil, (b) soil nailing.

If the crib wall is supported by an anchorage system its effectiveness becomes comparable with gravity and cantilever walls.

Sheet pile systems (Figure 8.18(e)) are mostly temporary retaining structures that are built to facilitate excavations or retain soils. However, the use of permanent sheet pile systems is common in both onshore and offshore structures. A sheet pile is a flexible beam constructed from concrete, timber or, most commonly, steel. Whilst precast concrete piles are quite heavy and difficult to handle and drive, they may be competitive with steel piles if they are cast close to the jobsite. Timber sheet piles are for low heights up to 3 m. The stability of a sheet pile wall is maintained by embedment depth using the mechanics of a cantilever beam. When the embedment depth is not adequate or decreases during excavation, anchors may be used to increase the stability.

8.5.2   *Improved earth walls or internally stabilized systems*

Improved earth walls consist of stabilized backfill soil and facing elements. Improvement of the soil (apart from compaction) is carried out either by means of chemicals or by using inclusions. Chemical stabilization of the soil may be realized by cement or lime or other chemicals. Soils stabilized by cement or lime normally fall into the category of a Mohr-Coulomb material with improved shear strength parameters. A mechanically stabilized soil is reinforced by strips or grids (Figure 8.19(a)) that may be metallic, polymeric or organic. A mixture of soil and polymeric elements of fine diameter and small length has also been used. The main objective is to transfer the tensile stresses to reinforcement elements. Anchored earth systems combined by soil reinforcement have been developed and applied successfully in highway construction. The in-situ reinforcement includes soil nailing and dwelling by means of grouted bars as shown in Figure 8.19(b).

In the selection of a retaining wall for a specific project, consideration should be given to the type of soil, its deformation compatibility with the retaining wall, the height of the retained soil, ground water level, construction and environmental aspects, time and cost.

8.5.3   *Factor of safety in retaining walls*

For externally stabilized systems the factors of safety against sliding and overturning are considered separately. The factor of safety against sliding is defined as the ratio of the sum of the resisting forces to the sum of the disturbing forces along the sliding surface:

$$F_S = \frac{\sum F_r}{\sum F_d} \tag{8.33}$$

Forces that resist sliding include of the passive thrust at the front of the wall, and adhesive and frictional forces mobilized on the sliding surface. The disturbing forces are the components of the active thrust in the direction of the sliding and the force due to any water pressure behind the wall.

For overturning, the factor of safety is defined as the ratio of the sum of resisting moments to the sum of disturbing moments about the toe of the retaining wall.

$$F_V = \frac{\sum M_r}{\sum M_d} \tag{8.34}$$

In most cases the adhesive and frictional forces mobilized under the base have no effect as they pass through the toe. The resultant of the contact pressure under the base N, which is equal to the total vertical forces applied to the wall, has no moment about toe because at this limiting state it approaches the toe as the wall loses contact with the foundation soil along the base. For both sliding and overturning, the factor of safety must be not less than 1.5; reference should be made to the standard codes of the relevant country.

For a reinforced earth wall a simplified traditional analysis assumes that the active thrust in the soil is fully mobilized through a linear distribution along the back face of the wall. This lateral active pressure is resisted solely by the strips or geogrids so (theoretically) there is no pressure on the facing elements. The factor of safety is applied to the tensile strength of the strips and geogrids takes due account of environmental and construction factors.

The factor of safety in a sheet pile system is applied to c', ϕ', the passive forces or k_p. For strength data of moderate reliability, a factor of safety of 1.3 to 1.5 is applied to both the cohesion and internal friction angle but for less reliable data a factor of safety of 2 is normally adopted.

8.6   STATIC ANALYSIS OF CANTILEVER AND GRAVITY RETAINING WALLS

8.6.1   *Soil pressure distribution under the footing*

The contact pressure under the footing is assumed to be linear and, ideally, compressive throughout. Maximum contact pressure occurs under the toe, while the minimum pressure occurs under the heel. If the contact pressure becomes negative at the heel then the factor of safety against overturning must be calculated ignoring unreliable properties such as wall friction and adhesion. In this case a part of the contact area becomes ineffective and the computed toe pressure increases and may exceed the allowable bearing pressure.

8.6.2   *Basic concepts of static analysis*

Typical sections of cantilever and gravity walls are shown in Figure 8.20, together with the forces applied to the walls. The weight of the wall W is comprised from the materials within the dotted lines. Thus the active thrust is calculated on the imaginary wall ab assuming that the soil at the left of ab is a rigid material attached to the rigid wall. Similarly the passive thrust is calculated on dc. The shear force T represents the adhesion and frictional forces mobilized under the base whilst the normal force N is equal to the sum of the vertical forces applied to the wall system:

$$T = \frac{1}{F_S}(Bc_b + N\tan\delta_b), \quad N = W + P_{av} + \frac{P_{pv}}{F_S} + Q \tag{8.35}$$

where c_b and δ_b are the maximum values of adhesion and friction angle mobilized under the base respectively, F_S is the factor of safety against sliding, P_{av} is the vertical component of the active thrust, P_{pv} is the vertical component of the passive thrust and Q is the resultant of the vertical surface load applied at the upper boundary of the dotted lines.

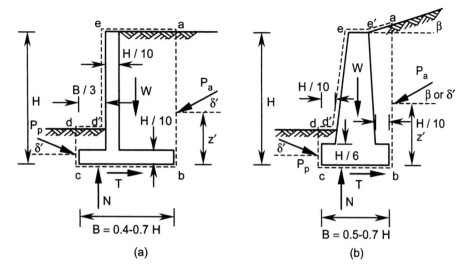

Figure 8.20. Sections and free-body diagrams of typical cantilever and gravity retaining walls.

It is assumed that all the resisting forces have been decreased by the ratio $1 / F_S$ to maintain equilibrium.

To calculate the active thrust on the imaginary wall *ab*, the concept of a linear earth pressure distribution can be employed in the following form:

$$p_a = (\gamma z + q)k_a - 2c'\sqrt{k_a(1+\frac{c_w}{c'})} \qquad (8.36)$$

where k_a is defined by Equation 8.19 and c_w is the cohesion mobilized on *ab*. In the calculation of k_a a reasonable value for δ' on *ab* must be assumed. For the passive case the linear pressure distribution is given by:

$$p_p = (\gamma z + q)k_p + 2c'\sqrt{k_p(1+\frac{c_w}{c'})} \qquad (8.37)$$

where k_p is defined by Equation 8.21. In a $c' = 0$, ϕ' soil and with the assumption of $\delta' = \beta$, k_a and k_p are calculated from Equations 8.14 and 8.17; and the term $(\gamma z + q)$ must be multiplied by $\cos\beta$. The point of application of the active or passive thrust is determined from the relevant earth pressure diagrams. Alternatively, a Coulomb wedge analysis may be used to obtain the active and passive thrusts. The point of application of the active thrust can be determined using the method of Section 8.4.5 or may be taken as $0.4H$ above the base. Stability criteria for both analyses include:

1. N must be located within the middle third of the base to avoid tensile stress under the heel.

2. The contact pressure at the toe must be equal to or less than the allowable bearing pressure.

3. The settlement of the toe must be within the tolerable limits.

4. The wall system must be safe against sliding and overturning.

The long-term stability of a wall may be investigated using the concept of at-rest earth pressure. In this instance, the total thrust on the wall is calculated using k_o from Equation 8.4. For walls with a granular backfill and founded on rock, Duncan et al. (1990) suggested $k_o = 0.45$ for a compacted backfill, and $k_o = 0.55$ for an uncompacted backfill. Stability criteria were the same as above but with N permitted to be in the middle half of the base.

Regardless of the linear stress distribution along the line ab, the point of application of the active thrust may be assumed to be $0.4H$ above the base. For the walls with clayey soils in the backfill and foundation, none of the methods above give reliable predictions due to creep of the material. As a result, the earth pressure charts given by Terzaghi & Peck (1967) are more reliable as they are based on field experiments. However, the chart solution does not provide a value for vertical component of the active thrust.

Example 8.7

For the cantilever retaining wall shown in Figure 8.21(a) determine: (a) the soil pressure distribution along the vertical plane ab, (b) the factors of safety against sliding and overturning and (c) the maximum and minimum contact pressures under the base. For the backfill soil: $c' = 0$, $\phi' = 30°$, and $\gamma = 17$ kN/m^3. The angle of friction mobilized under the base is $20°$ and there is no adhesion. The unit weight of the wall material is 24 kN/m^3.

Solution:

(a) Calculate k_a from Equation 8.14:

$$k_a = \frac{\cos 20.0° - \sqrt{\cos^2 20.0° - \cos^2 30.0°}}{\cos 20.0° + \sqrt{\cos^2 20.0° - \cos^2 30.0°}} = 0.441.$$ Using Equation 8.36:

At $z = 0$, $p_a = 0$. At $z = 4.5$ m, $p_a = \gamma z k_a \cos\beta = 17.0 \times 4.5 \times 0.441 \times \cos 20.0° = 31.7$ kPa.
$P_a = 31.7 \times 4.5/2 \times 1.0 = 71.3$ kN, and its location is $4.5/3 = 1.5$ m above the base.

$P_{a(horizontal)} = P_{ah} = P_a \cos\beta = 71.3 \times \cos 20.0° = 67.0$ kN.

$P_{a(vertical)} = P_{av} = P_a \sin\beta = 71.3 \times \sin 20.0° = 24.4$ kN.

Calculate the passive force in the front of the wall by assuming $\delta' = 0$:

$k_p = (1 + \sin 30.0°)/(1 - \sin 30.0°) = 3.0$. At $z = 0$ (in the front of the wall), $p_p = 0$.

At $z = 1.1 + 0.4 = 1.5$ m, $p_p = \gamma z k_p + 2c' \sqrt{k_p} = 1.5 \times 17.0 \times 3.0 = 76.5$ kPa.

$P_p = 1.5 \times 76.5/2 \times 1.0 = 57.4$ kN and is located 0.5 m above the base of the wall.

Figure 8.21(b) illustrates the results.

(b) Calculate the total weight W (see Figure 8.21(b) for a definition of W_1 to W_4):

$W_1 = 1.3 \times 4.1 \times 1.0 \times 17.0 = 90.6$ kN at 1.85 m from the toe.

$W_2 = 0.4 \times 4.1 \times 1.0 \times 24.0 = 39.4$ kN at 1.0 m from the toe.

$W_3 = 0.4 \times 2.5 \times 1.0 \times 24.0 = 24.0$ kN at $2.5/2 = 1.25$ m from the toe.

$W_4 = (2.5 - 1.3 - 0.4) \, 1.1 \times 1.0 \times 17.0 = 15.0$ kN at 0.4 m from the toe.

Thus: $W = 90.6 + 39.4 + 24.0 + 15.0 = 169.0$ kN.

$N = W + $ vertical component of the active thrust $= 169.0 + 24.4 = 193.4$ kN.

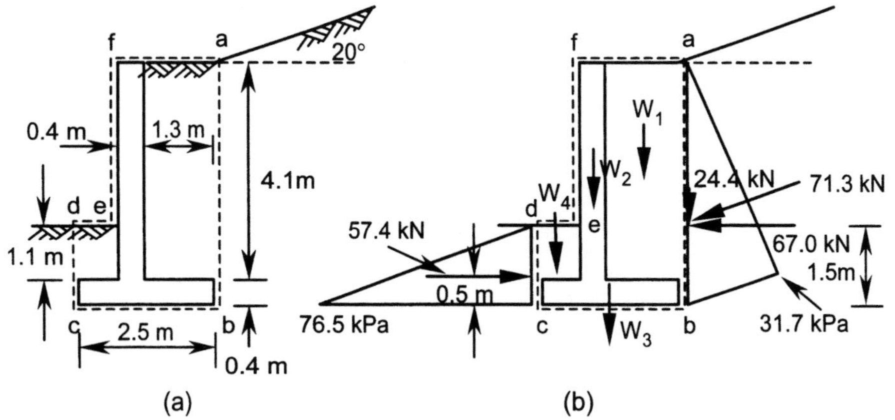

Figure 8.21. Example 8.7.

Shear resistance due to friction between the base and soil $= 193.4 \times \tan 20.0° = 70.4$ kN.
Total force resisting the active thrust $= 70.4 + 57.4 = 127.8$ kN, and the factor of safety against sliding is:
$F_S = 127.8 / 67.0 = 1.91$. From Equation 8.34:
$F_V = (57.4 \times 0.5 + 90.6 \times 1.85 + 39.4 \times 1.0 + 24.0 \times 1.25 + 15.0 \times 0.4 + 24.4 \times 2.5) / 67.0 \times 1.5$
$F_V = 3.31$.
(c) Moment equilibrium about the heel:
$193.4x = 67.0 \times 1.5 + 90.6(2.5 - 1.85) + 39.4(2.5 - 1.0) + 24.0(2.5 - 1.25)$
$+15.0(2.5 - 0.4) - (57.4 / 1.91) \times 0.5 = 264.96 \rightarrow x = 1.37, e = 1.37 - 1.25 = 0.12$ m.
Making use of Equations 5.98:
$q_{max} = (193.4 / 2.5 \times 1.0)(1.0 + 6 \times 0.12 / 2.5) = 99.6$ kPa (at point c, Figure 8.21(b)).
$q_{min} = (193.4 / 2.5 \times 1.0)(1.0 - 6 \times 0.12 / 2.5) = 55.1$ kPa (at point b, Figure 8.21(b)).

Example 8.8

A gravity retaining wall 7 m high retains a backfill soil with $c' = 20$ kPa, $\phi' = 18°$, and $\gamma = 18$ kN/m^3. A surcharge load of $q = 30$ kPa is applied to the horizontal upper ground surface. The thickness of the soil in the front of the wall to provide the passive resistance is 3 m (Figure 8.22). Determine: (a) the soil pressure distribution along the wall, (b) the factors of safety against sliding and overturning, and (c) the maximum and minimum contact pressures under the base. Re-solve parts (b) and (c) if the tension cracks are filled with rainwater. Assume no base friction and adhesion. The unit weight of the gravity wall is 24 kN/m^3.

Solution:

(a) For a smooth wall $k_a = (1 - \sin 18.0°) / (1 + \sin 18.0°) = 0.528$, $k_p = 1 / k_a = 1.894$.
At $z = 0$, $p_a = qk_a - 2c'\sqrt{k_a} = 30.0 \times 0.528 - 2 \times 20.0 \times \sqrt{0.528} = -13.2$ kPa.
From Equation 8.12, the position of the point of zero pressure is:
$z_o = 2 \times 20.0 \tan(45.0° + 18.0° / 2) / 18.0 - 30.0 / 18.0 = 1.39$ m;
the tensile force above this point is ignored. At $z = 7.0$ m,
$p_a = (7.0 \times 18.0 + 30.0) \times 0.528 - 2 \times 20.0 \times \sqrt{0.528} = 53.3$ kPa.

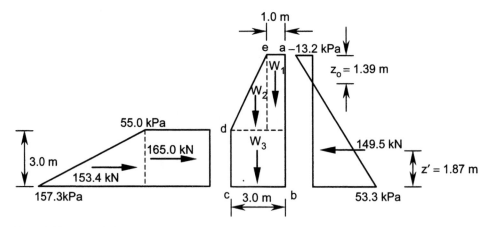

Figure 8.22. Example 8.8.

$P_a = (7.0 - 1.39) \times 53.3/2 \times 1.0 = 149.5$ kN and its location is $(7.0 - 1.39)/3 = 1.87$ m above the base of the wall. Calculate the passive force in the front of the wall:

At $z = 0$ (in the front of the wall), $p_p = 2 \times 20.0 \times \sqrt{1.894} = 55.0$ kPa.

At $z = 3.0$ m, $p_p = 3.0 \times 18.0 \times 1.894 + 2 \times 20.0 \times \sqrt{1.894} = 157.3$ kPa.

$P_{p1} = 3.0 \times 55.0 \times 1.0 = 165.0$ kN, and is located 1.5 m above the base.

$P_{p2} = 3.0(157.3 - 55.0)/2 \times 1.0 = 153.4$ kN, and is located 1.0 m above the base.

Figure 8.22 illustrates the results.

(b) The weight of the concrete wall has the three components shown in Figure 8.22. Using the geometry of the wall section we find that $W_1 = W_2 = 96.0$ kN, and $W_3 = 216.0$ kN. The horizontal distances of W_1, W_2 and W_3 from the heel of the wall are 0.5 m, 1.67 m and 1.5 m respectively. Thus $W = 96.0 + 96.0 + 216.0 = 408.0$ kN.

Total resistance force against active thrust $= 165.0 + 153.4 = 318.4$ kN,

$F_S = 318.4 / 149.5 = 2.13$. From Equation 8.34:

$F_V = 165.0 \times 1.5 + 153.4 \times 1.0 + 96.0 \times 2.5 + 96.0 \times 1.33 + 216.0 \times 1.50 / (149.5 \times 1.87) = 3.91$.

(c) Mobilized passive forces:

$P_{p1} = 165.0/2.13 = 77.5$ kN and $P_{p2} = 153.4/2.13 = 72.0$ kN.

This ensures horizontal equilibrium of the wall. Moment equilibrium about the heel:

$408.0x = 149.5 \times 1.87 + 96.0 \times 0.5 + 96.0 \times 1.67 + 216.0 \times 1.50 - 77.5 \times 1.5 - 72.0 \times 1.0 = 623.6$

$x = 623.6/408.0 = 1.528$ m, $e = 1.528 - 1.50 = 0.028$ m.

Note that the resultant of the contact pressure is located on the left side of the centroid and thus the maximum contact pressure will be under the toe. Making use of Equation 5.98:

$q_{max} = (408.0/3.0 \times 1.0)(1.0 + 6 \times 0.028/3.0) = 143.6$ kPa (at point c, Figure 8.22).

$q_{min} = (408.0/3.0 \times 1.0)(1.0 - 6 \times 0.028/3.0) = 128.4$ kPa (at point b, Figure 8.22).

$P_w =$ force due to water $= \gamma_w \times z_o \times z_o/2 \times 1.0 = 9.81 \times 1.39 \times 1.39/2 \times 1.0 = 9.5$ kN.

Total thrust on the back of the wall $= P_a + P_w = 149.5 + 9.5 = 159.0$ kN.

$F_S = 318.4/159.0 = 2.0$.

$$F_V = \frac{165.0 \times 1.5 + 153.4 \times 1.0 + 96.0 \times 2.5 + 96.0 \times 1.33 + 216.0 \times 1.50}{149.5 \times 1.87 + 9.5(7.0 - 1.39 \times 2/3)} = 3.24.$$

The mobilized passive forces are:

$P_{p1} = 165.0/2.0 = 82.5$ kN and $P_{p2} = 153.4/2.0 = 76.7$ kN.

Similarly: $x = 1.640$ m, $e = 1.64 - 1.50 = 0.14$ m, $q_{max} = 174.1$ kPa, and $q_{min} = 97.9$ kPa.

8.7 STATIC ANALYSIS OF SHEET PILE WALLS

8.7.1 *Basic concepts*

A cantilever sheet pile wall is a flexural structure and its stability depends entirely on the mobilized passive resistance within the embedment depth. The recommended maximum height for a cantilever sheet pile is 8 m. Figure 8.23(a) shows a mechanism of failure due to the rotation of the sheet pile about point s. This point is close to the end point and can be determined by static equilibrium. As a result of the rotation, the state of the stress below point s changes on both sides and there is a transition from active to passive at the back and from passive to active at the front as idealized in Figure 8.23(b). The two unknowns of embedment depth D and z_s can be found by force and moment equilibrium providing that the mobilized shear strength parameters on both sides of the sheet pile are known. In a c', ϕ' soil this requires the iterative solution of two sets of rather complex equations in terms of D and z_s. Alternatively the simplified distribution of Figure 8.23(c) can be adopted and the stability can be formulated to yield the embedment depth. The formulations for D in both $c' = 0$, ϕ' and c_u, $\phi_u = 0$ soils are simple and will be discussed in the following sections.

An anchored sheet pile wall is shown in Figure 8.24(a). The recommended maximum height for an anchored sheet pile is 15 m. The anchor rod is either horizontal or inclined and its depth from the ground surface is 1 m to 2 m. These rods are usually spaced 2 m to 4 m apart. The active earth pressure in the backfill is resisted by both the passive earth pressure in the front and the anchor rod. This allows a reduction in the embedment or

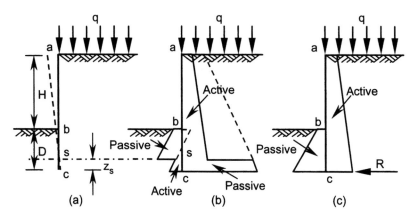

Figure 8.23. Cantilever sheet pile: (a) failure mechanism, (b) idealized lateral earth pressure distribution, (c) simplified lateral earth pressure distribution.

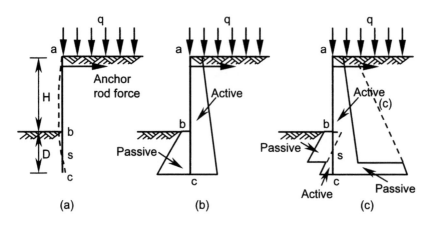

Figure 8.24. Anchored sheet pile: (a) failure mechanism, (b) lateral earth pressure distribution: free earth support method, (c) lateral earth pressure distribution: fixed earth support method.

an increase in the height and, as a consequence, this technique is used extensively in excavations and waterfront constructions.

The simple mechanism of Figure 8.24(a) allows rotation about the anchor rod level without any lateral displacement at this point. Active earth pressure is fully mobilized along the depth of the backfill. At the front of the sheet pile there is only the passive resistance (Figure 8.24(b)). As there is no point of contraflexure the equivalent beam analysis, which is referred to as *free earth support*, is statically determinate.

In dense granular materials the anchored sheet pile tends to rotate about point *s*, which is usually taken as the point where the bending moment is zero. This rotation causes passive and active states at the back and the front of the wall respectively similar to the cantilever sheet pile (Figure 8.23(b)). The idealized earth pressure distribution is shown in Figure 8.24(c). Having the three unknowns of D, z_s and the anchor rod force, the problem is statically indeterminate unless the position of point *s* is estimated from the equivalent beam method solved by the theory of elasticity (Terzaghi, 1966). This analysis is referred to as *fixed earth support* method. If the number of anchors down the depth exceeds one then the analysis is statically indeterminate and the stiffness properties of the sheet pile and soil have to be included in the analysis.

The active and passive earth pressures are computed from the general linear earth pressure distribution expressed by Equations 8.36 and 8.37. It is convenient to combine the lateral earth pressures of both sides and construct a net pressure diagram. The weight of the sheet pile is ignored and only the horizontal components of the active and passive thrusts and anchor rod force are considered.

An analysis of both cantilever and anchored sheet piles can be carried out by assuming that the full active and passive states are mobilized. The computed embedment depth may be increased by 20% to 40% (this must comply with the standard codes) to ensure safety, or alternatively the shear strength parameters or k_p can be reduced. Any change in the embedment depth will change the lateral stress distribution below the dredge line. The factor of safety applied to k_p in a c', ϕ' soil must be in the range 1.5 to 2.25 depending on the magnitude of the effective internal friction angle ϕ' (Fleming et al., 1992). Lower values of the factor of safety correspond to lower values of ϕ'. For a c_u, $\phi_u = 0$ soil (under the dredge line), Burland et al. (1981) suggested a factor of safety of 2 should be applied to c_u on both sides under the dredge line. In both cantilever and anchored sheet piles the application of the factor of safety to the shear strength parameters or, alternatively, increasing the length of the embedment, will result in the mobilization of only part of the passive resistance. Consequently, the ideal condition of a linear earth pressure distribution, which applies to a material on the verge of failure, is no longer valid. This will cause an overestimation of the bending moments and therefore the maximum bending moment has to be reduced.

The traditional correction method is based on the moment reduction factors suggested by Rowe (1952 and 1957). In this method, which is based on experimental results, a relationship was developed between the embedment ratio $\alpha = H / (H + D)$, the anchor rod position, a flexibility number $\rho = (H + D)^4 / EI$ and the bending moment reduction factor. It is recommended that the method be used only if a factor of safety is applied to the passive resisting forces. For more flexible walls and lower values of k_o, the results of the traditional methods become more reliable. Potts & Fourie (1984 and 1985) applied a numerical analysis to investigate a single anchored retaining wall using finite elements that obeyed an elasto-plastic constitutive law. The position of the anchor rod was selected at the top of the sheet pile and the elastic properties that varied linearly with depth were known. It was found that a traditional method using the concept of a linear lateral earth pressure distribution or simple limit equilibrium calculations provide a reliable estimate of the embedment depth. Although the k_o of the soil had little effect in backfill walls, higher values of k_o dominated the behaviour of excavated walls. Lower values of k_o correspond to a reduction in the bending moments and anchor rod force. Figure 8.25 shows the effects of wall stiffness, expressed in terms of the flexibility number ρ, on the bending moments and anchor rod force for the case where the anchor rod is positioned at the top of the sheet pile. In general the major shortcomings of traditional analyses are the uncertainty in the displacement behaviour, and the distribution of the earth pressure. If the cost can be justified a numerical approach should be considered.

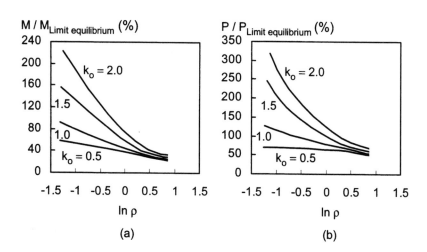

Figure 8.25. Correction of bending moments and anchor rod force (Potts & Fourie, 1985).

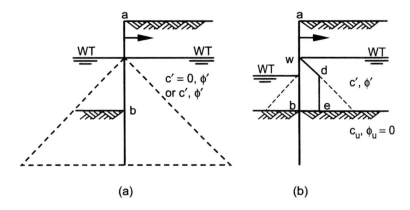

(a)   (b)

Figure 8.26. Hydrostatic pore pressure distribution in sheet pile walls.

8.7.2   *Pore pressure distribution*

In the cases when the water table has different levels in the backfill and at the front of the sheet pile, approximations are made to include the unbalanced force due to this difference together with the seepage pressure. In a simplified analysis it is assumed that in the immediate vicinity of the wall water moves vertically downward in the backfill and upward at the front of the wall. The hydraulic gradients at both sides are assumed to be equal and constant with depth, which means equal velocity and seepage pressure on both sides. Alternatively, a flow net can be constructed to evaluate the pore pressure and seepage pressure distribution. For equal levels of the water table (Figure 8.26(a)) the active and passive earth pressures are calculated using the submerged unit weight γ' and there is no unbalanced pore pressure. If the soil under the dredge line is in undrained conditions then the hydrostatic pressures are distributed according to Figure 8.26(b). The unbalanced net pore pressure diagram is the area of *wdeb*. The pore pressure distribution shown in Figure 8.27(a) is linear on both sides creating equal hydraulic gradients of i_R and i_L on the right and the left respectively:

$$i_R = \frac{(h+d)-u_c/\gamma_w}{h+d}, \quad i_L = \frac{u_c/\gamma_w - d}{d}.$$

where u_c is the pore pressure at the base and parameters h and d are defined in Figure 8.27(a). The pore pressure at the base is found by equating the two equations above:

$$u_c = \frac{2d(h+d)\gamma_w}{h+2d} \tag{8.38}$$

The unbalanced pore pressure diagram is the area *wdc*. The net pore pressure at point *d* is:

$$u_d = \frac{2hd\gamma_w}{h+2d} \tag{8.39}$$

On the backfill side u_c is lower than the hydrostatic pressure and causes an increase in the submerged unit weight equal to the average seepage pressure along the wall. At the front of the sheet pile the submerged unit weight decreases by the same amount. The seepage pressure is defined by Equation 3.10:

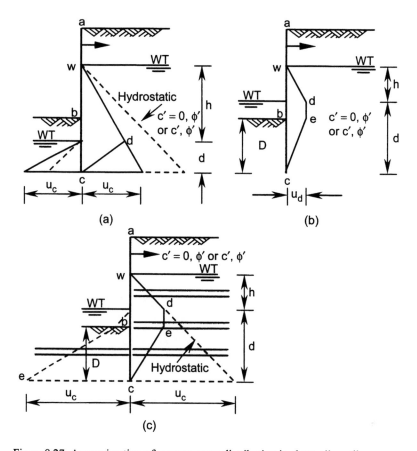

Figure 8.27. Approximation of pore pressure distribution in sheet pile walls.

$$j = i\gamma_w = \frac{h}{h+2d}\gamma_w \tag{8.40}$$

The effective unit weights used in the calculation of the lateral earth pressures are γ_R for the backfill and γ_L at the front of the wall:

$$\gamma_R = \gamma' + J, \; \gamma_L' = \gamma' - J \tag{8.41}$$

If, at the front of the wall, the water table is above the ground surface (Figure 8.27(b)), the unbalanced pore pressure diagram is the area of *wdec*, where:

$$u_d = \frac{h(d+D)}{h+d+D}\gamma_w \tag{8.42}$$

Using a similar approach, the average seepage pressure within the thickness of h is:

$$J = \frac{h}{h+d+D}\gamma_w \tag{8.43}$$

Figure 8.27(c) shows a section of a clayey soil with thin layers of a granular material on both sides of the sheet pile wall.

The pore pressure u_c is assumed to be equal to the hydrostatic pressure calculated from the backfill whilst the seepage pressure is applied only to the soil at the front of the sheet pile. The unbalanced pore pressure is represented by *wdec* where u_d is the hydrostatic pressure at *d* and the point *e* is at the level of the lower ground surface. This pressure dissipates linearly at the front of the sheet pile within the embedment depth so that:

$$J = \frac{h\gamma_w}{D}$$
(8.44)

8.7.3 Cantilever sheet pile in $c' = 0$, ϕ' soil

A cantilever sheet pile in a $c' = 0$, ϕ' soil is shown in Figure 8.28. Under the dredge line the active and passive pressures are combined to represent the net pressure diagram. All the active forces above the zero pressure point *o* are represented by their resultant R_a located at a distance $\bar{z}$ above this point. At point *o* the active and passive pressures acting from opposite sides are equal. If there is no water in the vicinity of the wall, the depth of this point from the dredge line z_o is given by:

$$z_o = \frac{p_d}{\gamma(k_p - k_a)}$$
(8.45)

where p_d is the active earth pressure at the level of the dredge line (point *b*). If water exists on both sides of the sheet pile, then:

$$z_o = \frac{p_d}{\gamma'(k_p - k_a)}$$
(8.46)

where $\gamma' = \gamma_{sat} - \gamma_w$ and k_a, k_p are the earth pressure coefficients corresponding to the drained internal friction angle. Horizontal equilibrium of the forces and moment equilibrium about point *c* yields the following equation in terms of *z*:

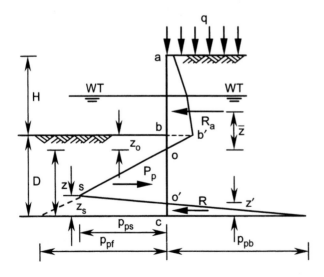

Figure 8.28. Net pressure diagram for a cantilever sheet pile wall in $c' = 0$, ϕ'soil.

$$z^4 + \frac{p_o}{C}z^3 - \frac{8R_a}{C}z^2 - \left[\frac{6R_a}{C^2}(2\bar{z}C + p_o)\right]z - \frac{6R_a\bar{z}p_o + 4R_a^2}{C^2} = 0 \tag{8.47}$$

where p_o and C are defined by:

$$p_o = \gamma(H + z_o)k_p - \gamma z_o k_a, \quad C = \gamma(k_p - k_a) \tag{8.48}$$

and z_s is calculated from:

$$z_s = (p_{pf}z - 2R_a)/(p_{pf} + p_{pb}) \tag{8.49}$$

A trial and error method usually provides a rapid solution. The total required length of the pile is: $L = H + D$, where $D = z_o + z$. A simplified solution is obtained if the net passive pressure R is replaced by a concentrated horizontal force acting at the base of the sheet pile ($z_s = 0$).

$$z^3 - \frac{6z_o R_a}{p_d}z - \frac{6R_a\bar{z}z_o}{p_d} = 0 \tag{8.50}$$

Example 8.9

A sheet pile supports 6 m of soil with the following properties: $c' = 0$, $\phi' = 30°$, $\delta' = 17°$, $\gamma = 17.3$ kN/m^3, and $\gamma' = 9.5$ kN/m^3. The water table is on both sides of the sheet pile and is located 3 m below the upper ground surface. Find the length of the embedment D by applying a factor of safety of 1.5 to k_p at both sides. For the computed embedment depth determine the location and magnitude of the maximum bending moment.

Solution:

For $\phi' = 30°$ and $\delta' = 17°$, $k_a = 0.299$, $k_p = 5.385$. The earth pressure along the wall is inclined at 17° to the horizontal. We can introduce the horizontal earth pressure coefficients of k_{ah} and k_{ph} providing that $c' = 0$: $k_{ah} = 0.299 \times \cos 17.0° = 0.286$, $k_{ph} = 5.385 \times \cos 17.0° = 5.150$.

Construct the soil pressure diagram above the dredge line and compute z_o, R_a, and $\bar{z}$:
$dd' = \gamma z k_{ah} = 17.3 \times 3.0 \times 0.286 = 14.8$ kPa.
$bb' = dd' + 9.5 \times 3.0 \times 0.286 = 23.0$ kPa. $F_1 = (14.8 \times 3.0)/2 \times 1.0 = 22.2$ kN.
$F_2 = 14.8 \times 3.0 \times 1.0 = 44.4$ kN. $F_3 = (23.0 - 14.8)/2 \times 3.0 \times 1.0 = 12.3$ kN.
The mobilized passive pressure coefficient $k_{pm} = 5.150 / 1.5 = 3.433$.

$$z_o = \frac{p_d}{C} = \frac{23.0}{9.5(3.433 - 0.286)} = 0.77 \text{ m}, \quad F_4 = (23.0 \times 0.77)/2 \times 1.0 = 8.8 \text{ kN}.$$

$R_a = F_1 + F_2 + F_3 + F_4 = 22.2 + 44.4 + 12.3 + 8.8 = 87.7$ kN.
Taking moments about point o:
$R_a \times \bar{z} = F_1(1.0 + 3.0 + 0.77) + F_2(1.5 + 0.77) + F_3(1.0 + 0.77) + F_4(0.77 \times 2/3)$,
$87.7 \times \bar{z} = 22.2 \times 4.77 + 44.4 \times 2.27 + 12.3 \times 1.77 + 8.8 \times 0.77 \times 2/3 = 232.97$,
$\bar{z} = 232.97/87.7 = 2.66$ m. Find p_o (from Equation 8.48) and necessary coefficients:
$p_o = (17.3 \times 3.0 + 9.5 \times 3.77) \times 3.433 - 9.5 \times 0.77 \times 0.286 = 299.0$ kPa.
$p_o/C = 299.0/9.5(3.433 - 0.286) = 10.0$, $8R_a/C = 8 \times 87.7/9.5(3.433 - 0.286) = 23.47$.

$6R_a/C^2 = 6\times 87.7/[9.5(3.433-0.286)]^2 = 0.5887.$

$2\bar{z}C+p_o = 2\times 2.66\times 9.5(3.433-0.286)+299.0 = 458.05.$

$6R_a(2\bar{z}C+p_o)/C^2 = 0.5887\times 458.05 = 269.65.$

$(6R_a\bar{z}p_o+4R_a^2)/C^2 = (6\times 87.7\times 2.66\times 299.0+4\times 87.7^2)/[9.5(3.433-0.286)]^2,$

$(6R_a\bar{z}p_o+4R_a^2)/C^2 = 502.65.$

Substituting into Equation 8.47: $z^4+10.0z^3-23.47z^2-269.65z-502.65 = 0,$

Iteration yields: $z = 5.61$ m. Thus $D = z+z_o = 5.61+0.77 = 6.38$ m,

$L = 6.00+6.38 = 12.38$ m. Calculate P_p and R:

$p_{pf} = Cz = 9.5(3.433-0.286)\times 5.61 = 167.7$ kPa,

$p_{pb} = p_o+Cz = 299.0+167.7 = 466.7$ kPa.

$z_s = (p_{pf}z-2Ra)/(p_{pf}+p_{pb}) = (167.7\times 5.61-2\times 87.7)/(167.7+466.7) = 1.21$ m.

$p_{ps} = C(z-z_s) = 9.5(3.433-0.286)(5.61-1.21) = 131.5$ kPa.

To find $o'c$ we use the geometry of Figure 8.28 (or 8.29):

$z_s/o'c = (p_{ps}+p_{pb})/p_{pb}$, $1.21/o'c = (131.5+466.7)/466.7 \rightarrow o'c = 0.94$ m.

$P_p = p_{ps}(5.61-0.94)/2\times 1.0 = 131.5\times 4.67/2 = 307.0$ kN.

$R = 0.94\times 466.7/2\times 1.0 = 219.3$ kN.

Note that the distance of R from the end point is $0.94/3 = 0.31$ m. Figure 8.29 shows the net pressure distribution. The maximum bending moment occurs at a point where shear force is zero. The net lateral stress below point o is represented by line os and is equal to Cx, where x is the vertical distance of the point of interest blow point o.

Thus the shear force $SF = -R_a+Cx\times x/2\times 1.0 = -R_a+Cx^2/2,$

$SF = -87.7+9.5(3.433-0.286)x^2/2 = 0 \rightarrow x = \sqrt{2\times 87.7/[9.5(3.433-0.286)]} = 2.42$ m.

The magnitude of bending moment at this point is:

$BM = -87.7(2.66+2.42)+[9.5(3.433-0.286)2.42^2/2](2.42\times 1/3) = -374.9$ kN.m.

This means that the back of the sheet pile at this point will be in tension, while its front will be in compression.

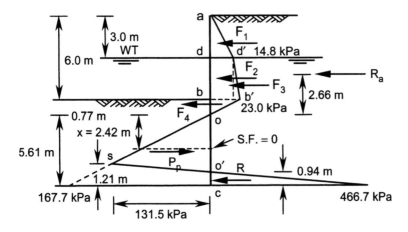

Figure 8.29. Example 8.9.

Example 8.10

Re-work Example 8.9 assuming that the passive force R behind the wall acts as a concentrated force at the bottom of the sheet pile.

Solution:

Calculate z from Equation 8.50:

$$z^3 - 6 \times 0.77 \times 87.7 / 23.0 z - 6 \times 87.7 \times 2.66 \times 0.77 / 23.0 = 0,$$

$z^3 - 17.62z - 46.86 = 0$, $z = 5.17$ m. $D = 5.17 + 0.77 = 5.94$ m, take $D = 6.0$ m.
$P_{pf} = Cz = 9.5(3.433 - 0.286) \times 5.17 = 154.6$ kPa, $P_p = 154.6 \times 5.17 / 2 \times 1.0 = 399.6$ kN.

$R = P_p - R_a = 399.6 - 87.7 = 311.9$ kN. Maximum bending moment remains the same.

8.7.4 *Cantilever sheet pile in c_u, $\phi_u = 0$ soil*

The backfill material is a c', ϕ' or preferably $c' = 0$, ϕ' soil but the soil below the dredge line is a purely cohesive soil with c_u and $\phi_u = 0$. The net pressure diagram is shown in Figure 8.30. Noting that in the clay layer $k_a = k_p = 1$, the net pressure at b becomes: $4c_u - q_d$, where q_d is the effective vertical stress at the dredge line level at the back of the sheet pile. As the active and passive pressure coefficients under the dredge line are equal, the effect of the weight disappears and the net pressure below point b remains constant up to the point of rotation. It can be shown that the net pressure at the end point at the back of the wall is $4c_u + q_d$. For equilibrium of the horizontal forces and moments about the base:

$$z_s = \frac{D(4c_u - q_d) - R_a}{4c_u} \tag{8.51}$$

$$D^2 - \frac{2R_a}{4c_u - q_d} D - \frac{R_a(12c_u \bar{z} + R_a)}{(4c_u - q_d)(2c_u + q_d)} = 0 \tag{8.52}$$

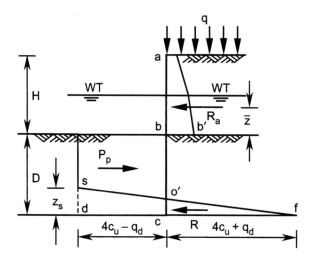

Figure 8.30. Net pressure diagram for cantilever sheet pile wall in c_u, $\phi_u = 0$ soil.

For the simplified pressure diagram ($z_s = 0$, and R acting as a concentrated force at c):

$$D^2 - \frac{2R_a}{4c_u - q_d}D - \frac{2R_a\bar{z}}{4c_u - q_d} = 0 \tag{8.53}$$

The computed D values are increased by 20% to 40% or, alternatively, the mobilized cohesion $c_{um} = c_u / F$ is substituted for c_u where $F = 1.5$ to 2.0. For a stable sheet pile $4c_{um} - q_d > 0$ or:

$$\frac{4c_u}{F} > q_d \tag{8.54}$$

Example 8.11

Re-solve Example 8.9 assuming that the soil under the dredge line is in undrained conditions with $c_u = 60$ kPa and $\phi_u = 0$. Assume full passive resistance is mobilized.

Solution:

Calculate the resultant of the active forces above the dredge line (see Example 8.9):
$R_a = F_1 + F_2 + F_3 = 22.2 + 44.4 + 12.3 = 78.9$ kN. Distance of R_a from the dredge line:
$78.9 \times \bar{z} = 22.2 \times 4.0 + 44.4 \times 1.5 + 12.3 \times 1.0 = 167.7$, $\bar{z} = 2.12$ m.

Calculate the necessary coefficients:
$4c_u - q_d = 4 \times 60.0 - (17.3 \times 3.0 + 9.5 \times 3.0) = 159.6$ kPa, $2R_a = 157.8$ kN/m,
$4c_u + q_d = 320.4$ kPa.

$$\frac{R_a(12c_u\bar{z} + R_a)}{2c_u + q_d} = \frac{78.9(12 \times 60.0 \times 2.12 + 78.9)}{2 \times 60.0 + (17.3 \times 3.0 + 9.5 \times 3.0)} = 632.0.$$

Substituting into Equation 8.52: $159.6D^2 - 157.8D - 632.0 = 0$, $D = 2.54$ m.
Using Equation 8.51, $z_s = (2.54 \times 159.6 - 78.9)/(4 \times 60.0) = 1.36$ m.
From similarity concept:

$$\frac{o'c}{4c_u + q_d} = \frac{1.36}{(4c_u + q_d) + (4c_u - q_d)}, \quad \frac{o'c}{320.4} = \frac{1.36}{320.4 + 159.6}, \quad o'c = 0.91 \text{ m}.$$

$P_p = 159.6 \times [(2.54 - 1.36) + (2.54 - 0.91)]/2 \times 1.0 = 224.2$ kN.
$R = (320.4 \times 0.91)/2 \times 1.0 = 145.8$ kN. The net pressure distribution along the wall is shown in Figure 8.31. Apply a 30% increase on D: total $D = 2.54 \times 1.30 = 3.3$ m.

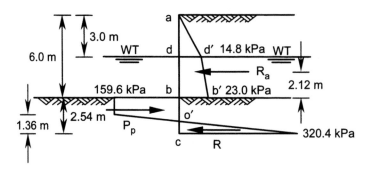

Figure 8.31. Example 8.11.

Example 8.12

Re-work Example 8.11 assuming that the passive force R behind the wall acts as a con-centrated force at the base of the sheet pile. Increase the computed embedment depth by 30% and calculate the factor of safety in terms of the mobilized cohesion.

Solution:

Calculate the necessary coefficients:

$$2R_a/(4c_u - q_d) = 2 \times 78.9/159.6 = 0.99, \ 2R_a \bar{z}/(4c_u - q_d) = 2 \times 78.9 \times 2.12/159.6 = 2.10.$$

Substituting into Equation 8.53:

$$D^2 - 0.99D - 2.10 = 0, D = 2.03 \text{ m}.$$

Calculation of P_p and R:

$$P_p = 159.6 \times 2.03 \times 1.0 = 324.0 \text{ kN}, R = 324.0 - 78.9 = 245.1 \text{ kN}.$$

$$D = 2.03 \times 1.3 = 2.65 \text{ m}, L = 6.0 + 2.65 = 8.65 \text{ m}, \text{ For } D = 2.65 \text{ m}:$$

$$P_p = (4c_u/F - q_d)2.65 \times 1.0 = [4 \times 60.0/F - (17.3 \times 3 + 9.5 \times 3)]2.65,$$

$$P_p = 636.0/F - 213.1 \text{ kN}.$$

Taking moments about point c:

$$(636.0/F - 213.1)2.65/2 - 78.9(2.12 + 2.65) = 0 \rightarrow F = 1.28.$$

The mobilized cohesion is $60.0 / 1.28 = 46.9$ kPa.

$$P_p = 636.0/1.28 - 213.1 = 283.8 \text{ kN}, R = 283.8 - 78.9 = 204.9 \text{ kN}.$$

8.7.5   *Anchored sheet pile in c', ϕ' soil*

Most of the sheet piles in this category are constructed from steel and are supported by one or more anchor rods (or tie rods), which transfer the load to isolated anchorages. A typical net pressure distribution for a free earth support is shown in Figure 8.32. Summing moments about the anchor rod and simplifying we obtain:

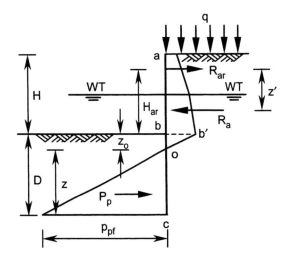

Figure 8.32. Net pressure diagram for an anchored sheet pile wall in c', ϕ' soil.

$$z^3 + 1.5(H_{ar} + z_0)z^2 - \frac{3R_a z'}{\gamma(k_p - k_a)} = 0 \tag{8.55}$$

When the water table is at the same level on both sides the effective unit weight and effective earth pressure coefficients are used. The embedment depth of the pile is $D = z + z_0$ and is computed using either a value of k_p suitably reduced by an appropriate factor of safety or by using the normal value of k_p and simply increasing the resulting D by 20% to 40%. In a fixed earth support an estimate of the depth of the point of contraflexure (zero bending moment), with the distance x measured from the dredge line (positive downwards) based on Table 8.1, makes the analysis statically determinate (Scott, 1980). Tschebotarioff (1973) proposed that the point of contraflexure be taken at the level of the dredge line together with a 33% increase in the allowable stresses of the sheet pile material. Williams & Waite (1993) suggested that the point of contraflexure be taken at the zero pressure point under the dredge line.

Table 8.1. Position of the point of contraflexure in the fixed earth support method.

ϕ'	15°	20°	25°	30°	35°	40°
x/H	0.37	0.25	0.15	0.08	0.033	−0.06

Example 8.13

The side of a 10 m excavation in sand is to be supported by an anchored sheet pile wall in which the anchor rod is 1.3 m below the ground surface. The water table is 2.6 m below the ground surface and there is a uniform vertical surface load of 24 kPa. Soil properties are: $\phi' = 30°$, $\delta' = 17°$, $c' = 0$, $\gamma = 16.5$ kN/m³ and $\gamma' = 10.4$ kN/m³. Determine: (a) the embedment depth and the anchor rod force when k_p is fully mobilized, (b) the embedment depth and the anchor rod force when a factor of safety of 2 is applied to k_p, (c) the location and magnitude of the maximum bending moment for case (b), (d) re-work case (c) using the fixed earth method assuming that the point of contraflexure coincides with the point of zero lateral earth pressure.

Solution:

(a) For $\phi' = 30°$, $k_{ah} = 0.286$, $k_{ph} = 5.150$ (similar to Example 8.9).
Referring to Figure 8.33:
$aa' = qk_a = 24.0 \times 0.286 = 6.9$ kPa, $dd' = (24.0 + 16.5 \times 2.6) \times 0.286 = 19.1$ kPa.
$bb' = (24.0 + 16.5 \times 2.6 + 10.4 \times 7.4) \times 0.286 = 41.1$ kPa.
$z_0 = p_d / \gamma'(k_p - k_a) = 41.1/10.4(5.150 - 0.286) = 0.81$ m.
$F_1 = 6.9 \times 2.6 \times 1.0 = 17.9$ kN, $F_2 = (19.1 - 6.9) \times 2.6/2 \times 1.0 = 15.9$ kN.
$F_3 = 19.1 \times 7.4 \times 1.0 = 141.3$ kN, $F_4 = (41.1 - 19.1) \times 7.4/2 \times 1.0 = 81.4$ kN.
$F_5 = (41.1 \times 0.81)/2 \times 1.0 = 16.6$ kN, $R_a = 273.1$ kN.

By taking moments about the dredge line we find R_a is 3.68 m above the dredge line. Therefore $z' = 10.0 - 3.68 - 1.3 = 5.02$ m. Find the necessary coefficients:

$1.5(H_{ar} + z_0) = 1.5(7.4 + 1.3 + 0.81) = 14.265$.

$\frac{3R_a z'}{\gamma'(k_p - k_a)} = \frac{3 \times 273.1 \times 5.02}{10.4(5.150 - 0.286)} = 81.305$.

Substituting the coefficients into Equation 8.55: $z^3 + 14.265z^2 - 81.305 = 0$.

Using a trial and error method we find: $z = 2.22$ m, $D = z + z_o = 2.22 + 0.81 = 3.03 \approx 3$ m.

Compute cc' and F_6: $cc' = \gamma'z(k_p - k_a) = 10.4 \times 2.22 \times (5.150 - 0.286) = 112.3$ kPa.

$F_6 = (cc' \times z)/2 \times 1.0 = 112.3 \times 2.22/2 \times 1.0 = 124.6$ kN.

$R_{ar} + F_6 = R_a, R_{ar} = 273.1 - 124.6 = 148.5$ kN.

(b) $z_o = p_d / \gamma'(k_p - k_a) = 41.1/10.4(5.150/2 - 0.286) = 1.73$ m.

$F_5 = (41.1 \times 1.73)/2 \times 1.0 = 35.6$ kN. $R_a = 17.9 + 15.9 + 141.3 + 81.4 + 35.6 = 292.1$ kN.

Taking the moments of F_1 to F_5 about the dredge line we find R_a is 3.39 m above the dredge line, and therefore $z' = 10.0 - 3.39 - 1.3 = 5.31$ m. Find the necessary coefficients:

$1.5(H_{ar} + z_o) = 1.5(7.4 + 1.3 + 1.73) = 15.645$.

$$\frac{3R_a z'}{\gamma'(k_p - k_a)} = \frac{3 \times 292.1 \times 5.31}{10.4(5.150/2 - 0.286)} = 195.465.$$

Substituting the coefficients into Equation 8.55: $z^3 + 15.645z^2 - 195.465 = 0$, $z = 3.22$ m.

$D = z + z_o = 3.22 + 1.73 = 4.95$ m ≈ 5.0 m. Compute cc' and F_6:

$cc' = \gamma'z(k_p - k_a) = 10.4 \times 3.22 \times (5.150/2 - 0.286) = 76.6$ kPa.

$F_6 = (cc' \times z)/2 \times 1.0 = 76.6 \times 3.22/2 \times 1.0 = 123.3$ kN.

Anchor rod force per metre run of the sheet pile: $R_{ar} = 292.1 - 123.3 = 168.8$ kN.

(c) The maximum bending moment occurs at a point where the shear force is zero. As:

$F_1 + F_2 = 17.9 + 15.9 = 33.8$ kN $< R_{ar}$, then this point is below the water table.

The lateral stress below point d is represented by the line $d'b'$ and is equal to $19.1 + \gamma'xk_a$, where x is the vertical distance of the point of interest below d. Thus the shear force SF is:

$SF = R_{ar} - F_1 - F_2 - 19.1x \times 1.0 - \gamma'xk_a \times x/2 \times 1.0 = 0$,

$SF = 168.8 - 17.9 - 15.9 - 19.1x - 10.4 \times 0.286x^2/2 = 0$, $1.487x^2 + 19.1x - 135.0 = 0$,

$x = 5.07$ m < 7.4 m. The magnitude of the bending moment at this point is:

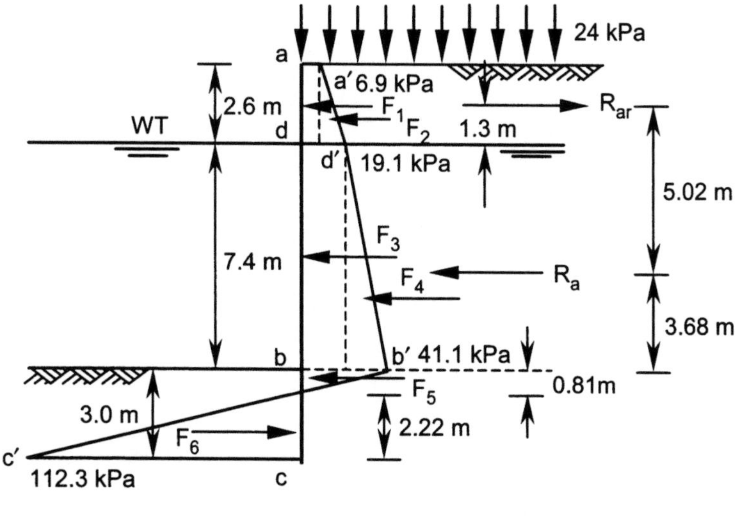

Figure 8.33. Example 8.13.

$BM = 168.8(1.3+5.07) - 17.9(1.3+5.07) - 15.9(2.6/3+5.07) - 19.1 \times 5.07^2/2$

$-10.4 \times 0.286 \times 5.07^3/6 = 556.7$ kN.m. This means that the back of the sheet pile at this point will be in compression while its front will be in tension.

(d) Bending moment at a depth of $z_o = 1.73$ m is zero. Thus:

$R_{ar}(H_{ar} + z_o) - R_a(z_1' + z_o) = 0$,

$R_{ar}(8.7+1.73) - 292.1(3.39+1.73) = 0$, $R_{ar} = 143.4$ kN.

$SF = 143.4 - 17.9 - 15.9 - 19.1x - 10.4 \times 0.286x^2/2 = 0$, $1.487x^2 + 19.1x - 109.6 = 0$,

$x = 4.30$ m. The magnitude of bending moment at this point is:

$BM = 143.4(1.3+4.30) - 17.9(1.3+4.30) - 15.9(2.6/3+4.30) - 19.1 \times 4.30^2/2$

$-10.4 \times 0.286 \times 4.3^3/6 = 404.6$ kN.m.

8.7.6 Anchored sheet pile in c_u, $\phi_u = 0$ soil

In this case the backfill is taken to be a c', ϕ' soil whilst the soil below the dredge line is purely cohesive with c_u and $\phi_u = 0$. With the net pressure diagram shown in Figure 8.34 and full mobilization of c_u, the sum of the moments about the anchor rod level yields:

$$D^2 + 2H_{ar}D - \frac{2R_a z'}{4c_u - q_d} = 0 \qquad (8.56)$$

Equilibrium of the horizontal forces is used to determine the force in the anchor rod. The general expression for the mobilized passive thrust is:

$$P_p = D(\frac{4c_u}{F} - q_d) \qquad (8.57)$$

where F is the factor of safety applied to c_u (replace c_u by c_u/F in Equation 8.56). If the water levels on both sides are different the area $wdeb$ of Figure 8.26(b) must be taken into account in the force and moment equilibrium equations.

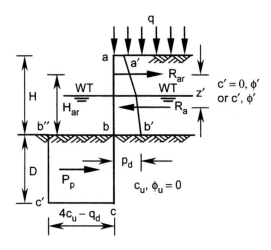

Figure 8.34. Net pressure diagram for an anchored sheet pile wall in c_u, $\phi_u = 0$ soil.

Example 8.14

In an anchored sheet pile the backfill has the same profile as in Example 8.13. The soil below the dredge line has $c_u = 75$ kPa and $\phi_u = 0$. Determine: (a) the embedment depth and the anchor rod force when a factor of safety of 2 is applied to c_u, (b) the factor of safety when there is a 3 m sudden drop of the water table at the front of the sheet pile.

Solution:

(a) $R_a = 17.9 + 15.9 + 141.3 + 81.4 = 256.5$ kN.

Taking moments about the dredge line we find R_a is 3.94 m above the dredge line. Thus

$z' = 10.0 - 3.94 - 1.3 = 4.76$ m.

$q_d = 2.6 \times 16.5 + 7.4 \times 10.4 = 119.9$ kN. Use Equation 8.56:

$D^2 + 2 \times 8.7D - 2 \times 256.5 \times 4.76/(4 \times 75.0/2 - 119.9) = 0$,

$D^2 + 17.4D - 81.12 = 0$, $D = 3.82$ m.

$P_p = 3.82(4 \times 75.0/2 - 119.9) = 115.0$ kN, $R_{ar} = R_a - P_p = 256.5 - 115.0 = 141.5$ kN.

(b) Calculate the magnitudes of the horizontal forces due to backfill water and the water at the front and their moments about the level of the anchor rod:

$P_{w(back)} = \gamma_w H^2_{w(back)}/2 = 9.81 \times 7.4^2/2 \times 1.0 = 268.6$ kN.

$M_{w(back)} = 268.6(8.7 - 7.4/3) = 1674.3$ kN.m.

$P_{w(front)} = \gamma_w H^2_{w(front)}/2 = 9.81(7.4 - 3.0)^2/2 \times 1.0 = 95.0$ kN.

$M_{w(front)} = 95.0[8.7 - (7.4 - 3.0)/3] = 687.2$ kN.m.

Sum moments about the anchor rod: $R_a z' + M_{w(back)} - M_{w(front)} = P_p(H_{ar} + D/2)$,

$256.5 \times 4.76 + 1674.3 - 687.2 = P_p(8.7 + 3.82/2) \rightarrow P_p = 208.1$ kN.

$P_p = D(4c_u/F - q_d) = 208.1$, $3.82(4 \times 75.0/F - 119.9) = 208.1 \rightarrow F = 1.72$.

The new anchor rod force is:

$R_{ar} = 256.5 - 208.1 + 268.6 - 95.0 = 222.0$ (a 57% increase).

8.7.7   Earth pressure on sheet piles in an excavation

A general arrangement of a braced cut is shown in Figure 8.35(a) in which horizontal struts support the sheet pile system. A traditional approach is to use the apparent earth pressure profiles of Figures 8.35(b) - 8.35(d) that are based on the work of Terzaghi & Peck (1967), Peck (1969) and Flaate & Peck (1973). These diagrams do not represent the actual lateral earth pressure but are an envelope to the approximate lateral stress distribution due to the force in each strut and is constructed from the measurement of the strut forces. For sands the uniform lateral pressure of Figure 8.35(b) is used, where k_a is calculated from Rankine's theory. For a purely cohesive soil a trapezoidal distribution is used that depends on the magnitude of the stability number defined by $n = \gamma H / c_u$. For $n > 4$ the lateral distribution of Figure 8.35(c) is more appropriate, where the parameter m varies from 0.4 to 1.0, with lower values of m for soft clays. For $n < 4$ the clay soil is in elastic equilibrium and it is convenient to use the distribution of Figure 8.35(d). As the stability number approaches 8, plastic equilibrium dominates the backfill and the soil underneath

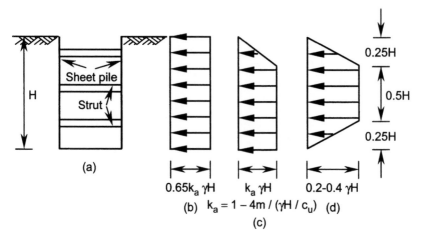

Figure 8.35. Apparent earth pressure profiles on sheet piles in an excavation.

and the heave of the bottom of the excavation will lead to a shear failure of the system. A slightly different profile of the apparent pressure diagram in sand and clay may be found in Tschebotarioff (1973). Note that the traditional method does not account for the effective stress analysis and multi-layer soils. Bowles (1996) suggested the use of an average of the two pressure diagrams obtained from Rankine's theory and the at-rest condition. Alternatively, either the active earth pressure coefficient or the magnitudes of the lateral stresses may be increased by a factor in the range 1.0 to 1.3. The analysis of the sheet pile is carried out using the following methods:

1. Calculate the force at each strut by assuming that the area of the pressure diagram between the two struts is equally shared by the adjacent struts. Note that this satisfies equilibrium of the horizontal forces, but moment equilibrium is not satisfied.

2. Assume the top and the bottom struts have fixed connections with the sheet pile, while the other struts have hinge connections. Use the above mentioned pressure diagrams to analyse the sheet pile beam. The problem is statically determinate.

3. Using the average pressure diagrams obtained from k_a and k_o as the actual response of the backfill soil, treat the sheet pile as a statically indeterminate continuous beam.

In a slurry excavation the force P due to hydrostatic pressure from the slurry material made of water and bentonite (montmorillonite group) supports the cast in-situ concrete sheet pile. Generally bentonite can be used in combination with soil and other additives (e.g. cement) to improve the flexibility and decrease the permeability. During the excavation a thin impermeable layer will be formed within the slurry at the face of the excavation. This layer prevents the loss of slurry into the soil and allows the construction of the concrete elements to cover the face of the excavated soil. A Coulomb wedge analysis can be adopted with $\alpha = 45° + \phi' / 2$ to determine the maximum value of P provided by the slurry. This allows the unit weight of the slurry to be obtained and thereby the proportion of bentonite that should be used.

8.7.8 *Common types of plate, concrete, and ground anchors*

Common types of sheet pile anchorages are shown in Figure 8.36. The vertical plate anchor shown in Figure 8.36(a) may be made of steel plates, precast or in-situ concrete. The anchor (or tie) rods are spaced (preferably) equidistantly and are either horizontal or slightly below the horizontal. Each rod is secured to a plate anchor that may be a strip (thereby facilitating plane strain conditions) or a plate of finite dimensions. Its holding capacity is mainly derived from the passive resistance at the front of the plate, which is reduced by the active thrust created at the back of the plate. The ultimate value of holding capacity is determined by introducing suitable failure mechanisms based on the development of a passive failure surface from the tip of the plate. This type of failure mechanism (general or local) has been verified by laboratory tests on models (Hueckel, 1957; Das, 1990). In a general failure mechanism the ratio of the depth to the height of the plate is small and the failure surface intersects the ground surface. In a localized failure a cylinder of soil with a radius equal to the height of the plate is considered. The length of the anchor rod must exceed the active zone at the back of the sheet pile and the passive zone at the front of the plate anchor.

A block concrete (or dead man concrete) anchor is shown in Figure 8.36(b). Its holding capacity relies mostly on the passive resistance at the front of the block. The weight of the block and the soil resting on its upper surface mobilizes shear resistance on the upper and lower contact surfaces, further increasing its holding capacity. If, for a specified geometry, the analysis shows a possibility of sliding or overturning, then the block could be supported by a group of piles as shown in which the piles at the front will be in compression while those at the back will be in tension.

A ground anchor consists of an anchor rod grouted in a drilled hole as shown in Figure 8.36(c). Typically, these drilled holes have diameters between 100 mm and 300 mm over a length of some 10 m to 30 m. A ground anchor depends entirely upon the shear resistance over its grouted length. This resistance is a function of the method of construction, the pressure by which the grout is inserted into the hole, the average effective vertical stress in the vicinity of the grouted portion and the dimensions of the anchor. In Europe a Coulomb wedge-type analysis is normally used to evaluate the ultimate holding capacity of a ground anchor (e.g. BS 8081, 1989). This method allows the sheet pile and the anchor to be analysed as a unified system. Anchor rods are often inclined at 15° to 25° below the horizontal to increase the effective vertical stress and thereby the corresponding shear resistance. Steep slopes, up to 2 vertical to 1 horizontal, are used to provide sufficient effective vertical stress, facilitate gravity grouting and possibly avoid underground utilities.

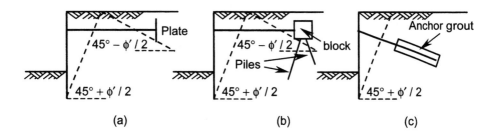

Figure 8.36. Common types of sheet pile anchorages.

The anchor rod must meet the requirements of the relevant standards. Generally anchor rods are made from hot-rolled alloy steel bars strengthened by cold working that produces a high tensile strength. Anchor rods may be housed in pipes to prevent possible damage caused by backfill settlement or ground heave, and must be covered by a material specified in the relevant code in order to prevent contamination and corrosion by minerals and domestic waste.

The factor of safety is defined as the ratio of the ultimate holding capacity to the anchor rod force calculated from equilibrium of the sheet pile. The factor of safety is normally in the range 1.5 to 2.5 depending on the type of the anchor, the soil and the design life span of the system. For temporary anchorages a factor of safety of 1.5 can be used but for permanent anchors higher values of the factor of safety should be adopted. In ground anchors where the anchor has a steep slope (e.g. 2 vertical, 1 horizontal) the factor of safety must be increased (from 1.5 for horizontal) to 2.

8.7.9 *Holding capacity of plate anchors*

The design of a plate anchor relies upon an understanding of its behaviour, the shear strength properties of the soil and the magnitudes of the angles of friction mobilized on both sides of its surfaces. It is normal practice to assume that the plate is rigid and undergoes sufficient translation towards the retaining structure to create both the passive and active states. Figure 8.37 illustrates the vertical section of a typical plate anchor system in sand. Based on the design methods suggested by Teng (1962), Ovesen & Stromann (1972), Tschebotarioff (1973) and Meyerhof (1973), a number of approaches are available. In Teng's method the active and passive thrusts are calculated using Rankine's earth pressure coefficients. If the depth ratio $h/B > 2$ to 3, the ultimate holding capacity in plane strain conditions is obtained from equilibrium of the horizontal forces (Das, 1990):

$$R_u = \frac{1}{\cos\eta}\left[\frac{\gamma h^2}{2}k_p - \frac{\gamma h^2}{2}k_a - qk_a h\right] \tag{8.58}$$

For safety reasons, the effect of any surface loading is usually ignored in the calculation of the passive thrust. For h/B values smaller than 2 the lateral soil pressure is estimated according to that given previously in Section 8.3. For a plate of finite width the effect of

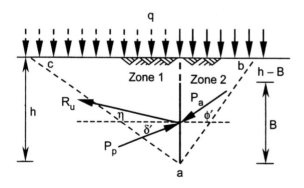

Figure 8.37. Ultimate holding capacity of a vertical plate anchor.

the shear resistance developed on the vertical faces of the failure surface (plane strain in the previous case) is taken into account, thus:

$$R_u = \frac{1}{\cos\eta}\left[L(\frac{\gamma h^2}{2}k_p - \frac{\gamma h^2}{2}k_a - qk_aH) + \frac{\gamma h^3}{3}k_o(\sqrt{k_p}+\sqrt{k_a})\right] \tag{8.59}$$

where k_o is the coefficient of earth pressure at-rest that may be assumed to be 0.4 and L is the width of the plate. In this analysis vertical equilibrium may not be satisfied.

In the original Ovesen & Stromann (1972) method, equilibrium of the forces both horizontally and vertically is considered in conjunction with the following assumptions:

1. The anchor rod is horizontal.
2. The passive and active failure surfaces are combinations of planes and spiral surfaces and reflect the actual failure mechanism.
3. The mobilized friction angle on the plate on the active side (zone 2, Figure 8.37) is equal to ϕ'.
4. The mobilized friction angle on the plate on the passive side (zone 1, Figure 8.37) is $\delta' < \phi'$ and is computed from equilibrium of vertical forces. The weight of the plate is ignored.
5. Moment equilibrium is not satisfied.

For an inclined anchor rod the following analysis takes account of equilibrium in both horizontal and vertical directions. For horizontal equilibrium:

$R_u \cos\eta + P_a\cos\phi' - P_p\cos\delta' = 0$, $R_u = (P_p\cos\delta' - P_a\cos\phi')/\cos\eta$, thus:

$$R_u = \frac{1}{\cos\eta}\left[\frac{\gamma h^2}{2}k_p\cos\delta' - \frac{\gamma h^2}{2}k_a\cos\phi'\right] \tag{8.60}$$

For vertical equilibrium, neglecting the weight of the vertical plate,

$$R_u = \frac{1}{\sin\eta}\left[\frac{\gamma h^2}{2}k_a\sin\phi' - \frac{\gamma h^2}{2}k_p\sin\delta'\right] \tag{8.61}$$

Equating Equations 8.60 and 8.61, δ' may be obtained by a trial and error method. By substituting δ' into Equations 8.60 or 8.61, the ultimate holding capacity can be calculated. For a horizontal anchor rod the right-hand term of Equation 8.61 is zero:
$k_a\sin\phi' - k_p\sin\delta' = 0$,

$$k_a = \frac{\sin\delta'}{\sin\phi'}k_p \tag{8.62}$$

Substituting k_a and k_p from Equations 8.20 and 8.22 we find:

$$\sin\delta' = \left[\frac{\sqrt{\cos\delta'} - \sqrt{\sin(\phi'+\delta')\sin\phi'}}{\sqrt{\cos\phi'} + \sqrt{\sin 2\phi'\sin\phi'}}\right]^2\sin\phi' \tag{8.63}$$

The mobilized friction angle δ' is found by a trial and error method. Substituting Equation 8.62 into Equation 8.60 the ultimate holding capacity R_u may be expressed as:

$$R_u = \frac{\gamma h^2}{2} k_p \frac{\sin(\phi' - \delta')}{\sin \phi'} \tag{8.64}$$

For a horizontal anchor rod in an undrained soil, a breakout factor in dimensionless form F_c is defined by Tschebotariff (1973) to give the ultimate holding capacity as:

$$R_u = F_c c_u h \tag{8.65}$$

According to laboratory small-scale model tests, the magnitude of F_c increases with the depth ratio h / B and remains constant after a critical depth ratio is achieved. This latter ratio defines a shallow anchorage condition. The following empirical equations were proposed by Meyerhof (1973):

$$F_c = 1.2 h / B \quad \text{square anchor, critical } h / B = 7.5 \tag{8.66}$$

$$F_c = h / B \quad \text{strip anchor, critical } h / B = 8.0 \tag{8.67}$$

Further tests showed that the critical depth ratio depends upon the magnitude of the undrained cohesion; for square anchors this ratio is approximately 7.4 (Das et al., 1985). It has also been suggested that Equations 8.66 and 8.67 underestimate the actual ultimate holding capacity (Das, 1990). The breakout factor can also be determined using a suitable upper bound failure mechanism. Ignoring the active failure at the back of the anchor plate, a plane strain mechanism, similar to the tunnel heading mechanism of Figure 7.26, can be applied. With sufficient accuracy Equation 7.61, suggested by the author for a plane strain tunnel heading, is rearranged to represent the breakout factors for depth ratios up to 10:

$$F_c = 2.64 + 2\ln\frac{h}{B}, \ (h/B:1 \text{ to } 2); \quad F_c = 4.0 + 2\ln\frac{h-B}{B}, \ (h/B:2 \text{ to } 10) \tag{8.68}$$

It is common to apply a factor of safety of 3 to the ultimate holding capacity. In the analysis of a sheet pile system, the anchor rod force is calculated by using a factor of safety against the passive state at the front of the sheet pile, thus for the plate anchor a reduced factor of safety of 1.5 to 2 can be used.

Example 8.15

Calculate the ultimate holding capacity of a vertical anchor plate in sand using the following data: $\eta = 0$, $\phi' = 35°$, $\gamma = 17.5$ kN/m^3, $B = 2$ m, $L = 2.5$ m, and $h = 2$ m.

Solution:

From Equations 8.6 and 8.9:

$k_a = \tan^2(45.0° - 35.0°/2) = 0.271$,

$k_p = \tan^2(45.0° + 35.0°/2) = 3.690$.

$P_a = \gamma h^2 k_a / 2 = 17.5 \times 2.0^2 \times 0.271 / 2 = 9.5$ kN/m.

$P_p = \gamma h^2 k_p = 17.5 \times 2.0^2 \times 3.690 / 2 = 129.1$ kN/m.

$\gamma h^3 k_o (\sqrt{k_p} + \sqrt{k_a})/3 = 17.5 \times 2.0^3 \times 0.4(\sqrt{3.690} + \sqrt{0.271})/3 = 45.6$ kN.

From Equation 8.59:

$R_u = 2.5(129.1 - 9.5) + 45.6 = 344.6$ kN.

Example 8.16

Re-work Example 8.15 assuming that at the active side, the angle of friction between the soil and the vertical plate is mobilized to ϕ'.

Solution:

Using Equation 8.63: $\sin \delta' = \left(\dfrac{\sqrt{\cos \delta'} - \sqrt{\sin(35.0° + \delta')\sin 35.0°}}{\sqrt{\cos 35.0°} + \sqrt{\sin 70.0° \sin 35.0°}} \right)^2 \sin \phi'.$

From a trial and error procedure δ' is found to be 2.07°. From Equations 8.22:

$k_p = [\cos 35.0° /(\sqrt{\cos 2.07°} - \sqrt{\sin(35.0° + 2.07°)\sin 35.0°})]^2 = 3.96.$ Using Equation 8.64:

$R_u = (17.5 \times 2.0^2 \times 3.96 \times 2.5/2)\sin(35.0° - 2.07°)/\sin 35.0° = 328.4$ kN.

8.7.10 *Holding capacity of block concrete anchors*

An idealized free body diagram of a block concrete anchor is shown in Figure 8.38. The factor of safety may be applied to either R_u or k_p. Equilibrium of horizontal and vertical forces allows the unknown magnitudes of R_u and N_2 to be computed whilst the point of application of the base reaction N_2 is found by moment equilibrium.

Example 8.17

For a strip concrete block anchor the following data are known: $B' = 1$ m, $B = 1.3$ m, $h = 2.5$ m, $q = 30$ kPa, $\gamma = 18$ kN/m^3, $\gamma_c = \gamma_{concrete} = 23.5$ kN/m^3, $c' = 0$, $\phi' = 34°$, and $\delta' = 17°$. Determine the ultimate holding capacity of the anchor assuming a horizontal anchor rod and a factor of safety of 2 for the mobilized k_p.

Solution:

From Equations 8.20 and 8.22: $k_a = 0.256$ $k_p = 6.767$, and therefore (as $c' = 0$):

$k_{ah} = 0.256 \times \cos 17.0° = 0.245$ and $k_{ph} = 6.767 \times \cos 17.0° = 6.471.$

$a'a = (30.0 + 18.0 \times 1.2) \times 0.245 = 12.6$ kPa, $b'b = (30.0 + 18.0 \times 2.5) \times 0.245 = 18.4$ kPa.

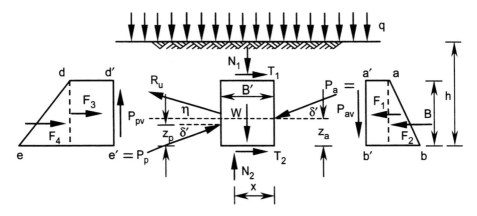

Figure 8.38. Ultimate holding capacity of a concrete block anchor.

$P_{ah} = F_1 + F_2 = 12.6 \times 1.3 \times 1.0 + (18.4 - 12.6) \times 1.3 \times 1.0/2 = 16.4 + 3.8 = 20.2$ kN.

$z_a = (F_1 z_1 + F_2 z_2)/P_{ah} = (16.4 \times 1.3/2 + 3.8 \times 1.3/3)/20.2 = 0.61$ m.

$d'd = 18.0 \times 1.2 \times 6.471 = 139.8$ kPa, $e'e = 18.0 \times 2.5 \times 6.471 = 291.2$ kPa.

$P_{ph} = F_3 + F_4 = 139.8 \times 1.3 \times 1.0 + (291.2 - 139.8) \times 1.3 \times 1.0/2 = 181.7 + 98.4 = 280.1$ kN.

$z_p = (F_3 z_3 + F_4 z_4)/P_{ph} = (181.7 \times 1.3/2 + 98.4 \times 1.3/3)/280.1 = 0.57$ m.

$N_1 = B' \times 1.0 (\gamma h + q) = 1.0 \times 1.0 (18.0 \times 1.2 + 30.0) = 51.6$ kN.

$P_{av} = P_{ah} \times \tan \delta' = 20.2 \times \tan 17.0° = 6.2$ kN.

$P_{pv} = P_{ph} \times \tan \delta' = 280.1 \times \tan 17.0° = 85.6$ kN.

$W = B' \times B \times 1.0 \times \gamma_c = 1.0 \times 1.3 \times 1.0 \times 23.5 = 30.5$ kN. From vertical equilibrium:

$N_2 = N_1 + W + P_{av} - P_{pv} = 51.6 + 30.5 + 6.2 - 85.6 = 2.7$ kN.

$T_1 = N_1 \tan \delta' = 51.6 \times \tan 17.0° = 15.8$ kN, $T_2 = N_2 \tan \delta' = 2.7 \times \tan 17.0° = 0.8$ kN.

For horizontal equilibrium:

$R_u = P_{ph} - P_{ah} + T_1 + T_2 = 280.1 - 20.2 + 15.8 + 0.8 = 276.5$ kN.

Applying a factor of safety of 2 for the mobilized k_p:

$N_{2m} = N_1 + W + P_{av} - P_{pv}/2 = 51.6 + 30.5 + 6.2 - 85.6/2 = 45.5$ kN.

$T_{2m} = N_2 \tan \delta' = 45.5 \times \tan 17° = 13.9$ kN.

$R_{ar} = P_{phm} - P_{ah} + T_1 + T_{2m} = 280.1/2 - 20.2 + 15.8 + 13.9 = 149.6$ kN.

Factor of safety against holding capacity = 276.5 / 149.6 = 1.85.

Assume the anchor rod is 0.75 m above the base and check that the force N_{2m} is located within the base ($0 < x < 1.0$ m). Taking moments to the heel (right corner of the base):

$x = [P_{ah} \times z_a + (W + N_1)B'/2 + R_{ar} \times z_{ar} - T_1 \times B - P_{phm} \times z_p - P_{pvm} \times B']/N_{2m}$,

from which $x = 0.49$ m.

8.7.11 *Holding capacity of ground anchors*

Based on the soil-grout ultimate bond stress τ_{ult}, the length of the grouted zone l_r is:

$$l_r = \frac{R_u}{\pi D_h \tau_{ult}} = \frac{F \times s R_{ar}}{\pi D_h \tau_{ult}} \qquad (8.69)$$

where F is the factor of safety, s is the ground anchor spacing, R_{ar} is the anchor rod force and D_h is the diameter of the hole. The minimum grout length normally lies between 3 m to 4.5 m (Littlejohn & Bruce, 1977; Munfakh et al., 1987) and must comply with the relevant code. The average ultimate bond stress depends on the type of the soil and the grout pressure (PTI, 1996). In cohesive soils, this bond stress is between 30 kPa to 380 kPa for high grout pressures and between 30 kPa to 70 kPa for low grout pressures. In cohesionless soils the ultimate bond stress for a low pressure grout is 70 kPa to 140 kPa and for high grout pressure is 80 kPa to 1380 kPa. For soft and hard rocks a low grout pressure is used and the ultimate bond stress varies from 150 kPa to 3100 kPa with 1400 kPa as a maximum value for soft rocks. The average ultimate bond stress can be obtained by in-situ testing that includes pullout, performance and proof tests. In the latter two tests the ground anchor is loaded up to between 1.3 and 1.5 times of the design load. Cheney (1988) recommended test procedures and the requirements to be used in the selection of the anchors. Further information about design considerations, methods and testing procedures may be found in Xanthakos (1991), Xanthakos et al. (1994) and Schaefer et al. (1997). Alternatively, the following equation can be used to estimate the average ultimate bond stress:

$$\tau_{ult} = \sigma'_{zm}k \tan \delta' + c_a \tag{8.70}$$

where σ'_{zm} is the effective vertical stress at the midpoint of the grouted length, k is the lateral earth pressure coefficient between k_a and k_o, $\delta' \le \phi'$ is the friction angle mobilized on the surface of the grout cylinder, and $c_a \le c'$ is the cohesion mobilized on the soil-grout surface.

8.7.12 *Numerical analysis of sheet pile walls*

Sheet pile walls can be analysed using the beam-foundation model discussed previously in Chapter 5 by assuming a suitable embedment depth. Lateral earth pressure is applied to the vertical beam only above the zero pressure point. The anchor rod and the soil resistance below the zero pressure point are idealized by a set of springs. The spring stiffness for the anchor rod (per unit width of sheet pile) is given by:

$$K_{ar} = \frac{AE \cos \eta}{sL} \tag{8.71}$$

where E is the Modulus of Elasticity of the anchor rod material, A is the cross-sectional area, L is the length of the anchor rod, s is the anchor rod spacing and η is the slope of the anchor rod with horizontal. For a unit width of soil, the spring stiffness under the dredge line is:

$$K_i = \frac{a_i + a_{i-1}}{2} k_{si} \tag{8.72}$$

where a_i and a_{i-1} are the distances to the springs adjacent to spring i and k_{si} is the modulus of subgrade reaction at spring i. Bowles (1996) suggested the use of the following equations for the estimation of modulus of subgrade reaction at a depth z:

$$k_s = A_s + B_s z^n \qquad n > 0 \tag{8.73}$$
$$k_s = 40 q_u \qquad q_u \text{ in kPa, } k_s \text{ in kN/m}^3 \tag{8.74}$$

where A_s, B_s and n are coefficients and z is the depth of the point. The term q_u in Equation 8.74 represents the ultimate bearing capacity of the ground at a certain depth that can be estimated theoretically if the shear strength parameters c' and ϕ' are known (Chapter 10) – it may also be obtained from field experiments. Using Equation 8.74 at several specified depths allows the unknown parameters of A_s, B_s and n to be obtained. If x_i is the horizontal deformation of the ith spring, the force in the spring is:

$$R_i = K_i \times x_i .$$

The solution of the equivalent sheet pile beam can be arranged in terms of the unknown x_i values. By dividing the corresponding spring force by the effective area of:
$1.0 \times (a_i + a_{i-1}) / 2$, the average normal stress acting on the vertical interface can be evaluated. If any of these normal stresses or the spring deformations are higher than specified allowable values, the embedment depth has to be increased.

8.8   INTERNALLY STABILIZED EARTH RETAINING WALL

8.8.1   *Geosynthetic reinforcement*

Factory manufactured geosynthetic products are gradually replacing conventional reinforcement materials such as metal strips, metal bars and welded wire grids as soil reinforcement materials. Major applications of geosynthetics in ground improvement, reinforcement and treatment have been given by Koerner (1994) and include earth walls, steep slopes, clay fill slopes, embankments on soft soils, pile embankments, unpaved road stabilization, and treatment of areas prone to subsidence. In retaining walls economic benefits and a reduced construction time may be realized by using a poor quality soil reinforced by geosynthetics as backfill. In comparison with reinforced concrete walls, a 50% reduction in cost is predicted for a 5 m high wall; for a 10 m high wall the reduction in cost could be as much as 250%. Geosynthetics are manufactured polymeric materials that are currently available in the following forms (Schaefer et al., 1997):

1. Geotextiles (GT); permeable polymeric materials comprised of textile yarns used for filtration, drainage, separation or reinforcement.
2. Geogrids (GG); grid-like sets of interconnected polymer ribs used for reinforcement.
3. Geonets (GN); net-like sets of interconnected polymer ribs used for the transmission of liquids within the plane of their structure.
4. Geomembranes (GM); impermeable barriers used to contain various solids and liquids.
5. Geosynthetic clay liners (GCL); factory manufactured hydraulic barriers consisting of a thin layer of bentonite supported by geotextiles and/or geomembranes.
6. Geocomposites (GC); a combination of any of the above geosynthetics that sometimes includes natural soils.

Since the basic objective of the reinforcement in the backfill is to transfer the active thrust to the reinforcement elements, the working tensile strength of the element must be established accurately. Due to the high extensibility of the most geosynthetics they are subject to installation damage and creep during the design life. Construction damage may lower the tensile strength by as much as 30% which is equivalent to a factor of safety of $F_c = 1.3$ to 1.4 on the initial (nominal) tensile strength. The magnitude of F_c may vary between 1.1 and 1.6 (Task force 27, 1990) and its minimum value must comply with the relevant design code.

In general, the load-extension behaviour is a function of time. Laboratory results are commonly presented in two sets of curves showing: (1) the relationship between time and ultimate tensile strength; (2) the relationship between extension and tensile strength for specified time intervals. The time-tensile strength relationship is approximated by a line if time is plotted on a logarithmic scale and is constructed for the time period over which the creep data is available (10000 hours according to ASTM D-5262, 1997). As time increases, the ultimate tensile strength decreases. To find the ultimate tensile strength for the required design life (say 120 years), extrapolation of the linear relationship is permitted providing that a reduction factor of about 70% is applied to the computed tensile strength. This is equivalent to a creep reduction factor of $CRF = 40\%$ (higher for woven geotextiles) applied to the nominated ultimate tensile strength. In the absence of any creep tests, the default values of the corresponding code must be applied. In addition, a general factor of safety of approximately $F_s = 1.2$ (slightly higher for retaining walls) is applied to take account of the possibility of reaching a limit state and other uncertainties during the design life (Task force 27, 1990).

If necessary, factors of safety against chemical damages and biological degradations (F_d) must be considered. Consequently, the allowable tensile strength may be computed from the following equation:

$$T_{all} = \frac{T_{ult} \times CRF}{F_c \times F_d \times F_s} \tag{8.75}$$

where T_{all} and T_{ult} represent the allowable and ultimate tensile strengths respectively. The corresponding extension (tensile strain) to the design load can be determined from the set of the load-extension-time relationship constructed in the laboratory. For specified time intervals such as one hour, one month, one year and 5 to 10 years, the stress-strain type curves are determined by conducting tensile tests on the geotextile materials. Extrapolation is used to estimate the extension during the design life; however, with the above factors of safety considered, the initial and final extensions become nearly identical because the time isochrones share a common slope up to 40% of the ultimate tensile strength (Exxon, 1992). The tensile force in the reinforcement is redistributed in the soil through the surface area of the reinforcement. Laboratory tests are used to determine the bond stress between the soil and the reinforcement. The friction angle mobilized between the geosynthetic reinforcement and the adjacent soil δ_b may be related to the internal friction angle of soil ϕ' using the following equation:

$$\tan \delta_b = f_b \tan \phi' \tag{8.76}$$

The magnitude of f_b depends on the type of the reinforcement material and varies from 0.5 to 1.0. It is determined from the recommended standard laboratory tests such as a direct shear test or a full-scale pullout test. In the direct shear test, the reinforcement material is placed on the enforced failure plane. In a pullout test the reinforcement material is pulled through the soil along the separation of the two half boxes of the shear box. For granular soils, the apparent friction angle from pullout test is lower than the results obtained from the direct shear test.

8.8.2 Reinforced earth retaining walls

Depending on the type of the reinforcement, two design methods are available. For a reinforcement with high extensibility such as geotextiles, the active state in the earth wall is fully mobilized and the active Rankine earth pressure coefficient k_a is used. In a granular backfill with $c' = 0$ and ϕ', the position of the potential failure plane may be computed using the Coulomb wedge method. For a vertical wall, horizontal ground surface and with $\delta' = 0$, the failure plane makes an angle of $\alpha = 45° + \phi' / 2$ with the horizontal. With a sloping ground surface and $\delta' = \beta$, the position of the failure plane may be formulated from the Coulomb wedge analysis. Using the same terms given in Equation 8.19, the magnitude of α is given by:

$$\tan(\alpha - \phi') = \frac{-\tan(\phi' - \beta) + \sqrt{A}}{B} \tag{8.77}$$

$$A = \tan(\phi' - \beta)[\tan(\phi' - \beta) + \cot(\phi' + \theta - 90°)] \ [1 + \tan(\delta' + 90° - \theta)\cot(\phi' + \theta - 90°)],$$

$$B = 1 + \tan(\delta' + 90° - \theta)[\tan(\phi' - \beta) + \cot(\phi' + \theta - 90°)].$$

In the absence of hydrostatic water in the soil, the axial tensile load transferred to each horizontal reinforcement element (Figure 8.39) can be calculated as:

$$p_r = p_{ah} \times s_x \times s_z = (k_a \sigma_z \cos\beta) s_x s_z \tag{8.78}$$

where p_r is the tensile force in the element, σ_z is the vertical stress at depth z within the active zone, s_x and s_z are the horizontal and vertical spacing the of the reinforcement elements respectively. The embedment length l_r is calculated to resist the tensile force p_r. For a strip element of width w, the maximum tensile load p_t due to the frictional bond on both sides is:

$$p_t = 2\sigma_{zr} \cos\beta l_r w \tan\delta_b = 2\sigma_{zr} \cos\beta l_r w f_b \tan\phi' \tag{8.79}$$

where σ_{zr} is the vertical stress on the reinforcement element (at depth z_r) in the vicinity of the resisting zone. If the ground surface is horizontal and there is no surface loading, or the surface loading is uniform and covers a large area behind the facing panel, we may assume $\sigma_z = \sigma_{zr}$. Thus, equating Equations 8.78 and 8.79 we obtain:

$$l_r \geq \frac{k_a s_x s_z}{2w f_b \tan\phi'} \tag{8.80}$$

The tensile force in the reinforcement must be equal to or smaller than T_{all} (Equation 8.75). For continuous grid reinforcement, s_x and w are unity and p_t represents the tensile force per unit length of the wall. For reinforcement of low extensibility, Juran & Schlosser (1978) suggested an active zone that is separated from the resisting zone by a logarithmic spiral surface and represents the maximum states of tensile stress in the horizontal reinforcement. The assumed failure surface intersects the ground surface at a distance of $0.3H$ from the facing panel. AASHTO (1997) suggested that the logarithmic spiral may be

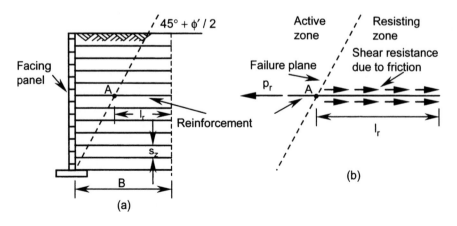

Figure 8.39. Soil-reinforcement bond due to friction.

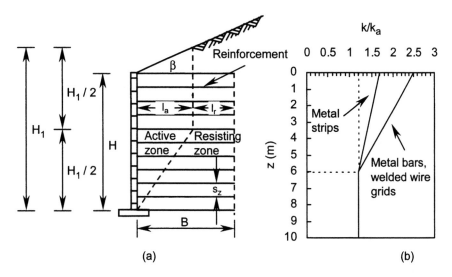

Figure 8.40. Reinforcement of low extensibility: (a) active zone, (b) variation of k / k_a with depth.

idealized by two linear failure planes (Figure 8.40(a)). The distribution of the appropriate earth pressure coefficient is shown in Figure 8.40(b). A slightly different arrangement is suggested by BS 8006 (1995).

In the sloping backfill the magnitude of H_1 defined in Figure 8.40(a) is:

$$H_1 = H + \frac{0.3H \tan\beta}{1 - 0.3\tan\beta}$$

(8.81)

In the presence of surface loading, the arrangement of the failure planes will change and this can be determined from the relevant code.

8.8.3 Soil-nailed earth retaining walls

Nailed soils represent a recent development in ground improvement and the technology of earth retaining structures. They can be classified into two categories of driven and grouted nails. Driven nails are small diameter rods that are driven into the soil or drilled holes by means of hammers or (more recently) by compressed air launchers. In the latter category, the nail is subjected to the tensile stress while being fired into the ground thereby eliminating the possibility of nail buckling. In the former category, the nails are not pretensioned but are grouted with or without a grout pressure similar to ground anchors. The main advantages of nailed retaining walls and slopes are their stability against dynamic loading (Felio et al., 1990) and their use in excavations as well as retaining walls. Methods of analysis include using an empirical method that assumes an idealized active zone of locus of maximum tensile stresses (Bowles, 1996), a kinematic limit analysis (Juran et al., 1988) and the finite element method (Unterreiner et al., 1997).

In the empirical method, the active zone is defined by Figure 8.40(a) and Equation 8.81 applies. The distribution of the lateral earth pressure is trapezoidal (Figure 8.35(d)) and the maximum k_a that occurs at mid-height is given by:

$$k_a = 0.65 \tan^2(45° - \phi'/2) \qquad \text{for sands with } \gamma H / c' \geq 20 \qquad (8.82)$$

$$k_a = \left[1 - \frac{4c'}{\gamma H \tan(45° - \phi'/2)}\right] \tan^2(45° - \phi'/2) \quad \text{for } c', \phi' \text{ soils} \qquad (8.83)$$

Equation 8.70 is used to estimate the maximum bond stress developed on the grouted nails. Both the finite element and kinematic methods have shown that the application of the trapezoidal pressure diagrams (used in braced excavations) provides rational estimates of the working tensile forces in the nails (Schaefer et al., 1997).

Example 8.18

It is required to design a 9 m high vertical face with a horizontal ground surface and using a geotextile reinforcement strip of width $w = 55$ mm and $T_{ult} = 150$ kN.
The soil properties are: $c' = 0$, $\phi' = 35°$ and $\gamma = 18$ kN/m^3.
The fill is placed in 0.3 m thick layers. Other known data are:
$CRF = 0.40$, $F_c = 1.0$, $F_d = 1.2$, $F_s = 1.3$, $f_b = 0.8$, and $s_x = s_z = 0.9$ m.
Solution:

From Equation 8.75: $T_{all} = \dfrac{T_{ult} \times CRF}{F_c \times F_d \times F_s} = \dfrac{150.0 \times 0.4}{1.0 \times 1.2 \times 1.3} = 38.5$ kN.

Using Equation 8.6 we obtain: $k_a = 0.271$. The position of the potential failure plane is:
$\alpha = 45.0° + \phi'/2 = 45.0° + 35.0°/2 = 62.5°$ from the horizontal.

Using Equation 8.80 we have:
$l_r = 0.271 \times 0.9 \times 0.9/2 \times 0.055 \times 0.8 \times \tan 35.0° = 3.56$ m.

The first reinforcement may be placed 0.45 m below the ground surface, and thus the last reinforcement at the bottom will be at a depth of 8.55 m. The tensile load transferred to this reinforcement is calculated from Equation 8.78:
$p_r = 0.271 \times 18.0 \times 8.55 \times 1.0 \times 0.9 \times 0.9 = 33.8$ kN $< T_{all} = 38.5$ kN.

Total length of reinforcement (first from the top):
$B = 8.55/\tan 62.5° + 3.56 = 8.0$ m.

One may use a variable length by considering the linear variation of the width of the wedge by depth.

8.9 THE OVERALL STABILITY OF RETAINING STRUCTURES

8.9.1 *Slope stability analysis*

The overall stability of a gravity or a cantilever retaining wall may be evaluated by a slope stability analysis (Chapter 9) with (sliding) trial circles that are outside the wall system and possible anchorages (Figure 8.41(a)). A similar method is used for a sheet pile and anchor system as shown in Figure 8.41(b) where the (sliding) trial circles pass through the tip of the sheet pile. The minimum requirements for the factor of safety must comply with the relevant codes.

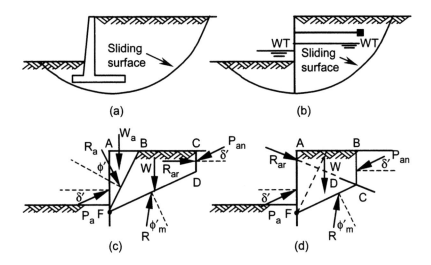

Figure 8.41. Evaluation of the overall factor of safety for retaining structures.

Bowles (1996) suggested a minimum factor of safety of 1.2 for the mobilization of the shear strength on the trial circle. For sheet pile walls, failure modes due to the loss of bond between the rod and grout, and excessive deformation of the sheet pile, must also be considered.

8.9.2   *Sliding block analysis*

Evaluation of the overall stability of anchored sheet piles using the method of the sliding block was first suggested by Kranz (1953) and subsequently modified by Locher (1969) and Littlejohn (1970). The method considers the equilibrium of a wedge type block located at the back of the sheet pile that includes the anchor system, as shown in Figures 8.41(c), (block *FBCD, FB* being the failure plane of the active zone), and 8.41(d), (block *FABC*). The lower sliding surface passes through the zero shear point *F* in the equivalent sheet pile beam. This sliding surface terminates at the tip of the vertical anchor or at the mid-point of the grouted anchor (point *C* in Figure 8.41(d)). A failure mode in which the sliding surface terminates at the top of the grouted portion (point *D* in Figure 8.41(d)) also has to be considered. This method, with some modifications to the geometry of the sliding block, can also be applied to multiple anchors (Hanna, 1982).

The application of the sliding block for overall stability analysis has been accepted by the codes of several European countries (e.g. BS 8081, 1989). The sheet pile system is assumed to be a part of the block so that the anchor rod force and the active thrust (immediately behind the sheet pile) are treated as internal forces that do not affect the stability of the block. Relevant forces are the effective weight of the block, the soil reaction R on the sliding surface, and the active thrust P_{an} on the right side of the anchor which is calculated using a linear distribution of earth pressure. Force equilibrium yields both the magnitude and direction of the reaction force R. The overall factor of safety is defined by the following equation where ϕ'_m is the angle between the reaction force R and a line normal to the sliding surface.

$$F_{overall} = \frac{\tan \phi'}{\tan \phi'_m} \geq 1.5 \tag{8.84}$$

In the revised version by Ostermayer (1977), the passive force acting between point F and the dredge line (if point F is under the dredge line) was included in the force equilibrium. The overall factor of safety is calculated from Equation 8.84 with a minimum value of 1.2. BS 8081 (1989) suggests that the active thrust above point F and anchor rod force be included in the force equilibrium equation. Note that the horizontal component of P_a (above point F) is equal and in the opposite direction to the horizontal component of the anchor rod force as the point F represents the zero shear point. In this case the minimum requirement for the factor of safety defined by Equation 8.84 is 1.3 for non-critical applications and 1.5 for critical applications. In a stratified soil the sliding surface passes through different types of soil and the problem is indeterminate unless a simplifying idealization is made. The sliding block is divided into vertical slices by drawing vertical lines through the points of intersection of the sliding surface and layers of the soil. From the force equilibrium of each slice, together with the assumption that the angle ϕ' (for each soil) is fully mobilized on the sliding surface, the corresponding soil reaction R is computed. The factor of safety is defined as the ratio of the computed anchor rod force from the force equilibrium of the block to the anchor rod force calculated from the statics of the sheet pile.

Example 8.19

Calculate the overall factor of safety for the sheet pile system of Example 8.13. Assume the length of the anchor rod is 13 m and is supported by a plate anchor of length 2.6 m with its mid-point 1.3 m below the ground surface.

Solution:

The point of zero shear (F) is located 4.3 m below the water table (for factor of safety of 2 applied to k_p and fixed earth support method: see Example 8.13). Referring to Figure 8.42, the forces applied to the sliding block are calculated as follows:
The lateral active pressure at point F is:

$19.1 + 10.4 \times 4.3 \times 0.286 = 31.9$ kPa.

Horizontal active thrust above point F:

$$P_{ah} = \frac{6.9 + 19.1}{2} \times 2.6 \times 1.0 + \frac{19.1 + 31.9}{2} \times 4.3 \times 1.0 = 143.4 \text{ kN},$$

(equal to the anchor rod force). The vertical component of the active thrust is:

$$P_{av} = P_{ah} \times \tan \delta' = 143.4 \times \tan 17.0° = 43.8 \text{ kN}.$$

The horizontal active thrust acting at the back of anchor plate is:

$$P_{anh} = \frac{6.9 + 19.1}{2} \times 2.6 \times 1.0 = 33.8 \text{ kN}.$$

The vertical component is:

$$P_{anv} = 33.8 \times \tan 17.0° = 10.3 \text{ kN}.$$

Calculate the weight of the sliding block including the vertical load due to the surcharge:

$W = 16.5(2.6 \times 13.0 \times 1.0) + 10.4(4.3 \times 13.0 / 2) \times 1.0 + 24.0 \times 13.0 \times 1.0 = 1160.4 \text{ kN}.$

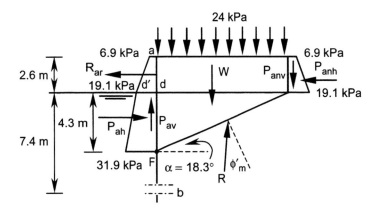

Figure 8.42. Example 8.19.

The sliding block and the corresponding forces are shown in Figure 8.42:
$$\alpha = \tan^{-1}(4.3/13.0) = 18.3°.$$
Horizontal equilibrium:
$$R\sin(\phi'_m - 18.3°) - 33.8 + 143.4 - 143.4 = 0.0.$$
Vertical equilibrium:
$$R\cos(\phi'_m - 18.3°) - 10.3 + 43.8 - 1160.4 = 0.0. \text{ Solving for } R \text{ and } \phi'_m:$$
$$R = 1127.4 \text{ kN}, \ \phi'_m = 20.0°.$$
From Equation 8.84:
$$F_{overall} = \tan 30.0° / \tan 20.0° = 1.59.$$

8.10 PROBLEMS

8.1 An 8 m high retaining wall retains a soil with two 4 m thick layers that have the following properties:
Upper layer: $c' = 10$ kPa, $\phi' = 18°$, $\gamma = 18$ kN/m³;
lower layer: $c' = 0$, $\phi' = 35°$, $\gamma = 18$ kN/m³. For a surface load $q = 50$ kPa, determine the active thrust and its distance from the base of the wall.

Answers: 295 kN, 3.42 m

8.2 A retaining wall of 5 m height retains a sloping backfill with $\beta = 20°$. The properties of backfill are:
$c' = 0$, $\phi' = 35°$, and $\gamma = 17$ kN/m³.
Determine the active thrust on the wall and its horizontal and vertical components.

Answers: 68.3 kN, 64.2 kN, 23.4 kN

8.3 Re-work Problem 8.1 assuming that the water table is located 2 m below the ground surface. The saturated unit weight for both layers is 19.5 kN/m³.

Answers: 250 kN (note this is not the total horizontal load applied to the wall), 3.62 m

8.4 A 10 m retaining wall retains soil with the following properties:
$c_u = 40$ kPa, $\gamma = 17.5$ kN/m^3, $c_w = 17.6$ kPa, and $\delta' = 0$.
Determine the magnitude of the active thrust:
(a) when the surface carries no surcharge,
(b) when a surface surcharge of 50 kPa is applied.
In both cases the ground surface is horizontal.

Answers: 178.5 kN, 475.4 kN

8.5 A retaining wall of height 12 m retains a two-layer soil having the following properties:
0 m - 5 m below the surface:
$c_u = 12$ kPa, $\phi_u = 0$, $\gamma = 17$ kN/m^3.
Below 5 m:
$c_u = 35$ kPa, $\phi_u = 0$, and $\gamma = 18$ kN/m^3.
Calculate the magnitude of the total active thrust and the critical value of α. For this purpose formulate P_a in terms of the angle α using the force diagram and set:
$\partial P_a / \partial \alpha$ to zero. $\delta' = 0$ and $c_w = 10$ kPa.

Answers: $\alpha = 40.63°$, $P_a = 578$ kN

8.6 A gravity concrete retaining wall is 6.6 m high and 3.2 m wide. If the thickness of the soil at the front of the wall is 2 m, determine the maximum and minimum base pressures assuming no base friction or adhesion. The soil has the following properties:
$c' = 0$, $\phi' = 35°$, ρ (for soil) = 1.8 Mg/m^3.
ρ (for concrete) = 2.4 Mg/m^3.

Answers: 61.6 kPa (at the heel), 249.2 kPa (at the toe)

8.7 For the gravity concrete retaining wall of Problem 8.6, determine the maximum and minimum base pressures assuming a base friction angle of $\delta' = 15°$, and a base adhesion of 10 kPa. A surcharge load of 20 kPa is applied vertically to the ground surface on the backfill. The soil has the following properties:
$c' = 10$ kPa, $\phi' = 25°$, ρ (for soil) = 1.8 Mg/m^3.
ρ (for concrete) = 2.4 Mg/m^3.

Answers: 35.9 kPa (at the heel), 274.8 kPa (at the toe)

8.8 A gravity concrete retaining wall with a vertical back supports the soil of Example 8.4. The width of the wall is 3.5 m at the base and 2 m at the top and has a 2 m vertical front supported by soil with:
$c' = 20$ kPa, $\phi' = 25°$ and $\gamma = 18$ kN/m^3.
The rest of the front face is sloping upwards, terminating at the ground surface. Determine:
(a) the factor of safety against sliding (the cohesion between the base of the wall and the soil is 20 kPa, and the mobilized friction angle on this interface is 25°),
(b) the factor of safety against overturning,
(c) the distribution of the contact pressure under the base of the wall.
Take the unit weight of the concrete as 24 kN/m^3.
Assume the back and front faces of the wall are smooth ($c_w = 0$, $\delta' = 0$).

Answers: 1.65, 1.61, −91.5 kPa (at the heel), 338.4 kPa (at the toe)

8.9 A cantilever sheet pile supports a 6 m high backfill with the following properties:

0 m - 2 m: $c' = 0$, $\phi' = 30°$, $\gamma = 16.5$ kN/m^3,

2 m - 4 m: $c' = 0$, $\phi' = 35°$, $\gamma = 17$ kN/m^3,

4 m - 6 m: $c' = 15$ kPa, $\phi' = 20°$, $\gamma = 17$ kN/m^3.

$D = 3.5$ m and the soil under the dredge line (at both sides) is the same as the soil under the 4 m depth.

Determine the factor of safety (in terms of c' and k_p) assuming a simplified net pressure diagram.

Note that the case of the cantilever sheet pile in c', ϕ' soil is not formulated in this chapter.

Answer: 1.28

8.10 A cantilever sheet pile supports a 9 m high backfill with the following properties:

0 m - 3 m: $c' = 0$, $\phi' = 35°$, $\gamma = 18$ kN/m^3,

3 m - 6 m: $c' = 15$ kPa, $\phi' = 20°$, $\gamma_{sat} = 20.3$ kN/m^3,

6 m - 9 m: $c' = 0$, $\phi' = 35°$, $\gamma_{sat} = 21.1$ kN/m^3.

There is a vertical surface load of 20 kPa applied at the backfill. The water table is 3 m below the ground surface of the backfill and has the same level at the front of the sheet pile. The soil under the dredge line (at both sides) is purely cohesive soil with: $c_u = 100$ kPa, and $\phi_u = 0$.

(a) Determine the embedment depth D assuming that the full passive resistance is mobilized,

(b) add a horizontal anchor rod at a depth of 1.5 m and with the same embedment depth of part (a) calculate the factor of safety in terms of mobilized cohesion under the dredge line and the anchor rod force,

(c) determine the location and magnitude of the maximum bending moment,

(d) if the anchor rod is supported by a concrete block anchor with thickness of 0.5 m, width of 2 m (parallel to the sheet pile), and height of 1.5 m, calculate the distance between anchor rods along the sheet pile (anchor rod is anchored at the centre point of the concrete block).

Assume no cohesion and friction resistance along the surfaces of the anchor. Include the surface load in both the active and passive thrusts and assume a factor of safety of 1.5 for the mobilized k_p at the front of the anchor.

Answers: 4.1 m, 2.49, 119.2 kN/m, $x = 0.41$ m (below the second layer),

$BM = 286.5$ kN.m/m, 2.6 m

8.11 An earth retaining wall 12 m high is reinforced with metal strips of width 100 mm, the first row of which is at a depth of 0.5 m. The strips are spaced at $s_x = 1$ m and $s_z = 1$ m, and the allowable tensile strength of the metal strip is 140 MPa. The thickness of the strip is 5 mm. The properties of backfill soil are:

$c' = 0$, $\phi' = 35°$, and $\gamma = 18$ kN/m^3.

The friction angle mobilized in the soil-strip contact area is $\delta_b = 23°$. Calculate the length of the reinforcements at depths 0.5 m, 5.5 m and 11.5 m.

Answers: 8.9 m, 7.6 m, 4.1 m

8.11 REFERENCES

AASHTO. 1997. *Standard specifications for highway bridges*. Section 5.8. Washington D.C.

Alpan, I. 1967. The empirical evaluation of the coefficient k_o and $k_{o,OCR}$. *Soils and foundations* 7(1): 31-40, Tokyo.

ASTM D-5262. 1997. *Standard test method for evaluating the unconfined tension creep behavior for geosynthetics*. Philadelphia: ASTM.

Benoit, J. & Lutenegger, A.J. 1993. Determining lateral stress in soft clays. In G.T. Houlsby & A.N. Schofield (eds), *Predictive soil mechanics*. London: Thomas Telford.

Bishop, A.W. 1958. Test requirements for measuring the coefficient of earth pressure at rest. *Proc. Brussels conf. on earth pressure problems*.

Bowles, J.E. 1996. *Foundation analysis and design*. New York: McGraw-Hill.

BS 8006. 1995. *Code of practice for strengthened/reinforced soils and other fills*. London: British Standard Institution.

BS 8081. 1989. *Ground anchorages*. London: British Standard Institution.

Burland, J.B., Potts, D.M. & Walsh, N.M. 1981. The overall stability of free and propped embedded cantilever retaining walls. *Ground engineering* 14(5): 28-38.

Cheney, R.S. 1988. *Permanent ground anchors*. Report FHWA/DP-68/IR U.S: 136. Washington D.C.: Department of Transportation Federal Highway Administration.

Das, B.M. 1990. *Earth anchors*. Amsterdam: Elsevier.

Das, B.M., Tarquin, A.J. & Moreno, R. 1985. Model tests for pullout resistance of vertical anchors in clay. *Civil eng. for practising and design engineers* 4(2): 191-209. New York: Program Press.

Duncan, M., Clough, G.W. & Ebeling, R.M. 1990. Behavior and design of gravity earth retaining structures. In P.C. Lambe & L.A. Hansen (eds), *Design and performance of earth retaining structures*: 251-277. New York: ASCE.

Duncan, J.M. & Seed, R.B. 1986. Compaction-induced earth pressure under k_o-conditions. *Journal SMFE, ASCE* 112(1): 1-22.

Exxon. 1992. *Designing for soil reinforcement*. 2nd edition, UK: Exxon Chemical Geopolymers.

Felio, G.Y., Vucetic, M., Hudson, M., Barar, O. & Chapman, R. 1990. Performance of soil nailed walls during the October 17, 1989 Loma Prieta earthquake. *Proc. 43rd Canadian geotechnical conf.*: 165-173. Quebec.

Flaate, K. & Peck, R.B. 1973. Braced cuts in sand and clay. *Norwegian geotechnical institute*, Publication 96.

Fleming, W.G.K., Weltman, A.J., Randolph, M.F. & Elson, W.K. 1992. *Piling engineering*. Surrey University Press: Halsted Press.

Hanna, T.H. 1982. *Foundations in tension-ground anchors*. 1st edition: 269-274. USA: McGraw – Hill.

Hueckel, S. 1957. Model tests on anchoring capacity of vertical and inclined plates. *Proc. 4th intern. conf. SMFE* 2: 203-206. London.

Jáky, J. 1948. Earth pressure in soils. *Proc. 2nd intern. conf. SMFE* 1: 103-107. Rotterdam.

Juran, I. & Schlosser, F. 1978. Theoretical analysis of failure in reinforced earth structures. *Proc. symp. on earth reinforcement, ASCE convention, Pittsburg*: 528-555. New York: ASCE.

Juran, I., Baudrand, G., Farrag, K. & Elias, V. 1988. Kinematical limit analysis for design of nailed structures. *Journal GED, ASCE* 116(1): 54-72.

Koerner, R.M. 1994. *Designing with geosynthetics*. 3rd edition. Englewood, New Jersey: Prentice Hall.

Kranz, E. 1953. *Über die verankerung von spundwänden*. Berlin: Verlag Ernst & Sohn.

Kulhawy, F.H., Jackson, C.S. & Mayne, P.W. 1989. First order estimation of k_o in sands and clays. *Foundation engineering: current principles and practices, Journal GED, ASCE* 1: 121-134.

Littlejohn, G.S. 1970. Soil anchors. *Proc. conf. ground engineering*: 33-44. London: Institution of Civil Engineers.

Littlejohn, G.S. & Bruce, D.A. 1977. *Rock anchors-state of the arts*: 50. UK: Foundation Publications Ltd..

Locher, H.G. 1969. *Anchored retaining walls and cut-off walls*. Berne: Losinger & Co..

Mayne, P.W. & Kulhawy, F.H. 1982. k_o-OCR relationships in soil. *Journal GED, ASCE* 108(GT6): 851-872.

Meyerhof, G.G. 1973. Uplift resistance of inclined anchors and piles. *Proc. intern. conf. SMFE* 2.1: 167-172. Moscow.

Munfakh, G.A. 1990. Innovative earth retaining structures: Selection, design, & performance. In P.C. Lambe & L.A. Hansen (eds), *Design and performance of earth retaining structures*: 85-118. New York: ASCE.

Munfakh, G.A., Abramson, L.W., Barksdale, R.D. & Juran, I. 1987. Soil improvement-a ten year update. In J.P. Welsh (ed), *ASCE geotechnical special publication* (12): 59. New York: ASCE.

O'Rourke, T.D. 1987. Lateral stability of compressible walls. *Geotechnique* 37(1): 145-149.

O'Rourke, T.D. & Jones, C.J.F.P. 1990. Overview of earth retention systems: 1970-1990. In P.C. Lambe & L.A. Hansen (eds), *Design and performance of earth retaining structures*: 22-51. New York: ASCE.

Ostermayer, H. 1977. *Practice on the detail design application of anchorages-A review of diaphragm walls*: 55-61. London: Institution of Civil Engineers.

Ovesen, N.K. & Stromann, H. 1972. Design methods for vertical anchor slabs in sand. *Proc. speciality conf. on performance of earth and earth supported structures* 2(1): 1481-1500. New York: ASCE.

Pálossy, L., Scharle, L. & Szalatkay, I. 1993. *Earth walls*. New York: Ellis Horwood.

Peck, R.B. 1969. Deep excavation and tunnelling in soft ground. *Proc. 7th intern. conf. SMFE* 225-290. Mexico.

Peck, R.B. 1990. Fifty years of lateral earth support. In P.C. Lambe & L.A. Hansen (eds), *Design and performance of earth retaining structures*. New York: ASCE.

Post-tensioning Institute (PTI). 1996. *Recommendations for prestressed rock and soil anchors*. Phoenix, Arizona.

Potts, D.M. & Fourie, A.B. 1984. The behaviour of a propped retaining wall: results of a numerical experiment. *Geotechnique* 34(3): 383-404.

Potts, D.M. & Fourie, A.B. 1985. The effect of wall stiffness on the behaviour of a propped retaining wall. *Geotechnique* 35(3): 347-352.

Rowe, P.W. 1952. Anchored sheet pile walls. *Proc. institution of civil engineers,* 1(1): 27-70.

Rowe, P.W. 1957. Sheet pile walls in clay. *Proc. institution of civil engineers,* 1(7): 629-654.

Schaefer, V.R., Abramson, L.W., Drumheller, J.C., Hussin, J.D. & Sharp, K.D. 1997. Ground improvement, ground reinforcement, ground treatment-developments 1987-1997. *ASCE geotechnical special publication* (69): 178-200. New York: ASCE.

Scott, C.R. 1980. *An introduction to soil mechanics and foundations*. 3rd edition. London: Applied Science Publishers.

Task force 27 (Federal Highway Administration). 1990. *In-situ soil improvement techniques: design guidelines for use of extensible reinforcements (geosynthetics) for mechanically stabilized earth walls in permanent applications*. Joint committee of AASHTO, AGC, ARTBA. Washington D.C.

Teng, W.C. 1962. *Foundation design*. Englewood Cliffs, New Jersey: Prentice-Hall.

Terzaghi, K. 1966. *Theoretical soil mechanics*. 14th edition. New York: John Wiley & Sons.

Terzaghi, K. & Peck, R. B. 1967. *Soil mechanics in engineering practice*. 2nd edition, New York: John Wiley & Sons.

Terzaghi, K., Peck, R. B. & Mesri, G. 1996. *Soil mechanics in engineering practice*. 3rd edition, New York: John Wiley & Sons.

Tschebotarioff, G.P. 1973. *Foundations, retaining and earth structures*. 2nd edition. New York: McGraw-Hill.

Unterreiner, P., Benhamida, B. & Schlosser, F. 1997. Finite element modelling of the construction of a full-scale experimental soil-nailed wall. French national research project CLOUTERRE. *Journal of ground improvement* 1(1): 1-8.

Williams, B.P. & Waite, D. 1993. *The design and construction of sheet-piled cofferdams*. Special publication 95. London: Construction Industry Research and Information Association.

Williams, G.W., Duncan, J.M. & Sehn, A.L. 1987. Simplified chart solution of compaction- induced earth pressures on rigid structures. *Geotechnical engineering report*. Blackburg, VA.: Virginia Polytechnic Institute and State University.

Wroth, C.P. & Houlsby, G.T. 1985. Soil mechanics-property characterization and analysis. *Proc. intern. conf. SMFE* 1: 1-56.

Xanthakos, P.P. 1991. *Ground anchors and anchored structures*. New York: John Willey & Sons.

Xanthakos, P.P., Abramson, L.W. & Bruce, D.A. 1994. *Handbook on ground control and improvement*. New York: John Willey & Sons.

CHAPTER 9

Stability of Earth Slopes

9.1 INTRODUCTION

This chapter examines the stability of earth slopes in two-dimensional space using a limit equilibrium method. Figures 9.1(a) and 9.1(b) illustrate a rotational failure where a mass of soil rotates on a circular or non-circular failure surface at its limit state. In the presence of a hard layer, a combined translational and rotational failure could occur (Figure 9.1(c)). An infinite slope is normally associated with a translational failure parallel to the ground surface (Figure 9.1(d)). The failure of a reinforced slope, however, is rather different and is idealized either by a combination of wedges (Department of Transport UK, 1994) or by a circular failure surface (Koerner & Robins, 1986; Koerner, 1994).

In a rotational circular failure, the mass of soil is treated as a rigid body sliding on the failure surface (Fellenius, 1927; Bishop, 1955). On this surface the Mohr-Coulomb failure criterion applies and the shear strength parameters used correspond to the peak strength obtained by a total or effective stress analysis. For a frictional soil, the mass is divided into vertical slices to facilitate the application of the force and moment equilibrium requirements. In Fellenius' method, only moment equilibrium is satisfied, but Bishop's simplified method satisfies both moment equilibrium and vertical force equilibrium. In non-circular failure surfaces, all of the methods introduced by Morgenstern & Price (1965), Spencer (1967) and Janbu (1968 and 1973) satisfy both force and moment equilibrium,

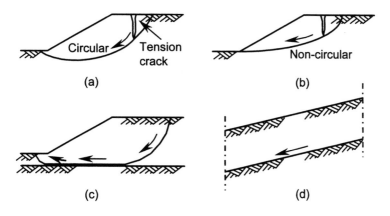

Figure 9.1. Idealized types of earth slope failure in two dimension.

but differ slightly in the assumptions made for the side forces acting on a slice. Circular failure surface analyses yield results that are sufficiently accurate for many practical purposes.

The factor of safety is defined as the ratio of the shear strength to the mobilized shear stress on the sliding surface required for equilibrium, and is assumed to be constant along the surface. Whilst the factor of safety actually varies along the sliding surface (Tavenas et al., 1980); the average value obtained from a traditional circular analysis is a reliable indication of overall slope stability.

The choice of drained or undrained conditions depends upon the magnitude of the time factor T_v defined by Equation 6.20. For the construction time t substituted into this equation, if the value of T_v exceeds 3, it is reasonable to treat the material as drained. If the value of T_v is less than 0.01, the material is treated as undrained (Duncan, 1996). For the values of T_v between the above limits both drained and undrained analyses must be considered.

In the absence of consolidation data, consideration should be given to the magnitude of the coefficient of permeability k. Soils with values of k greater than 10^{-6} m/s can be assumed to be drained, whilst those with values of k less than 10^{-9} m/s can be considered to be undrained. In both cut and fill slopes, it is customary to study their short-term stability in undrained conditions with shear strength parameters c_u, ϕ_u; their long-term stability, however, should be examined using the effective shear strength parameters c', ϕ' (Atkinson, 1993).

For homogenous soils, general solutions for the critical values (at failure) of the dimensionless parameters $\gamma H / c'$ or $c' / \gamma H$ have been obtained for specific values of ϕ' and the slope angle β (Taylor, 1948; Bishop & Morgenstern, 1960; Janbu, 1968 and 1973; Cousins, 1978; Duncan et al., 1987). For slopes reinforced by geosynthetics, general solutions and charts provided by Jewell et al. (1985), Jewell (1991 and 1996) can be used. In engineering practice however, an individual solution is usually required to accommodate geometric irregularities, variations in the shear strength parameters and different layers of soils. In a slope stability analysis the objective is to locate the circle (or any type of the sliding surface) that yields the minimum factor of safety. This usually requires that an iterative procedure be employed.

9.2   STABILITY OF SLOPES IN c_u, $\phi_u = 0$ SOIL – CIRCULAR FAILURE SURFACE

9.2.1   *Factor of safety in c_u, $\phi_u = 0$ soil using circular failure surface*

A plane strain slope of height H and angle β is shown in Figure 9.2(a). The trial circular failure surface is defined by its centre C, radius R and central angle θ. Shear stresses along the trial surface are due only to undrained cohesion and are mobilized to c_u / F (to maintain the equilibrium of the sliding block), where F is the factor of safety. The weight of the sliding block W acts at a distance d from the centre of the circle. Taking moments of the forces about the centre of the circular arc, and noting that the normal stresses on the arc pass through the centre, then:

$$W \times d - \int_0^\theta dl \times 1.0 \frac{c_u}{F} \times R = W \times d - \frac{c_u R}{F} \int_0^\theta dl = W \times d - \frac{c_u R^2 \theta}{F} = 0, \text{ thus:}$$

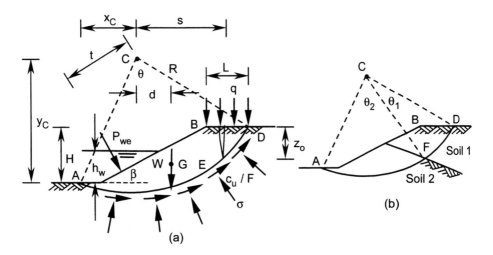

Figure 9.2. Slope failure in c_u, $\phi_u = 0$ soil.

$$F = \frac{c_u R^2 \theta}{Wd} = \frac{c_u L_a R}{Wd} \qquad (9.1)$$

where L_a is the length of the circular arc. If a surcharge loading q is applied to the upper ground surface and external water exists at the front of the slope (Figure 9.2(a)):

$$F = \frac{c_u R^2 \theta}{Wd + qLs - P_{we}t} \qquad (9.2)$$

If the soil is composed of two different layers (Figure 9.2(b)), F is obtained from:

$$F = \frac{R^2(c_{u1}\theta'_1 + c_{u2}\theta'_2)}{Wd} \qquad (9.3)$$

where c_{u1} and c_{u2} are the undrained cohesions of soil 1 and soil 2 respectively and θ'_1 and θ'_2 are the corresponding central angles.

The stability of an earth slope in undrained conditions can be expressed in terms of a dimensionless parameter N called the *stability number*:

$$N = \frac{\gamma H}{c_u} \qquad (9.4)$$

It can be shown that for a specified value of β, the magnitude of N at failure has a constant value (N_f) and, as a consequence, the factor of safety (Equations 9.1) may be presented by:

$$F = \frac{N_f}{N_d} \qquad (9.5)$$

where N_d is the stability number corresponding to the design values of γ, H and c_u. The centroid of the sliding mass is obtained using a mathematical procedure based on the geometry or the sub-division of the sliding mass into narrow vertical slices.

Example 9.1

Find the factor of safety of a 1 vertical to 1.5 horizontal slope that is 6 m high. The centre of the trial circle is located 2.5 m to the right of and 9.15 m above the toe of the slope. $c_u = 25$ kPa, and $\gamma = 18$ kN/m^3. The circle passes through the toe of the slope.

Solution:

Geometrical data are: $\theta = 85.9°$, area of the sliding mass $= 29.87$ m^2 and $d = 3.85$ m.

$$R = \sqrt{9.15^2 + 2.5^2} = 9.48 \text{ m}, \ W = 29.87 \times 1.0 \times 18.0 = 537.7 \text{ kN}.$$

$$F = \frac{c_u R^2 \theta}{Wd} = \frac{25.0 \times 9.48^2 \times (85.9°/180.0°)\pi}{537.7 \times 3.85} = 1.63.$$

9.2.2   The effect of tension cracks

Tension cracks may develop from the upper ground surface to a depth z_o that can be estimated using Equation 8.12. The effect of a tension crack can be taken into account by assuming that the trial failure surface terminates at the depth z_o, thereby reducing the weight W and central angle θ. Any external water pressure in the crack creates a horizontal force that must be included in equilibrium considerations.

Example 9.2

Re-work Example 9.1 by taking into account tension cracks.

Solution:

Use Equation 8.12: $z_o = 2c_u/\gamma = 2 \times 25.0/18.0 = 2.78$ m.

From the geometry of the circle and slope (calculations omitted): $\theta = 66.6°$, area of the sliding mass $= 27.46$ m^2 and $d = 3.48$ m.

$W = 27.46 \times 1.0 \times 18.0 = 494.3$ kN. Calculate the horizontal force (P_w) due to the water pressure in the tension crack and its vertical distance (y_1) from the centre of the circle:

$$P_w = 9.81 \times 2.78^2/2 = 37.9 \text{ kN}, \ y_1 = y_C - H + 2z_o/3 = 9.15 - 6.0 + 2 \times 2.78/3 = 5.0 \text{ m}.$$

Parameters z_o and y_C are defined in Figure 9.2(a). Taking moments about the centre:

$$F = c_u R^2 \theta /(Wd + P_w y_1),$$

$$F = (25.0 \times 9.48^2 \times 66.6° \times \pi/180°)/(494.3 \times 3.48 + 37.9 \times 5.0) = 1.37.$$

9.2.3   Location of the critical circle

Cousins (1978) developed a series of charts for homogeneous soils using extensive computer analyses to investigate the effects of pore pressure and a hard stratum (located horizontally under the slope) on the magnitude of the stability number N. Figure 9.3 is reconstructed for a c_u, $\phi_u = 0$ soil where the hard layer is deep and has no effect on the mode of failure. The coordinate of the centre of the critical circle through the toe (x_C, y_C) in the xAy plane (with its origin A at the toe) is normalized in terms of height and slope angle. Note that the most critical circle may or may not be a toe circle. Selecting a trial circle using Figure 9.3 or other available charts can reduce the number of iterations in the search for critical circle.

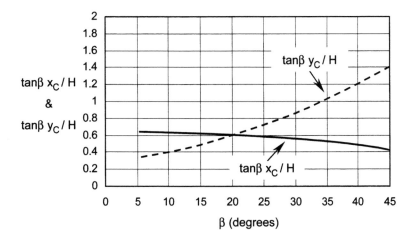

Figure 9.3. The position of the critical toe circle in c_u, $\phi_u = 0$ soil (Cousins, 1978).

Example 9.3

A cut 10 m deep is to be made in a stratum of cohesive soil for which $c_u = 45$ kPa. The slope angle $\beta = 40°$ and the soil has a unit weight of 17 kN/m^3. Using Cousins's chart of Figure 9.3, find the factor of safety for the critical toe circle.

Solution:

From Figure 9.3: $\tan\beta x_C / H \approx 0.47$, $\tan\beta y_C / H \approx 1.20$ (selected with eye accuracy).

$x_C = 0.47 \times 10.0 / \tan 40.0° \approx 5.6$ m, $y_C = 1.20 \times 10.0 / \tan 40.0° \approx 14.3$ m.

The central angle, area of the sliding block and the position of its centroid are found to be:

$\theta = 95.1°$, $S = 120.4$ m^2, $d = 6.21$ m.

$R = \sqrt{5.6^2 + 14.3^2} = 15.4$ m, $W = 120.4 \times 1.0 \times 17.0 = 2046.8$ kN. From Equation 9.1:

$F = (45.0 \times 15.4^2 \times 95.1° \times \pi / 180°) / (2046.8 \times 6.21) = 1.39 \approx 1.4$.

9.2.4   Taylor's stability charts

In undrained conditions, a horizontal hard stratum located $n_d H$ below the upper ground surface affects the critical stability number N_f. The stability number increases as n_d decreases. For $\beta > 53°$, the critical circle is a toe circle and the hard stratum has no effect on the stability number. The reconstructed slope stability chart developed by Taylor (1948) is shown in Figure 9.4, where the dashed curves represent the undrained conditions.

Example 9.4

Using Taylor's stability chart of Figure 9.4, determine the factor of safety for the slope of Example 9.3.

Solution:

For $\beta = 40°$, the stability number from the chart is 5.52.

$N_d = \gamma H / c_u = 17.0 \times 10.0 / 45.0 = 3.78$, thus $F = 5.52 / 3.78 = 1.46$.

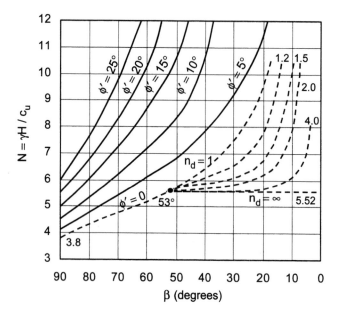

Figure 9.4. Relationship between stability number and slope angle.

9.3 STABILITY OF SLOPES IN c', ϕ' SOIL – THE METHOD OF SLICES

9.3.1 *Forces acting on a slice and moment equilibrium*

The soil mass above a trial circle is divided into a series of vertical slices of width b as shown in Figure 9.5(a). For each slice, its base is assumed to be a straight line defined by its angle of inclination α with the horizontal whilst its height h is measured along the centreline of the slice. The forces acting on a slice shown in Figure 9.5(b) are as follows:

w = total weight of the slice = $h \times b \times \gamma$ where γ is the unit weight.

N = total normal force on the base = $N' + U$,

where N' is the effective normal force and $U = ul$ is the force due to the pore pressure at the midpoint of the base of length l.

T = mobilized shear force on the base = $\tau_m l$,

where τ_m is the shear stress required for equilibrium and is equal to the shear strength divided by the factor of safety:

$$\tau_m = \frac{\tau_f}{F}.$$

X_1, X_2 = shear forces on sides of slice.

E_1, E_2 = normal forces on sides of slice.

Due to the internal nature of the side forces the sum of their moments about the centre is zero. Thus for moment equilibrium about C:

$$\sum_{i=1}^{i=n} T_i R = R \sum_{i=1}^{i=n} (\tau_m l)_i = R \sum_{i=1}^{i=n} \frac{(\tau_f l)_i}{F} = \sum_{i=1}^{i=n} (w \sin \alpha)_i R, \quad i = 1, 2, \ldots n, \text{ and } n \text{ is the total number}$$

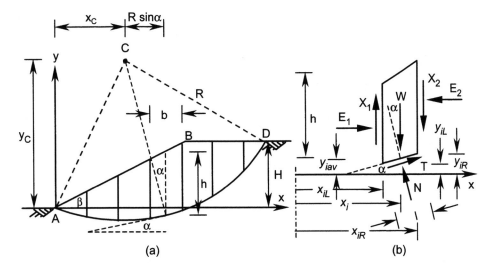

Figure 9.5. The method of slices for c', ϕ' soil.

of slices. Replacing τ_f by $c' + \sigma'_n \tan\phi'$ (Mohr-Coulomb failure criterion) we obtain:

$$F = \frac{\sum\limits_{i=1}^{i=n}\left[(c' + \sigma'_n \tan\phi')l\right]_i}{\sum\limits_{i=1}^{i=n}(w\sin\alpha)_i} = \frac{\sum\limits_{i=1}^{i=n}(c'l + N'\tan\phi')_i}{\sum\limits_{i=1}^{i=n}(w\sin\alpha)_i} \qquad (9.6)$$

The term $c'l$ may be replaced by $c'b / \cos\alpha$. For uniform c', the algebraic summation of $c'l$ is replaced by $c'L_a$, where L_a is the length of the circular arc. The value of N' must be determined from the force equilibrium equations. As the problem is statically indeterminate, some simplifying assumptions have to be made. Two common methods that apply different simplifying assumptions are Fellenius' method and Bishop's simplified method.

9.3.2   *Fellenius' method*

This method assumes that for each slice $X_1 = X_2$ and $E_1 = E_2$. Resolving the system of forces normal to the base of each slice:

$N = w\cos\alpha$, $N' + ul = w\cos\alpha$, and $N' = w\cos\alpha - ul$.

Substituting N' into Equation 9.6 and assuming $r_u = ub/w$ gives:

$$F = \frac{\sum\limits_{i=1}^{i=n}\left[c'l + (w\cos\alpha - ul)\tan\phi'\right]_i}{\sum\limits_{i=1}^{i=n}(w\sin\alpha)_i} = \frac{\sum\limits_{i=1}^{i=n}\left[c'l + w(\cos\alpha - r_u \sec\alpha)\tan\phi'\right]_i}{\sum\limits_{i=1}^{i=n}(w\sin\alpha)_i} \qquad (9.7)$$

The parameter r_u is dimensionless because the term $ub = \gamma_w \times h_w \times b \times 1$ represents the weight of water with a volume of $h_w \times b \times 1.0$, where h_w is the height of the water above the midpoint of the base. Furthermore, r_u can be simplified as follows:

$$r_u = \frac{ub}{w} = \frac{ub}{bh\gamma} = \frac{u}{\gamma h} = \frac{\gamma_w h_w}{\gamma h} \tag{9.8}$$

In the case of steady seepage, the height of water above the midpoint of the base is obtained by constructing the flow net. Alternatively, hydrostatic conditions may be assumed with an average r_u value for the slope. By doing so it is assumed that the height of the water above the base of each slice is a constant fraction of the average height of each slice. If the height of the water and the average height of the slice are equal, the maximum value of r_u becomes γ_w / γ which is approximately 0.5. Note that the effective normal force N' acting on the base is equal to $N' = w\cos\alpha - ul$ or $w(\cos\alpha - r_u\sec\alpha)$. If the term $\cos\alpha - r_u\sec\alpha$ is negative, N' is set to zero because effective stress cannot be less than zero. The whole procedure must be repeated for number of trial circles until the minimum factor of safety corresponding to the critical circle is determined. The accuracy of the predictions depends on the number of slices, position of the critical circle (shallow or deep) and the magnitude of r_u. For high values of r_u and circles with a relatively long radius, a conservative value for the factor of safety is obtained that could be 20% or more on the low side.

Example 9.5

Using Fellenius' method of slices, determine the factor of safety for a slope with the same geometry of Example 9.1 for $r_u = 0$ and 0.4. Take the number of slices as 8, each having 1.5 metre width (check the width of the last slice). Soil properties are:
$c' = 10$ kPa, $\phi' = 29°$, and $\gamma = 18$ kN/m^3.

Solution:

From the calculations in Example 9.1 and the geometry of the circle, it can be shown that $R = 9.48$ m and $x_D = 11.44$ m (point D is defined in Figure 9.5(a)). The width of slice 8 is: $11.44 - 7 \times 1.5 = 0.94$ m. Equation of the trial circle in xAy coordinate system is: $(x - 2.50)^2 + (y - 9.15)^2 = 9.48^2$. Differentiating this equation with respect to x: $2(x - 2.50) + 2(y - 9.15)(dy/dx) = 0$, $dy/dx = \tan\alpha = -(x - 2.50)/(y - 9.15)$.

For each slice, average values of y, α, and h are tabulated below along with the other terms needed for the conventional Fellenius' method. Sample calculation for slice 6:

Slice	y_{av} (m)	h (m)	α (deg.)	w (kN)	$r_u = 0$ $w\cos\alpha$ (kN)	$r_u = 0.4$ $w(\cos\alpha - r_u\sec\alpha)$ (kN)	$w\sin\alpha$ (kN)
1	-0.14	0.64	-10.6	17.28	16.98	9.95	-3.18
2	-0.30	1.80	-1.5	48.60	48.58	29.14	-1.27
3	-0.22	2.72	7.6	73.44	72.79	43.16	9.71
4	0.10	3.40	16.9	91.80	87.83	49.46	26.68
5	0.71	3.79	26.6	102.33	91.50	45.72	45.82
6	1.66	3.84	37.3	103.68	82.47	30.34	62.83
7	3.15	2.85	49.9	76.95	49.56	1.78	58.86
8	5.03	0.97	63.2	16.41	7.40	0.00	14.65
Total:					457.11	209.55	214.10

x_6 (x coordinate at the mid point of the arc at slice 6) $= 5\times1.5+1.5/2 = 8.25$ m, substituting into the equation of the circle we find: $y_6 = 1.60$ m, thus:

$\alpha_6 = \tan^{-1}[-(8.25-2.5)/(1.60-9.15)] = 37.3°$. Find the y value at the middle of the base (not arc) y_{6av} by taking the average values of y at the two ends of the base, thus:

$h = x_i \tan\beta - y_{iav} = x_i/1.5 - y_{iav}$ for $x \le x_B = 9.0$ m, and

$h = 6.0 - y_{iav}$ for $x \ge x_B = 9.0$ m, where point B is defined in Figure 9.5(a).

For x_{6L} (x coordinate at the left end of the base) $= 5 \times 1.5 = 7.5$ m, $y_{6L} = 1.09$ m.

For x_{6R} (x coordinate at the right end of the base) $= 6 \times 1.5 = 9.0$ m, $y_{6R} = 2.24$ m.

$y_{6av} = (1.09+2.24)/2 = 1.66$ m, $h_6 = 8.25/1.5 - 1.66 = 3.84$ m.

$w_6 = 1.5\times3.84\times1.0\times18.0 = 103.68$ kN.

$\sum c'l = c'\sum l = 10.0(85.9°\times\pi/180.0°)9.48 = 142.13$. Using Equation 9.7 for $r_u = 0$:

$F = (142.13+457.11\times\tan 29.0°)/214.10 = 1.85$.

For $r_u = 0.4$: $F = (142.13+209.55\times\tan 29.0°)/214.10 = 1.21$.

9.3.3   Bishop's simplified method

This method assumes that for each slice $X_1 = X_2$ but $E_1 \ne E_2$. These assumptions are considered to make this method more accurate than Fellenius' method. An increase of 5% to 20% in the factor of safety over Fellenius' method is usually realised. Referring to Figure 9.5(b) and writing force equilibrium in vertical (in order to eliminate E_1 and E_2), an equation for N' can be found and substituted into Equation 9.6. Putting $l = b / \cos\alpha$ and $ub = wr_u$, we obtain:

$N' = (w-ul-c'l\sin\alpha/F)/(\cos\alpha+\sin\alpha\tan\phi'/F)$, which in turn gives:

$$F = \frac{1}{\sum\limits_{i=1}^{i=n}(w\sin\alpha)_i}\sum_{i=1}^{i=n}\left[\frac{c'b+w(1-r_u)\tan\phi'}{m_\alpha}\right]_i \tag{9.9}$$

where m_α is defined by:

$$m_\alpha = \cos\alpha+\sin\alpha\tan\phi'/F \tag{9.10}$$

Equation 9.9 is non-linear in F and is solved by fixed-point iteration. An initial value for F is guessed (slightly greater than the F obtained by Fellenius' method) and substituted into the appropriate terms of Equation 9.9 to compute a new value for F. This procedure is repeated until the difference between the assumed and computed F becomes negligible. Convergence is normally rapid and only 2 or 3 iterations are required. The procedure is repeated for number of trial circles to locate the critical failure surface with the lowest factor of safety. Due to the computational nature of this method, it can be readily programmed on a computer and extended to deal with layered soils, irregular slope geometry (Bromhead, 1992) and variable r_u (King, 1989).

Example 9.6

Re-work Example 9.5 for $r_u = 0.4$ using Bishop's simplified method.

Solution:

A summary of the computations is tabulated below where 3 iterations have been carried out. The initial value for F was taken as 1.2. For subsequent iterations, the initial value of F is that computed from the previous iteration. The final factor of safety for the selected trial circle is 1.34.

Slice	(a) $c' \times b$ (kN)	(b) $w \times (1-r_u) \times$ $\tan\phi'$ (kN)	(c) $(a)+(b)$ (kN)	$F = 1.200$ $\dfrac{(d)}{m_a}$	$(c)/(d)$	$F = 1.319$ $\dfrac{(d)}{m_a}$	$(c)/(d)$	$F = 1.339$ $\dfrac{(d)}{m_a}$	$(c)/(d)$
1	15.0	5.75	20.75	0.898	23.11	0.906	22.90	0.907	22.88
2	15.0	16.16	31.16	0.987	31.57	0.989	31.51	0.989	31.51
3	15.0	24.42	39.42	1.052	37.47	1.047	37.65	1.046	37.69
4	15.0	30.53	45.53	1.091	41.73	1.079	42.20	1.077	42.27
5	15.0	34.03	49.03	1.101	44.53	1.082	45.31	1.080	45.40
6	15.0	34.48	49.48	1.075	46.03	1.050	47.12	1.046	47.30
7	15.0	25.59	40.59	0.997	40.71	0.966	42.02	0.961	42.24
8	9.40	5.46	14.86	0.863	17.22	0.826	17.99	0.820	18.12
				Total = 282.37		Total = 286.70		Total = 287.41	
				$F = 1.319$		$F = 1.339$		$F = 1.342$	

9.3.4   Stability coefficients by Bishop & Morgenstern (1960)

Bishop & Morgenstern (1960) developed a method using two stability coefficients, m and n, that satisfy the following equation:

$$F = m - nr_u \qquad (9.11)$$

Whitlow (1990) recalculated the m and n values over an extended range and these are given in Table 9.1. The stability number is defined as $c' / \gamma H$, which is the inverse of the definition expressed in Equation 9.4.

Using Bishop's simplified method, Chandler & Peiris (1989) published results for the coefficients m and n that are in good agreement with the values in Table 9.1. To estimate the factor of safety, the procedure is as follows:

1. Calculate $c' / \gamma H$ from the soil and slope data.
2. For a value of $c' / \gamma H$ just greater than that found in step 1, use the corresponding section of Table 9.1 and find m and n for $n_d = 1$. Use linear interpolation (for ϕ' values) if necessary.
3. If n is underlined the critical circle is at a greater depth. Use the next higher value of n_d to find a non-underlined n. Use linear interpolation (for ϕ' values) if necessary.
4. Repeat steps 2 and 3 for values of $c' / \gamma H$ just less than that found in step 1.
5. Use Equation 9.11 to obtain two factors of safety for the upper and lower values of $c' / \gamma H$.
6. Calculate the final factor of safety by interpolating between the above two values.

Table 9.1 Stability coefficients by Bishop & Morgenstern (1960) recalculated by Whitlow (1990).

$c'/\gamma H = 0.000$

cot β		0.5: 1		1: 1		2: 1		3: 1		4: 1		5: 1	
n_d	ϕ'	m	n	m	n	m	n	m	n	m	n	m	n
All	20°	0.18	0.90	0.36	0.72	0.73	0.90	1.08	1.21	1.45	1.54	1.81	1.88
	25°	0.23	1.16	0.47	0.92	0.92	1.16	1.40	1.55	1.86	1.97	2.32	2.41
	30°	0.29	1.43	0.58	1.15	1.15	1.43	1.72	1.91	2.30	2.44	2.88	2.98
	35°	0.35	1.74	0.70	1.39	1.39	1.74	2.10	2.32	2.79	2.97	3.48	3.62
	40°	0.42	2.09	0.83	1.67	1.67	2.09	2.51	2.79	3.34	3.55	4.18	4.34

$c'/\gamma H = 0.025$

cot β		0.5: 1		1: 1		2: 1		3: 1		4: 1		5: 1	
n_d	ϕ'	m	n	m	n	m	n	m	n	m	n	m	n
1.00	20°	0.52	0.72	0.70	0.76	1.11	1.01	1.53	1.34	1.95	1.69	2.37	2.04
	25°	0.59	0.79	0.83	0.96	1.35	1.27	1.87	1.69	2.39	2.13	2.91	2.59
	30°	0.67	0.88	0.97	1.19	1.60	1.56	2.23	2.07	2.86	2.61	3.41	3.17
	35°	0.76	1.00	1.13	1.44	1.87	1.88	2.63	2.50	3.38	3.15	4.14	3.83
	40°	0.86	1.17	1.30	1.72	2.18	2.24	3.07	2.98	3.95	3.76	4.85	4.56
1.25	20°	1.00	0.93	1.07	1.02	1.29	1.20	1.60	1.45	1.93	1.76	2.30	2.06
	25°	1.22	1.18	1.31	1.30	1.60	1.53	1.97	1.87	2.42	2.25	2.87	2.65
	30°	1.46	1.47	1.59	1.62	1.95	1.91	2.41	2.33	2.93	2.80	3.49	3.28
	35°	1.74	1.76	1.90	1.96	2.32	2.31	2.89	2.83	3.50	3.38	4.17	3.98
	40°	2.04	2.11	2.23	2.35	2.74	2.75	3.43	3.39	4.14	4.04	4.93	4.75

$c'/\gamma H = 0.050$

cot β		0.5: 1		1: 1		2: 1		3: 1		4: 1		5: 1	
n_d	ϕ'	m	n	m	N	m	n	m	n	m	n	m	n
1.00	20°	0.69	0.78	0.90	0.83	1.37	1.06	1.83	1.38	2.32	1.77	2.77	2.08
	25°	0.80	0.98	1.05	1.03	1.61	1.33	2.18	1.75	2.77	2.20	3.33	2.64
	30°	0.91	1.21	1.21	1.24	1.88	1.62	2.56	2.15	3.24	2.68	3.91	3.24
	35°	1.02	1.40	1.37	1.46	2.17	1.95	2.99	2.78	3.58	3.25	4.57	3.96
	40°	1.14	1.61	1.55	1.71	2.50	2.32	3.44	3.06	4.40	3.91	5.30	4.64
1.25	20°	1.16	0.98	1.24	1.07	1.50	1.26	1.82	1.48	2.22	1.79	2.63	2.10
	25°	1.40	1.23	1.50	1.35	1.81	1.59	2.21	1.89	2.70	2.28	3.19	2.67
	30°	1.65	1.51	1.77	1.66	2.14	1.94	2.63	2.33	3.20	2.81	3.81	3.30
	35°	1.93	1.82	2.08	2.00	2.53	2.33	3.10	2.84	3.78	3.39	4.48	4.01
	40°	2.24	2.16	2.42	2.38	2.94	2.78	3.63	3.38	4.41	4.07	5.22	4.78
1.50	20°	1.48	1.28	1.55	1.33	1.74	1.49	2.00	1.69	2.33	1.98	2.68	2.27
	25°	1.82	1.63	1.90	1.70	2.13	1.89	2.46	2.17	2.85	2.52	3.28	2.88
	30°	2.18	2.01	2.28	2.09	2.56	2.33	2.95	2.69	3.42	3.10	3.95	3.56
	35°	2.57	2.42	2.68	2.52	3.02	2.82	3.50	3.25	4.05	3.75	4.69	4.31
	40°	3.02	2.91	3.16	3.02	3.55	3.37	4.11	3.90	4.77	4.48	5.50	5.12

Example 9.7

For a slope of 1 vertical to 3 horizontal and of height 17 m, compute the factor of safety for $r_u = 0.2$. Relevant soil properties are: $c' = 12$ kPa, $\phi' = 22°$ and $\gamma = 18$ kN/m^3.

$c'/\gamma H = 0.075$

cot β		0.5:1		1:1		2:1		3:1		4:1		5:1	
n_d	ϕ'	m	n	m	n	m	n	m	n	m	n	m	n
1.00	20°	0.85	0.80	1.09	0.84	1.61	1.10	2.14	1.44	2.66	1.80	3.17	2.13
	25°	0.95	1.01	1.25	1.05	1.86	1.38	2.50	1.80	3.13	2.26	3.74	2.72
	30°	1.06	1.24	1.42	1.30	2.14	1.69	2.88	2.20	3.62	2.76	4.36	3.33
	35°	1.19	1.49	1.61	1.56	2.44	2.03	3.31	2.66	4.18	3.33	5.02	4.00
	40°	1.33	1.76	1.80	1.82	2.77	2.39	3.78	3.15	4.79	3.95	5.78	4.76
1.25	20°	1.34	1.02	1.39	1.09	1.69	1.29	2.07	1.54	2.49	1.82	2.95	2.17
	25°	1.58	1.28	1.66	1.39	2.00	1.64	2.47	1.96	2.97	2.32	3.52	2.73
	30°	1.83	1.56	1.94	1.70	2.35	2.01	2.89	2.39	3.50	2.86	4.15	3.36
	35°	2.11	1.87	2.25	2.03	2.73	2.39	3.36	2.87	4.08	3.46	4.83	4.04
	40°	2.42	2.21	2.58	2.40	3.15	2.84	3.89	3.43	4.73	4.13	5.60	4.83
1.50	20°	1.64	1.31	1.71	1.35	1.92	1.51	2.20	1.73	2.55	1.99	2.93	2.27
	25°	1.98	1.66	2.05	1.71	2.31	1.91	2.66	2.20	3.08	2.53	3.55	2.92
	30°	2.34	2.04	2.43	2.10	2.74	2.36	3.16	2.71	3.66	3.13	4.22	3.59
	35°	2.74	2.46	2.84	2.54	3.21	2.85	3.71	3.29	4.30	3.79	4.96	4.34
	40°	3.19	2.93	3.31	3.03	3.74	3.40	4.33	3.93	5.03	4.53	5.79	5.19

$c'/\gamma H = 0.100$

cot β		0.5:1		1:1		2:1		3:1		4:1		5:1	
n_d	ϕ'	m	n	m	n	m	n	m	n	m	n	m	n
1.00	20°	0.98	0.80	1.25	0.86	1.83	1.13	2.41	1.46	2.97	1.83	3.53	2.15
	25°	1.10	1.02	1.41	1.07	2.09	1.42	2.78	1.84	3.36	2.29	4.09	2.72
	30°	1.21	1.25	1.58	1.30	2.37	1.72	3.17	2.25	3.91	2.80	4.71	3.34
	35°	1.34	1.50	1.77	1.57	2.68	2.08	3.59	2.71	4.49	3.34	5.39	4.03
	40°	1.48	1.78	1.99	1.87	3.01	2.44	4.07	3.21	5.10	3.97	6.14	4.80
1.25	20°	1.48	1.03	1.52	1.09	1.86	1.29	2.27	1.55	2.74	1.83	3.23	2.15
	25°	1.72	1.29	1.79	1.38	2.19	1.63	2.67	1.96	3.21	2.32	3.81	2.74
	30°	1.99	1.59	2.08	1.73	2.53	2.00	3.09	2.41	3.73	2.84	4.42	3.35
	35°	2.27	1.90	2.40	2.07	2.91	2.41	3.58	2.90	4.30	3.44	5.10	4.04
	40°	2.58	2.23	2.74	2.44	3.33	2.85	4.09	3.44	4.96	4.11	5.88	4.84
1.50	20°	1.77	1.30	1.85	1.36	2.07	1.52	2.38	1.73	2.76	2.00	3.14	2.28
	25°	2.11	1.66	2.20	1.72	2.47	1.93	2.83	2.21	3.28	2.53	3.78	2.91
	30°	2.48	2.05	2.58	2.11	2.90	2.38	3.33	2.72	3.86	3.12	4.44	3.59
	35°	2.88	2.47	2.98	2.54	3.37	2.86	3.88	3.28	4.49	3.78	5.17	4.34
	40°	3.33	2.94	3.45	3.03	3.90	3.42	4.49	3.92	5.21	4.51	5.99	5.16

Solution:

Step 1: calculate the dimensionless parameter $\dfrac{c'}{\gamma H} = \dfrac{12.0}{18.0 \times 17.0} = 0.0392$.

Step 2: for $c'/\gamma H = 0.050$, $\cot\beta = 3.0$ and $D = 1$, n is underlined for the range of ϕ' from 20° to 25°.

Step 3: select $D = 1.25$ for a deeper critical circle.

For $\phi' = 20°$, $m = 1.82$, $n = 1.48$ and for $\phi' = 25°$, $m = 2.21$, $n = 1.89$.

Interpolating linearly for $\phi' = 22°$:

$c'/\gamma H = 0.125$

cot β		0.5:1		1:1		2:1		3:1		4:1		5:1	
n_d	φ'	m	n	m	n	m	n	m	n	m	n	m	n
1.00	20°	1.13	0.81	1.43	0.88	2.04	1.15	2.69	1.54	3.26	1.78	3.87	2.12
	25°	1.25	1.04	1.60	1.11	2.32	1.45	3.06	1.91	3.74	2.27	4.45	2.72
	30°	1.38	1.27	1.77	1.34	2.62	1.78	3.46	2.30	4.25	2.81	5.07	3.37
	35°	1.50	1.51	1.96	1.59	2.93	2.12	3.88	2.71	4.82	3.41	5.77	4.05
	40°	1.61	1.75	2.17	1.89	3.27	2.48	4.36	3.18	5.46	4.06	6.55	4.89
1.25	20°	1.64	1.06	1.67	1.10	2.05	1.32	2.49	1.58	2.98	1.86	3.50	2.17
	25°	1.89	1.33	1.94	1.40	2.38	1.67	2.89	1.99	3.48	2.38	4.08	2.75
	30°	2.16	1.63	2.23	1.73	2.73	2.04	3.32	2.43	4.01	2.92	4.71	3.41
	35°	2.45	1.95	2.56	2.09	3.11	2.45	3.80	2.93	4.59	3.50	5.41	4.13
	40°	2.77	2.30	2.92	2.49	3.54	2.91	4.33	3.49	5.24	4.16	6.21	4.95
1.50	20°	1.92	1.32	2.02	1.39	2.23	1.55	2.57	1.75	2.96	2.00	3.40	2.29
	25°	2.26	1.68	2.37	1.75	2.64	1.97	3.03	2.23	3.50	2.55	4.02	2.91
	30°	2.63	2.07	2.75	2.15	3.07	2.43	3.53	2.75	4.08	3.15	4.69	3.60
	35°	3.04	2.50	3.16	2.58	3.55	2.92	4.08	3.32	4.73	3.81	5.44	4.36
	40°	3.50	2.98	3.63	3.07	4.09	3.49	4.71	3.98	5.46	4.57	6.28	5.23

$c'/\gamma H = 0.150$

cot β		0.5:1		1:1		2:1		3:1		4:1		5:1	
n_d	φ'	m	n	m	n	m	n	m	n	m	n	m	n
1.00	20°	1.25	0.81	1.58	0.89	2.25	1.16	2.89	1.44	3.57	1.80	4.21	2.15
	25°	1.37	1.02	1.75	1.12	2.53	1.45	3.24	1.80	4.01	2.27	4.78	2.77
	30°	1.50	1.25	1.93	1.36	2.83	1.78	3.64	2.24	4.54	2.79	5.41	3.39
	35°	1.65	1.53	2.12	1.61	3.14	2.14	4.09	2.71	5.10	3.38	6.09	4.09
	40°	1.80	1.82	2.33	1.89	3.49	2.53	4.57	3.24	5.74	4.05	6.86	4.85
1.25	20°	1.79	1.07	1.80	1.10	2.22	1.32	2.69	1.59	3.22	1.86	3.77	2.17
	25°	2.03	1.33	2.07	1.40	2.55	1.68	3.09	2.01	3.71	2.37	4.33	2.76
	30°	2.30	1.63	2.37	1.74	2.90	2.06	3.51	2.44	4.22	2.92	4.96	3.38
	35°	2.60	1.96	2.69	2.08	3.28	2.47	4.00	2.94	4.81	3.50	5.66	4.10
	40°	2.92	2.33	3.05	2.44	3.72	2.92	4.53	3.48	5.46	4.17	6.44	4.92
1.50	20°	2.05	1.33	2.15	1.39	2.38	1.54	2.74	1.75	3.15	2.01	3.63	2.30
	25°	2.39	1.68	2.51	1.76	2.77	1.97	3.19	2.23	3.67	2.55	4.23	2.90
	30°	2.76	2.07	2.89	2.16	3.22	2.43	3.70	2.75	4.26	3.14	4.90	3.57
	35°	3.16	2.50	3.30	2.59	3.69	2.92	4.24	3.31	4.90	3.79	5.64	4.33
	40°	3.62	2.98	3.76	3.07	4.23	3.48	4.87	3.95	5.63	4.54	6.47	5.19

$$m = 1.82 + \frac{2.21 - 1.82}{(25.0° - 20.0°)}(22.0° - 20.0°) = 1.976,$$

$$n = 1.48 + \frac{1.89 - 1.48}{(25.0° - 20.0°)}(22.0° - 20.0°) = 1.644.$$

Step 4: For $c'/\gamma H = 0.025$, $\cot\beta = 3.0$ and $D = 1$, for $\phi' = 20°$, $m = 1.53$, $n = 1.34$ and for $\phi' = 25°$, $m = 1.87$, $n = 1.69$. Interpolating linearly for $\phi' = 22°$:

$$m = 1.53 + \frac{1.87 - 1.53}{(25.0° - 20.0°)}(22.0° - 20.0°) = 1.666,$$

$$n = 1.34 + \frac{1.69 - 1.34}{(25.0° - 20.0°)}(22.0° - 20.0°) = 1.480.$$

Step 5: $F_1 = 1.976 - 1.644 \times 0.2 = 1.647$, $F_2 = 1.666 - 1.480 \times 0.2 = 1.370$.

Step 6: find F by interpolation: $F = 1.370 + \dfrac{1.647 - 1.370}{(0.050 - 0.025)}(0.0392 - 0.025) = 1.53$.

9.3.5 *Determination of pore pressure ratio r_u*

One of the major factors causing instability and subsequent failure of slopes is the increase of the pore pressure due to natural causes or construction. An increase in pore pressure reduces the effective normal stress on the potential failure plane, which in turn reduces the shear strength on this surface. Values of r_u used in design must represent those states that are created during and at the end of construction, and later in the lifetime of the slope. In the construction process of an embankment, pore pressures are increased above the original level as a result of loading. Depending on the type of the material, an undrained condition may prevail until the excess pore pressures dissipate over time. An estimate of r_u can be made by utilising Equation 4.26. Consider u_i to be the initial pore pressure at the base of slice. As a result of the loading equivalent to the weight of the slice, the pore pressure increases to $u_i + \gamma hB$, where B is the pore pressure coefficient obtained in the laboratory triaxial tests that simulate field conditions. The design value of r_u may therefore be taken as:

$$r_u = \frac{u_i + u_e}{\gamma h} = \frac{u_i + \gamma hB}{\gamma h} = \frac{u_i}{\gamma h} + B \tag{9.12}$$

In earth dams subjected to an increase in the height of the external water in the upstream side, the pore pressure will increase and may make the downstream side unstable. To estimate the r_u vales, it is necessary to construct the corresponding flow net representing the seepage through the earth dam from the upstream side to the downstream side. Any rapid drawdown in the upstream side will eliminate the contribution of the external water pressure and may cause the failure of that side. At the same time, the effective stresses will increase due to a decrease in pore pressure which is equal to the change in the total vertical stress $\gamma_w h_w$ times the pore pressure coefficient B. Therefore, the new r_u value is:

$$r_u = \frac{u_i - u_e}{\gamma h} = \frac{\gamma_w h_w - \gamma_w h_w B}{\gamma h} = \frac{\gamma_w h_w}{\gamma h}(1 - B), \text{ thus:}$$

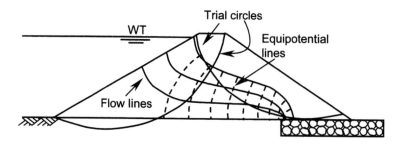

Figure 9.6. Determination of pore pressure ratio at steady flow.

$$r_u = \frac{u_i}{\gamma h}(1 - B) \tag{9.13}$$

In general with the change in the level of the external water, the flow net within the embankment will change and it is necessary to search for a critical flow net to compute the corresponding pore pressures. An example of flow net is shown in Figure 9.6 for the case where the water moves from the upstream side to the downstream side. In the case of a cut slope, pore pressures will decrease but, depending on the drainage conditions, will recover over time and a new equilibrium will be achieved. The evaluation of slope stability in undrained conditions and the application of the effective stress analysis with a critical r_u value may closely bracket the above different cases.

The determination of an average value of r_u for a specified trial circle can be justified by a mathematical model. Bromhead (1992) suggests systematic calculations to define an average value for r_u. According to this method, the soil mass between the slope boundaries and the hard stratum underneath is divided into vertical strips of equal width. Each strip is then divided vertically into three equal portions and r_u is determined at the centroid of each portion. The average of the r_u values of the middle portions (weighted by the area of each portion) represents, the average r_u for a general failure. For shallow failures, the average of the upper portions and for deep failures the average of the lower portions can be used.

9.4   STABILITY OF INFINITELY LONG EARTH SLOPES

9.4.1   *Planar failure mechanism*

A two-dimensional infinitely long earth slope is shown in Figure 9.7. For the purpose of analysis it is assumed that the material is homogeneous. The failure mechanism is assumed to be a plane parallel to the ground surface on which the normal and shear stresses are $\gamma H \cos^2\beta$ and $\gamma H \sin\beta\cos\beta$ respectively as was shown in Section 8.3.3 for a rhombic element. In practice, this mechanism is applied to the case when a soft material (e.g. clay) of very long length with constant slope may slide on a hard material (e.g. rock) having the same slope. The objective of the analysis is to determine either the critical slope or the critical height or, alternatively, the factor of safety based on a concept similar to that of the circular failure surface described previously. Steady seepage conditions in the slope will reduce the factor of safety.

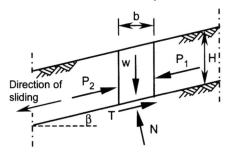

Figure 9.7. Planar failure mechanism for an infinitely long slope.

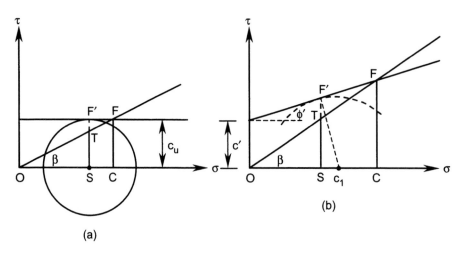

Figure 9.8. State of stress in an infinitely long slope: (a) c_u, $\phi_u = 0$ soil, (b) c', ϕ' soil.

9.4.2   Stability of infinitely long earth slopes in c_u, $\phi_u = 0$ soil

On any plane parallel to the ground surface the ratio of the shear stress to the normal stress is constant and equal to $\tan\beta$. Thus, the states of stress on these planes are located on the line OF in the σ, τ coordinate system, as shown in Figure 9.8(a). If the normal stress is less than OC, (point T on line OF) the shear stress will be less than c_u and the factor of safety is:

$$F = \frac{SF'}{ST}, F = \frac{c_u}{\gamma H \sin\beta\cos\beta} \qquad (9.14)$$

For normal stresses greater than OC the material will (theoretically) fail. The corresponding critical height H_c for a specified angle β is found by setting F to 1 in Equation 9.14:

$$H_c = \frac{c_u}{\gamma \sin\beta\cos\beta} = \frac{2c_u}{\gamma \sin 2\beta} \qquad (9.15)$$

Similarly, a critical slope angle β_c may be defined for a specified value of H as follows:

$$\beta_c = 0.5\sin^{-1}\frac{2c_u}{\gamma H} \qquad (9.16)$$

9.4.3   Stability of infinitely long earth slopes in c', ϕ' soil – no seepage

The states of stress on planes parallel to the ground surface and with $\beta > \phi'$ is represented by line OF in Figure 9.8(b). Defining the factor of safety as the ratio of SF' (shear strength) to ST (shear stress), then:

$$F = \frac{SF'}{ST} = \frac{c' + OS\tan\phi'}{OS\tan\beta} = \frac{c'}{\gamma H\cos^2\beta\tan\beta} + \frac{\tan\phi'}{\tan\beta},$$

$$F = \frac{c'}{\gamma H\sin\beta\cos\beta} + \frac{\tan\phi'}{\tan\beta} \qquad (9.17)$$

As the depth of the sliding plane increases, the normal stress on the plane approaches its limiting value *OC*. The critical height H_c is defined by setting Equation 9.17 to unity:

$$H_c = \frac{c'}{\gamma}\left[\frac{\sec^2 \beta}{\tan \beta - \tan \phi'}\right] \tag{9.18}$$

For the case where $\beta < \phi'$, the factor of safety is always greater than 1 and is computed from Equation 9.17. This means that there is no limiting value for *H*, and at an infinite depth the factor of safety approaches:

$$F = \frac{\tan \phi'}{\tan \beta} \tag{9.19}$$

For a granular material with $c' = 0$ and $\beta < \phi'$, the factor of safety is computed from Equation 9.17 (or 9.19). The case where $\beta > \phi'$ and $c' = 0$ is always unstable and cannot be applied to practical situations. This means that the critical value of the slope angle is:

$$\beta_c = \phi' \tag{9.20}$$

Example 9.8

An infinitely long slope is resting on a shale formation with the same inclination. The height of the slope is 3.2 m. Determine the factor of safety, the shear stress on the sliding surface and the critical height. $\beta = 25°$, $\gamma = 17.5$ kN/m^3, $c' = 12$ kPa, and $\phi' = 20°$.

Solution:

Using Equation 9.17:
$F = 12.0/(17.5 \times 3.2 \times \sin 25.0° \cos 25.0°) + \tan 20.0°/\tan 25.0° = 1.34$.
$\tau = \gamma H \sin \beta \cos \beta = 17.5 \times 3.2 \times \sin 25° \times \cos 25.0° = 21.4$ kPa.
The critical height is calculated from Equation 9.18:
$H_c = (12.0 \times \sec^2 25.0°/17.5)/(\tan 25.0° - \tan 20.0°) = 8.16$ m.

9.4.4 Stability of infinitely long earth slopes in c', φ' soil – steady state seepage

The case of steady state flow parallel to the slope angle β and with the water table at the ground surface is shown in Figure 9.9(a). From the geometry of the flow net, the pore pressure at point *M* of the base of an arbitrary vertical element is given by:

$$u = MN \times \gamma_w = MV \times \cos\beta \times \gamma_w = H\cos^2\beta\gamma_w.$$

The factor of safety is: $F = \dfrac{\tau_f}{\tau} = \dfrac{c' + \sigma'\tan\phi'}{\gamma H \sin\beta\cos\beta} = \dfrac{c' + (\gamma H \cos^2\beta - \gamma_w H \cos^2\beta)\tan\phi'}{\gamma H \sin\beta\cos\beta}$

$$F = \frac{c'}{\gamma H \sin\beta\cos\beta} + \frac{\tan\phi'}{\tan\beta} - \frac{\gamma_w \tan\phi'}{\gamma \tan\beta} \tag{9.21}$$

The first two terms of Equation 9.21 are identical to Equation 9.17; the third term indicates the reduction in the factor of safety due to steady state flow. If the ground water level is at some depth but parallel to the ground surface, Equation 9.21 is modified to:

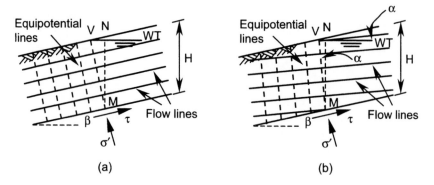

Figure 9.9. Seepage in an infinitely long slope with water at the ground surface.

$$F = \frac{c'}{\gamma H \sin\beta \cos\beta} + \frac{\tan\phi'}{\tan\beta} - \frac{\gamma_w h_w \tan\phi'}{\gamma H \tan\beta} \qquad (9.22)$$

where h_w is the height of the water above the base of the slice.
Substituting r_u for $\gamma_w h_w / \gamma H$ (Equation 9.8) into Equation 9.22 we obtain:

$$F = \frac{c'}{\gamma H \sin\beta \cos\beta} + \frac{\tan\phi'}{\tan\beta} - \frac{\tan\phi'}{\tan\beta} r_u = m - nr_u \qquad (9.23)$$

which is the fundamental relationship of the Bishop and Morgenstern approach to the stability of shallow slopes. If the unit weights of the saturated zone and the zone above the water table are not the same, the term γH in Equations 9.22 and 9.23 must be replaced by $\Sigma\gamma H$. Equation 9.21 may be conveniently presented in the following form:

$$F = \frac{c'}{\gamma H \sin\beta \cos\beta} + \frac{\gamma' \tan\phi'}{\gamma \tan\beta} \qquad (9.24)$$

By setting Equation 9.24 to unity the critical height is calculated as:

$$H_c = \frac{c' \sec^2\beta}{\gamma \tan\beta - \gamma' \tan\phi'} \qquad (9.25)$$

For the values of $\tan\beta < (\gamma' / \gamma) \tan\phi'$, the factor of safety expressed by Equation 9.24 is always greater than 1.0. At infinite depth the factor of safety is given by:

$$F = \frac{\gamma' \tan\phi'}{\gamma \tan\beta} \qquad (9.26)$$

From Equation 9.26 we can also calculate the factor of safety for a granular material with $c' = 0$. In this case $\tan\beta$ must be less than $(\gamma' / \gamma) \tan\phi'$, otherwise the slope will not be stable. By setting Equation 9.26 to unity a critical slope angle is defined:

$$\beta_c = \tan^{-1}\frac{\gamma' \tan\phi'}{\gamma} \qquad \text{for granular materials} \qquad (9.27)$$

The case where the steady flow is not parallel to the ground surface is shown in Figure 9.9(b). It can be shown that the pore pressure at point M is defined by:

$$u = \gamma_w MN = \frac{\gamma_w H}{1 + \tan\alpha\tan\beta} \tag{9.28}$$

where α is the angle of the flow lines to the horizontal. The corresponding factor of safety is derived in a similar way to that undertaken for Equation 9.21:

$$F = \frac{c'}{\gamma H \sin\beta\cos\beta} + \frac{\tan\phi'}{\tan\beta} - \frac{\gamma_w \tan\phi'(1 + \tan^2\beta)}{\gamma\tan\beta(1 + \tan\alpha\tan\beta)} \tag{9.29}$$

The critical height for a c', ϕ' soil and critical slope angle for $c' = 0$, ϕ' soil are as follows:

$$H_c = \frac{c'\sec^2\beta}{\gamma\left[\tan\beta - \tan\phi'(1 - \dfrac{\gamma_w}{\gamma}\dfrac{1 + \tan^2\beta}{1 + \tan\alpha\tan\beta})\right]} \tag{9.30}$$

$$\beta_c = \tan^{-1}\left[(1 - \frac{\gamma_w}{\gamma}\frac{1 + \tan^2\beta_c}{1 + \tan\alpha\tan\beta_c})\tan\phi'\right] \tag{9.31}$$

It is seen that Equation 9.31 must be solved using iteration.

Example 9.9

A long slope of $H = 4.5$ m and $\beta = 22°$ is to be constructed of material having the following properties:

$\gamma_{sat} = 20$ kN/m^3, $\gamma_{dry} = 17.5$ kN/m^3, $c' = 10$ kPa, and $\phi' = 32°$.

Determine the factor of safety when:
(a) the slope is dry,
(b) there is steady state seepage parallel to the surface with the water level 2 m above the base, and
(c) the water level is at the ground surface.

Solution:

(a) From Equation 9.17:

$$F = \frac{10.0}{17.5 \times 4.5 \times \sin 22.0° \times \cos 22.0°} + \frac{\tan 32.0°}{\tan 22.0°} = 1.91.$$

(b) Using Equation 9.22 and replacing γH with $\Sigma\gamma H$:

$$F = \frac{10.0}{(17.5 \times 2.5 + 20.0 \times 2.0)\sin 22.0° \cos 22.0°} +$$

$$\frac{\tan 32.0°}{\tan 22.0°} - \frac{9.81 \times 2.0 \times \tan 32.0°}{(17.5 \times 2.5 + 20.0 \times 2.0)\tan 22.0°} = 0.34 + 1.55 - 0.36 = 1.53.$$

(c) Use Equation 9.22:

$$F = \frac{10.0}{(20.0 \times 4.5)\sin 22.0° \cos 22.0°} + \frac{\tan 32.0°}{\tan 22.0°} - \frac{9.81 \times 4.5 \times \tan 32.0°}{(20.0 \times 4.5)\tan 22.0°} = 1.11.$$

9.5 STABILITY OF REINFORCED AND NAILED EARTH SLOPES

9.5.1 *Circular failure surfaces for slopes reinforced or nailed by geosynthetics*

Earth slopes can be stabilized by geosynthetic reinforcement with various deployment schemes and spacing (Schaefer et al., 1997). Examples of applications of geosynthetics in geotechnical engineering may be found in Koerner (1991) and Ingold (1994). Normally the reinforcement is placed horizontally with uniform vertical spacing. However, reduced vertical spacing may be used in the deeper regions where there are higher lateral stresses due to gravity. If the foundation underneath is a weak saturated soil, the reinforcement may be applied only at the base of the slope. The length of the reinforcement must extend by the length l_r beyond the failure surface as described in Section 8.8.2. Its vertical spacing is designed by taking into account the thickness of the compacted layers and the weight of the soil above the reinforcement that provides frictional resistance for the embedded length l_r. For a reinforced slope, the concept of a circular sliding surface may be applied by modifying Fellenius' or Bishop's simplified methods to account for the moment contributions of the tensile forces developed in the layers of reinforcement. The problem becomes statically determinate by making appropriate assumptions, which include a linear distribution of the lateral stresses resisted by reinforcement. In the case of parallel reinforcement in a c_u, $\phi_u = 0$ soil, adding the moments of the geosynthetic tensile forces to the resisting moments results in the following equation:

$$F = \frac{c_u L_a R + \sum_{j=1}^{j=m} [T(y_C - y)]_j}{Wd} \tag{9.32}$$

where T_j is the geosynthetic tensile force, y_j is the vertical distance of the reinforcement from the x-axis (Figure 9.10(a)), and m is the number of the geosynthetic layers. Note that by including the additional moment in the denominator of Equation 9.32, the tensile force required for equilibrium is T_j / F. In the presence of a surcharge load on the upper ground surface and external water pressure at the front of the slope, the corresponding additional moments must be taken into account. For the slope in the c', ϕ' soil shown in Figure 9.10(a), Bishop's simplified method is modified as follows:

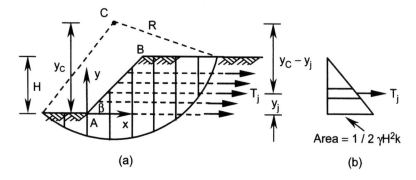

(a)   (b)

Figure 9.10. Parallel geosynthetic reinforcement.

$$F = \frac{\sum_{i=1}^{i=n}\left\{[c'b+w(1-r_u)\tan\phi']/m_\alpha\right\}_i + \frac{1}{R}\sum_{j=1}^{j=m}[T(y_C-y)]_j}{\sum_{i=1}^{i=n}(w\sin\alpha)_i} \qquad (9.33)$$

The total reinforcement force T_{total} is the sum of the tensile forces in the reinforcement; it is equivalent to the integral of the lateral soil pressure σ_h with the pressure coefficient k less than k_a (for $F = 1$) due to the slope angle. It is convenient to assume a linear distribution for lateral stress with the depth as suggested by Jewell (1991). Equation 9.33 can be solved for a specified F to yield T_{total}. The procedure is repeated until a maximum value of T_{total}, corresponding to the critical circle, is obtained. This type of analysis may overestimate the reinforcement forces and the current trend is to assume that the full strength of the soil on the critical circle is mobilized ($F = 1$). A final check on the overall stability must be carried out using trial circles that are deep and well beyond the reinforcement. This will eliminate the possibility of sliding where the foundation soil has poor strength properties. In order to evaluate the tensile forces in the reinforcement, the following procedure should be followed:

1. Specify a factor of safety F not less than 1.4. Substitute F into Equation 9.33 and compute the sum of the moments contributed by the reinforcement: $\Delta M = \Sigma[T(y_C-y)]_j$.
2. Assume a linear distribution for σ_h in the form of: $\sigma_h = \gamma(H-y)k$, thus $T_{total} = \gamma H^2 k/2$. Equate the moment of the resultant of the linear distribution to the sum of the moments contributed by the reinforcement (step 1) and compute T_{total} and k. Repeat the procedure for a number of trial circles until a maximum value for T_{total} (or k) is obtained. T_{total} is located $H/3$ above the base of the slope. Note that for $F > 1$, the magnitude of k may be greater than k_a.

$$\Delta M = \sum_{j=1}^{j=m}[T(y_C-y)]_j = T_{total}(y_C-\frac{H}{3}) = \frac{\gamma H^2}{2}k(y_C-\frac{H}{3}) \qquad (9.34)$$

3. Divide the lateral stress distribution into horizontal strips. The area of each strip is equal to the tensile force of the reinforcement (per metre run) as shown in Figure 9.10(b).
4. If a surface loading q exists: $\sigma_h = [\gamma(H-y)+q]k$, $T_{total} = \gamma H^2 k/2 + qHk$ and:

$$\Delta M = \sum_{i=1}^{i=m}[T(y_C-y)]_i = \frac{\gamma H^2}{2}k(y_C-H/3)+qHk(y_C-\frac{H}{2}) \qquad (9.35)$$

5. Calculate the required reinforcement length according to Section 8.8.2.

In the construction of Equations 9.32 and 9.33, the reinforcement contribution in penetrating the failure surface has been ignored. Thus, the shear strength parameters and the factor of safety may be multiplied by factors greater than 1 (Koerner & Robins, 1986).

Soil nailing is applied in cut slopes where rods are inserted into the slope as excavation proceeds. Using the method suggested by Koerner (1984), it is possible to construct fill slopes with geosynthetic nailing. This is a surface-deployed reinforcement (Figure 9.11(a)) that is nailed into the slope as the construction proceeds. Assuming that each slice contains a reinforcement element, Bishop's simplified method can be modified to take into account the moment and vertical force equilibrium contributions of the reinforcement element:

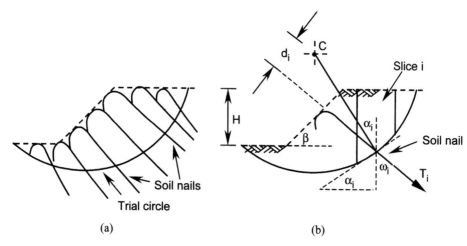

Figure 9.11. Application of the slice method to a slope nailed by geosynthetics.

$$F = \frac{\sum\limits_{i=1}^{i=n}\left\{\dfrac{c'b+\left[w(1-r_u)+T\cos\omega\right]\tan\phi'}{m_\alpha}\right\}_i}{\sum\limits_{i=1}^{i=n}(w\sin\alpha - Td/R)_i} \qquad (9.36)$$

where ω_i is the angle of the reinforcement from the vertical and d_i is its distance from the centre of the trial circle. This equation may be modified to include the contributions of nails penetrating the failure surface. Koerner & Robins (1986) suggested to replace F with $F/(1+f)$ where $f > 0$, and use improved shear strength parameters $c'_m \geq c'$ and $\phi'_m \geq \phi'$. The tensile force at each layer of reinforcement is expressed in terms of its horizontal projection T_{ih}. Each horizontal projection is then expressed in terms of the (unknown) lateral soil pressure coefficient k, obtained by estimating a reasonable width for the stress strip shown in Figure 9.10(b). The maximum value of k is determined by searching for a critical sliding circle. The horizontal component of each reinforcement force can then be found which in turn is used to compute T_i.

Example 9.10

A 7.2 m high slope, 1.0 horizontal to 1.8 vertical, is to be reinforced with horizontal geo-synthetic elements. $c' = 0$, $\phi' = 35°$, $\gamma = 19$ kN/m^3 and $r_u = 0.4$. Two trial toe circles ($F = 1.4$) are to be considered, with x_C, y_C coordinates (0, 13.60 m) and (0, 8.04 m). For both circles compute the total tensile force in the reinforcement. Relevant details of the geometry are given in the first three columns of the following two tables.

Solution:

For the trial circle with $R = 13.6$ m using Equation 9.33:
$F = 1.4 = (340.2 + \Delta M/13.6)/371.0 \rightarrow \Delta M = 2437.12$ kN.m. From Equation 9.34:
$\Delta M = 2437.12 = T_{total}(13.6 - 7.2/3) \rightarrow T_{total} = 2437.12/11.2 = 217.6$ kN.

For $R = 8.04$ m, $F = 1.4 = (209.9 + \Delta M/8.04)/275.8 \rightarrow \Delta M = 1416.81$ kN.m, and
$\Delta M = 1416.81 = T_{total}(8.04 - 7.2/3) \rightarrow T_{total} = 1416.81/5.64 = 251.2$ kN > 217.6 kN.

$R = 13.6$ m, $n = 6$, $b = 2.0$ m for all slices, $F = 1.4$, $r_u = 0.4$.

Slice	h (m)	α (deg.)	w (kN)	$w\sin\alpha$ (kN)	$w(1 - r_u)\tan\phi' / m_\alpha$ (kN)
1	1.7	4.2	64.6	4.7	26.2
2	5.0	12.7	190.0	41.8	73.5
3	6.2	21.6	235.6	86.7	88.9
4	5.2	31.0	197.6	101.8	74.5
5	3.7	41.4	140.6	93.0	54.6
6	1.4	54.0	53.2	43.0	22.5
Total:				371.0	340.2

$R = 8.04$ m, $n = 4$, $b = 2.0$ m for all slices, $F = 1.4$, $r_u = 0.4$.

Slice	h (m)	α (deg.)	w (kN)	$w\sin\alpha$ (kN)	$w(1 - r_u)\tan\phi' / m_\alpha$ (kN)
1	1.7	7.1	64.6	8.0	25.7
2	4.7	21.9	178.6	66.6	67.3
3	5.3	38.4	201.4	125.1	77.3
4	2.3	60.5	87.4	76.1	39.6
Total:				275.8	209.9

9.5.2 *Application of a two-part wedge mechanism to slopes reinforced by geosynthetics*

The stability of a reinforced slope may be assessed by the two-part wedge mechanism shown in Figure 9.12(a). With a vertical inter-wedge boundary, the mechanism is defined by three independent variables h, θ_1 and θ_2. If the inter-wedge boundary is not vertical then an additional variable is needed to specify the boundary. In the evaluation of stability, only force equilibrium is used. The failure criterion is assumed to apply on the three sliding surfaces, which in turn implies that the shear strength on these surfaces is fully mobilized. In the two-part wedge method adopted by the Department of Transport, UK (1994), it is assumed that the friction angle on the inter-wedge sliding surface is zero; which results in reduced computational effort. A base-sliding factor to modify the ϕ' of the lower wedge may also be applied. Free body diagrams of the wedges are shown in Figure 9.12(b) where the total reinforcement force T_{total} is the sum of T_1 and T_2 corresponding to wedges 1 and 2 respectively.

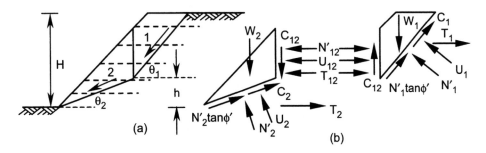

Figure 9.12. Application of a two-part wedge analysis for slopes reinforced by geosynthetics.

The force system is statically determinate and T_{total} can be formulated by considering horizontal and vertical equilibrium of both wedges:

$$T_{total} = \frac{W_1(\tan\theta_1 - \tan\phi') + (U_1\tan\phi' - C_1)\sec\theta_1}{1 + \tan\theta_1\tan\phi'} +$$

$$\frac{W_2(\tan\theta_2 - \tan\phi') + (U_2\tan\phi' - C_2)\sec\theta_2}{1 + \tan\theta_2\tan\phi'} \tag{9.37}$$

where W_1, W_2 are weights of the wedges 1 and 2, C_1, C_2 are the forces due to cohesion acting on the sliding bases, and U_1, U_2 are the forces due to water pressure acting on the sliding bases. The resultant of any surcharge loading acting on the slope surface or upper ground surface must be added to W_1 or W_2 where appropriate. By optimising (usually by iteration) the three variables of the mechanism, the critical two-part wedge corresponding to maximum T_{total} can be obtained. Similar to the circular method, the computed value of T_{total} is assumed to be linearly distributed along the slope height.

Jewell et al. (1985) and Jewell (1991) used a two-part wedge mechanism (as well as a log-spiral failure surface) to estimate the soil pressure coefficient k for a $c' = 0$, ϕ' soil with different values of r_u. Depending on the height of the water in the slope, an expression for an average value of r_u was defined. The k values so obtained were nearly identical to those obtained from circular failure surfaces. A typical chart for $r_u = 0$ is shown in Figure 9.13.

Example 9.11

For the two-part wedge mechanism shown in Figure 9.14, calculate the total force in the horizontal reinforcement and equivalent soil pressure coefficient k. $H = 10$ m, $\beta = 60°$, $\gamma = 19$ kN/m^3, $c' = 0$ kPa, and $\phi' = 35°$. BD is vertical and, $\theta_1 = 60°$, $\theta_2 = 30°$.

Solution:

From the geometry of the slope in Figure 9.14: $AD = 11.55$ m, $AB = 6.67$ m, $BD = 6.67$ m,

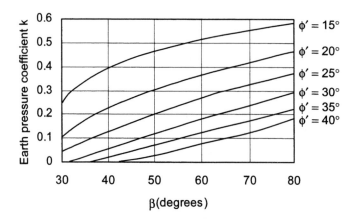

Figure 9.13. Variation of k with slope angle β for $c' = 0$, ϕ' soil and $r_u = 0$ (Exxon chemicals, 1992).

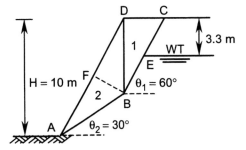

Figure 9.14. Example 9.11.

$DC = 3.85$ m, BF (normal to AD) $= 3.33$ m, $BE = 3.89$ m,

h_B = depth of water at point $B = 3.37$ m, and h_A = depth of water at point $A = 6.7$ m.

Note that BC is parallel to AD.

$W_1 = (BD \times DC \times 1.0)\gamma/2 = 6.67 \times 3.85 \times 19.0/2 = 243.9$ kN.

$W_2 = (AD \times BF \times 1.0)\gamma/2 = 11.55 \times 3.33 \times 19.0/2 = 365.4$ kN.

$u_B = \gamma_w h_B = 9.81 \times 3.37 = 33.0$ kPa, $u_A = \gamma_w h_A = 9.81 \times 6.7 = 65.7$ kPa.

$U_1 = (BE \times u_B \times 1.0)/2 = 3.89 \times 33.0/2 = 64.2$ kN.

$U_2 = AB(u_B + u_A)1.0/2 = 6.67(33.0 + 65.7)/2 = 329.2$ kN. Using Equation 9.37:

$T_{total} = [243.9(\tan 60.0° - \tan 35.0°) + (64.2 \tan 35.0°) \sec 60.0°]/(1 + \tan 60.0° \tan 35.0°) +$

$[365.4(\tan 30.0° - \tan 35.0°) + (329.2 \tan 35.0°) \sec 30.0°]/(1 + \tan 30.0° \tan 35.0°) =$

$154.4 + 157.6 = 312.0$ kN. $T_{total} = 312.0 = \gamma H^2 k/2 = 19.0 \times 100 \times k/2 \rightarrow k = 0.33$.

9.5.3 Slopes constructed on soft clay

Soft clay is a soil with a significant volume change beyond the preconsolidation pressure (Ladd, 1991). Normally and lightly consolidated clays with $c_u < 30$ kPa are considered to be soft (Jewell, 1996). If the foundation soil underneath a slope with no internal reinforcement is soft clay, geosynthetic reinforcement is placed on the interface of the two soils to increase stability against possible horizontal sliding. In a $c' = 0$, ϕ' fill, the failure mechanism shown in Figure 9.15(a) may be used to assess the stability against sliding. The lateral active thrust P_a on the vertical inter-wedge boundary BE is resisted by the shear stresses developed on the reinforcement along AB. Shear stresses are mobilized by the weight of the triangular block ABE and the maximum value of their resultant is: $Wf_b \tan\phi'$, where W is the weight of the block ABE, and f_b is defined by Equation 8.76. The factor of safety against sliding is $Wf_b \tan\phi'/P_a$. To calculate P_a it is convenient to assume the full active state on the vertical boundary BE. This model implies that no shear stress is transferred to the surface of the soft clay. However, depending on the extensibility of the reinforcement, a part of the shear force may transfer to the surface of the soft clay. This will reduce the bearing capacity of the soft clay (Chapter 10) and may cause horizontal sliding of the soft clay, particularly if it has a finite thickness.

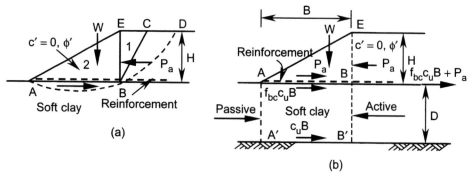

Figure 9.15. Outward sliding of a slope on soft soil.

The proportion of the shear force so transferred is estimated from the properties of the reinforcement and empirical equations suggested by the relevant codes. The length of the embedment of the reinforcement beyond point B is calculated in usual way.

For cases where $\beta < 30°$ the use of a two-part wedge analysis is not recommended and the force in the reinforcement and the equivalent k may be calculated using the circular failure surface method. The solution can be extended to the case where the soft clay has a finite thickness D and is resting on a hard horizontal stratum (Figure 9.15(b)). Active and passive states are created over the boundaries BB' and AA' respectively. On the boundary AB, which is the interface of the soft soil and the underside of the reinforcement, the resisting shear stress is $f_{bc} \times c_u$, where f_{bc} must be determined by a modified direct shear test. On the lower boundary $A'B'$ full mobilization of undrained cohesion may be assumed. Knowing all forces applied to the sliding mechanism, an estimate of the factor of safety against sliding can be made. The force in the reinforcement will be the sum of shear forces acting on both surfaces of reinforcement. Note that the calculation is carried out for a unit thickness of the slope and this must be taken into account if strip-type reinforcement is used.

Partial factors of safety may be applied to the strength properties of slope fill and soft clay. These are of the order of 1.2 for drained conditions and 1.5 for undrained conditions (Jewell, 1996). The soft clay may be in undrained, partially drained or drained conditions, the latter occurring after a long time under load. The undrained foundation is probably the most critical condition. To estimate the elastic and consolidation settlement of the soft clay, an estimate of the contact pressure on the surface of the soft clay must first be made. One (versatile) approach is to use the linear contact pressure concept under the length AB caused by the weight of the triangular block of ABE and the horizontal force P_a. Making use of Equation 5.98:

$$q_{max} = q_R = \gamma H (1 - k_a \tan^2 \beta), \quad q_{min} = q_L = \gamma H k_a \tan^2 \beta > 0 \qquad (9.38)$$

Example 9.12

A fill slope of $H = 6$ m, $\beta = 25°$, $c' = 0$, $\phi' = 35°$, and $\gamma = 19$ kN/m^3, is constructed on a soft clay soil with geotextile reinforcement on the horizontal interface of two soils. From laboratory test results, $f_b = 0.7$ and $f_{bc} = 0.6$. The thickness of the soft clay is 3 m, $c_u = 15$ kPa and $\gamma = 17$ kN/m^3.

Determine: (a) the factor of safety against outward sliding of the side of the slope, (b) the factor of safety against outward sliding of the soft soil, (c) the tensile force in the reinforcement, and (d) the contact pressure at points A and B of Figure 9.15(b).

Solution:

(a) $W_{ABE} = H \times H \cot \beta \times 1.0 \times \gamma / 2 = 6.0 \times 6.0 \cot 25.0° \times 1.0 \times 19.0 / 2 = 733.4$ kN.

Using Equation 8.6:

$k_a = (1 - \sin 35.0°) / (1 + \sin 35.0°) = 0.271$.

From Equation 8.10:

$P_a(fill) = \gamma H^2 k_a / 2 = 19.0 \times 6.0^2 \times 0.271 / 2 = 92.7$ kN.

$P_{t1} = W f_b \tan \phi' = 733.4 \times 0.7 \times \tan 35° = 359.5$,

$F_S = 359.5 / 92.7 = 3.9$.

(b) The active force acting along BB' is in accordance with Equation 8.10 with $k_a = 1$:

$P_a = 0.5 \gamma H^2 + qH - 2c_u H =$

$0.5 \times 17.0 \times 3.0^2 + (19.0 \times 6.0) \times 3.0 - 2 \times 15.0 \times 3.0 = 328.5$ kPa.

The passive force on the boundary AA' is calculated from Equation 8.11:

$P_p = 0.5 \times 17.0 \times 3.0^2 + 2 \times 15.0 \times 3.0 = 166.5$ kPa.

The shear force developed beneath the reinforcement is:

$P_{t2} = f_{bc} c_u B \times 1.0 = 0.6 \times 15.0 \times 6.0 \cot 25.0° = 115.8$ kN.

On the sliding surface between the soft clay and the hard stratum, the mobilized shear force is:

$P_{t3} = c_u B \times 1.0 = 15.0 \times 6.0 \cot 25° = 193.0$ kN.

The factor of safety against sliding of the block $BAA'B'$:

$F_S = (166.5 + 115.8 + 193.0) / 328.5 = 1.45$.

(c) The total tensile force in the reinforcement is:

$P_a(fill) + P_{t2} = 92.7 + 115.8 = 208.5$ kN.

(d) From Equation 9.38:

$q_{max} = 19.0 \times 6.0(1 - 0.271 \times \tan^2 25.0°) = 107.3$ kPa, (right edge).

$q_{min} = 19.0 \times 6.0 \times 0.271 \times \tan^2 25.0° = 6.7$ kPa, (left edge).

9.6   GENERAL SLOPE STABILITY ANALYSIS

9.6.1   *Principles of a general slope stability analysis*

The conventional limit equilibrium methods described above do not satisfy all the conditions of equilibrium. Fellenius' method assumes the tangential and normal forces on the right and the left of a slice are equal and have identical points of application. Bishop's simplified method assumes the tangential forces on both sides of a slice are equal. A general stability analysis must satisfy all conditions of equilibrium and brief descriptions of several general methods follow.

9.6.2   Bishop's method

Referring to Figure 9.5(b), vertical and horizontal equilibrium of each slice gives:

$$\Delta X = X_2 - X_1 = N'(\cos\alpha + \sin\alpha\tan\phi'/F) + c'b\tan\alpha/F - w(1 - r_u) \tag{9.39}$$

$$\Delta E = E_2 - E_1 = N'(\cos\alpha\tan\phi'/F - \sin\alpha) + c'b/F - wr_u\tan\alpha \tag{9.40}$$

For the specified values of F and ΔX, the magnitude of ΔE is obtained by calculating N' from Equation 9.39 and substituting the result into Equation 9.40. Both the X and E forces are internal and the sum of these forces must satisfy the following two conditions:

$$\sum \Delta X = 0, \ \sum \Delta E = 0 \tag{9.41}$$

The general expression for the factor of safety is obtained by adding the difference in the interslice shear forces ΔX to the vertical forces in Equation 9.9:

$$F = \frac{1}{\sum\limits_{i=1}^{i=n}(w\sin\alpha)_i} \sum_{i=1}^{i=n}\left\{\frac{c'b + [w(1 - r_u) + \Delta X]\tan\phi'}{m_\alpha}\right\}_i \tag{9.42}$$

The solution can be obtained by taking a set of X forces and a value of F that satisfy Equations 9.41 and 9.42. This requires increased iteration (in comparison with the simplified method) and computational effort. Different assumptions for the lines of action of the E forces may improve the factor of safety at the range of 1%.

9.6.3   Janbu's method

Janbu (1968 and 1973) proposed a general slope stability analysis that permits any shape of failure surface to be considered. The analysis is based on the slice method and all equilibrium conditions are fully satisfied. It also allows for variable lines of action for the interslice forces. Similar to Bishop's simplified method, a minimum factor of safety is obtained for a critical non-circular failure surface, but it requires comparatively more computational effort.

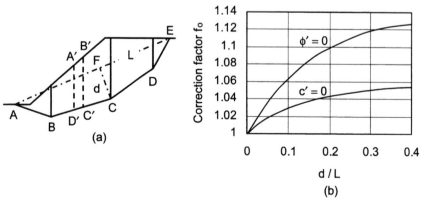

Figure 9.16. Janbu's general slope stability analysis.

While the search for a critical circular failure surface is relatively straightforward, the search for a critical non-circular failure surface is considerably more complex and guidance may be obtained from Celestino & Duncan (1981) and Chen & Shao (1988). Janbu's general equation for the factor of safety is expressed by:

$$F = \frac{1}{\sum\limits_{i=1}^{i=n}(w\tan\alpha)_i} \sum_{i=1}^{i=n}\left[\frac{c'b+(w-ub+\Delta X)\tan\phi'}{m_{\alpha(j)}}\right]_i \tag{9.43}$$

where all parameters have definitions identical to Bishop's method and $m_{\alpha(j)}$ is:

$$m_{\alpha(j)} = \cos^2\alpha(1+\frac{\tan\alpha\tan\phi'}{F}) \tag{9.44}$$

A method suitable for routine applications is based on neglecting the ΔX terms and increasing the number of slices. To simplify calculations, the sliding mass can be divided into several vertical blocks, as shown in Figure 9.16(a), which are in turn further subdivided into finer slices. Consequently all the slices within a block will have a common α value but different pore pressures. Application of the routine method will cause a reduction in the factor of safety corresponding to the critical failure surface. Janbu's proposal to increase the factor of safety for $c' = 0$, ϕ' and c', $\phi' = 0$ soils is shown in Figure 9.16(b). The correction factor f_o varies with the ratio of d / L, where L is the length AE shown in Figure 9.16(a) and d is the maximum distance of the non-circular failure surface from this line. The increased factor of safety is given by:

$$F_{Cor.} = f_o F \tag{9.45}$$

Example 9.13

For a slope of $H = 5$ m, $\beta = 45°$, $c' = 15$ kPa, $\phi' = 20°$, $\gamma_{sat} = 20.7$ kN/m^3 and $\gamma_{dry} = 17.5$ kN/m^3, apply Janbu's routine method to the assumed failure surface and with the piezometric level shown in Figure 9.17. Number of slices = 14, with $b = 1$ m. For the first trial assume the factor of safety is 1.4.

Solution:

A summary of the computations for two trial values of the factor of safety is given in the table. The final factor of safety for the given non-circular failure surface (without using Equation 9.45) is 1.85.

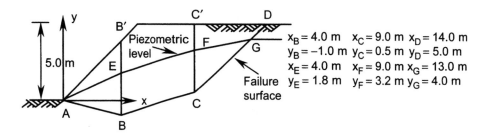

Figure 9.17. Example 9.13.

Slice	α (deg.)	$\tan\alpha$	h (m)	h_w (m)	w (kN)	$w\tan\alpha$ (kN)	ub (kN)	$(w-ub)\tan\phi'$ (kN)	$c'b$ (kN)
1	−14.0	−0.25	0.625	0.35	12.06	−3.01	3.43	3.14	15.0
2	−14.0	−0.25	1.875	1.05	36.17	−9.04	10.30	9.41	15.0
3	−14.0	−0.25	3.125	1.75	60.29	−15.07	17.17	15.69	15.0
4	−14.0	−0.25	4.375	2.45	84.40	−21.10	24.03	21.97	15.0
5	16.7	0.30	5.35	2.79	102.55	30.77	27.37	27.36	15.0
6	16.7	0.30	5.55	2.77	105.99	31.80	27.17	28.69	15.0
7	16.7	0.30	5.25	2.75	100.67	30.20	26.98	26.82	15.0
8	16.7	0.30	4.95	2.73	95.36	28.61	26.78	24.96	15.0
9	16.7	0.30	4.65	2.71	90.05	27.01	26.58	23.10	15.0
10	42.0	0.90	4.05	2.35	78.39	70.55	23.05	20.14	15.0
11	42.0	0.90	3.15	1.65	60.40	54.36	16.19	16.09	15.0
12	42.0	0.90	2.25	0.95	42.41	38.17	9.32	12.04	15.0
13	42.0	0.90	1.35	0.25	24.42	21.98	2.45	7.99	15.0
14	42.0	0.90	0.45	0.00	7.87	7.08	0.00	2.86	15.0
Total:						292.31			

	$F = 1.4$			$F = 1.85$
$m_{\alpha(j)}$	$[c'b + (w - ub)\tan\phi']/m_{\alpha(j)}$		$m_{\alpha(j)}$	$[c'b + (w - ub)\tan\phi']/m_{\alpha(j)}$
0.880	20.61		0.895	20.27
0.880	27.74		0.895	27.27
0.880	34.87		0.895	34.29
0.880	42.01		0.895	41.31
0.989	42.83		0.972	43.58
0.989	44.17		0.972	44.95
0.989	42.28		0.972	43.02
0.989	40.40		0.972	41.11
0.989	38.52		0.972	39.20
0.682	51.52		0.650	54.06
0.682	45.59		0.650	47.83
0.682	39.65		0.650	41.60
0.682	33.71		0.650	35.37
0.682	26.19		0.650	27.48
	Total = 530.09			Total = 541.34
	$F = 530.09 / 292.31 = 1.81$			$F = 541.34 / 292.31 = 1.85$

9.6.4   Morgenstern and Price's method

In this method the sliding mass above the non-circular failure surface is first divided into wide vertical slices (Figure 9.18(a)). Equilibrium of a vertical element of width dx within each slice (Figure 9.18(b)) is then investigated by introducing following five functions:

1. $y = s(x)$ defines the shape of the slope and its boundaries. For each slice this function is given as data and may be represented by a linear equation.
2. $y = y(x)$ defines the selected failure surface. For each slice a linear equation is assumed.

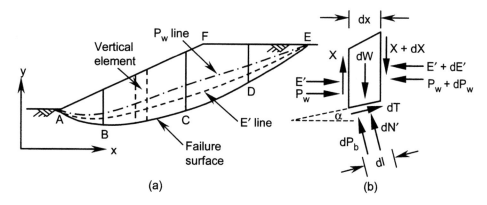

Figure 9.18. Morgenstern and Price's method: (a) slope section and wide slices, (b) forces acting on a vertical element within a wide slice.

3. $y = h(x)$ defines the line of action of the forces due to water pressure P_w acting on the vertical planes of each slice. This function is constructed from the position of the piezometric level.

4. $y = y'_t(x)$ defines the line of action of the effective normal force E' acting on the vertical planes.

5. $X = \lambda f(x) E'$ defines the relationship between the shear force X and the effective normal force E' on the vertical planes of each slice. This makes the slope stability problem statically determinate. The function $f(x)$ is a given relationship that represents the ratio of the shear force to the effective normal force on each vertical plane; λ is a scale factor and is determined within the solution. Force equilibrium equations, constructed parallel with and normal to the sliding direction, are combined with an application of the Mohr-Coulomb failure criterion on the failure plane to yield the following differential equation:

$$\frac{c'}{F}\sec^2\alpha - \frac{\tan\phi'}{F}(\tan\alpha\frac{dE'}{dx} + \tan\alpha\frac{dP_w}{dx} + \frac{dX}{dx} - \frac{dW}{dx} + \sec\alpha\frac{dP_b}{dx}) =$$

$$\frac{dE'}{dx} + \frac{dP_w}{dx} - \tan\alpha\frac{dX}{dx} + \tan\alpha\frac{dW}{dx} \tag{9.46}$$

For convenience moment equilibrium is considered at the mid point of the base of the element and yields the following differential equation:

$$y(\frac{dE'}{dx} + \frac{dP_w}{dx}) - \left[\frac{d(E'y'_t)}{dx} + \frac{d(P_w h)}{dx}\right] + X = 0 \tag{9.47}$$

Difficulties in the integration of the pair of the differential equations above may be overcome by assuming linear equations for the five basic functions within the integration limits (wide slices). This facilitates the integration of Equation 9.46 for selected values of F and λ and gives the E' forces as a function of x. The solution must satisfy the specified boundary conditions in terms of the E' forces at the first and last slices. Using the same basic five linear functions, Equation 9.47 is integrated to yield the change in the moment, which is essentially zero or as specified at the first and last slices. If the solution does not satisfy the moment equilibrium conditions another set of values for F and λ must be tried. Estimates of these new values can be obtained using a two variable Newton approximation method that gives the increments ΔF and $\Delta \lambda$ to be added to the previous values of F and λ.

9.6.5 *Spencer's method*

This analysis is based on the slice method and is applicable to failure surfaces of any shape and satisfies all conditions of equilibrium. Its application to circular failure surfaces has shed some light on explaining the accuracy of Bishop's simplified method. Referring to Figure 9.5(b), the inter-slice forces X_1, X_2, E_1 and E_2 can be represented by a single force Z with its point of application on the centre line of the slice and making an angle of θ with the horizontal. For the general formulation it may be assumed that the force Z is a disturbing force in a direction opposite to the force T. Vertical and horizontal force equilibrium equations are combined (to eliminate N') and result in the following expression:

$$Z = \frac{c'l/F + (W \cos\alpha - ul)\tan\phi'/F - W\sin\alpha}{\cos(\alpha - \theta)[1 + \tan(\alpha - \theta)\tan\phi'/F]} \tag{9.48}$$

From the definition of the force Z, the difference in the inter-slice forces at sides of the slice is expressed by the following two equations:

$$\Delta X = Z \sin\theta, \quad \Delta E = Z \cos\theta \tag{9.49}$$

Substituting the expressions above into Equations 9.41 we obtain:

$$\sum Z \sin\theta = 0, \quad \sum Z \cos\theta = 0 \tag{9.50}$$

Moment equilibrium of the Z forces about the centre of rotation of the slice gives:

$$\sum Z \cos(\alpha - \theta) = 0 \tag{9.51}$$

An iterative method is used to solve Equations 9.48 and 9.50. An equal θ value for all slices implies that the resultants of the side forces are parallel. Equation 9.50 then reduces to:

$$\sum Z = 0 \tag{9.52}$$

In the case of a constant θ, the solution may be obtained as follows:

1. Assume a value for the unknown angle θ and find the Z force for each slice using Equation 9.48.
2. Substitute the Z values into Equation 9.51 and evaluate F using a trial and error method.
3. If the assumed value of θ and the computed value of F satisfy Equation 9.52, the solution is correct; otherwise a new value for θ must be selected.

It may be shown that the factor of safety is not very sensitive to variations in θ which explains the accuracy of Bishop's simplified method. For $\theta = 0$ the factor of safety is identical to that predicted by Bishop's method.

9.7 APPLICATION OF THE WEDGE METHOD TO UNREINFORCED SLOPES

9.7.1 *Hard stratum in the vicinity of an earth slope*

Single or multiple wedge mechanisms are applied to unreinforced slopes where there is the possibility of translational sliding on the surface of a hard stratum close to the ground surface. Possible sliding mechanisms of an earth slope in the presence of a hard stratum are shown in Figure 9.19. In Figures 9.19(a) - 9.19(c), the surface of the hard stratum is inclined in the direction of the slope and may intersect the sloping ground, or pass through the toe or be located at some depth below the toe. Whilst the first two cases may be analysed using a single wedge, a multiple wedge mechanism is more appropriate for the third case. Figures 9.19(d) - 9.19(f) show the mechanisms for the particular instance of a horizontal hard stratum. The first two cases can be analysed using a single wedge but the accuracy of the result depends on the slope angle. The third case is analysed using a multiple wedge mechanism or a circular method. For a single wedge the stability of the slope can be formulated in terms of the stability number $\gamma H / c'$, critical height or critical slope angle. In the case of a multiple wedge mechanism the factor of safety is defined as the ratio

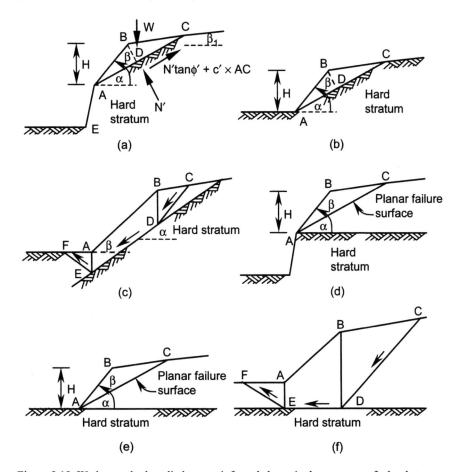

Figure 9.19. Wedge method applied to unreinforced slopes in the presence of a hard stratum.

of the shear strength on the sliding surface to the shear stress on that surface required to maintain equilibrium. In a $c' = 0$, ϕ' soil this ratio becomes:

$$F = \tau_f / \tau = N' \tan\phi' / N' \tan\phi'_m = \tan\phi' / \tan\phi'_m \tag{9.53}$$

where ϕ'_m is the mobilized friction angle on the interface of the soil and the hard stratum that satisfies the force equilibrium equations. It is assumed that the maximum friction angle that can be mobilized on the sliding surface is equal to the internal friction angle of the slope material. To ensure a conservative factor of safety, the friction angle on the vertical inter-wedge surfaces may be assumed zero. For a given mechanism, the geometry of the wedges must be optimised to obtain the minimum factor of safety.

9.7.2   *Single wedge mechanism*

By assuming full mobilization of the shear strength parameters for a specified slope angle β, the stability number corresponding to the critical height can be evaluated. Referring to Figure 9.19(a) and considering force equilibrium in the *AC* and *BD* (normal to *AC*) directions we obtain:

$$N' \tan\phi' + c' \times AC - W \sin\alpha = 0, \quad N' - W \cos\alpha = 0 \tag{9.54}$$

Combining the two above equations and replacing W with $H_c \sin(\beta - \alpha)AC\gamma/(2\sin\beta)$:

$$\frac{\gamma H_c}{c'} = \frac{2\sin\beta\cos\phi'}{\sin(\beta - \alpha)\sin(\alpha - \phi')} \tag{9.55}$$

It is seen that the stability number is independent of the slope of the upper ground surface. If a factor of safety is specified for the shear strength parameters, then:

$$\frac{\gamma H}{c'_m} = \frac{2\sin\beta\cos\phi'_m}{\sin(\beta - \alpha)\sin(\alpha - \phi'_m)} \tag{9.56}$$

where $c'_m =$ mobilized effective cohesion $= c' / F$, and ϕ'_m is the mobilized friction angle on the sliding surface so that $\tan\phi'_m = (\tan\phi') / F$. For the case where $H < H_c$, a factor of safety can be derived by replacing $N'\tan\phi' + c' \times AC$ with $(N'\tan\phi' + c' \times AC)/F$ in Equation 9.54 and rearranging the result thus:

$$F = \frac{2c'\sin\beta}{\gamma H \sin(\beta - \alpha)\sin\alpha} + \frac{\tan\phi'}{\tan\alpha} \tag{9.57}$$

In general the analysis of a single wedge is mathematically identical to an upper bound analysis where force equilibrium and energy equilibrium both yield the same result. In a $c' = 0$, ϕ' soil, the factor of safety is expressed by the second term of Equation 9.57, which is identical to Equation 9.19 and is independent of the height and the slope angle. In this case, the value of $F > 1$ means the slope will not slide on the hard stratum but it may slide on another plane or failure surface within the cohesionless material. In undrained conditions Equation 9.57 is rearranged (with $F = 1$) to the following form:

$$N_f = \frac{\gamma H_c}{c_u} = \frac{2\sin\beta}{\sin(\beta - \alpha)\sin\alpha} \tag{9.58}$$

In this case Equation 9.5 can be used to evaluate the factor of safety.

9.7.3   *Multiple wedge mechanism*

Multiple wedge mechanisms are used for the cases shown in Figures 9.19(c) and 9.19(f) by considering the general definition of the factor of safety for a limit equilibrium method and maximum values of the mobilized shear strength parameters. To begin with, an iterative method is used to verify the assumed factor of safety for the mechanism. The next step is to optimise the variables of the mechanism to obtain the minimum factor of safety. On the inter-wedge surfaces the mobilized cohesion and friction angle must also be specified. A common approach to the solution is to guess the factor of safety and evaluate the force equilibrium of each wedge starting from one side of the mechanism and moving towards the last wedge on the other side. The interface forces calculated for the last wedge must satisfy equilibrium of this wedge. After the first iteration, the residual interface forces for the last wedge can be used to formulate the variation of the mobilized strength parameters in terms of the computed reactions on the sliding surfaces. The mobilized strength parameters should be changed in a systematic way until force equilibrium of the last wedge is satisfied.

Example 9.14

A slope with $\beta = 45°$ is to be constructed on the surface of a hard stratum that has an angle of $30°$ to the horizontal. The soil properties are: $c' = 10$ kPa, $\phi' = 20°$ and $\gamma = 18$ kN/m^3. Assuming that the soil above the surface of the hard stratum slides on this surface, determine: (a) the critical height and (b) the corresponding height for a factor of safety of 2.

Solution:

Use Equations 9.55 for case (a) and 9.56 for case (b):

(a) $N_f = \dfrac{\gamma H_c}{c'} = \dfrac{18.0 H_c}{10.0} = \dfrac{2 \times \sin 45.0° \times \cos 20.0°}{\sin(45.0° - 30.0°)\sin(30.0° - 20.0°)} = 29.57 \rightarrow H_c = 16.4 \text{ m.}$

(b) $c'_m = c'/F = 10.0/2 = 5.0$ kPa,

$\tan \phi'_m = \tan \phi'/F = \tan 20.0°/2 \rightarrow \phi'_m = 10.3°.$

$N = \dfrac{\gamma H}{c'_m} = \dfrac{18.0 H_c}{5.0} = \dfrac{2 \times \sin 45.0° \times \cos 10.3°}{\sin(45.0° - 30.0°)\sin(30.0° - 10.3°)} = 15.95,$

$H_c = 4.4$ m.

Example 9.15

Figure 9.20(a) shows a slope with $\beta = 45°$ constructed on a hard stratum inclined at $\alpha = 31°$ to the horizontal. Calculate the factor of safety for the selected three-wedge mechanism. Lines *FB* and *EC* are vertical. The soil properties are: $c' = 0$, $\phi' = 36°$ and $\gamma = 18$ kN/m^3. Assume there are no frictional forces on the vertical interfaces of the wedges. $FB = 2.0$ m, $\angle ABF = 63°$, $\angle ECD = 27°$.

Solution:

The areas of the wedges are calculated from their geometry:

$A_1 = 4.076$ m^2, $A_2 = 15.0$ m^2, $A_3 = 3.925$ m^2.

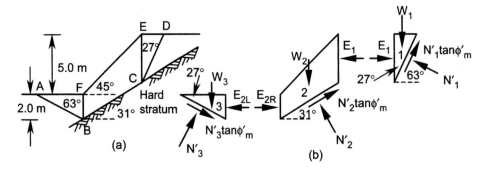

Figure 9.20. Example 9.15.

Thus the corresponding weights are:

$W_1 = 4.076 \times 1.0 \times 18.0 = 73.37$ kN, $W_2 = 15.0 \times 1.0 \times 18.0 = 270.0$ kN,

$W_3 = 3.925 \times 1.0 \times 18.0 = 70.65$ kN.

A summary of the computations is tabulated below, where for $\phi'_m = 24.5°$, $E_{2R} \approx E_{2L}$ and the factor of safety for the assumed mechanism is 1.59.

ϕ'_m (deg.)	E_1 (kN)	E_{2R} (kN)	E_{2L} (kN)	F
36.0	37.38	13.76	138.66	1.00
25.0	57.32	85.70	90.43	1.56
24.5	58.36	89.12	88.82	1.59

Sample calculation for $\phi'_m = 25.0°$:

$F = \tan \phi' / \tan \phi'_m = \tan 36.0° / \tan 25.0° = 1.56$.

Force equilibrium of wedge 1 in the directions CD and normal to CD gives:

$N'_1 \tan 25.0° - W_1 \sin 63.0° + E_1 \cos 63.0° = 0$,

$N'_1 - W_1 \cos 63.0° - E_1 \sin 63.0° = 0$.

Solving for E_1:

$$E_1 = \frac{W_1(\tan 63.0° - \tan 25.0°)}{1 + \tan 63.0° \tan 25.0°} = \frac{73.37(\tan 63.0° - \tan 25.0°)}{1 + \tan 63.0° \tan 25.0°} = 57.32 \text{ kN, similarly:}$$

$$E_{2R} = \frac{270.0(\tan 31.0° - \tan 25.0°)}{1 + \tan 31.0° \tan 25.0°} + E_1 = 28.38 + 57.32 = 85.70 \text{ kN.}$$

In the third wedge force equilibrium yields:

$$E_{2L} = \frac{W_3(\tan 27.0° + \tan 25.0°)}{1 - \tan 27.0° \tan 25.0°} = \frac{70.65(\tan 27.0° + \tan 25.0°)}{1 - \tan 27.0° \tan 25.0°} = 90.43 \text{ kN} > 85.70 \text{ kN.}$$

Thus the mobilized friction angle on the sliding surfaces must be slightly decreased to maintain force equilibrium.

9.8   CONCLUDING REMARKS

9.8.1   *Estimation of settlement in slopes and earth dams*

In slope stability analyses that use limit equilibrium methods, the state of stress within the sliding mass is unknown. Estimation of settlements in slopes and earth dams may be made using methods based on continuum mechanics (e.g. ABAQUS finite element code by Hibbit, Karlsson & Sorensen, Inc., 1995). Simplifying assumptions, such as a linear distribution of vertical stress with depth, facilitates estimation of the deformation and settlement of the sliding mass. For one-dimensional displacements (oedometer type), the distribution of vertical displacement can be formulated by assuming a linear distribution of vertical stress. For the plane strain case the semi-numerical method described in Chapter 5 (Equation 5.75) may be used. In this case the slope is divided into horizontal layers and estimates are made of the vertical and lateral stresses at the centre of each layer. Hooke's law can then be used to estimate of the vertical deformation of each layer.

A simplified one-dimensional analysis, applicable to the construction of an earth dam with symmetrical slopes, is given by Marsal (1958) who suggested using a parabolic shape for the elastic vertical displacement under the centre line. At elevation z measured from the foundation level, the vertical stress is $\gamma(H-z)$ where γ is the unit weight of the slope material. Considering the simplified definition of strain, the vertical displacement $v(z)$ is given by:

$$\varepsilon_z = \frac{\sigma_z}{E_s} = \frac{\gamma(H-z)}{E_s} = \frac{v(z)}{z} \rightarrow v(z) = \frac{\gamma}{E_s} z(H-z) \tag{9.59}$$

where E_s is the Modulus of Elasticity of the slope material. From the parabolic function expressed by Equation 9.59, it can be seen that the vertical displacement is maximum at mid-height of the dam and is zero value at the upper ground surface and the foundation level. Substituting $H/2$ for z we obtain:

$$v_{max} = \frac{\gamma H^2}{4E_s} \tag{9.60}$$

Although the conditions assumed in this model are unrealistic, this simplified analysis can be used to estimate vertical displacements in the early stages of construction and assist in the interpretation of results obtained from the plane strain models (Naylor, 1991). A modified method assumes a linear variation for E_s and γ with depth (Pagano et al., 1998). For a constant γ, a linear expression for E_s may be defined by: $E_s = E_o + \eta z$, where E_o is the Modulus of Elasticity at the foundation level. The vertical displacement at elevation z is calculated from:

$$v(z) = [\frac{\gamma(H-z)}{\eta}] \ln(1 + \frac{\eta z}{E_o}) \tag{9.61}$$

For constant E_s, a linear expression for γ is defined by: $\gamma = \gamma_o - \lambda z$, where γ_o is the unit weight at the foundation level. The vertical displacement at elevation z is then given by:

$$v(z) = \frac{1}{2E_s} [2\gamma_o (H-z)z - \lambda(H^2 - z^2)z] \tag{9.62}$$

For positive values of η and λ, the maximum vertical displacement occurs below mid-height whilst for negative values it occurs above mid-height.

9.8.2 *Application of numerical methods to slope stability*

With the advent of the new generation of powerful computers, the finite element and explicit finite difference methods have been applied successfully to a variety of slope stability problems. A comprehensive history of the finite element analysis of slope stability is presented by Duncan (1996) and the following conclusions were drawn:

1. An elastic stress-strain model (linear, multilinear, and hyperbolic) represents soil behaviour well at low levels of stress but is unable to predict the plastic deformation. Difficulties in the determination of Poisson's ratio and the Modulus of Elasticity are the major problems.
2. The total deformation (elastic plus plastic) can be predicted with reasonable accuracy using elastic-plastic models or non-linear stress-strain models.
3. The finite element method overestimates the deformation because soil properties from laboratory tests are used. Ageing of the soil in the field and disturbance of the sample in the test are the major reasons for this outcome.
4. The cost of using finite element method is about 10% of the total cost of an analysis.

The stability of a vertical cut for $r_u = 0$ is shown in Tables 7.1(a) and 7.1(b), (Chapter 7). It is seen that, for ϕ' values less than 40°, there is little difference between the slice solution and the two numerical techniques of finite element method (FEM: AFENA package by Carter & Balaam, 1990) and explicit finite difference method (FDM: FLAC package version 3.2, 1993 by Cundall, 1980 and 1987). Results obtained from the mechanism and slice methods are almost identical. All results fall within the upper bound and the lower bound solutions obtained by finite element formulations (Aysen & Sloan, 1992). An elastic-plastic stress strain model has been used in both FEM and FDM applications to study the stability of slopes. For slopes with small angles, the FEM/FDM results always exceed those of the slice solution (higher stability number or load parameter). Typically these differences are negligible for $\phi' < 20°$ but significant (say 25% or more) for $\phi' > 20°$. But this generalization must be viewed with caution as the results from numerical methods depend upon the selected model, its corresponding parameters and the technique in which the ultimate failure load is determined in the incremental process of the computations (Carter & Balaam, 1990).

In the case where $\gamma H / c' = 0$ (a weightless material), a uniformly distributed load q applied to the upper ground surface is increased incrementally using a dummy time variable which calculates the proportion of the specified loads acting on the top of the slope at each computation step. The accuracy of the method depends on the number of increments and interpretation of the stress-strain (or load-deformation) behaviour of some selected elements. Usually the failure load for this case corresponds to an element located at the intersection of the upper ground surface and the inclined face of the slope. A similar procedure must be carried out for some other elements that are thought to be in a state of failure. In the case of self-weight with a zero surface load, techniques are adopted to simulate excavation and fill processes. For example, to simulate the fill slope case, a progressive increase in $\gamma H / c'$ may be achieved by adding to the flat layer a set of 6-8 rows of elements, with each set forming a uniform thickness of soil layer to be added as an increment of fill (Aysen & Loadwick, 1995). Displacement behaviour of the toe element may be used to calculate the final value for the stability number at failure.

In the numerical application of FEM or FDM, the initial state of stress within the domain of the problem is required as input data. The vertical stress can be assumed to be equal to the overburden pressure while the horizontal normal stress is estimated using the concept of the at-rest condition (Chapter 8). If required the initial strains are set to zero.

9.8.3   *Three-dimensional slope stability analysis*

Whilst the slope stability analyses in this chapter are based on the plane strain conditions, many practical cases should be treated as a three-dimensional stability problem. Such analyses include limit equilibrium and variational methods (assuming a polynomial equation for the failure surface) such as extended slice methods (Fellenius, Bishop's simplified method, Janbu and Spencer), extended Morgenstern and Price method, upper bound solutions and numerical methods. Reference may be made to Hovland (1977: extended Fellenius), Hunger (1987: extended Bishop's simplified method), Chen & Chameau (1982: extended Spencer method and finite elements), Chen & Chameau (1983: extended Spencer method), Leshchinsky & Huang (1992: limit equilibrium and variational method). Generally the three-dimensional factor of safety F_3 is greater than the two-dimensional factor of safety F_2, where the difference depends on the shear strength parameters and the length of the extension of the failure surface from the plane strain into the third dimension. The ratio F_3 / F_2 in a $c' = 0$, ϕ' soil is normally in the range of 1.0 to 1.08 but may rise to 1.30 as the cohesion increases (Azzoz et al., 1981).

In some cases the ratio has been found to be less than 1.0 (Chen & Chameau, 1983; Seed et al., 1990). This is believed to be a result of simplifying assumptions made in special cases but this has yet to be proven.

9.8.4   *General observations*

The results obtained by circular methods for a c_u, $\phi' = 0$ material may differ from those of more sophisticated analyses by some 5% to 8% (Bromhead, 1992). Duncan (1996) reported a maximum variation of 12% between the results obtained for rotational sliding using the different approaches presented in this chapter. Generally it seems reasonable to use Bishop's simplified method with a large number of slices (say 30) and start the iterative process with a factor of safety slightly higher than the one obtained (for the same circular failure surface) from Fellenius' method.

Settlement of the slope under its own weight and any boundary loading can be evaluated using the method of Section 9.8.1. Alternatively, a numerical method with an appropriate stress-strain model obtained in the laboratory condition may be adopted. If settlement is not an important factor and it is required to evaluate the stability using FEM or FDM, an elastic-plastic stress-strain model can be used. In cut slopes the heave of the lower ground surface, which occurs due to unloading, must be modelled using a stress-strain relationship that includes both loading and unloading cycles. Whilst an estimate of the heave is possible using the elastic methods presented in Chapter 5, the use of FEM or FDM analyses is preferred.

Consolidation of the foundation soil may be estimated using the principles of Chapter 6. A coupled solution, where the stability of slope and consolidation of the foundation soil are analysed under a time-related loading, must be carried out by numerical methods. A crude estimate, however, can be obtained by applying the principles of Chapter 6 with time related loading. It is preferable to obtain pore pressures from the flow net, otherwise an average pore pressure must be evaluated.

9.9 PROBLEMS

9.1 Determine the factor of safety for a 1 vertical to 2 horizontal slope 5 m high using a trial toe circle for which $x_C = 4.5$ m and $y_C = 8$ m. The cross-sectional area of the sliding mass is 40.22 m^2 and its centroid is located 2.69 m from the centre of the trial circle.
The soil properties are:
$c_u = 18$ kPa, $\phi_u = 0$, and $\gamma = 18$ kN/m^3.

Answer: 1.36

9.2 For a 45° plane strain slope 30 m high (Example 7.8), determine: (a) the factor of safety for a toe circle for which $x_C = 12.5$ m and $y_C = 42$ m if there is no surcharge load on the upper ground surface, and (b) find the maximum surcharge load q that will cause the failure of the slope on the same slip circle.
The soil properties are: $c_u = 100$ kPa, $\phi_u = 0$, and $\gamma = 18$ kN/m^3.

Answers: 1.09, 44.7 kPa.

9.3 Using Taylor's stability chart (Figure 9.4), re-solve part (a) of Problem 9.2.

Answer: 1.02

9.4 Using Fellenius' method determine the factor of safety for a slope of 1 vertical to 2 horizontal and height $H = 4.5$ m using a trial toe circle for which $x_C = 4.5$ m and $y_C = 6.25$ m. The soil mass is divided into 3 m wide slices whose average height and angle α are tabulated below.
$c' = 6.75$ kPa, $\phi' = 17°$, and $\rho = 1.96$ Mg/m^3.

Slice no.	1	2	3	4
h (m)	1.6	3.7	4.6	3.0
α (deg.)	−23	0	23	51

Answer: 1.49

9.5 A 5 m high slope has an angle of $\beta = 45°$. Data on the 1 m wide slices are given in the table below where h_w is the height of water measured from the midpoint of the base of each slice. The trial circle is not a toe circle and slices 1 and 2 are located to the left of the toe. Using both Fellenius' method, and Bishop's method determine the factor of safety for this trial circle.
$c' = 15.0$ kPa, $\phi' = 20°$, $\gamma_{sat} = 20.7$ kN/m^3, and $\gamma_d = 17.5$ kN/m^3.

Answer: 1.465, 1.57

Slice no.	h (m)	h_w (m)	α (deg.)
1	0.20	0.20	−24.0
2	0.60	0.60	−14.0
3	1.35	1.35	−11.0
4	2.40	2.40	−3.0
5	3.40	3.20	0.0
6	4.35	3.60	5.5
7	5.25	3.80	11.5
8	5.60	3.80	14.0
9	5.25	3.70	24.0
10	4.75	3.40	29.0
11	4.20	3.10	32.5
12	3.50	2.60	38.5
13	2.50	1.70	46.0
14	1.25	0.60	57.0

9.6 An 8 m slope for which $\beta = 35°$ is composed of two layers of soil with the following properties:

Upper layer:

5 m thick, $c'_1 = 25.0$ kPa, $\phi'_1 = 12°$, $\gamma_1 = 18$ kN/m^3.

Lower layer:

3 m thick, $c'_2 = 7.0$ kPa, $\phi'_2 = 25°$, $\gamma_2 = 19.5$ kN/m^3.

The centre of the trial circle is located 5 m to the right and 12 m above the toe and its radius is 13.89 m. Consequently the failure surface is not a toe circle but passes from a point 2 m to the left of the toe. The total central angle is 103.6° for which the central angles corresponding to the upper and lower layers are 23.7° and 79.9° respectively. Data on the slices are tabulated below. Slices 1 to 9 are 2 m wide whilst the width of slice 10 is 2.4 m. Calculate the weight of each slice from the equation:

$$w = b\,(\gamma_1 h_1 + \gamma_2 h_2),$$

where h_1, h_2 are the heights corresponding to layers 1 and 2 respectively. The angle α for each slice is to be computed from:

$$\alpha = \sin^{-1}(x\,/\,R),$$

where x is the horizontal distance from the midpoint of the base to the centre of the circle. Using Fellenius' method, calculate the factor of safety for the given trial circle. $r_u = 0.3$.

(Source: Whitlow, 1990)

Slice no.	1	2	3	4	5	6	7	8	9	10
h_1 (m)	0.00	0.00	0.00	0.55	2.00	3.30	4.70	5.00	5.00	2.90
h_2 (m)	0.60	2.10	3.90	5.00	4.80	4.40	3.50	2.50	0.70	0.00

Answer: 1.25

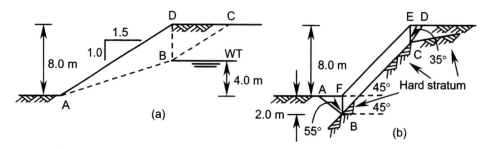

Figure 9.21. (a) Problem 9.10, (b) Problem 9.11.

9.7 A slope of 1 vertical to 2 horizontal and height of 7.5 m has the following soil properties:

$c' = 15$ kPa, $\phi' = 25°$, and $\gamma = 20$ kN/m^3.

Using the stability coefficients of Bishop & Morgenstern (1960), compute the factor of safety for $r_u = 0.0$, 0.2, and 0.4.

Answers: 2.19, 1.86, 1.54

9.8 A long slope is to be constructed using a material with:

$c' = 0$, $\phi' = 35°$, and $\gamma_{sat} = 20$ kN/m^3.

Determine the critical angles for both the dry condition and steady state flow parallel to the surface. Calculate the factor of safety for both cases if the selected slope angles are equal to $\beta_c / 1.5$.

Answers: 35.0°, 19.6°, 1.62, 1.54

9.9 Re-work Example 9.10 using a toe circle of radius 10.54 m tangent to the base at the toe. Relevant data are given in the table below.

Slice no.	1	2	3	4	5
h (m)	1.7	4.9	5.9	4.4	1.8
α (deg.)	5.4	16.6	28.4	41.8	59.5

Answer: $T_{total} = 250.8$ kN

9.10 For the reinforced slope shown in Figure 9.21(a), calculate the total force in the reinforcement for the trial two-part wedge shown. BC is parallel to AD. The soil properties are:

$c' = 0$, $\phi' = 29°$, and $\gamma = 18$ kN/m^3.

Answer: 59.5 kN

9.11 For the multiple-wedge mechanism shown in Figure 9.21(b), calculate the factor of safety for the slope assuming that no cohesion or friction is mobilized on the vertical inter-wedge planes of CE and BF.

$c' = 9.5$ kPa, $\phi' = 30.8°$, and $\gamma = 18$ kN/m^3.

Answer: 1.9

9.10 REFERENCES

Atkinson, J.H. 1993. *An introduction to the mechanics of soils and foundations*. London: McGraw-Hill.

Aysen, A. & Sloan, S.W. 1992. Stability of slopes in cohesive frictional soil. *Proc. 6th Australia-New Zealand conf. on geomechanics*: Geotechnical risk-identification, evaluation and solutions: 414-419. New Zealand: New Zealand Geomechanics Society.

Aysen, A. & Loadwick, F. 1995. Stability of slopes in cohesive frictional soil using upper bound collapse mechanisms and numerical methods. *Proc. 14th Australasian conf. on the mechanics of structures and materials* 1: 55-59. Hobart, Australia: University of Hobart.

Azzoz, A.S., Baligh, M.M. & Ladd, C.C. 1981. Three-dimensional stability analysis of four embankment failures. *Proc. 10th intern. conf. SMFE* 3: 343-346. Rotterdam: Balkema.

Bishop, A.W. 1955. The use of slip circle in the stability analysis of slopes. *Geotechnique* 5(1): 7-17.

Bishop, A.W. & Morgenstern, N.R. 1960. Stability coefficients for earth slopes. *Geotechnique* 10(4): 129-147.

Bromhead, E.N. 1992. *The stability of slopes*. 2nd edition, Surrey: Surrey University Press.

Carter, J.P. & Balaam, N.P. 1990. *Program AFENA: A general finite element algorithm*. University of Sydney, Australia: Centre for Geotechnical Research.

Celestino, T.B. & Duncan, J.M. 1981. Simplified search for non-circular slip surface. *Proc. 10th intern. conf. SMFE* 3: 391-394. Rotterdam: Balkema.

Chandler, R.J. & Peiris, T.A. 1989. Further extensions to the Bishop and Morgenstern slope stability charts. *Ground engineering*, May: 74-91.

Chen, R.H. & Chameau, J.L. 1982. Three-dimensional slope stability analysis. *Proc. 4th intern. conf. numer. meth. in geomech.* 2: 671-677. Rotterdam: Balkema.

Chen, R.H. & Chameau, J.L. 1983. Three-dimensional limit equilibrium analysis of slopes. *Geotechnique* 33(1): 31-40.

Chen, Z. & Shao, C. 1988. Evaluation of minimum factor of safety in slope stability analysis. *Canadian geotechnical journal* 25(4): 735-748.

Cousins, B.F. 1978. Stability charts for simple earth slopes. *Journal GE, ASCE* 104.

Cundall, P.A. 1980. *UDEC - A generalized distinct element program for modelling jointed rock*. Peter Cundall Associates, Report PCAR-1-80, US Army, European research office.

Cundall, P.A. 1987. Distinct element models of rock and soil structure. In E. T. Brown (ed), *Analytical and computational methods in engineering rock mechanics*: 129-139. London: George Allen & Unwin.

Duncan, J.M. 1996. State of the art: Limit equilibrium and finite element analysis of slopes. *Journal GE, ASCE* 122(7): 577-596.

Duncan, J.M., Buchignani, A.L. & De Wet, M. 1987. *An engineering manual for slope stability studies*. Blacksburg, VA: Virginia Tech.

Exxon Chemicals. 1992. *Geotextiles: design for soil reinforcement*. 2nd edition: 58. UK: Exxon Chemicals Geopolymers Ltd.

Fellenius, W. 1927. *Erdstatische berechnungen mit reibung und kohasion*. Berlin: Ernst, (in German).

Hibbit, Karlsson & Sorensen, Inc. 1995. *ABAQUS/ Standard - user manual - version 5.5*. Pawtucket, R.I.: Hibbit, Karlsson & Sorensen, Inc..

Hovland, H.J. 1977. Three-dimensional slope stability analysis method. *Journal GE, ASCE* 103(9): 971-986. ASCE.

Hunger, O. 1987. An extension of Bishop's simplified method of slope stability analysis to three dimensions. *Geotechnique* 37(1): 113-117.

Ingold, T.S. 1994. *The geotextiles and geomembranes manual*. Oxford, UK: Elsevier Advanced Technology.

Janbu, N. 1968. Slope stability computations. *Soil mechanics and foundation engineering report.* Trondheim: Norway, The Technical University of Norway.

Janbu, N. 1973. Slope stability computations. In E. Hirschfield & S. Poulos (eds), *Embankment dam engineering, Casagrande memorial volume*: 47-86. New York: John Wiley.

Jewell, R.A. 1991. Application of revised design charts for steep reinforced slopes. *Geotextiles geomembranes* 10(3): 203-233. UK: Elsevier.

Jewell, R.A. 1996. *Soil reinforcement with geotextiles.* Construction industry research and information association (CIRIA). Special publication 123. UK: Thomas Telford.

Jewell, R.A., Paine, N. & Woods, R.I. 1985. *Design methods for steep reinforced embankments, polymer grid reinforcement*: 70-81. UK: Thomas Telford.

King, C.J.W. 1989. Revision of effective stress method of slices. *Geotechnique* 39(3): 497-502.

Koerner, R.M. 1984. Slope stabilization using anchored geotextiles: Anchored spider netting. *Proc. special geotechnical engineering for roads and bridges conf.*: 1-11.Harrisburg, PA.: Penn DOT.

Koerner, R.M. 1991. Geomembranes overview, significance and background. In A. Rollin & J.M. Rigo (eds), *Geomembranes identifications and performance testing.* Chapter 1: 355. London: Chapman & Hall.

Koerner, R.M. 1994. *Designing with geosynthetics.* 3rd edition. London: Prentice Hall.

Koerner, R.M. & Robins, J.C. 1986. In-situ stabilization of soil slopes using nailed geosynthetics. *Proc. 3rd conf. on geosynthetics*: 395-399. Vienna.

Ladd, C.C. 1991. Stability evaluation during staged construction. *Journal GE, ASCE* 117(4): 537-615.

Leshchinsky, D. & Huang, C. 1992. Generalized three-dimensional slope stability analysis. *Journal GE, ASCE* 118(11): 1748-1764. ASCE.

Marsal, R.J. 1958. Analisis de asentamientos en la presa Presidente Aleman. *Oaxaca (5). Instituto de Ingenieria.* Mexico City, Mexico: UNAM.

Morgenstern, N.R. & Price, V.E. 1965. The analysis of the stability of general slip surfaces. *Geotechnique* 15(1): 79-93.

Naylor, D.J. 1991. Finite element methods for fills and embankment dams. In M. das Neves (ed), *Advances in rockfill structures.* North Atlantic Treaty Organization advanced study institute series: 291-339. Dordrecht, The Netherlands: Kluver Academic Publisher.

Pagano, L., Desideri, A. & Vinale, F. 1998. Interpreting settlement profiles of earth dams. *Journal GGE, ASCE* 124(10): 923-932.

Schaefer, V.R., Abramson, L.W., Drumheller, J.C., Hussin, J.D. & Sharp, K.D. 1997. Ground improvement, ground reinforcement, ground treatment-developments 1987-1997. *ASCE Geotechnical special publication*, (69): 263-270. New York: ASCE.

Seed, R.B., Mitchell, J.K. & Seed, H.B. 1990. Kettlemen Hills waste landfill slope failure, II: stability analysis. *Journal GED, ASCE* 116(4): 669-690.

Spencer, E. 1967. A method of analysis of the stability of embankments assuming parallel interslice forces. *Geotechnique* 17(1): 11-26.

Tavenas, F., Trak, B. & Leroueil, S. 1980. Remarks on the validity of stability analyses. *Canadian geotechnical journal* 17(1): 61-73.

Taylor, D.W. 1948. *Fundamentals of soil mechanics.* New York: Wiley.

UK Department of Transport. 1994. Design methods for the reinforcement of highway slopes by reinforced soil and soil nailing techniques. *Design manual 4, section 1, HA 68/94.*

Whitlow, R. 1990. *Basic soil mechanics.* 2nd edition, New York: Longman Scientific & Technical.

CHAPTER 10

Bearing Capacity of Shallow Foundations and Piles

10.1 INTRODUCTION

Shallow foundations are comprised of footings and rafts, while the deep foundations include piles that are used when the soil near the ground surface has not adequate strength to stand the applied loading.

The *ultimate bearing capacity* (in kPa) is the load that causes the shear failure of the soil immediately underneath and adjacent to the footing. A relatively high factor of safety (3 or more) is applied to allow for uncertainties in soil properties, inaccuracies in the method of analysis and allowable settlement recommended by the building codes or the designer.

To calculate the ultimate bearing capacity for a *general shear failure* a failure mechanism is introduced that satisfies equilibrium, stress boundary conditions and the failure criterion. The failure criterion is satisfied on the sliding surface(s) as well as inside the individual blocks of the mechanism creating a plastic equilibrium in the shearing zone. In practice this failure occurs suddenly and is catastrophic. Sometimes the failure is associated by a significant vertical settlement and failure planes do not reach to the ground surface. This is referred to as a *punching shear failure* because the soil around the footing is relatively uninvolved in the failure. In a *local shear failure* a significant settlement immediately beneath the footing is associated with the upward movement of soil at both sides of the footing and the failure planes may not reach the ground surface. Thus the failure pattern has the characteristics of general and punching shear representing a transitional mode (Ramiah & Chickanagappa, 1982). It is proposed that the ultimate bearing capacity for local shear to be computed based on 1/3 reduction of the shear strength parameters. In-situ tests such as plate bearing test, Standard Penetration Test (*SPT*), and Cone Penetration Test (*CPT*) are performed to estimate the *allowable bearing pressure* and ultimate bearing capacity.

10.2 ULTIMATE BEARING CAPACITY OF SHALLOW FOUNDATIONS

10.2.1 *Ultimate bearing capacity of frictionless shallow strip footing due to Parndtl*

A failure mechanism for a shallow strip footing proposed by Prandtl in early 1920s is shown in Figure 10.1. The footing is assumed frictionless and thus no shear stress can develop on the contact area. This mechanism with some modifications has become the basic

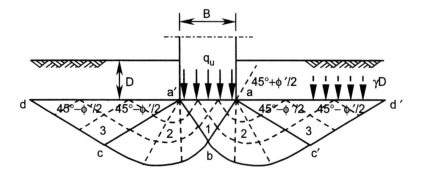

Figure 10.1. Failure mechanism for frictionless strip footing.

tool for evaluation of the ultimate bearing capacity. The mechanism is comprised of three zones. In zone 1, immediately under the footing the major principal stress q_u is in vertical direction. The minor principal stress is horizontal and can be found (in terms of q_u) from Equation 4.14. In zone 3 the minor principal stress is equal to the overburden pressure γD and the major principal stress can be found from Equation 4.13. In both zones the states of stress are assumed to be uniform. The state of stress at zone 2 obeys a logarithmic function, which satisfies the equilibrium, boundary conditions and the failure criterion, thus creating a relationship between two states of stress of zone 1 and zone 3. In this zone the soil is in plastic equilibrium and failure planes pass trough point a'. Another set of failure surfaces are parallel to the logarithmic spiral bc. For a weightless soil (in the zones 1 to 3) it can be shown that the ultimate bearing capacity is:

$$q_u = c'N_c + \gamma DN_q \tag{10.1}$$

$$N_q = \exp(\pi \tan \phi') \tan^2(45° + \phi'/2), \; N_c = \cot \phi'(N_q - 1) \tag{10.2}$$

where N_q and N_c are bearing capacity factors.
For c_u, $\phi_u = 0$ soil $N_q = 1$, $N_c = (\pi + 2) = 5.14$:

$$q_u = 5.14 c_u + \gamma D \tag{10.3}$$

The magnitude of q_u at the ground surface is $5.14 c_u$.

Example 10.1

Determine the ultimate bearing capacity of a frictionless strip footing of width 1.5 m at a depth of 1 m in a soil with $c' = 10$ kPa, $\phi' = 28°$, $c_u = 105$ kPa, $\phi_u = 0$ and $\gamma = 19$ kN/m³.
Solution:
$\gamma D = 19.0 \times 1.0 = 19.0$ kPa for both drained and undrained conditions.
Using Equations 10.2:
$N_q = e^{\pi \tan 28.0°} \tan^2(45.0° + 28.0°/2) = 14.72$, $N_c = \cot 28.0°(14.72 - 1) = 25.80$.
Thus from Equation 10.1: $q_u = 10.0 \times 25.80 + 19.0 \times 14.72 = 538$ kPa.
For the undrained conditions use Equation 10.3: $q_u = (\pi + 2)105.0 + 19.0 = 559$ kPa.

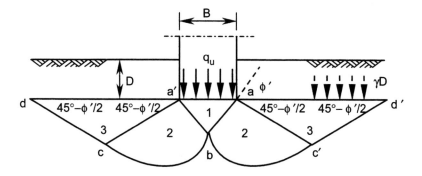

Figure 10.2. Failure mechanism for Terzaghi's bearing capacity solution.

10.2.2 *Terzaghi's bearing capacity equations*

Terzaghi (1943) improved the Prandtl solution to include the roughness of the footing, and the weight of the failure zone. The failure mechanism for a general shear failure in a c', ϕ' soil is shown in Figure 10.2. The ultimate bearing capacity for a strip, square and circular footing and the corresponding bearing capacity factors are:

$$q_u = c'N_c + \gamma D N_q + 0.5 B \gamma N_\gamma \tag{10.4}$$

$$q_u = 1.3 c'N_c + \gamma D N_q + 0.4 B \gamma N_\gamma \tag{10.5}$$

$$q_u = 1.2 c'N_c + \gamma D N_q + 0.3 B \gamma N_\gamma \tag{10.6}$$

$$N_q = \frac{e^{(3\pi/2-\phi')\tan\phi'}}{2\cos^2(45°+\phi'/2)}, \;\; N_c = \cot\phi'(N_q - 1), \;\; N_\gamma = 0.5\tan\phi'(\frac{k_{p\gamma}}{\cos^2\phi'} - 1) \tag{10.7}$$

The parameter $k_{p\gamma}$ is evaluated from the equilibrium of the failure mechanism. From the

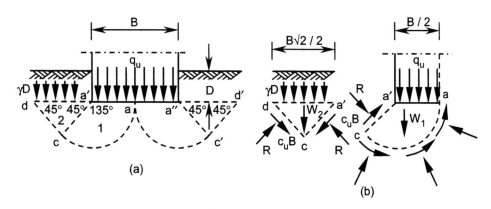

Figure 10.3. Failure mechanism for Terzaghi's bearing capacity solution in c_u, $\phi_u = 0$ soil.

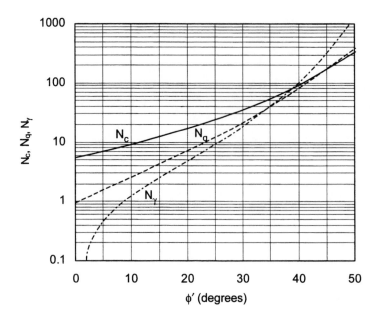

Figure 10.4. Bearing capacity factors using Terzaghi's equations.

given values of N_γ the following matching empirical equation may be proposed:

$$k_{p\gamma} = (8\phi'^2 - 4\phi' + 3.8)\tan^2(60° + \phi'/2)$$ (10.8)

where ϕ' (in the first term) is in radians. In the undrained conditions with c_u and $\phi_u = 0$, the mechanism of Figure 10.3 yields: $N_c = 1.5\pi + 1 = 5.71$, $N_q = 1$ and $N_\gamma = 0$.

Figure 10.4 shows the variation of the bearing capacity factors with the effective internal friction angle.

Example 10.2

Re-do Example 10.1 using Terzaghi's bearing capacity factors.

Solution:

Using Equations 10.8 and 10.7: $k_{p\gamma} = 45.68$, $N_q = 17.81$, $N_c = 31.61$ and $N_\gamma = 15.31$.

$q_u = c'N_c + \gamma DN_q + 0.5B\gamma N_\gamma =$
$10.0 \times 31.61 + 19.0 \times 17.81 + 0.5 \times 1.5 \times 19.0 \times 15.31 = 873$ kPa.

For the undrained conditions $N_c = 5.71$, $N_q = 1$, and $N_\gamma = 0$, thus:

$q_u = 5.71c_u + \gamma D = 5.71 \times 105.0 + 19.0 \times 1.0 = 619$ kPa.

Example 10.3

Determine the ultimate bearing capacity of a square footing 1.5 m, at a depth of 1 m in a soil with: $c' = 10$ kPa, $\phi' = 28°$, $c_u = 105$ kPa, $\phi_u = 0$, and $\gamma = 19$ kN/m^3.

Solution:

$\gamma D = 19.0 \times 1.0 = 19.0$ kPa for both drained and undrained conditions.
From Example 10.2:

$N_q = 17.81$, $N_c = 31.61$ and $N_\gamma = 15.31$. Using Equation 10.5:

$q_u = 1.3 \times 10.0 \times 31.61 + 19.0 \times 17.81 + 0.4 \times 1.5 \times 19.0 \times 15.31 = 924 \text{ kPa}$.

For undrained conditions, $N_c = 5.71$, $N_q = 1$, and $N_\gamma = 0$, thus:

$q_u = 1.3 \times 105.0 \times 5.71 + 19.0 \times 1.0 = 798 \text{ kPa}$.

10.2.3 *Meyerhof's bearing capacity equations*

Meyerhof (1951, 1953, 1963, 1965 and 1976) developed the bearing capacity equations by extending the Terzaghi's mechanism to the soil above the base of the footing for both shallow and deep foundations, where in the latter a local shear failure was also considered. The bearing capacity factors N_c and N_q are identical to Parandtl analysis (Equations 10.2). Furthermore the shape of the footing, inclination of the applied load and the depth of the footing were taken into account by introducing the corresponding factors of s, i, and d. For a rectangular footing of L by B ($L > B$) and inclined load:

$$q_u = c'N_c s_c i_c d_c + \gamma D N_q s_q i_q d_q + 0.5B\gamma N_\gamma s_\gamma i_\gamma d_\gamma \tag{10.9}$$

For vertical load: $i_c = i_q = i_\gamma = 1$ and:

$$q_u = c'N_c s_c d_c + \gamma D N_q s_q d_q + 0.5B\gamma N_\gamma s_\gamma d_\gamma \tag{10.10}$$

The bearing capacity factor N_γ is defined by:

$$N_\gamma = (N_q - 1)\tan(1.4\phi') \tag{10.11}$$

The bearing capacity factors are graphically presented in Figure 10.5. The shape, inclination and depth factors are according to:

$$s_c = 1 + 0.2(B/L)\tan^2(45° + \phi'/2), s_q = s_\gamma = 1 + 0.1(B/L)\tan^2(45° + \phi'/2) \tag{10.12}$$

For c_u, $\phi_u = 0$ soil $s_q = s_\gamma = 1$.

$$i_c = i_q = (1 - \alpha°/90°)^2, \; i_\gamma = (1 - \alpha/\phi')^2 \tag{10.13}$$

For c_u, $\phi_u = 0$ soil $i_\gamma = 1$.

$$d_c = 1 + 0.2(D/B)\tan(45° + \phi'/2), d_q = d_\gamma = 1 + 0.1(D/B)\tan(45° + \phi'/2) \tag{10.14}$$

For c_u, $\phi_u = 0$ soil $d_q = d_\gamma = 1$. The equivalent plane strain ϕ' is related to triaxial ϕ' by:

$$\phi'_{ps} = \phi'_{tri}(1.1 - 0.1B/L) \tag{10.15}$$

For the eccentric load the length and width of the footing rectangle are modified to:

$$L' = L - 2e_L, \; B' = B - 2e_B \tag{10.16}$$

where e_L and e_B represent the eccentricity along the appropriate directions.

Example 10.4

Re-do example 10.3 using Meyerhof's bearing capacity factors.

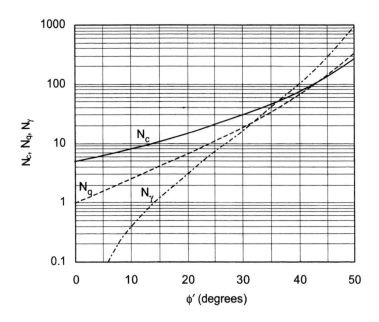

Figure 10.5. Bearing capacity factors using Meyerhof's equations.

Solution:

From Equations 10.2 $N_q = 14.72$, $N_c = 25.80$ (Example 10.1), using Equation 10.11:
$N_\gamma = (14.72 - 1)\tan(1.4 \times 28.0°) = 11.19$. The load is vertical, thus $i_c = i_q = i_\gamma = 1$.
Calculate shape factors from Equations 10.12 and depth factors from Equations 10.14:

$s_c = 1 + 0.2(1.5/1.5)\tan^2(45.0° + 28.0°/2) = 1.55$,

$s_q = s_\gamma = 1 + 0.1(1.5/1.5)\tan^2(45.0° + 28.0°/2) = 1.28$.

$d_c = 1 + 0.2(1.0/1.5)\tan(45.0° + 28.0°/2) = 1.22$,

$d_q = d_\gamma = 1 + 0.1(1.0/1.5)\tan(45.0° + 28.0°/2) = 1.11$. From Equation 10.10:

$q_u = 10.0 \times 25.80 \times 1.55 \times 1.22 + 19.0 \times 14.72 \times 1.28 \times 1.11 + 0.5 \times 1.5 \times 19 \times 11.19 \times 1.28 \times 1.11$,

$q_u = 1112$ kPa.

For the undrained conditions, $N_c = 5.14$, $N_q = 1$, and $N_\gamma = 0$.

$s_c = 1 + 0.2(1.5/1.5)\tan^2(45.0° + 0/2) = 1.2$, $s_q = s_\gamma = 1.0$,

$d_c = 1 + 0.2(1.0/1.5)\tan(45.0° + 0/2) = 1.13$, $d_q = d_\gamma = 1.0$,

$q_u = 105.0 \times 5.14 \times 1.2 \times 1.13 + 19.0 \times 1.0 \times 1.0 \times 1.0 = 751$ kPa.

Example 10.5

Re-do example 10.4 assuming that the central load applied to the footing is inclined at an angle $\alpha = 10°$ to the vertical. The load is located in a plane parallel to the side of the square footing. What is the ultimate value of this load?

Solution:

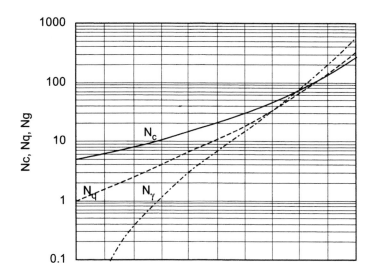

Figure 10.6. Bearing capacity factors using Hansen's equations.

From Equations 10.13: $i_c = i_q = (1-10.0°/90.0°)^2 = 0.79$, $i_\gamma = (1-10.0°/28.0°)^2 = 0.41$.
The shape and depth factors are the same as in Example 10.4. From Equation 10.9:
$q_u = 10.0 \times 25.80 \times 1.55 \times 1.22 \times 0.79 + 19.0 \times 14.72 \times 1.28 \times 1.11 \times 0.79 +$
$0.5 \times 1.5 \times 19 \times 11.19 \times 1.28 \times 1.11 \times 0.41 = 792$ kPa.
The ultimate value of the vertical load is:
$Q_{uv} = 792 \times 1.5^2 = 1782$ kN, $Q_u = 1782/\cos 10.0° = 1810$ kN.
For undrained conditions the shape and depth factors are identical to the values calculated
in Example 10.4. $i_c = i_q = 0.79$ (as above) and $i_\gamma = 1.0$, thus:
$q_u = 105.0 \times 5.14 \times 1.2 \times 1.13 \times 0.79 + 19.0 \times 1.0 \times 1.0 \times 1.0 \times 0.79 = 593$ kPa.
$Q_{uv} = 593 \times 1.5^2 = 1334$ kN, $Q_u = 1334/\cos 10.0° = 1355$ kN.

10.2.4 *Hansen's bearing capacity equations*

Hansen (1961 and 1970) extended Meyerhof's solutions by considering the effects of sloping ground surface and tilted base (Figure 10.7) as well as modification of N_γ and other
factors. For a rectangular footing of L by B ($L > B$), sloping ground surface, tilted base,
and inclined load:

$$q_u = c'N_c s_c i_c d_c b_c g_c + \gamma D N_q s_q i_q d_q b_q g_q + 0.5 B \gamma N_\gamma s_\gamma i_\gamma d_\gamma b_\gamma g_\gamma \qquad (10.17)$$

For horizontal ground surface and base: $g_c = g_q = g_\gamma = b_c = b_q = b_\gamma = 1$, the general bearing
capacity equation becomes the same as Equation 10.9. For the undrained conditions:

$$q_u = 5.14c_u \left(1 + s_c + d_c - i_c - b_c - g_c\right) + \gamma D \qquad (10.18)$$

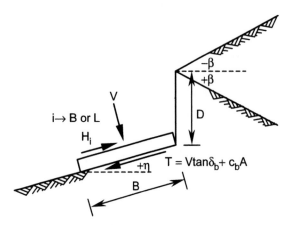

Figure 10.7. Identification of the terms in Hansen's bearing capacity equations.

The bearing capacity factors of N_q and N_c are defined by Equations 10.2; N_γ is defined by:

$$N_\gamma = 1.5(N_q - 1)\tan\phi' \qquad (10.19)$$

$$s_{c,B} = 1 + \frac{N_q}{N_c}\frac{B}{L}i_{c,B}, \; s_{q,B} = 1 + \frac{B}{L}i_{q,B}\sin\phi', s_{\gamma,B} = 1 - 0.4\frac{B}{L}i_{\gamma,B} \geq 0.6 \qquad (10.20)$$

$$s_{c,L} = 1 + \frac{N_q}{N_c}\frac{L}{B}i_{c,L}, \; s_{q,L} = 1 + \frac{L}{B}i_{q,L}\sin\phi', \; s_{\gamma,L} = 1 - 0.4\frac{L}{B}i_{\gamma,L} \geq 0.6 \qquad (10.21)$$

For $c_u, \phi_u = 0$ soil $\quad s_{c,B} = 0.2(B/L)i_{c,B}, \; s_{c,L} = 0.2(L/B)i_{c,L} \qquad (10.22)$

$$i_{c,i} = i_{q,i} - \frac{1 - i_{q,i}}{N_q - 1}, \; i_{q,i} = (1 - \frac{0.5H_i}{V + Ac_b\cot\phi'})^{\alpha_1}, \; i_{\gamma,i} = (1 - \frac{0.7H_i}{V + Ac_b\cot\phi'})^{\alpha_2}$$

$$(10.23)$$

where i (in Equations 10.23) = B or L, $2 \leq \alpha_1 \leq 5$, $2 \leq \alpha_2 \leq 5$, A is the area of the footing base and c_b is the cohesion mobilized in the footing-soil contact area. For the tilted base:

$$i_{\gamma,i} = \left[1 - \frac{(0.7 - \eta°/450°)H_i}{V + Ac_b\cot\phi'}\right]^{\alpha_2} \qquad (10.24)$$

For $c_u, \phi_u = 0$ soil $i_{c,i} = 0.5 - \sqrt{1 - H_i/Ac_b} \quad i = B$ or $L \qquad (10.25)$

In the above equations B and L may be replaced by their effective values expressed by Equations 10.16. The depth factors are specified in two sets. For $D/B \leq 1$, $D/L \leq 1$:

$$d_{c,B} = 1 + 0.4D/B, \; d_{q,B} = 1 + 2\tan\phi'(1 - \sin\phi')^2 D/B \qquad (10.26)$$

$$d_{c,L} = 1 + 0.4D/L, \; d_{q,L} = 1 + 2\tan\phi'(1 - \sin\phi')^2 D/L \qquad (10.27)$$

For $D/B > 1$, $D/L > 1$:

$$d_{c,B} = 1 + 0.4\tan^{-1}(D/B), \; d_{q,B} = 1 + 2\tan\phi'(1-\sin\phi')^2 \tan^{-1}(D/B) \quad (10.28)$$

$$d_{c,L} = 1 + 0.4\tan^{-1}(D/L), \; d_{q,L} = 1 + 2\tan\phi'(1-\sin\phi')^2 \tan^{-1}(D/L) \quad (10.29)$$

For both sets $d_\gamma = 1.0$ \hfill (10.30)

For c_u, $\phi_u = 0$ soil $d_{c,B} = 0.4(D/B)$, $d_{c,L} = 0.4(D/L)$ \hfill (10.31)

For the sloping ground and tilted base the ground factor g_i and base factor b_i are proposed by Equations 10.32 to 10.35. The angles β and η are at the plane either parallel to B or L:

$$g_c = 1 - \beta^\circ/147^\circ, \; g_q = g_\gamma = (1-0.5\tan\beta)^5 \quad (10.32)$$

For c_u, $\phi_u = 0$ soil $g_c = \beta^\circ/147^\circ$ \hfill (10.33)

$$b_c = 1 - \eta^\circ/147^\circ, \; b_q = e^{-2\eta\tan\phi'}, \; b_\gamma = e^{-2.7\eta\tan\phi'} \quad (10.34)$$

For c_u, $\phi_u = 0$ soil $b_c = \eta^\circ/147^\circ$ \hfill (10.35)

Example 10.6

Re-do example 10.4 using Hansen's bearing capacity factors.

Solution:

From Equations 10.19: $N_\gamma = 1.5(14.72-1.0)\tan 28.0^\circ = 10.94$.

Shape factors are calculated from Equations 10.20 by setting the inclination factors to 1.0.
$s_c = 1 + (14.72/25.80)(1.5/1.5) = 1.57$, $s_q = 1 + (1.5/1.5)\sin 28.0^\circ = 1.47$,
$s_\gamma = 1 - 0.4(1.5/1.5) = 0.6$. From Equations 10.26 and 10.30:
$d_c = 1 + 0.4(1.0/1.5) = 1.27$,
$d_q = 1 + 2\tan 28.0^\circ(1-\sin 28.0^\circ)^2(1.0/1.5) = 1.20$ and $d_\gamma = 1.0$.
Using Equation 10.17:
$q_u = 10.0 \times 25.80 \times 1.57 \times 1.27 + 19.0 \times 14.72 \times 1.47 \times 1.20 + 0.5 \times 1.5 \times 19 \times 10.94 \times 0.6 \times 1.0$,
$q_u = 1101$ kPa. For the undrained conditions $N_\gamma = 0$. The shape factors s_q and s_γ are calculated from the same equations for $\phi' > 0$ case, but s_c is calculated from Equations 10.22:
$s_c = 0.2(1.5/1.5) = 0.2$, $s_q = 1.0$, $s_\gamma = 0.6$.
Using Equations 10.31 for d_c ($\phi' = 0$), 10.26 for d_q and 10.30 for d_γ:
$d_c = 0.4(1.0/1.5) = 0.27$, $d_q = 1.0$, $d_\gamma = 1.0$. Substituting in Equation 10.18:
$q_u = 5.14 \times 105.0(1 + 0.2 + 0.27) + 19.0 = 812$ kPa.
The results of Examples 10.3, 10.4 and 10.6 are tabulated below:

Soil condition	Terzaghi	Meyerhof	Hansen
q_u (kPa): $c' = 10.0$ kPa, $\phi' = 28^\circ$	921	1112	1101
q_u (kPa): $c_u = 105$ kPa, $\phi_u = 0$	797	751	812

10.2.5 *Vesić's bearing capacity equations*

Vesić (1973) suggested a procedure for ultimate bearing capacity that is similar to Hansen's method with minor modifications in N_γ and some selected factors. This method seems easier to use, as there is no interrelationships between different factors. In this method Equations 10.17 and 10.18 are used with the following modifications:

$$N_\gamma = 2(N_q + 1)\tan\phi' \tag{10.36}$$

For c_u, $\phi_u = 0$ soil with the sloping ground (Figure 10.7):

$$N_\gamma = -2\sin\beta \tag{10.37}$$

The other terms are as follows:

$$s_{c,B} = 1 + (N_q/N_c)(B/L),\ s_{q,B} = 1 + (B/L)\tan\phi',\ s_{\gamma,B} = 1 - 0.4(B/L) \geq 0.6 \tag{10.38}$$

$$s_{c,L} = 1 + (N_q/N_c)(L/B),\ s_{q,L} = 1 + (L/B)\tan\phi',\ s_{\gamma,L} = 1 - 0.4(L/B) \geq 0.6 \tag{10.39}$$

For c_u, $\phi_u = 0$ soil $s_{c,B} = 0.2(B/L),\ s_{c,L} = 0.2(L/B)$ $\tag{10.40}$

For $i_{c,i}$ use Hansen's equation; other terms are defined by:

$$i_{q,i} = (1 - \frac{H_i}{V + Ac_b\cot\phi'})^{m_i},\ i_{\gamma,i} = (1 - \frac{H_i}{V + Ac_b\cot\phi'})^{m_i+1}\quad i = B\ \text{or}\ L \tag{10.41}$$

For c_u, $\phi_u = 0$ soil $i_{c,i} = 1 - m_i H_i/Ac_b N_c,\ i = B\ \text{or}\ L$ $\tag{10.42}$

$$m_B = (2 + B/L)/(1 + B/L),\ m_L = (2 + L/B)/(1 + L/B) \tag{10.43}$$

For the sloping ground and tilted base (Figure 10.7):

$$g_c = i_q - (1 - i_q)/(5.14\tan\phi'),\ g_q = g_\gamma = (1 - \tan\beta)^2 \tag{10.44}$$

For c_u, $\phi_u = 0$ soil $g_c = \beta/5.14$ $\tag{10.45}$

$$b_c = 1 - 2\beta/(5.14\tan\phi'),\ b_q = b_\gamma = (1 - \eta\tan\phi')^2 \tag{10.46}$$

For c_u, $\phi_u = 0$ soil $b_c = \beta/5.14$ $\tag{10.47}$

The depth factors are the same as Hansen's method. For circular bases in both Hansen's and Vesić's methods the dimensions of an equivalent square may be used.

Example 10.7

A square footing of 1.5 m is to be constructed in sand with $c' = 0$, $\phi' = 40°$. The thickness of the footing is 0.45 m and its top surface is level with the horizontal ground surface. The footing is subjected to a central vertical force of 700 kN and a central horizontal force (parallel to the side) of 210 kN. Find the ultimate bearing capacity by the (a) Meyerhof's method, (b) Hansen's method and (c) Vesić's method. The unit weight of the sand is 18 kN/m³.

Solution:

$\gamma D = 0.45 \times 18.0 = 8.1$ kPa. Find N_q from Equation 10.2:

$N_q = \exp(\pi \tan \phi') \tan^2(45° + \phi'/2) = e^{\pi \tan 40.0°} \tan^2(45.0° + 40.0°/2) = 64.19$.

(a) Meyerhof's method: find N_γ from Equation 10.11:

$N_\gamma = (N_q - 1) \tan(1.4\phi') = (64.19 - 1) \tan(1.4 \times 40.0°) = 93.68$.

Using Equations 10.12 and 10.14:

$s_q = s_\gamma = 1 + 0.1(B/L) \tan^2(45° + \phi'/2) = 1 + 0.1(1.5/1.5) \tan^2(45.0° + 40.0°/2) = 1.46$.

$d_q = d_\gamma = 1 + 0.1(D/B) \tan(45° + \phi'/2) = 1.0 + 0.1(0.45/1.5) \tan(45.0° + 40.0°/2) = 1.06$.

Inclination angle $\alpha = \tan^{-1}(210.0/700.0) = 16.7°$, thus from Equations 10.13:

$i_q = (1 - \alpha°/90°)^2 = (1 - 16.7°/90.0°)^2 = 0.66$, $i_\gamma = (1 - 16.7°/40.0°)^2 = 0.34$.

Using Equation 10.9:

$q_u = 8.1 \times 64.19 \times 1.46 \times 1.06 \times 0.66 + 0.5 \times 1.5 \times 18.0 \times 93.68 \times 1.46 \times 1.06 \times 0.34 = 1196$ kPa.

(b) Hansen's method: find N_γ from Equation 10.19:

$N_\gamma = 1.5(N_q - 1) \tan \phi' = 1.5(64.19 - 1.0) \tan 40.0° = 79.53$.

The inclination factors are calculated from Equations 10.23 with $\alpha_1 = \alpha_2 = 3$.

$i_q = [1 - 0.5 H_i /(V + A c_b \cot \phi')]^{\alpha_1} = [1 - 0.5 \times 210.0/(700.0 + 0.0)]^3 = 0.61$,

$i_\gamma = [1 - 0.7 H_i /(V + A c_b \cot \phi')]^{\alpha_2} = [1 - 0.7 \times 210.0/(700.0 + 0.0)]^3 = 0.49$.

Calculate shape and depth factors from Equations 10.20, 10.26 and 10.30:

$s_q = 1 + (B/L) i_q \sin \phi' = 1 + (1.5/1.5) \times 0.61 \times \sin 40.0° = 1.39$,

$s_\gamma = 1 - 0.4(B/L) i_\gamma = 1 - 0.4(1.5/1.5) \times 0.49 = 0.80$.

$d_q = 1 + 2 \tan \phi'(1 - \sin \phi')^2 (D/B) = 1 + 2 \tan 40.0°(1 - \sin 40.0°)^2 (0.45/1.5) = 1.06$,

$d_\gamma = 1.0$.

Thus from Equation 10.17 and with $c' = 0$ and $b_c = b_q = b_\gamma = g_c = g_q = g_\gamma = 1$:

$q_u = \gamma D N_q s_q i_q d_q + 0.5 B \gamma N_\gamma s_\gamma i_\gamma d_\gamma = 8.1 \times 64.19 \times 1.39 \times 0.61 \times 1.06 +$

$0.5 \times 1.5 \times 18.0 \times 79.53 \times 0.8 \times 0.49 \times 1.0 = 888$ kPa.

(c) Vesić's method: from Equation 10.36:

$N_\gamma = 2(N_q + 1) \tan \phi' = 2(64.19 + 1) \tan 40.0° = 109.40$.

Shape factors are calculated from Equations 10.38:

$s_q = 1 + (B/L) \tan \phi' = 1 + (1.5/1.5) \tan 40.0° = 1.84$,

$s_\gamma = 1 - 0.4(B/L) = 1 - 0.4(1.5/1.5) = 0.60$.

Depth factors are the same as Hansen's method: $d_q = 1.06$, $d_\gamma = 1.0$. From Equation 10.43:

$m = (2 + B/L)/(1 + B/L) = (2 + 1.5/1.5)/(1 + 1.5/1.5) = 1.5$, using Equations 10.41:

$i_q = [1 - H_i /(V + A c_b \cot \phi')]^m = [1 - 210.0/(700.0 + 0.0)]^{1.5} = 0.58$,

$i_\gamma = [1 - H_i /(V + A c_b \cot \phi')]^{m+1} = [1 - 210.0/(700.0 + 0.0)]^{2.5} = 0.41$, thus:

$q_u = \gamma D N_q s_q i_q d_q + 0.5 B \gamma N_\gamma s_\gamma i_\gamma d_\gamma = 8.1 \times 64.19 \times 1.84 \times 0.58 \times 1.06 +$

$0.5 \times 1.5 \times 18.0 \times 109.40 \times 0.6 \times 0.41 \times 1.0 = 951$ kPa.

10.2.6 Skempton's N_c value

For c_u, $\phi_u = 0$ soil Skempton (1951) suggested the following formulas:

$$q_u = c_u N_c + \gamma D \tag{10.48}$$

$$N_c = 5.14(1 + \frac{0.2B}{L})(1 + \frac{0.24D}{B} - \frac{0.031D^2}{B^2}) \tag{10.49}$$

For circular base $B = L$ = diameter of the footing. The values of N_c for strip and square (or circular) footing are presented graphically in Figure 10.8.

Example 10.8

Determine the ultimate bearing capacity of a square footing 1.2 m, at a depth of 2.5 m in a soil with $c_u = 105$ kPa, $\phi_u = 0$, $\gamma = 19$ kN/m^3.

Solution:

$\gamma D = 19.0 \times 2.5 = 47.5$ kPa.

Using Equation 10.49:

$N_c = 5.14(1 + 0.2 \times 1.2/1.2)(1 + 0.24 \times 2.5/1.2 - 0.031 \times 2.5^2/1.2^2) = 8.42$, thus:

$q_u = c_u N_c + \gamma D = 105.0 \times 8.42 + 47.5 = 932$ kPa.

10.2.7 Effect of water table on bearing capacity

At the presence of hydrostatic water an effective unit weight γ_e must be substituted in the bearing capacity expressions. The effective unit weight is 50% to 100% of the unit weight of the soil above the water table. The following equation is a reasonable estimation for γ_e under the footing level:

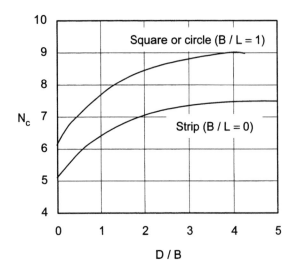

Figure 10.8. Skempton's N_c values.

$$\gamma_e = \gamma' + \frac{z_w}{B}(\gamma - \gamma') \qquad\qquad (z_w \le B) \qquad\qquad (10.50)$$

where z_w is the depth of water level from the base, γ is the unit weight above the water level and γ' is the submerged unit weight of the soil. If water table is on the foundation base ($z_w = 0$), $\gamma_e = \gamma'$; for $z_w = B$, $\gamma_e = \gamma$. An alternative equation can be obtained by calculating an average unit weight within the zone 1 of failure mechanism of Figure 10.1:

$$\gamma_e = \gamma\left[1 - (\frac{H - z_w}{H})^2\right] + \gamma'(\frac{H - z_w}{H})^2 \quad (z_w \le H) \qquad\qquad (10.51)$$

where $H = B\tan(45° + \phi'/2)$. Note that in most cases the third term of the bearing capacity expression ($0.5\, B\, \gamma\, N_\gamma$) has not a significant value and its further reduction by water will have little effect on the bearing capacity. However in granular materials this term is the main contributor (apart from $\gamma D N_q$) and its reduction due to the water table could be of significant value. For the buried footings similar reduction factors in the form of effective unit weight are applied for the surcharge load γD.

Example 10.9

Re-do Example 10.7 assuming that the water table is at the footing level (0.45 m below the ground surface). The saturated unit weight is 21 kN/m^3.

Solution:

Using Equation 10.50 or 10.51:
γ_e (under the footing) $= \gamma' = 21.0 - 9.81 = 11.19$ kN/m^3.

(a) Meyerhof's method:

$N_q = 64.19$, $N_\gamma = 93.68$, $s_q = s_\gamma = 1.46$, $d_q = d_\gamma = 1.06$, $i_q = 0.66$, and $i_\gamma = 0.34$.

From Equation 10.10:
$q_u = 8.1 \times 64.19 \times 1.46 \times 1.06 \times 0.66 + 0.5 \times 1.5 \times 11.19 \times 93.68 \times 1.46 \times 1.06 \times 0.34 = 945$ kPa.

(b) Hansen's method:

$N_q = 64.19$, $N_\gamma = 79.53$, $s_q = 1.39$, $s_\gamma = 0.80$, $d_q = 1.06$, $d_\gamma = 1.0$, $i_q = 0.61$, and $i_\gamma = 0.49$.

From Equation 10.17:
$q_u = 8.1 \times 64.19 \times 1.39 \times 1.06 \times 0.61 + 0.5 \times 1.5 \times 11.19 \times 79.53 \times 0.80 \times 1.0 \times 0.49 = 729$ kPa.

(c) Vesić's method:

$N_q = 64.19$, $N_\gamma = 109.40$, $s_q = 1.84$, $s_\gamma = 0.60$, $d_q = 1.06$, $d_\gamma = 1.0$, $i_q = 0.58$, and $i_\gamma = 0.41$.

From Equation 10.17:
$q_u = 8.1 \times 64.19 \times 1.84 \times 0.58 \times 1.06 + 0.5 \times 1.5 \times 11.19 \times 109.40 \times 0.60 \times 0.41 \times 1.0 = 814$ kPa.

10.2.8 *Allowable bearing pressure and factor of safety*

Allowable bearing (or soil) pressure q_a is the contact pressure beneath the footing which (a) satisfies the settlement requirements, (b) provides safety against shear failure. Thus the ultimate bearing capacity must be divided by a desired factor of safety F to provide a *safe bearing capacity* equal to or greater than the allowable bearing pressure.

The settlement requirements depend on the type of the structure; however a maximum differential settlement (between adjacent columns of the structure) of 25 mm is accepted universally by geotechnical and structural engineers. For sands a maximum settlement of 40 mm for isolated footings and up to 65 mm for raft foundations is recommended. The corresponding maximum settlements for clay are 65 mm and 100 mm respectively (Mac-Donald & Skempton, 1955; Walsh, 1981).

The magnitude of factor of safety applied to the ultimate bearing capacity depends on the type of the foundation. For spread footings it is between 2 to 3, whilst for mat foundations it may selected within the range 1.7 to 2.5 (Bowles, 1996). The factor of safety in terms of actual contact pressure $q \leq q_a$ and footing depth D may be expressed by:

$$F = \frac{q_u - \gamma D}{q - \gamma D} \tag{10.52}$$

Most of the building codes recommend the *Ultimate Strength Design (USD)* for design and analysis of foundations. In this method the working loads are factorised and combined (according to the code) to give the ultimate load applied to the footing. This load in many cases is about 1.5 times of the working load; however higher or lower values may be obtained depending on the nature and critical combination of the involved loads. Under this ultimate load the footing material is in plastic state. While the initial plan section of a footing is estimated using working loads and allowable bearing pressure, the cross section and reinforcements are calculated by *USD* method. The contact pressure under the ultimate load is called *ultimate soil pressure*. If the allowable bearing pressure is obtained by application of a factor of safety of 3 to the ultimate bearing capacity, and the factorised ultimate load is 1.5 times of the working load, then the ultimate soil pressure is 50% of the ultimate bearing capacity. This ensures the factor of safety of 2 against shear failure of the foundation soil at the time when the foundation material has reached to its ultimate strength.

Example 10.10

Find the factor of safety against shear failure in Examples 10.7 and 10.9.

Solution:

Using Equation 10.52 for Example 10.7:

Note $\gamma D = 8.1$ kPa and,

$q =$ (contact pressure) $= 700$ (vertical load) $/ 1.5^2 = 311.1$ kPa.

(a) Meyerhof's method: $F = (1196 - 8.1)/(311.1 - 8.1) = 3.92$,

(b) Hansen's method: $F = (888 - 8.1)/(311.1 - 8.1) = 2.90$,

(c) Vesić's method: $F = (951 - 8.1)/(311.1 - 8.1) = 3.11$.

For the Example 10.9:

(a) Meyerhof's method: $F = (945 - 8.1)/(311.1 - 8.1) = 3.09$,

(b) Hansen's method: $F = (729 - 8.1)/(311.1 - 8.1) = 2.38$,

(c) Vesić's method: $F = (814 - 8.1)/(311.1 - 8.1) = 2.66$.

10.3 FIELD TESTS

10.3.1 *Plate loading test*

The plate-loading test is carried out to estimate the bearing capacity and settlement beneath the single footings (Section 5.4.6). A loading plate is circular or square in shape and is manufactured from machined steel plate with a minimum thickness of 25 mm. The diameter of circular plates ranges from 150 mm to 760 mm. The square or circular plate with 1 ft^2 area has many practical applications in geotechnical site investigations. The load is either of gravity type applied through a platform or reaction type applied by a hydraulic jack. The test is carried out in a pit with its bottom located at the actual footing level. The diameter of the pit is 4 to 5 times of the diameter of the plate to allow the developments of failure planes to the ground surface (bottom of the pit). The load is applied in increments each 1 / 5 to 1 / 10 of the predicted ultimate load. For each increment a time (logarithmic scale)-settlement plot is constructed from the recorded data. The next increment is applied after a specified time interval (not less than 1 hour) when no settlement is observed. The settlement-load plot is also constructed to establish the maximum contact pressure or the contact pressure needed for a specified settlement (commonly 25 mm). Equation 5.63 can be used to relate the test results to the actual footing; however the disadvantages of an extrapolation has been investigated by some authors (Section 5.4.6) and must not be used if the ratio of the width of the footing (B) to the width or diameter of the plate exceeds 3. In sands a screw-plate test is performed without a need for a pit excavation. The plate loading test procedure may be found in ASTM D-1194, D-1195, and D-1196.

10.3.2 *Standard penetration test*

Standard Penetration Test (*SPT*) is a dynamic in-situ test performed mostly in cohesionless soils to estimate the allowable bearing pressure as well as the internal friction angle and the ultimate bearing capacity. Furthermore the correlation of *SPT* results and elastic properties of soils has given useful practical equations (Section 5.4.7). In the absence of a sampling technique to provide undisturbed specimens of cohesionless soils *SPT* is considered a simple but efficient tool in site investigation of these soils. This test determines the resistance of the soil to the penetration of a split barrel sampler and obtains disturbed soil specimens for identification purposes. The details of basic equipment and test procedure may be found in ASTM D-1586, BS 1377 and AS 1289.6.3.1. The schematic view of the sampler is shown in Figure 10.9. The section is comprised of a driving shoe, a split barrel for collection of the disturbed sample and a sampler head for connection to the rod and drilling assembly. The central part of the sampler is of split type construction (to allow the removal of the recovered soil) with provision for a liner. The sampler is driven 450 mm through a bottom of a borehole (already drilled in the ground) using a hammer of 64 kg falling freely from a vertical distance of 760 mm. The number of blows are counted for each 150 mm penetration. The sum of number of blows for the last two 150 mm penetrations gives the *SPT* number N (also called penetration resistance). The *SPT* number for adjacent holes from equipment made by different manufacturers are not equal and must be normalized to an energy ratio recommended by the code or computed by the user (Seed et al., 1984; Skempton, 1986).

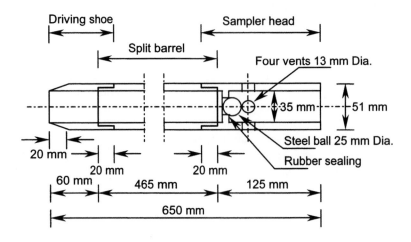

Figure 10.9. Schematic diagram of a split barrel sampler.

The energy ratio is defined by:

$$E_r = (E_{ac} / E_{in}) \times 100\% \tag{10.53}$$

where E_{in} is the input driving energy caused by 760 mm free dropping of 64 kg hammer (0.475 kN.m) and E_{ac} is the actual energy transferred to the sampler. The energy ratio may change from 50% to 80% and more depending on the type of the hammer. In some countries this ratio is standardized. For example in UK for rope-pulley or cathead lift hammer E_r is 50%, whilst for hammers with automatic trip it is 60%. A corrected N value for an apparatus of E_{r1} with N_1 count can be calculated from the following equation:

$$E_r \times N = E_{r1} \times N_1 \tag{10.54}$$

where E_r is the standardized energy ratio. Note that to establish E_{r1} we need to calibrate the apparatus (drilling rigs) on regular bases. Further adjustments are due to water table, overburden pressure, length of the drilling rod (reduction effects for lengths less than 10 m), sampler liner, and borehole diameter. The correction for water table is carried out using the following equation (Peck, et al., 1974) where the *SPT* number N must be multiplied by parameter C_w:

$$C_w = 0.5[1 + D_w / (D + B)] \tag{10.55}$$

where D_w and D are the depths of water table and footing base from the surface respectively. In the current practice the correction for water table is ruled out due to conservative nature of the above equation and that the *SPT* number obtained below the water table reflects the reduction in the N value. The early experimental results by Gibbs & Holtz (1957) and Meyerhof (1956) showed that the increase in overburden pressure influences the *SPT* number; thus the measured field N value must be multiplied by a correction factor C_N. In the Meyerhof's proposal the corrected N value depends on the square of the density index I_D (or relative density: Equation 1.13) and changes linearly with effective overburden pressure p'_o according:

$$N = I_D^2(a + bp_o')$$ (10.56)

Terzaghi et al. (1996) proposed the following table for sands:

Table 10.1. Relationship between *SPT* number and density index (Terzaghi et al., 1996).

Sand	SPT number N
Very loose	0-4
Loose	4-10
Medium	10-30
Dense	30-50
Very dense	Over 50

Peck et al. (1974) presented a correction factor that has a considerable applications in practise:

$$C_N = 0.77 \log(2000/p_o') \qquad (p_o' \text{ in kPa})$$ (10.57)

Based on Meyerhof's concept, Skempton (1986) proposed the following equation:

$$C_N = a/(b + p_o')$$ (10.58)

where p_o' is in (kPa); a and b are according to the table below:

Table 10.2. Parameters to obtain the correction factor C_N (Skempton, 1986).

Sands	Range of I_D	a	b
Fine-medium	0.35 to 0.65	200	100
Dense-coarse	0.65 to 0.85	300	200
Overconsolidated	0.85 to 1.00	170	70

Correction for overburden pressure may be carried out by the following equation, which covers most of the proposed corrections by different authors (Liao & Whitman, 1986):

$$C_N = \sqrt{95.76/p_o'}$$ (10.59)

where p_o' is the effective overburden pressure at the footing level in kPa.

Correction factors for the drill rod length and borehole diameter are given by Skempton (1986). For rods longer than 10 m no correction is necessary whilst for rods less than 10 m a reduction factor of 0.75 (short rods) to 1 (10 m rod) is applied. For boreholes 65 mm to 115 mm no correction factor is needed. For wider boreholes the correction factor is greater than 1 being 1.15 for 200 mm borehole. The existence of a liner in the split barrel increases the N value due to side friction, thus a reduction factor must be applied. The amounts of this factor for dense and loose materials are suggested 0.8 and 0.9 respectively. The correlations between shear strength parameters and *SPT* number N have been reported by number of authors. Peck et al. (1974) gave correlations between corrected N (for effective overburden pressure) and the bearing capacity factors. Furthermore the correlations were extended to estimate the ϕ' values of the normally consolidated sands (Schmertmann, 1975) and overconsolidated clays (Stroud, 1989).

Table 10.3 relates the *SPT* number to the consistency states of clayey soils (Terzaghi et al., 1996). While using Tables 10.1 to 10.3, one must take into account the energy ratio corresponding to *SPT* number.

Table 10.3. Bearing capacity in clay soils and the corresponding *SPT* numbers (Terzaghi et al., 1996).

Consistency	Very soft	Soft	Medium	Stiff	Very stiff	Hard
SPT number N	< 2	2-4	4-8	8-15	15-30	> 30
q_u (kPa)	< 25	25-50	50-100	100-200	200-400	> 400

Correlation between *SPT* number and allowable bearing pressure was investigated by Terzaghi & Peck (1967) for footings on sands relating the isobars of *SPT* number N to allowable bearing pressure q_a and width of the footing B. However it has been shown that these charts are very conservative being 50% or more on the safe side. Meyerhof (1956 and 1974) provided the following relationships:

$$q_a = \frac{12 S_e \overline{N}}{25} k_d \qquad\qquad B < 1.22 \text{ m} \qquad\qquad (10.60)$$

$$q_a = \frac{8 S_e \overline{N}}{25} (\frac{B + 0.305}{B})^2 k_d \qquad B > 1.22 \text{ m} \qquad\qquad (10.61)$$

where S_e is the elastic settlement of the layer in mm, $\overline{N}$ is the average *SPT* number within the influence depth of D (depth of the base) to $D + B$ and k_d is defined by:

$$k_d = 1 + 0.33 D / B \qquad\qquad (10.62)$$

Bowles (1996) improved Meyerhof's equations by 50% increase in the allowable bearing pressure according:

$$q_a = \frac{20 S_e \overline{N}}{25} k_d \qquad\qquad B < 1.22 \text{ m} \qquad\qquad (10.63)$$

$$q_a = \frac{12.5 S_e \overline{N}}{25} (\frac{B + 0.305}{B})^2 k_d \qquad B > 1.22 \text{ m} \qquad\qquad (10.64)$$

Figure 10.10 shows the allowable bearing pressure corresponding to 25 mm settlement constructed from the above equations for $D / B = 0$. In the above equations an energy ratio of 55% must be assumed. Burland & Burbidge (1985) suggested the following equations for sands (Section 5.4.6, Equations 5.69 to 5.71). For normally consolidated sands:

$$q_a = \frac{S_e \overline{N}^{1.4}}{1.7 B^{0.75}} \qquad\qquad (10.65)$$

For overconsolidated sands:

$$q_a = \frac{S_e \overline{N}^{1.4}}{1.7 B^{0.75}} + \frac{2}{3} p'_c \quad (q_a > p'_c) \text{ and } q_a = \frac{3 S_e \overline{N}^{1.4}}{1.7 B^{0.75}} \quad (q_a < p'_c) \qquad (10.66)$$

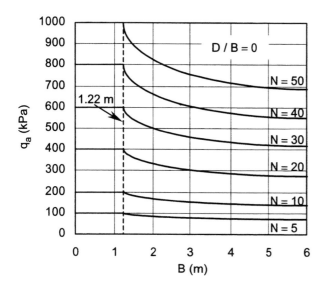

Figure 10.10. Allowable bearing pressure corresponding to 25 mm settlement (Bowles, 1996).

where p'_c is the preconsolidation pressure. The magnitude of $\overline{N}$ may be corrected according to Equations 5.67 and 5.68 (for saturated sands and gravelly sands respectively) including the various adjustments mentioned above. For rectangular footings the assumed settlement must satisfy the shape factor expressed by Equation 5.72. This means the allowable bearing pressure (including the term $2/3p'_c$) must be reduced by the shape factor (maximum 1.56). If an incompressible layer is located within the influence depth Z_l, the calculated q_a (including the term $2/3p'_c$) may be increased by the correction factor defined by Equation 5.73.

Example 10.11

For the *SPT* results given in Example 5.21 and a square footing with $B = 2$ m calculate the allowable bearing pressure corresponding to 25 mm settlement (a) if the base of the footing is on the ground surface, (b) if the base of the footing is 1 m below the ground surface.
Solution:

Meyerhof's method:

From Example 5.21 the average uncorrected (for water table) *SPT* values for cases (a) and (b) are 15 ($z = 0$ to $z = 2.0$ m) and 20 ($z = 1.0$ m to $z = 3.0$ m) respectively.

Case (a): $q_a = \dfrac{8S_e\overline{N}}{25}(\dfrac{B+0.305}{B})^2 k_d = \dfrac{8\times25.0\times15}{25}(\dfrac{2.0+0.305}{2.0})^2 \times 1.0 = 159.4$ kPa.

Case (b): $q_a = \dfrac{8\times25.0\times20}{25}(\dfrac{2.0+0.305}{2.0})^2 \times(1+0.33\dfrac{1.0}{2.0}) = 247.6$ kPa.

Using Bowles method the q_a values obtained from Meyerhof's method are multiplied by the ratio of 12.5 / 8. For case (a) $q_a = 249.0$ kPa and for case (b) $q_a = 386.9$ kPa.

Burland & Burbidge method: the average corrected value of *SPT* within the influence depth is 14 and 17 for cases (a) and (b) respectively.

Case (a) $q_a = \dfrac{S_e \overline{N}^{1.4}}{1.7 B^{0.75}} = \dfrac{25.0 \times 14^{1.4}}{1.7 \times 2.0^{0.75}} = 351.8 \text{ kPa}.$

Case (b) $q_a = \dfrac{25.0 \times 17^{1.4}}{1.7 \times 2.0^{0.75}} = 461.7 \text{ kPa}.$

10.3.3   *Cone penetration test*

Cone Penetration Test (*CPT*) is a static in-situ test used for subsurface exploration in fine and medium sands, soft silts and clays without taking soil samples. The term *static* implies that there is no impact on the penetrometer and the device is pushed into the soil by an external thrust. In this test the tip resistance of a steel cone of 35.7 mm end diameter with a projected area of 1000 mm^2 and 60° point angle is measured while advancing 80 mm into the soil at the test level at a constant rate of 10 mm/s to 20 mm/s. The force is applied through a thrust machine and is transferred to the cone by an assembly of the rods attached to the cone. This assembly is protected by an outer rod (or sleeve). The cone resistance q_c is the force required to push the cone divided by the end area of the cone. The test may be carried out in 1 m to 1.5 m depth intervals to give the distribution of cone resistance in depth. The data is used to classify the soil and estimate the allowable bearing pressure and elastic properties via empirical relationships. The test was originally developed in Netherlands (*Dutch Cone Test*) and was widely used in Europe. The developed versions of the cone, which are capable of measuring physical properties of soils including pore pressure, has gained favour in the USA and Canada. A detailed description of the test along with different types of the penetrometer is standardized by ASTM D-3441, BS 1377(Part 9) and AS 1289.6.5.1. The test procedures cover the determination of the components of the penetration resistance, which includes the cone resistance q_c and side friction f_s. The main types of cone penetrometers are as follows:

Mechanical cone penetrometer. The earliest type had a direct connection to a push rod with a diameter slightly less than that of the cone. The improved version has hollow outer and solid inner rods. The inner rod is pushed first and the required force is measured. Next the outer rod is pushed (with reading of the force) until it rests on the cone. The thrusting is continued on the outer rod pushing the cone into the soil and thus providing a third reading.

Mechanical friction cone penetrometer. This penetrometer has similar design to the mechanical cone penetrometer with a friction sleeve of surface area of 10000 mm^2 or 15000 mm^2 added at the top of the cone. The mechanism of the penetrometer allows the inner rod to push the cone engaging the friction sleeve after specified advance, thus measuring the cone resistance q_c in the first start and the total cone resistance plus side friction f_s (per unit area of sleeve) as the test proceeds.

Electrical cone penetrometer. A typical design has tubular push rods and a force transducer fitted to the cone, which is connected via a central cable (through the push rods) to a surface data acquisition system.

Electrical friction cone penetrometer. In this penetrometer the friction force on the friction sleeve is measured separately.

The push rods assembly form a rigid-jointed string with screwed flush joints, straight axis, smooth surface, and constant external diameter. The thrust machine should have a capacity of 40 kN to 200 kN and be capable of forcing the penetrometer at a constant rate. Depending on the mass of the thrust machine, additional reaction equipment may be required. Test intervals depend on the type of the project and are 150 mm to 200 mm. In a pavement project a shorter depth intervals may be used. The test report include calculation of cone resistance q_c at any depth with taking account the applied force and the weights of the push rods (in mechanical penetrometers) and calculation of side friction f_s. These are used to calculate the friction ratio (percentage) defined by:

$$F_r(\%) = \frac{f_s}{q_c} \times 100 \tag{10.67}$$

which is used for soil classification and determination of soil sensitivity. The soil sensitivity S_t is defined as the ratio of the undisturbed strength to the remoulded strength (see Section 4.3.7).

The correlations of *CPT* results to the engineering properties of soils have been developed in three major areas of soil classification, the ultimate bearing capacity (either by estimation of bearing capacity factors or the internal friction angle) and the Modulus of Elasticity. Figure 10.11(a) shows a soil classification chart reported by Robertson & Campanella (1983). The lowest curve representing the upper limit for clay soils is suggested by Bowles (1996).

The undrained cohesion c_u may be estimated by applying Equation 10.48:

$$c_u = (q_c - \gamma D)/N_k \tag{10.68}$$

where the term γD represents the total overburden pressure above the tip of the cone and N_k is analogous to the bearing capacity factor N_c. The magnitude of N_k depends on the clay plasticity and reduces with the increase of the plasticity index (Lunne & Edie, 1976). It varies from 14-21 for normally consolidated clay and 24-30 for overconsolidated clay. The above investigation also correlates the cone resistance to the vane shear test (Section 4.3.7) results. One must be aware that the vane shear test overestimates the undrained shear strength c_u and the results must be multiplied by a reduction factor 1 to 0.6 for the range of plasticity index 20% to 100% (Bjerrum, 1974). Bowles (1996) suggested the use of the following formula:

$$N_k = (13 + 0.11PI) \tag{10.69}$$

where *PI* is the plasticity index. Further improvements in c_u may be achieved by adding the measured pore pressure (using electrical type cone penetrometer) to q_c. The measured pore pressure must be multiplied by $1 - \chi$ where χ is the ratio of the internal cross-section of the outer push rod (in the vicinity of the cone) to the end area of the cone (Robertson & Campanella, 1983). In sands Meyerhof (1976) presented a chart relating the internal friction angle ϕ' (from 30° to 45°) to the cone resistance apparently without taking into account the effective overburden pressure. From the Begemann (1974) a series of equations are given below which evaluate an equivalent bearing capacity factor in the form of:

$$q_c = V_b \times p'_o \tag{10.70}$$

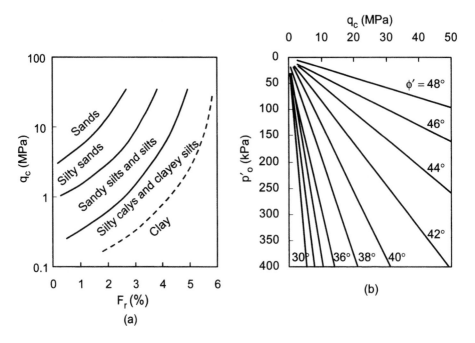

Figure 10.11. *CPT* correlations: (a) soil classification, (b) determination of effective internal friction angle (Robertson & Campanella, 1983).

and thus calculating the ϕ' from:

$$V_b = 10^{3.04 \tan \phi'} \qquad \text{(Caquot equation)} \qquad (10.71)$$

$$V_b = 1.3 \times e^{(2.5\pi - \phi') \tan \phi'} [(1 + \sin \phi')/(1 + \sin^2 \phi')] \quad \text{(Koppejan equation)} \quad (10.72)$$

$$V_b = 1.3 \times e^{2\pi \tan \phi'} \tan^2 (45° + \phi'/2) \qquad \text{(DeBeer equation)} \qquad (10.73)$$

Another correlation between ϕ' and cone resistance with taking into account of the effective overburden pressure is shown in Figure 10.11(b). In practice an evaluation of average ϕ' obtained from the available equations and charts is necessary if the soil deposit is deep and there is no possibility of obtaining undisturbed samples for laboratory testing. The correlation of *CPT* to the elastic properties of soils is discussed earlier in Section 5.4.7 where the Modulus of Elasticity was expressed in terms of cone resistance via empirical equations. However these equations are based on limited data as cone penetration test is more recent than the standard penetration test. With the availability of more data in the future the confidence in using the correspondent correlations will increase (Powrie, 1997). Based on the investigation by DeBeer & Martens (1957) a compressibility coefficient similar to compression index (Chapter 6) was proposed in the following form:

$$C = \frac{1.5 q_c}{p'_o} \qquad (10.74)$$

The settlement of a layer of finite thickness H is estimated from:

$$S_e = \frac{H}{C} \log(\frac{\Delta\sigma + p'_o}{p'_o})$$
(10.75)

where $\Delta\sigma$ is the increase in the vertical stress at the centre of the layer due to external loading and p'_o is the effective overburden pressure at that point. This method is applied by dividing the soil under the footing into finite layers (say up to the depth of 2B) and summing up the settlement of the layers. The above method overestimates the settlement, thus the methods based on the Modulus of Elasticity expressed by Equations 5.90 and 5.91 may be used. It is commonly accepted that the method proposed by Schmertmann (1970) and Schmertmann et al. (1978) (Section 5.4.6) is more reliable in sands and sandy soils.

The relationship between *CPT* resistance q_c and *SPT* number N, has been found to depend upon the average particle size D_{50}, and the q_c / N ratio (Robertson et al., 1983; Burland & Burbidge, 1985). The following equation is a reasonable correlation of the reported data by the above authors that is believed to be reliable data in use:

$$q_c = 800N(D_{50})^{0.3} \quad D_{50} \leq 1 \text{ mm}$$
(10.76)

where D_{50} is the particle size (in mm) corresponding to 50% finer and is determined from the particle size distribution curve obtained in the laboratory and q_c is in kPa. Note that the energy ratio corresponding to the N value is 55%. In general the ratio of q_c / N varies from 100 (silt) to 1000 (sand and gravel).

10.4 AXIAL ULTIMATE BEARING CAPACITY OF PILES

10.4.1 *Pile types*

Piles are slender structural members pushed or poured in soils where the strength and settlement criteria are not met using shallow foundations. The major purpose of piling is to transmit the loads from the superstructure to a hard stratum beneath the soil (*end bearing pile*) or to the surrounding soil (*friction pile*). Other applications include the cases of uplift and overturning forces, compaction of thick layers of cohesive or cohesionless materials and settlement control in slopes and landslides. Based on the method of installation and formation, the piles are classified into two categories of *displacement* and *non-displacement* piles (Weltman & Little, 1977). Displacement piles include preformed, screwed cast-in-place, and driven cast-in-place piles. In this group there is no removal of soil during installation. Preformed piles are constructed from concrete, steel, composite steel-concrete and timber. Driven cast-in-place piles include a temporary or permanent liner made from concrete or steel. In non-displacement piles the soil is removed during installation and the preformed pile is pushed into cavity or more commonly the cavity is filled with concrete. The cavity may be supported by permanent or temporary liner. Permanent liners are made of concrete or steel, whilst the temporary liners are provided by stabilizing fluid or by soil using continuous flight auger. A combination of both installation techniques can also be used to improve the performance of non-displacement piles.

Due to the expensive nature of the pile foundations, appropriate sufficient site investigations must be carried out according to the relevant codes. This investigation must provide strength and physical properties of the soil profile, the potential for heave, slope stability, damage to the adjacent structures and proposed technique for pile installation.

Timber piles. Timber piles have been used for centuries and are suitable for modest load and piling length of about 12 m. Timber piles (round or square) have cross sections from 250 mm (in diameter) to 500 mm (sometimes more). Working loads are unlikely to exceed 500 kN. Allowable stresses for compression and tension must be obtained according to the building codes and in particular the reduction factors for compression on the direction perpendicular to the grains must be investigated. In the *Ultimate Strength Design* (*USD*) the strength reduction factors of up to 0.8 are applied. In the *Working Stress Design* (*WSD*) the allowable compression in the direction of perpendicular to the grains may be adopted as 1 / 5 to 1 / 3 of the allowable tensile stress depending on the type of the tree used for the pile. The allowable tensile stress could be as high as 17 MPa, whilst the allowable compression is about 8 MPa. Note the average value of Modulus of Elasticity is at the range of 8000 MPa to 10000 MPa.

Timber piles are driven with pointed end, which may be protected, by a steel or a cast-iron shoe. Similarly a driving cap may be installed on driving end to protect the butt from hammering effects. Generally the code has restrictions on the size of the tip and butt as well as misalignment along the length of the pile. Timber piles have been used successfully in offshore structures. If the timber is treated with preservatives and is permanently under the water it will last more than the usual life spans of any structure; however the wetting and drying cycles will shorten its useful life to few years or less.

Concrete piles. Concrete piles are considered permanent, but necessary precautions have to be taken to protect the concrete from organic soils. They comprise precast and cast-in-situ piles. The precast group are displacement type piles and are driven into the ground with the flat or tapered point and generally equipped with steel or cast iron shoe. The precast concrete piles are reinforced and have solid square sections of 250 mm to 600 mm, but hexagonal, and hollow sections (octagonal and tubular) are also used. Hollow sections are commonly used in offshore structures, which have diameter in the excess of 0.9 m to stand high compressive and bending loads. The precast piles are usually of *jointed type* to ease the handling and transportation, but depending on the length of the piles and their numbers, full-length piles may also be cast in site. Sometimes the reinforcement is prestressed to reduce the sectional area for the same load capacity. In this way the compressive stresses generated during pile driving are better resisted, but the ultimate compression strength decreases. The maximum load is at the range of 900 kN for solid sections and 8500 kN for prestressed piles. The working load may be taken as the 1 / 3 of the maximum load. The hollow sections are designed for high loads at the range of 2000 kN or more. The reinforcement details may be found in the relevant concrete codes.

A concrete cast-in-situ pile is made by excavation of a borehole in the ground and filling the cavity by concrete. Reinforcement is placed centrally after completion of the borehole and casing is withdrawn during concreting. Normal diameters are 300 mm to 600 mm, but larger diameters of up to 2 m can be made. When there is more need to base resistance the cast-in-situ concrete piles are constructed with enlarged bases termed as *bell-bottom* piles (Mori & Inamura, 1989; Tomilinson, 1995). An economically cheaper replacement for a cast-in-situ concrete pile is a *grouted pile*. The cavity made by drilling is filled with sand-cement grout pumped through the hollow flight auger while withdrawing the auger (Gupte, 1984 and 1989). The main advantages of the grouted piles are minimal vibration and low-level noise. In large diameters pumpable concrete is preferable to cement-sand mortar.

Durability of concrete piles must comply with the durability requirements and *exposure classification* of the relevant concrete code. Durability requirements include adoption of minimum concrete strength and restriction on chemical content. Exposure classification investigates the chemical conditions of the soil surrounding the pile and commonly covers the determination of sulphates, chlorides and pH value. The details of the concrete pile characteristics, uses and installation are described by Fleming et al. (1992).

Steel piles. Steel piles comprise H, I and hollow sections. The hollow sections are of tubular or box type sections. Box sections are normally made by welding two sheet pile sections edge to edge termed as *Larssen double box* pile. Although they overcome the problems associated with the flexibility of the slender H sections they occupy more volume and produce large displacements. The hollow and tubular sections may be filled with concrete. Steel piles are used when the soil is free from contamination and the corrosion rate is low. Corrosion occurs in marine environment and disturbed soils (e.g. man made fills). Corrosion rate could range from non-aggressive to mild, to moderate, to severe to very severe. The corrosion allowances are categorized in the building codes under a common title of exposure classification and are at the range of less than 0.01 mm/year for non-aggressive exposure to 0.1- 0.5 mm/year for very severe conditions. In fresh water the corrosion rate is mild to moderate; in the seawater it can range from severe to very severe for submerging and tide or splash zones respectively. The acceleration of corrosion could occur at the presence of high levels of sulphates, domestic and industrial waste. Increase of chlorides in soil and water and decrease of pH accelerates the corrosion.

10.4.2 *Basics characteristics of axially loaded piles and pile test*

A pile subjected to a force along its central axis carries the load partly by shear stresses generated along the shaft (shaft or skin friction) and partly by normal stresses generated at the base of the pile (Figure 10.12(a)).

The ultimate bearing capacity P_u is the sum of the shaft capacity P_s and the base capacity P_b. Thus:

$$P_u = P_b + P_s \tag{10.77}$$

The allowable capacity of pile P_a is obtained by applying a suitable factor of safety on the above parts as:

$$P_a = \frac{P_b}{F_b} + \frac{P_s}{F_s} \tag{10.78}$$

A single factor of safety $F = F_b = F_s$ commonly within 2 to 4 is applied. Substituting values for the base and the shaft capacities in Equation 10.77 we obtain:

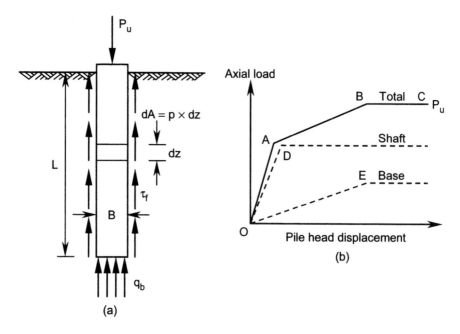

Figure 10.12. Axially loaded pile: (a) base and shaft resistance, (b) load-settlement behaviour.

$$P_u = q_b A_b + \int_0^L \tau_f dA = q_b A_b + \tau_s A_s \tag{10.79}$$

where q_b is the ultimate bearing capacity of the pile base, A_b is the sectional area of the base, τ_f is the limiting shear stress (skin friction) developed on the pile surface, dA is the element of the perimeter surface area defined by $p \times dz$ (p being the perimeter of the sectional area of the pile) and τ_s is the average limiting shear stress mobilized on the surface perimeter of the pile.

A typical load-settlement behaviour of a pile under axial loading is shown in Figure 10.12(b). This behaviour is consistent with the field data (Whitaker & Cooke, 1966; Bergdahl & Hult, 1981; Konrad & Roy, 1987) as well as with the small-scale laboratory tests (Anagnostpoulos & Georgiadis, 1993). The idealized response is comprised of three linear portions of *OA*, *AB* and *BC*. Initially there is a linear increase in the shaft resistance up to point *D* with a settlement at the range of 0.5% to 2% (or 5 mm to 10 mm) of the pile diameter or width (Fleming et al., 1992). This means that under a working loading below point *A* the applied load is resisted entirely by the shaft and little (or no) base resistance is present. In cohesive soils the shaft resistance is concentrated at the upper part of the pile as indicated by D'Appolonia & Romualdi (1963) and that the load-transfer mechanism has a time dependent behaviour (Frances et al., 1961). As the settlement continues (due to loading) the base capacity is mobilized with an increase in the rate of the settlement (Line *BC*). At point *C* the pile head displacement is at the range of 5% to 10% (more in bored piles) of the pile diameter (or width) and the base resistance is fully mobilized. Thus in a working condition the axial load P_a is shared between shaft and base, however the proportion of each share in terms of the total load cannot be defined. Some estimation is possible using a load-settlement model (Figure 10.12(b)) based on a pile load test. The limiting state of both base and shaft resistance may not occur at the same time and therefore the validity of Equation 10.77 may not be fully established. The mathematical models based on the load-settlement transfer curves have been successfully verified through closed form equations as well as numerical solutions. The closed form solutions (Cooke, 1974; Randolph & Wroth, 1978) are convenient for homogenous soils as well as piles with uniform sections. For layered soil and non-uniform (tapered) piles a numerical analysis is more suitable. The accuracy of the numerical analysis depends on the choice of the static stress-strain model or a dynamic (pile driving) model. DeNicola & Randolph (1993) and Ghazavi (1997) applied successfully an explicit finite difference method *FLAC* (Sections 7.3.4, 9.8.2), to verify a closed form equation and a new driving model respectively. Most of the recent computer programs use Smith's model (Smith, 1960), however new soil models developed by Goble & Rausche (1986) and Lee et al. (1988) and others have significant application in numerically based computer programs.

Pile load test is carried out either by static or dynamic loading and is the most reliable means of evaluating overall bearing capacity. The test procedure must comply with the relevant code (ASTM D-1143, BS 8004, AS 2159). Static load testing is used to evaluate the pile performance in the early or later stages of the construction work and to proof-test selected piles. Dynamic pile testing is used to measure the ultimate bearing capacity and its distribution, parameters to be used in the pile driveability equations, and to assess pile integrity.

Static load is usually applied to a pile by a jack supported by a reaction frame that consists of a structural beam supported by anchorage or existing coaxial piles. The distance of the anchorages and supporting piles from the testing pile must comply with the restrictions of the code; usually $3B$ for an anchorage system and $5B$ for a pile supporting system. The loading system must be capable of applying the load at a specified rate and maintaining the load for a specified duration. Loading times ranges from 5 minutes to 15 minutes and could reach up to 6 hours if the load is the representative of the design load. Unloading cycles may be introduced in the test program if required.

Dynamic testing may be carried out during pile installation or any time thereafter. The test is performed by using a hammer that is capable of mobilizing the pile strength requirements. In the pile integrity test the dynamic loading is provided by sonic impact test (*SIT*), sonic vibration test (*SVT*) and sonic logging test (*SLT*). In *SIT* the pile head is impacted by a small hammer to generate the stress wave whilst in the *SVT* an electrodynamic vibrator is used which generates sinusoidal stress waves. In the *SLT* the sonic pulses are provided by a transmitter to a receiver to determine the sonic profile. The results of the above tests are calibrated to assess the structural integrity and physical dimensions of the pile.

10.4.3   *Base capacity of piles*

For c', ϕ' soil the general form of the ultimate bearing capacity of the pile base is:

$$q_b = c'N_c d_c + \eta p_o' N_q d_q \qquad (10.80)$$

where η represents the effect of the at-rest lateral earth pressure. The N_γ term is negligible in comparison with the other terms. Bowles (1996) suggested the use of Hansen's bearing capacity equations where the pile length L replaces D in the depth factors and $\eta = 1$.

Alternatively, the values of N'_c and N'_q (instead of N_c and N_q in Equation 10.80) derived by Vesić (1975) may be used:

$$N'_c = (N'_q - 1)\cot\phi' \tag{10.81}$$

$$N'_q = \frac{3}{3 - \sin\phi'}\left\{\exp[(\pi/2 - \phi')\tan\phi']\tan^2(45.0° + \phi'/2)I_{rr}^{\frac{4\sin\phi'}{3(1+\sin\phi')}}\right\} \tag{10.82}$$

The term I_{rr} is called the reduced rigidity index and is defined by:

$$I_{rr} = I_r /(1 + \varepsilon_V I_r) \tag{10.83}$$

where ε_V is the volumetric strain of the soil at the vicinity of the pile base at failure and I_r is a rigidity index according:

$$I_r = G/(c' + p'_o \tan\phi') \tag{10.84}$$

The term G represents the shear modulus of the soil (Equation 5.7). Note that in undrained conditions $\varepsilon_V = 0$.

The parameter η is a function of the coefficient of soil pressure at-rest condition k_o in the following form:

$$\eta = (1 + 2k_o)/3 \tag{10.85}$$

For c_u, $\phi_u = 0$ soil (undrained conditions) the base capacity is reduced to:

$$q_b = c_u N_c d_c \tag{10.86}$$

The term $N_c d_c$ may be assumed 9 for most practical purposes, which is equal to the limiting value of N_c (derived from Equation 10.49) proposed by Skempton (1951) for square or circular bases. The Vesić's N'_c value for undrained conditions is:

$$N'_c = 4(\ln I_{rr} + 1)/3 + (\pi + 2)/2 \tag{10.87}$$

For piles in sands ($c' = 0$) Fleming et al. (1992) following Bolton (1986) proposed a solution based on the known values of density index I_D (Equation 1.13), the critical friction angle ϕ'_{cr} and the effective overburden pressure p'_o. Note that the maximum value of the density index where the field void ratio becomes equal to the minimum void ratio (densest state) obtained in the laboratory is 1. The minimum value of I_D is zero, which corresponds to the loosest possible state in the field. The critical friction angle ϕ'_{cr} is a constant property of the soil being independent from the initial void ratio and stress level. This is evident from Equation 4.33 ($q' = Mp'$) where q' equals to ($\sigma'_1 - \sigma'_3$) and p' is the effective mean stress ($\sigma'_1 + 2\sigma'_3$) / 3. From the geometry of Figure 4.11 or by taking the first term of Equation 4.17 (as there is no cohesion in the critical state) the magnitude of the critical friction angle is obtained:

$$\sin\phi'_{cr} = 3M/(6 + M).$$

The appropriate value of internal friction angle ϕ' under the pile base is estimated from the following equation:

$$\phi' = \phi'_{cr} + 3I_D[5.4 - \ln(p'/p_a)] - 3 \quad \text{(degrees)} \tag{10.88}$$

where p_a is the atmospheric pressure ($\approx$ 100 kPa). The base bearing capacity is defined by:

$$q_b = p'_o N_q \tag{10.89}$$

in which N_q is the bearing capacity factor corresponding to the internal friction angle ϕ' mobilized beneath the base (Equations 10.88). The effective mean stress p' in the vicinity of the pile base is taken as the geometric mean of the base bearing capacity and effective overburden pressure p'_o:

$$p' = \sqrt{N_q p'_o} \tag{10.90}$$

To evaluate N_q an iterative process is carried out by first assuming an initial value for N_q and calculating p' from Equation 10.90. The corresponding ϕ' is then calculated from Equation 10.88 and, using an appropriate ϕ'-N_q relationship, the value of N_q can be obtained. If this value is not sufficiently close to the assumed N_q, iteration is continued until the difference in N_q between successive cycles is insignificant. The ϕ'-N_q relationship used is taken from the theory of ultimate bearing capacity developed by Berezantzev et al. (1961) using a failure mechanism different to Terzaghi's, Meyerholf's and Hansen's. Whilst the failure mechanism under the base is similar to Terzaghi's mechanism shown in Figure 10.2, it extends vertically from points d and d' upwards until intersection with the ground surface, resulting in higher N_q values than the those from previously mentioned methods. A graphical presentation of N_q versus ϕ' is shown in Figure 10.13.

Correlations between field values of *SPT* and *CPT* and the base capacity are very useful, but the results must be assessed by the methods mentioned above. Meyerhof (1976) proposed the following equation for piles driven in sands:

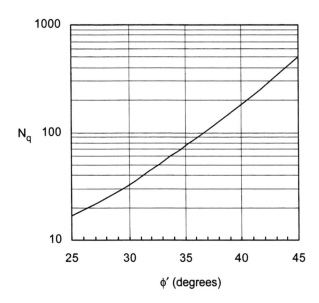

Figure 10.13. Bearing capacity factor N_q used for piles in sand.

$$q_b = 40\overline{N}L_b / B \le 400\overline{N} \quad \text{(kPa)} \tag{10.91}$$

where $\overline{N}$ is the average *SPT* number in the vicinity of $8B$ (pile diameter or width) above and $3B$ below the pile base, and L_b is the length of pile penetration into the sand layer. For fine sands and silts an upper limit of $300\,\overline{N}$ is taken. Note that in most cases the Equation 10.91 underestimates the base capacity. It is likely that the energy ratio in this equation is 55%, and therefore the $\overline{N}$ value must be corrected based on the specification of the equipment used. For bored piles a reduction factor in the range of 50% to 67% must be applied.

The average value of q_c within a zone similar to the *SPT* zone is assumed to be equal to the base capacity. Fleming & Thorburn (1983) suggested the use of the following equation:

$$q_b = (q_{c1} + q_{c2} + 2q_{c3})/4 \tag{10.92}$$

where q_{c1} is the average cone resistance over $2B$ below the pile base, q_{c2} is the minimum cone resistance over $2B$ below the pile base and q_{c3} is the average of the minimum values lower than q_{c1} over $8B$ above the pile base.

Example 10.12

Compute the base capacity of a round concrete pile that is 25 m long and has a diameter of 0.35 m. The shear strength parameter of base soil is: $c' = 20$ kPa, $\phi' = 26°$. The water table is 5 m below the ground surface. γ (above water table) $= 17$ kN/m³, $\gamma_{sat} = 20$ kN/m³.

Solution:

$A_b = \pi \times 0.35^2 / 4 = 0.0962$ m². Using Equations 10.2 we find $N_q = 11.85$, $N_c = 22.25$.

$p_o' = 17.0 \times 5.0 + (20.0 - 9.81) \times 20.0 = 288.8$ kPa. Depth factors from Equations 10.28:

$d_c = 1 + 0.4 \tan^{-1}(25.0/0.35) = 1.623$,

$d_q = 1 + 2 \tan 26.0°(1 - \sin 26.0°)^2 \tan^{-1}(25.0/0.35) = 1.479$. Thus using Equation 10.80:

$q_b = 20.0 \times 22.25 \times 1.623 + 288.8 \times 11.85 \times 1.479 = 5783.8$ kPa.

$P_b = q_b A_b = 5783.8 \times 0.0962 = 556.4$ kN.

Example 10.13

Re-work Example 10.12 using Vesić's bearing capacity factors with $I_{rr} = 10$.

Solution:

From Equations 10.81 and 10.82 we find: $N_q = 13.18$, $N_c = 24.98$.

$k_o = 1 - \sin\phi' = 1 - \sin 26.0° = 0.562$, and from Equation 10.85:

$\eta = (1 + 2 \times 0.562)/3 = 0.708$.

$q_b = 20.0 \times 24.98 \times 1.623 + 0.708 \times 288.8 \times 13.18 \times 1.479 = 4796.6$ kPa,

$P_b = 4796.6 \times 0.0962 = 461.4$ kN.

Example 10.14

A square pile of $B = 0.3$ m and length 18 m is embedded in a sand stratum; water table being 4 m below the ground surface. The properties of the sand layer are: $\phi'_{cr} = 30°$,

γ (above water table) = 16.5 kN/m^3, γ_{sat} = 19.2 kN/m^3, I_D = 0.5. Compute the base capacity of the pile.

Solution:

$p'_o = 16.5 \times 4.0 + (19.2 - 9.81) \times 14.0 = 197.5$ kPa.

Using Equation 10.90 with $N_q = 60$: $p' = \sqrt{60} \times 197.5 = 1529.8$ kPa. From Equation 10.88:

$\phi' = 30.0° + 3 \times 0.5[5.4 - \ln(1529.8/100.0)] - 3.0 = 31.0°$.

For this value of ϕ', Figure 10.13 yields a new value of 38 for N_q; trying $N_q = 42$:

$p' = \sqrt{42} \times 197.5 = 1279.9$ kPa,

$\phi' = 30.0° + 3 \times 0.5[5.4 - \ln(1279.9/100.0)] - 3.0 = 31.3°$.

From Figure 10.13 $N_q = 42$. Thus:

$q_b = p'_o N_q = 197.5 \times 42 = 8295$ kPa, $P_b = 8295 \times 0.3 \times 0.3 = 746.5$ kN.

10.4.4 *Shaft capacity of piles*

In the conventional analysis called α method (Skempton, 1959; Tomilinson, 1977) the average limiting shear stress τ_s mobilized on the shaft is estimated empirically as a fraction of the undrained cohesion c_u in the following form:

$$\tau_s = \alpha c_u \tag{10.93}$$

The magnitude of parameter α depends on the type of the soil, its past stress history, type of the pile and the method of installation and varies from 0.3 to 1 (and more). The generally accepted trend is higher values for soft clays (low shear strength) and lower values for stiff clays (high shear strength). This method has been improved (Fleming et al., 1992) relating α to the strength ratio defined by c_u / p'_o according:

$$\alpha = (c_u / p'_o)_{NC}^{0.5}(c_u / p'_o)^{-0.5} \qquad c_u / p'_o \le 1 \tag{10.94}$$

$$\alpha = (c_u / p'_o)_{NC}^{0.5}(c_u / p'_o)^{-0.25} \qquad c_u / p'_o > 1 \tag{10.95}$$

where the subscript *NC* represents the normally consolidated state. Equations 10.94 and 10.95 imply that for normally consolidated soil $\alpha = 1$ and the cohesion mobilized on the shaft is equal to the undrained shear strength. To apply the above equations in pile design it is necessary to obtain the undrained cohesion and effective overburden pressure profiles along the depth. Thus the shaft capacity may be expressed by:

$$P_s = \sum_{i=1}^{i=n} (\alpha c_u A_s)_i \tag{10.96}$$

where α_i, c_{ui} are the average values related to a finite length of a pile in specified depth, and A_{si} is the perimeter surface area of the finite length. It is recommended by number of authors that the shaft resistance to the depth of $3B$ to $4B$ from the ground surface to be ignored. Based on the experimental data for driven piles reported by Randolph & Murphy (1985) Equations 10.94 and 10.95 may be simplified to the following:

$$\alpha = 0.5(c_u / p_o')^{-0.5} \quad c_u / p_o' \le 1 \tag{10.97}$$

$$\alpha = 0.5(c_u / p_o')^{-0.25} \quad c_u / p_o' > 1 \tag{10.98}$$

An alternative effective stress analysis suggested by Burland (1973) assumes no effective cohesion on the pile shaft due to the remoulding effects of pile installation. The ultimate frictional shear stress mobilized at a specific depth on the shaft is:

$$\tau_s = \sigma_h' \tan \delta' = K p_o' \tan \delta' = \beta p_o' \tag{10.99}$$

where σ_h' is the effective normal stress horizontally applied by the soil on the pile, δ' is the effective friction angle mobilized on the pile surface, p_o' is the effective vertical stress, and K represents the lateral earth pressure coefficient. Whilst the value of K may be taken as the at-rest earth pressure coefficient k_o for a bored pile (Fleming et al., 1992), this will underestimate the mobilized shear stress on a driven pile. For bored piles in heavily con-solidated soils a value of $K = (1 + k_o) / 2$ is recommended. For driven piles experimental results give a range for K from 1.5 to 1.9 depending on the type of pile (Mansur & Hunter, 1970). For precast concrete and H section steel piles, an average value of 1.5 may be used (Meyerhof, 1976). The American Petroleum Institute's (*API*) 1984 guidelines recommend $K = 1$ for driven concrete piles, but may be assumed to be less for pipe type piles depend-ing on the end condition (open or closed). For open-end piles a value of 0.8 is appropriate. A 50% reduction in the K value may be applied where the axial force is tensile. Based on instrumented pile tests Francescon (1983) recommended $K = 1.5 k_o$. It was also suggested that δ' is mobilized to the soil's effective internal friction angle obtained from simple shear tests. Bowles (1996) suggested a range for δ' from 0.5 ϕ' to 0.75 ϕ' and emphasized that the method must be used only for granular materials. Note that the value of ϕ' may change with depth due to the increase in effective vertical stress and this has to be taken into account over the length of the pile.

In piles embedded in soft clay *negative skin friction* may develop when the consolida-tion settlement of the layer exceeds the pile settlement. The resulting force in the pile may be estimated using Equation 10.99 with $K = k_o$ where k_o is calculated from Equation 8.4. If the consolidation settlement is less than the pile settlement then the mobilized shear stress can be taken as a linear proportion to the limiting shear stress. The mobilized shear stress on the shaft is equal to the limiting shear stress calculated from Equation 10.99 mul-tiplied by the relative displacement (ratio or percentage) between the pile and the soil.

Empirical correlations between τ_s and the side friction f_s measured in a cone penetra-tion test may be used with care because of the radial consolidation associated with large diameter driven piles. Equating τ_s with f_s is acceptable only for slender piles with small volume displacement. Radial consolidation around large volume piles may increase τ_s up to twice the value of f_s.

Correlations between τ_s and the cone penetration resistance q_c are also used (Fleming et al., 1992). The magnitude of τ_s varies within the range $q_c / 10$ (Fleming & Thorburn, 1983) to $q_c / 40$ (Thorburn & McVicar, 1971). An empirical equation suggested by Meyerhof (1976) for piles driven in fine granular materials relates τ_s to the average *SPT* number over the length of the pile:

$$\tau_s = 2\overline{N} \quad (kPa) \tag{10.100}$$

Reduction factors for steel H piles of 50%, and for bored piles of 50% to 67%, must be applied.

Example 10.15

In an offshore structure a hollow pile of external diameter of 1.2 m and length 30 m is embedded in saturated soil with average $\gamma_{sat} = 20.5$ kN/m^3. The profile of undrained cohesion is shown in the second column of the table below. Calculate the shaft capacity ignoring the contribution of $3B$ of the length from the ground surface.

Solution:

The results of computations are tabulated where the total shaft capacity is 19.45 MN. Sample calculation for depth 15 m - 20 m:

$p'_o = 17.5(20.5 - 9.81) = 187.1$ kPa, $c_u / p'_o = 575/187.1 = 3.07$, from Equation 10.98:

$\alpha = 0.5 \times (3.07)^{-0.25} = 0.378,$

$\tau_s = 0.378 \times 575 = 217.3$ kPa.

$A_s = (\pi \times 1.2) \times 5.0 = 18.85$ m^2, $\tau_s \times A_s = 217.3 \times 18.85/1000 = 4.0961$ MN.

Depth (m)	c_u (kPa)	p'_o (kPa)	c_u / p'_o	α	τ_s (kPa)	A_s (m^2)	$\tau_s \times A_s$ (MN)
0-5	220	26.7	8.24	0.295	64.9	5.28	0.3427
5-10	490	80.2	6.11	0.318	155.8	18.85	2.9368
10-15	605	133.6	4.53	0.343	207.5	18.85	3.9113
15-20	575	187.1	3.07	0.378	217.3	18.85	4.0961
20-25	535	240.5	2.22	0.410	219.3	18.85	4.1338
25-30	485	294.0	1.65	0.441	213.9	18.85	4.0320
						Total	≈19.45

10.4.5 *Axial bearing capacity of piles using dynamic formulae and wave equation*

A pile driving formula is an idealized equation used to estimate the ultimate bearing capacity from the pile's resistance to penetration during driving. It is derived by equating the external energy induced by each blow of the hammer to the dissipated energy due to the permanent settlement per blow, including an allowance for energy loss. A part of the settlement is due to the elastic compression of the pile including the pile cap, cap block and ram assembly, and surrounding soil which will be recovered after each blow. There are numerous formulae in use including the following modified version of the Engineering News Record (*ENR*) formula that has been identified as reliable (Bowles, 1996):

$$P_u = \frac{1.25 e_h E_h}{s + C} \frac{W_r + n^2 W_p}{W_r + W_p}$$

(10.101)

where W_r is the weight of the ram, W_p is the weight of the pile, n is a coefficient of restitution (to be selected from the relevant code), E_h is the maximum rated energy of the ram (from manufacturer's catalogue), e_h is the hammer efficiency, s is the permanent settlement per blow, and C is the elastic compression of parts. The calculated P_u may be divided by 6 to yield the allowable bearing capacity of the pile. The magnitude of hammer efficiency e_h depends on the type of the hammer (single action, double action and drop hammers) and varies from 0.75 to 1. The elastic compression C represents the half of the recoverable compression and may be assumed 2.5 mm (0.1 inch) for most practical cases. The restitution parameter n is 0.25 for timber piles, 0.32 for compact cushion on steel pile and 0.5 for steel-on-steel anvil on either steel or concrete pile.

The dynamic response of a soil during a pile load test may be formulated more accurately using the one-dimensional pile-soil model originally suggested by Smith (1960). In this model the length of the pile is divided into discrete elements interconnected by springs. Smith's original model has been modified (Randolph & Simons, 1986) to include the damping effects of the surrounding soil. The governing differential equation, called the *wave equation*, is obtained from force equilibrium:

$$\frac{\partial^2 w}{\partial z^2} = \frac{1}{(E/\rho)_p}\frac{\partial^2 w}{\partial t^2} - \frac{T}{(AE)_p} \tag{10.102}$$

where w is the vertical displacement at depth z, E is the Modulus of Elasticity of the pile, ρ is the density of the pile material, A is the cross-sectional area of the pile, t is the time and T is the mobilized soil resistance per unit length of pile. This differential equation may be solved using finite differences (Fleming et al., 1992) or be transformed to a finite element form (Bowles, 1996) to yield P_u versus settlement as the load test proceeds.

Example 10.16

An H section pile (HP 310) has 12 m length and its weight per meter length is 93 kg. The properties of the hammer are: $W_r = 67$ kN, the equivalent stroke $= 0.39$ m. The pile has penetrated 300 mm in 20 blows. Calculate allowable bearing capacity of the pile using modified *ENR* formula. The weight of the cap assembly is 4.5 kN.

Solution:

$E_h = W_r \times h = 67.0 \times 0.39 = 26.13$ kN.m.

$W_p = (93.0 \times 12.0 \times 9.81/1000) + 4.5 = 15.45$ kN.

$s = (300.0/1000)/20 = 0.015$ m.

Take $e_h = 0.85$ and $n = 0.5$. Thus:

$$P_u = \frac{1.25e_h E_h}{s+C}\frac{W_r + n^2 W_p}{W_r + W_p} = \frac{1.25 \times 0.85 \times 26.13}{0.015 + 0.0025} \times \frac{67.0 + 0.5^2 \times 15.45}{67.0 + 15.45} = 1363.5 \text{ kN}.$$

10.4.6 *Pile settlement*

Settlement of a pile due to shear stresses developed on the shaft is calculated using two simplifying assumptions with regard to the distribution of the shear stress and the definition of shear strain. The shear stress at a distance r from the pile axis is assumed to be proportional to the average shear stress mobilized on the shaft (τ_{sm}):

$$\tau = \tau_{sm}\frac{r_o}{r} \tag{10.103}$$

where r_o is the radius of the pile section (or equivalent radius). The shear strain γ is approximated by:

$$\gamma \approx \frac{dw_s}{dr} = \frac{\tau_{sm} r_o}{rG} \tag{10.104}$$

where w_s is the settlement of the pile shaft, and G is the shear modulus of the soil. Integrating Equation 10.104 between $r = r_o$ and $r = r_m$ (where the settlement becomes insignificant):

$$w_s = \frac{\tau_{sm} r_o}{G} \ln(r_m / r_o) \tag{10.105}$$

The force resisted by the shaft equals its perimeter area multiplied by the average shear stress mobilized on the shaft:

$$P_{s(working)} = 2\pi r_o L \tau_o \tag{10.106}$$

The settlement under the base of the pile w_b is calculated from Equation 5.54 (Section 5.4.2). Alternatively, the base settlement may be calculated using the Mindlin stress distribution discussed in Section 5.3.11. After dividing the soil under the base of the pile into a finite number of layers (preferably of equal thickness), the vertical stress is computed using Equations 5.50 to 5.53 or Tables 5.2(a) to 5.2(c). The settlement of the base is approximately:

$$w_b = \sum_{i=1}^{i=n} (\frac{\sigma_z'}{E_s} l)_i \tag{10.107}$$

where l is the thickness of the finite layer, σ_z' is the increase in the effective vertical stress at the centre of the layer due to loading, and E_s is the Modulus of Elasticity of the soil layer. Substituting σ_z' from Equation 5.50 into Equation 10.107:

$$w_b = \frac{P_a}{L^2 E_s} \sum_{i=1}^{i=n} (lI_q)_i.$$

If $l_1 = l_2 = l_3 = \cdots = l_n$, and $l_1 + l_2 + l_3 + \cdots + l_n = L$ (length of the pile):

$$w_b = \frac{P_a}{LE_s} \sum_{i=1}^{i=n} (\frac{I_q}{n})_i = \frac{P_a}{LE_s} I_{q(average)} \tag{10.108}$$

which simplifies the use of Tables 5.2(a) to 5.2(c).

For stiff piles the total stiffness of the pile-soil system K_p is the sum of the shaft and base stiffness:

$$K_p = \frac{P_a}{w_a} = \frac{P_{s(working)}}{w_s} + \frac{P_{b(working)}}{w_b} \tag{10.109}$$

where w_a is the total settlement, and P_a is the working load applied to the pile (or allowable capacity defined by Equation 10.78). With known working values of the shaft and base capacities, the total settlement of the pile can be obtained using Equation 10.109. For piles that exhibit significant shortening under the working load, a characteristic solution reported by Fleming et al. (1992) defines the ratio $P_{b(working)} / P_a$ in terms of the elastic properties of the soil and the pile. Further information is available from Poulos (1982) and Poulos & Davis (1980). In order to obtain an estimate of the elastic shortening of the pile, it may be assumed that the pile force reduces linearly from P_a at the top to $P_a / 2$ at the base, which results in an elastic shortening of:

$$w_e = \frac{0.75 P_a L}{(AE)_p} \tag{10.110}$$

Example 10.17

An end bearing pile of 0.4 m diameter and 18 m length carries an axial load of 1500 kN. Calculate the settlement of the pile by dividing the underlying strata (down to the depth of 18 m from the base) into 5 layers of equal thickness. $\mu = 0.3$, $E_s = 28$ MPa.

Solution:

Thickness of each layer is $18.0 / 5 = 3.6$ m. Values of $m = z / L$ for $r = 0$ and the corresponding I_q values (From Table 5.2(a)) are tabulated from which the average I_q in the zone L to $2L$ is:

$I_{q(average)} = 23.1891 / 5 = 4.6378$. From Equation 10.108:

$w_b = [1500.0/(18.0 \times 28.0 \times 1000)]4.6378 = 0.0138\,\text{m} = 13.8\,\text{mm}$.

Layer	z (m)	$m = z / L$	I_q
1	19.8	1.1	19.3926
2	23.4	1.3	2.2222
3	27.0	1.5	0.8377
4	30.6	1.7	0.4500
5	34.2	1.9	0.2866
			Total: 23.1891

10.5 PILE GROUPS

10.5.1 *Force distribution using the rigid cap concept*

For the load distribution through a rigid pile cap Equation 5.95 (Section 5.5.1) may be used in the following form:

$$P_i = S_i(\frac{M_y}{I_y} x_i - \frac{M_x}{I_x} y_i + \frac{P}{S}) \tag{10.111}$$

where P is the total vertical load, M_x and M_y are the moments about the x and y axes (Figure 10.14), x_i, y_i are coordinates of the centriod of each pile, I_x and I_y are the second moments of area of the pile group about the x and y axes, S_i is the cross-sectional area of pile i, S is the total cross-sectional area of the pile group (excludes the area of the pile cap) and P_i is the vertical load taken by the pile i. The origin of the coordinate system is at the centriod of the pile group which may be different from the centriod of the rigid cap.

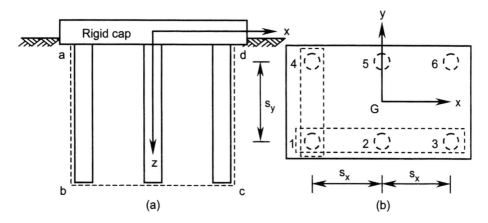

Figure 10.14. Rigid cap supported by a pile group.

For *n* piles of equal cross- sectional area:

$$P_i = \frac{M_y}{\sum x_i^2} x_i - \frac{M_x}{\sum y_i^2} y_i + \frac{P}{n} \qquad (10.112)$$

Whilst the magnitude of P_i calculated from the equations above is independent of the orientation of the *xy* coordinate system, it is prudent to orient the system so as to minimize the computational effort required.

Example 10.18

A pile group of Figure 10.14 carries 600 kN vertical force at $x = y = 0.7$ m. The piles are of equal diameter of 0.35 m. $s_x = 1.4$ m, $s_y = 1.2$ m. Calculate the vertical load at each pile.

Solution:

$|M_x| = |M_y| = 600 \times 0.7 = 420$ kN.m. Considering the right-hand rule sign convention:

$M_x = -420$ kN.m, $M_y = 420$ kN.m.

$\sum x_i^2 = 4 \times 1.4^2 = 7.84$ m^2, $\sum y_i^2 = 6 \times (1.2/2)^2 = 2.16$ m^2.

From Equation 10.112:

$P_1 = (420.0/7.84)(-1.4) - (-420.0/2.16)(-0.6) + 600.0/6 = -91.7$ kN.

$P_2 = (420.0/7.84)(0.0) - (-420.0/2.16)(-0.6) + 600.0/6 = -16.7$ kN.

$P_3 = (420.0/7.84)(1.4) - (-420.0/2.16)(-0.6) + 600.0/6 = 58.3$ kN.

Similarly, $P_4 = 141.7$ kN, $P_5 = 216.7$ kN, and $P_6 = 291.7$ kN.

10.5.2 *Ultimate bearing capacity and efficiency of a pile group*

In a group of closely spaced piles, or when the rigid cap rests on the ground surface as well as the piles, a block failure similar to block *abcd* of Figure 10.14(a) could occur under the applied loads. In a cohesionless soil the base bearing capacity of the group may be evaluated from the general bearing capacity equations for shallow footings by substituting the width of the block for *B*. The frictional resistance can be obtained using Equation

10.99 with $K = k_o$ and $\delta' = \phi'$. In sands the total ultimate bearing capacity of the block will be higher than the sum of the ultimate bearing capacities of the individual piles. In cohesive soils the base bearing capacity is found from Equations 10.48 and 10.49 (Skempton, 1951) by substituting the depth of the block for D. On all four sides of the block it is assumed that the whole shear strength of the soil is mobilized. In the case of clay soils the total block bearing capacity may be less than the sum of the ultimate bearing capacities of individual piles. Depending on the pile spacing a row type of block failure may also need to be considered (Figure 10.14(b)).

The ratio of the total block bearing capacity to the sum of the ultimate bearing capacity of individual piles is called the efficiency of the pile group. Jumikis (1971) defines the efficiency of pile group, independent of the type of the soil, as:

$$E_g = 1 - \frac{(n-1)m + n(m-1)}{90mn} \tan^{-1}(B/s)$$

(1.113)

where n in the number of piles in a row, m is the number of rows, B is the diameter of piles and s is the centre-to-centre spacing of the piles.

10.5.3 *Settlement of a pile group and Winkler model*

The traditional method of estimating the settlement of a pile group uses an equivalent raft at a specified depth with a nominated surface loading. However, the current method combines the stiffness of the adjacent piles to provide a reasonable soil-pile model (Butterfield & Douglas, 1981) that can be analysed by numerical methods. The settlement of a pile group may be estimated from Equation 10.107 where the vertical stress σ'_z due to loading is calculated at the centre of each finite layer (on each pile axis or any other point) considering the contributions of all piles. In order to compute σ'_z the axial force in each pile must be known. With a rigid cap the axial force taken by each pile may be estimated from Equation 10.111.

When an elastic (non-rigid) cap rests on a coaxial pile group, the concept of a Winkler foundation (Section 5.5.2) may be used to evaluate the load distribution and settlements (Figure 10.15(a)). If the elastic cap is not in contact with the ground surface then the following procedure may be carried out to determine the pile forces and corresponding settlements:

1. Calculate the vertical load in each pile using Equation 10.111 if the piles are of different sections. For piles of equal shape and cross-sectional area, use Equation 10.112.
2. Using the Mindlin solution calculate the settlement of each pile. Decide on the portions of load to be taken by the shaft and base. The effect of adjacent piles must be taken into account as well as the pile compression due to the working load and its own weight. Alternatively, any elastic method could be used to obtain the total displacement of the pile head. Care should be taken in the selection of the Modulus of the Elasticity of the soil as in most cases it is a function of the confining pressure and increases with depth.
3. Calculate the pile stiffness (or the spring constant) of the pile:

$K_p = P_a$ (pile force from step 1) / w_a (pile settlement from step 2).

4. Analyse the cap as a beam resting on finite number of springs with the K_p values calculated in step 3.

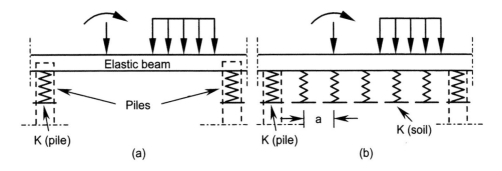

Figure 10.15. Winkler model for an elastic beam supported by a pile group.

If the cap is assumed to rest on the soil as well as the piles (Figure 10.15(b)), the K value of the soil has to be considered by replacing the soil between piles with additional springs.

Example 10.19

A rigid pile cap of 2 m by 2 m is resting on two piles and carries a vertical load of 1600 kN at its centre point. The piles are of diameter 0.4 m and length of 7 m with centre-to-centre distance of 1.4 m and are positioned on the centre line of the cap.
Compute the maximum probable settlement of the pile cap. Assume skin resistance load mechanism.
$\mu = 0.3$, $E_s = 28$ MPa, $E_p = 22000$ MPa.

Solution:

To calculate the settlement at the midpoint of two piles the 7 m thickness of the soil beneath the base is divided into 5 layers of thickness 1.4 m and I_q values for:

n_1 (under the pile) $= r / L = 0.0 / 7.0 = 0$,

n_2 (midpoint between two piles) $= r / L = 0.7 / 7.0 = 0.1$,

n_3 (under the other pile) $= r / L = 1.4 / 7.0 = 0.2$,

m_1 (centre of layer 1) $= z / L = 7.7 / 7 = 1.1$,

$m_2 = 1.3$, $m_3 = 1.5$, $m_4 = 1.7$, and $m_5 = 1.9$ are tabulated using Table 5.2(c).

Layer	z (m)	$m = z / L$	I_q ($n = 0.0$)	I_q ($n = 0.1$)	I_q ($n = 0.2$)
1	7.7	1.1	3.0612	1.6433	0.7680
2	9.1	1.3	0.8262	0.7411	0.5639
3	10.5	1.5	0.5189	0.4163	0.3679
4	11.9	1.7	0.3332	0.2739	0.2556
5	13.3	1.9	0.2654	0.1965	0.1887
		Total:	5.0049	3.2711	2.1441
		Average:	1.0010	0.6542	0.4288

The average I_q in the zone 7 m to 14 m is:

$I_q(pile) = 1.0010 + 0.4288 = 1.4298.$

$w_b(pile) = [(1600.0/2)/(7.0 \times 28.0 \times 1000)]1.4298 = 0.0058\text{m} = 5.8\,\text{mm}.$

$I_q(middle) = 0.6542 \times 2 = 1.3084.$

$w_b(middle) = [(1600.0/2)/(7.0 \times 28.0 \times 1000)]1.3084 = 0.0053\text{m} = 5.3\,\text{mm}.$

If the cap becomes in contact with the ground surface the settlement is calculated from Equation 5.54. The influence factor for square rigid footing is calculated using Equations 5.55 and 5.58 and is: $I_s = 0.88$. Thus:

$$S_e = (1600.0/2.0 \times 2.0)(2.0)(1 - 0.3^2) \times 0.88/28000.0 = 0.0229 \text{ m} = 22.9 \text{ mm}.$$

The elastic shortening of the piles are estimated from Equation 10.110:

$$w_e = 0.75 P_a L/(AE)_p = 0.75 \times 800.0 \times 7.0/[(\pi \times 0.4^2/4)22 \times 10^6] = 0.0015 \text{ m} = 1.5 \text{ mm}.$$

The maximum probable settlement $= 5.8 + 22.9 + 1.5 = 30.2 \approx 30$ mm.

10.6 PROBLEMS

10.1 A square footing of 1 m is located at a depth 1.5 m below the ground surface. The soil properties are:
$c' = 0$, $\phi' = 40°$, $\gamma = 16.7$ kN/m^3, $\gamma_{sat} = 20$ kN/m^3.
Using Terzaghi's bearing capacity factors calculate the ultimate bearing capacity
(a) the water table is well below the foundation level,
(b) the water table is at the ground surface.
Answers: 2706 kPa, 1651 kPa

10.2 Re-work Problem 10.1 using Meyerhof's bearing capacity equations.
Answers: 4606 kPa, 2811 kPa

10.3 Re-work problem 10.1 using Hansen's bearing capacity equations.
Answers: 3595 kPa, 2194 kPa

10.4 Re-work problem 10.1 using Vesić's bearing capacity equations.
Answers: 4126 kPa, 2518 kPa

10.5 Using Hansen's method calculate the base capacity of a square pile of 0.4 m width and 10 m length for the following two cases:
(a) the water table is well below the pile base with $\gamma = 16.7$ kN/m^3,
(b) the water table is at the ground surface with $\gamma_{sat} = 20$ kN/m^3. Ignore the N_γ term.
$c' = 0$, $\phi' = 40°$.
Answers: 2278 kPa, 1390 kPa

10.6 Re-work Problem 10.5 using Vesić's method with $I_{rr} = 20$.
Answers: 1177 kPa, 718 kPa

10.7 A pile of length 15 m is embedded in a sand layer with the following properties:
$I_D = 0.75$, $\phi'_{cr} = 33°$, $\gamma = 17$ kN/m^3, $\gamma_{sat} = 20$ kN/m^3.
Calculate the base bearing capacity if:

(a) there is no water in the vicinity of the pile base,

(b) the water table is at the ground surface.

Use the iteration based method of Fleming et al. (1992).

Answers: ≈19.1 MPa, 13.4 MPa

10.8 Estimate the ultimate pile capacity of a 30 m concrete pile with 0.4 m diameter in an offshore structure where the submerged unit weight is 8.3 kN/m^3. The profile of undrained shear strength, which changes linearly between the measured points, is:

Depth (m)	0	6	18	24	30
c_u (kPa)	200	440	440	220	220

Answer: 5073 kN

10.9 It is required to estimate the length of a frictional concrete pile of 0.4 m diameter embedded in cohesionless soil. The ultimate bearing capacity of the pile is 1850 kN. The soil is comprised of 10 m fine sand with $\gamma = 16.5$ kN/m^3 and $\phi' = 30°$ underlain by a course sand of $\gamma = 18.8$ kN/m^3 and $\phi' = 36°$.

Assume $K = 1.5$ for both layers.

Answer: 13.5 m

10.10 For the loading system of Example 10.18 a pile group of 3 rows with 3 piles in each row is designed.

Calculate the vertical force in each pile. $s_x = 1.4$ m, $s_y = 1.2$ m.

Answers: Left to right: lower row ($-41.7, 8.3, 58.3$), middle row ($16.7, 66.7, 116.7$), upper row ($75.0, 125.0, 175.0$), all in kN

10.11 A pile group has 3 coaxial piles all having equal length of 10 m and are spaced equally at 1.5 m. The central pile carries 600 kN; the side piles each has 450 kN. Calculate the settlement of the central pile assuming skin resistance load mechanism.

$\mu = 0.3$, $E_s = 8$ MPa.

Answer: 10 mm

10.7 REFERENCES

Anagnostopoulos, C. & Georgiadis, M. 1993. Interaction of axial and lateral pile responses. *Journal SMFED, ASCE* 119(4): 793-798.

API. 1984. *API recommended practice for planning, designing and constructing fixed offshore platforms*. 15th edition. API RP2A page 115. American Petroleum Institute.

Begemann, H. 1974. General report: central and western Europe. *Eropean symp. on penetration testing*. 2.1: 29-39. Stockholm.

Berezantzev, V.C., Khristoforov, V. & Golubkov, V. 1961. Load bearing capacity and deformation of piled foundations. *Proc. 5th intern. conf. SMFE* 2: 11-15.

Bergdahl, U. & Hult, G. 1981. Load tests on friction piles in clay. *Proc. 10th intern. conf. SMFE* 2: 625-630. Stockholm.

Bolton, M.D. 1986. The strength and dilatancy of sands. *Geotechnique* 36(1): 65-78.

Bowles, J.E. 1996. *Foundation analysis and design.* 5th edition. New York: Mc Graw-Hill.

Burland, J.B. 1973. Shaft friction piles in clay - a simple fundamental approach. *Ground engineering* 6(3): 30-42.

Burland, J.B. & Burbidge, M.C. 1985. Settlements of foundations on sands and gravel. *Proc.*, Part 1, 78: 1325-1381. UK: Institution of Civil Engineers.

Butterfield, R. & Douglas, R.A. 1981. Flexibility coefficients for the design of piles and pile groups. CIRIA technical note 108.

Cooke, R.W. 1974. Settlement of friction pile foundations. *Proc. conf. on tall buildings*: 7-19. Kuala Lumpur.

D'Appolonia, E. & Romualdi, J.P. 1963. Load transfer in end bearing steel H piles. *Journal SMFED, ASCE* 89(SM2): 1-26.

DeBeer, E.E. & Martens, A. 1957. Method of computation of an upper limit for the influence of heterogeneity of sand layers on the settlement of bridges. *Proc. 4th intern. conf. SMFE* 1: 275-281. London: Butterworths.

DeNicola, D.A. & Randolph, M.F. 1993. Tensile and compression shaft capacity of piles in sand. *Journal SMFED, ASCE* 119(12): 1952-1973.

Fleming, W.G.K. & Thorburn, S. 1983. Recent piling advances: state of the art report. *Proc. conf. on advances in piling and ground treatment for foundations.* London: Institute of Civil Engineers.

Fleming, W.G.K., Weltman, A.G., Randolph, M.F. & Elson, W.K. 1992. *Piling Engineering.* 2nd edition. New York: Blackie Academic & Professional.

Francescon, M. 1983. *Model pile tests in clay: stresses and displacements due to installation and axial loading.* Ph.D. Thesis, University of Cambridge.

Ghazavi, M. 1997. *Static and dynamic analysis of piled foundations.* Ph.D. Thesis, University of Queensland, Australia.

Gibbs, H.J. & Holtz, W.G. 1957. Reseach on determining the density of sands by spoon penetration testing. *Proc. 4th intern. conf. SMFE* 1: 35-39.

Goble, G.G. & Rausche, F. 1986. WEAP86 program documentation. FHWA, Washington D.C.

Gupte, A.A. 1984. Construction, design and application of post tensioned auger cast soil anchors. *Int. symp. of prestressed rock and soil anchors.* Post-Tensioning Institute, Des Plaines, IL, USA.

Gupte, A.A. 1989. Design, construction and applications of auger piling system for earth retention and underpinning of structures. In J. Burland & J. Mitchell (eds), *Piling and deep foundations* 1: 63-72. Rotterdam: Balkema.

Hansen, J.B. 1961. A general formula for bearing capacity. *Danish geotechnical institute.* Copenhagen, Denmark. Bulletin (11): 38-46.

Hansen, J.B. 1970. A revised and extended formula for bearing capacity. *Danish geotechnical institute.* Copenhagen, Denmark. Bulletin (28): 5-11.

Jumikis, A.R. 1971. *Foundation engineering.* Scranton, Pa: Intext Educational Publishers.

Konrad, J.M. & Roy, M. 1987. Bearing capacity of friction piles in marine clay. *Geotechnique* 37(2): 163-175.

Lee, S.L., Chow, Y.K. Karunarante, G.P. & Wong, K.Y. 1988. Rational wave equation model for pile driving analysis. *Journal SMFED, ASCE* 114(3): 306-325.

Liao, S.S. & Whitman, R.V. 1986. Overburden correction factors for sand. *Journal SMFED, ASCE* 112(3): 373-377.

Lunne, T. & Edie, O. 1976. Correlations between cone resistance and vane shear strength in some Scandinavian soft to medium stiff clays. *Canadian geotechnical journal* 13(4): 430-441.

MacDonald, D.H. & Skempton, A.W. 1955. A survey of comparisons between calculated and observed settlements of structures on clay. *Conf. on correlation of calculated and observed stresses and displacements*: 318-337. London: Institute of Civil Engineers.

Mansur, C.I. & Hunter, A.H. 1970. Pile tests-Arkansas river project. *Journal SMFD, ASCE* 96(SM5): 1545-1582.

Meyerhof, G.G. 1951. The ultimate bearing capacity of foundations. *Geotechnique* 2: 301-332.

Meyerhof, G.G. 1953. The bearing capacity of foundations under eccentric and inclined loads. *Proc. 3rd intern. conf. SMFE* 1: 440-445. Zurich.

Meyerhof, G.G. 1956. Penetration tests and bearing capacity of cohesionless soils. *Journal SMFED, ASCE* 82(SM1): 1-19.

Meyerhof, G.G. 1963. Some recent research on the bearing capacity of foundations. *Canadian geotechnical journal* 1(1): 16-26.

Meyerhof, G.G. 1965. Shallow foundations. *Journal SMFED, ASCE* 91(SM2): 21-31.

Meyerhof, G.G. 1974. General report: outside Europe. *Proc. 1st European symp. on penetration testing* 2.1: 40-48. Stockholm.

Meyerhof, G.G. 1976. Bearing capacity and settlement of pile foundations. *Journal SMFED, ASCE* 91(GT3): 195-228.

Mori, H & Inamura, T. 1989. Construction of cast-in-place piles with enlarged bases. In J. Burland & J. Mitchell (eds), *Piling and deep foundations* 1: 85-92. Rotterdam: Balkema.

Peck, R.B., Hanson, W.E. & Thornburn, T.H. 1974. *Foundation engineering*. New York: John Wiley & sons.

Powrie, W. 1997. *Soil mmechanics-concepts and applications*. London: E & FN Spon.

Ramiah, B.K. & Chickanagappa, L.S. 1982. *Handbook of soil mechanics and foundation engineering*. Rotterdam: Balkema.

Poulos, H.G. 1982. The influence of shaft length on pile load capacity in clays. *Geotechnique* 32(2): 145-148.

Poulos, H.G. & Davis, E.H. 1980. *Pile foundation analysis and design*. John Wiley & Sons.

Randolph, M.F. & Murphy, B.S. 1985. Shaft capacity of driven piles in clay. *Proc. 17th offshre technology conf.*: 371-378.

Randolph, M.F. & Simons, H.A. 1986. An improved soil model for one-dimensional pile driving analysis. *Proc. 3rd intern. conf. on numerical methods in offshore piling* 3-15. Nantes, France.

Randolph, M.F. & Wroth, C.P. 1978. Analysis of deformation of vertically loaded piles. *Journal SMFED, ASCE* 104(GT12): 1465-1488.

Robertson, P.K. & Camanella, R.G. 1983. Interpretation of cone penetration tests. *Canadian geotechnical journal* 20(4): 718-745.

Robertson, P.K., Campanella, R.G. & Wightman, A. 1983. SPT-CPT correlations. *Journal SMFED, ASCE* 109(11): 1449-1459.

Schmertmann, J.H. 1970. Static cone to compute static settlement over sand. *Journal SMFED, ASCE* 96(SM3): 1011-1043.

Schmertmann, J.H. 1975. Measurement of in-situ shear strength. *Proc. conf. on in-situ measurement of soil properties* 2: 57-138. New York: ASCE.

Schmertmann, J.H., Hartman, J.P. & Brown, P.R. 1978. Improved strain influence factor diagrams. *Journal SMFED, ASCE* 104(8): 1131-1135.

Seed, H.B., Tokimatsu, K., Harder, L.F. & Chung, R.M. 1984. The influence of SPT procedures in soil liquefaction evaluations. *Journal SMFED, ASCE* 111(12): 1425-1445.

Skempton, A.W. 1951. The bearing capacity of clays. *Proc. building research congress* 1: 180-189. London, UK.

Skempton, A.W. 1959. Cast in-situ bored piles in London Clay. *Geotechnique* 9: 153-173.

Skempton, A.W. 1986. Standard penetration test procedures and the effects in sands of overburden pressure, relative density, particle size, aging and overconsolidation. *Geotechnique* 36(3): 425-447.

Smith, E.A.L. 1960. Pile driving analysis by the wave equation. *Journal SMFED, ASCE* 86(SM4): 35-61.

Stroud, M.A. 1989. The standard penetration test: its application and interpretation. *Penetration testing in the UK* 29-49. London: Thomas Telford.

Terzaghi, K. 1943. *Theoretical soil mechanics*. New York: John Wiley & Sons.

Terzaghi, K. & Peck, R. B. 1967. *Soil mechanics in engineering practice*. New York: John Wiley.

Terzaghi, K., Peck, R. B. & Mesri, G. 1996. *Soil mechanics in engineering practice*. 3rd edition. New York: John Wiley & Sons.

Thorburn, S. & McVicar, R.S.L. 1971. Pile load tests to failure in the Clyde alluvium. *Proc. conf. on behaviour of piles*: 1-7, 53-54. London: Institute of Civil Engineers.

Tomilinson, M.J. 1977. *Pile design and construction practice*. 4th edition. London: E. & F.N. Spon.

Tomilinson, M.J. 1995. *Foundation design and construction*. 6th edition. Harlow, Essex: Longman Scientific & Technical.

Vesić, A.S. 1973. Analysis of ultimate loads of shallow foundations. *Journal SMFED, ASCE* 99(SM1): 45-73.

Vesić, A.S. 1975. Principles of pile foundation design. *Soil mechanics services*. School of Engineering, Duke University (38): page 48. Durham.

Walsh, H. 1981. Tolerable settlement of buildings. *Journal SMFED, ASCE* 107(ET11): 1489-1504.

Weltman, A.J. & Little, J.A. 1977. A review of bearing pile types. DoE/CIRIA piling development group, report PG1.

Whitaker, T. & Cooke, R.W. 1966. An investigation of the shaft and base resistance of large bored piles in London clay. *Proc. symp. on larged bored piles*: 7-49. London: Institution of Civil Engineers.

Index

GUILDFORD **college**

Learning Resource Centre

Please return on or before the last date shown.
No further issues or renewals if any items are overdue.

2 1 NOV 2007   2 NOV 2011

−4 JUN 2008
2 1 OCT 2009   − 7 MAR 2013

−4 NOV 2009

1 9 OCT 2010

− 3 NOV 2010

1 7 NOV 2010